P9-EJU-941

Praise for *The Singularity Is Near*

One of *CBS News*'s Best Fall Books of 2005

Among *St. Louis Post-Dispatch*'s Best Nonfiction Books of 2005

One of Amazon.com's Best Science Books of 2005

"Anyone can grasp Mr. Kurzweil's main idea: that mankind's technological knowledge has been snowballing, with dizzying prospects for the future. The basics are clearly expressed. But for those more knowledgeable and inquisitive, the author argues his case in fascinating detail. . . . *The Singularity Is Near* is startling in scope and bravado."
        —Janet Maslin, *The New York Times*

"Filled with imaginative, scientifically grounded speculation. . . . *The Singularity Is Near* is worth reading just for its wealth of information, all lucidly presented. . . . [It's] an important book. Not everything that Kurzweil predicts may come to pass, but a lot of it will, and even if you don't agree with everything he says, it's all worth paying attention to."
        —*The Philadelphia Inquirer*

"[An] exhilarating and terrifyingly deep look at where we are headed as a species. . . . Mr. Kurzweil is a brilliant scientist and futurist, and he makes a compelling and, indeed, a very moving case for his view of the future."
        —*The New York Sun*

"Compelling."
        —*San Jose Mercury News*

"Kurzweil links a projected ascendance of artificial intelligence to the future of the evolutionary process itself. The result is both frightening and enlightening. . . . *The Singularity Is Near* is a kind of encyclopedic map of what Bill Gates once called 'the road ahead.'"
        —*The Oregonian*

"A clear-eyed, sharply-focused vision of the not-so-distant future."
        —*The Baltimore Sun*

"This book offers three things that will make it a seminal document. 1) It brokers a new idea, not widely known, 2) The idea is about as big as you can get: the Singularity—all the change in the last million years will be superceded by the change in the next five minutes, and 3) It is an idea that demands informed response. The book's claims are so footnoted, documented, graphed, argued, and plausible in small detail, that it requires the equal in response. Yet its claims are so outrageous that if true, it would mean . . . well . . . the end of the world as we know it, and the beginning of utopia. Ray Kurzweil has taken all the strands of the Singularity meme circulating in the last decades and has united them into a single tome which he has nailed on our front door. I suspect this will be one of the most cited books of the decade. Like Paul Ehrlich's upsetting 1972 book *Population Bomb*, fan or foe, it's the wave at epicenter you have to start with."
    —Kevin Kelly, founder of *Wired*

"Really, really out there. Delightfully so."
    —Businessweek.com

"Stunning, utopian vision of the near future when machine intelligence outpaces the biological brain and what things may look like when that happens. . . . Approachable and engaging."
    —the unofficial Microsoft blog

"One of the most important thinkers of our time, Kurzweil has followed up his earlier works . . . with a work of startling breadth and audacious scope."
    —newmediamusings.com

"An attractive picture of a plausible future."
    —*Kirkus Reviews*

"Kurzweil is a true scientist—a large-minded one at that. . . . What's arresting isn't the degree to which Kurzweil's heady and bracing vision fails to convince—given the scope of his projections, that's inevitable—but the degree to which it seems downright plausible."
    —*Publishers Weekly* (starred review)

"[T]hroughout this tour de force of boundless technological optimism, one is impressed by the author's adamantine intellectual integrity. . . . If you are at all interested in the evolution of technology in this century and its consequences for the humans who are creating it, this is certainly a book you should read."

—John Walker, inventor of Autodesk, in Fourmilab Change Log

"Ray Kurzweil is the best person I know at predicting the future of artificial intelligence. His intriguing new book envisions a future in which information technologies have advanced so far and fast that they enable humanity to transcend its biological limitations—transforming our lives in ways we can't yet imagine."

—Bill Gates

"If you have ever wondered about the nature and impact of the next profound discontinuities that will fundamentally change the way we live, work, and perceive our world, read this book. Kurzweil's *Singularity* is a tour de force, imagining the unimaginable and eloquently exploring the coming disruptive events that will alter our fundamental perspectives as significantly as did electricity and the computer."

—Dean Kamen, recipient of the National Medal of Technology, physicist, and inventor of the first wearable insulin pump, the HomeChoice portable dialysis machine, the IBOT Mobility System, and the Segway Human Transporter

"One of our leading AI practitioners, Ray Kurzweil, has once again created a 'must read' book for anyone interested in the future of science, the social impact of technology, and indeed the future of our species. His thought-provoking book envisages a future in which we transcend our biological limitations, while making a compelling case that a human civilization with superhuman capabilities is closer at hand than most people realize."

—Raj Reddy, founding director of the Robotics Institute at Carnegie Mellon University and recipient of the Turing Award from the Association for Computing Machinery

"Ray's optimistic book well merits both reading and thoughtful response. For those like myself whose views differ from Ray's on the balance of promise and peril, *The Singularity Is Near* is a clear call for a continuing dialogue to address the greater concerns arising from these accelerating possibilities."

—Bill Joy, cofounder and former chief scientist, Sun Microsystems

ABOUT THE AUTHOR

Ray Kurzweil is one of the world's leading inventors, thinkers, and futurists, with a twenty-year track record of accurate predictions. Called "the restless genius" by *The Wall Street Journal* and "the ultimate thinking machine" by *Forbes* magazine, Kurzweil was selected as one of the top entrepreneurs by *Inc.* magazine, which described him as the "rightful heir to Thomas Edison." PBS selected him as one of "sixteen revolutionaries who made America," along with other inventors of the past two centuries. An inductee into the National Inventors Hall of Fame and recipient of the National Medal of Technology, the Lemelson-MIT Prize (the world's largest award for innovation), thirteen honorary doctorates, and awards from three U.S. presidents, he is the author of four previous books: *Fantastic Voyage: Live Long Enough to Live Forever* (coauthored with Terry Grossman, M.D.), *The Age of Spiritual Machines*, *The 10% Solution for a Healthy Life*, and *The Age of Intelligent Machines*.

RAY KURZWEIL

The Singularity Is Near

WHEN HUMANS TRANSCEND BIOLOGY

PENGUIN BOOKS

To my mother, Hannah,
who provided me with the courage to seek the ideas
to confront any challenge

PENGUIN BOOKS

Published by the Penguin Group

Penguin Group (USA) Inc., 375 Hudson Street, New York, New York 10014, U.S.A.

Penguin Group (Canada), 90 Eglinton Avenue East, Suite 700, Toronto,
Ontario, Canada M4P 2Y3 (a division of Pearson Penguin Canada Inc.)

Penguin Books Ltd, 80 Strand, London WC2R 0RL, England

Penguin Ireland, 25 St Stephen's Green, Dublin 2, Ireland (a division of Penguin Books Ltd)

Penguin Group (Australia), 250 Camberwell Road, Camberwell,
Victoria 3124, Australia (a division of Pearson Australia Group Pty Ltd)

Penguin Books India Pvt Ltd, 11 Community Centre, Panchsheel Park, New Delhi – 110 017, India

Penguin Group (NZ), cnr Airborne and Rosedale Roads, Albany,
Auckland 1310, New Zealand (a division of Pearson New Zealand Ltd)

Penguin Books (South Africa) (Pty) Ltd, 24 Sturdee Avenue,
Rosebank, Johannesburg 2196, South Africa

Penguin Books Ltd, Registered Offices:
80 Strand, London WC2R 0RL, England

First published in the United States of America by Viking Penguin,
a member of Penguin Group (USA) Inc. 2005
Published in Penguin Books 2006

10 9 8 7 6

Copyright © Ray Kurzweil, 2005
All rights reserved

Photograph on p. 368 by Helene DeLillo, 2005

Grateful acknowledgment is made for permission to reprint excerpts from the following copyrighted works: "Plastic Fantastic Lover" by Marty Balin, performed by Jefferson Airplane. Ice Bag Publishing Corp. "What I Am" by Edie Arlisa Brickell, Kenneth Neil Withrow, John Bradley Houser, John Walter Bush, Brandon Aly. © 1988 by Geffen Music, Edie Brickell Songs, Withrow Publishing, Enlightened Kitty Music, Strange Mind Productions. All rights reserved. Administered by Universal—Geffen Music (ASCAP). Used by permission. All rights reserved. "Season of the Witch" by Donovan Leitch. © 1966 by Donovan (Music) Limited. Copyright renewed. International copyright secured. Used by permission. All rights reserved. World rights administered by Peermusic (UK) Ltd. "Sailing to Byzantium" from *The Collected Works of W. B. Yeats, Volume I: The Poems, revised* edited by Richard J. Finneran. Copyright © 1928 by The Macmillan Company; copyright renewed © 1956 by Georgie Yeats. Reprinted with permission of Scribner, an imprint of Simon & Schuster Adult Publishing Group and A. P. Watt Ltd on behalf of Michael B. Yeats.

THE LIBRARY OF CONGRESS HAS CATALOGED THE HARDCOVER EDITION AS FOLLOWS:
    Kurzweil, Ray.
        The singularity is near : when humans transcend biology / Ray Kukrzweil.
            p.   cm.
        Includes bibliographical references (p. 497).
        ISBN 0-670-03384-7 (hc.)
        ISBN 0 14 30.3788 9 (pbk.)
        1. Brain—Evolution.   2. Human evolution.   3. Genetics.   4. Nanotechnology.   5. Robotics.   I.
    Title.
    QP376.K85 2005
    153.9—dc22   2004061231

Printed in the United States of America   •   Set in Minion   •   Designed by Amy Hill

Except in the United States of America, this book is sold subject to the condition that it shall not, by way of trade or otherwise, be lent, resold, hired out, or otherwise circulated without the publisher's prior consent in any form of binding or cover other than that in which it is published and without a similar condition including this condition being imposed on the subsequent purchaser.

The scanning, uploading and distribution of this book via the Internet or via any other means without the permission of the publisher is illegal and punishable by law. Please purchase only authorized electronic editions, and do not participate in or encourage electronic piracy of copyrighted materials. Your support of the author's rights is appreciated.

Contents

CHAPTER THREE

Achieving the Computational Capacity of the Human Brain

CHAPTER FOUR

Achieving the Software of Human Intelligence: How to Reverse Engineer the Human Brain

CHAPTER FIVE

GNR: Three Overlapping Revolutions

Life's Computer. Designer Baby Boomers. Can We Really Live Forever? RNAi (RNA Interference). Cell Therapies. Gene Chips. Somatic Gene Therapy. Reversing Degenerative Disease. Combating Heart Disease. Overcoming Cancer. Reversing Aging. DNA Mutations. Toxic Cells. Mitochondrial Mutations. Intracellular Aggregates. Extracellular Aggregates. Cell Loss and Atrophy. Human Cloning: The Least Interesting Application of Cloning Technology. Why Is Cloning Important? Preserving Endangered Species and Restoring Extinct Ones. Therapeutic Cloning. Human Somatic-Cell Engineering. Solving World Hunger. Human Cloning Revisited.

The Biological Assembler. Upgrading the Cell Nucleus with a Nano-computer and Nanobot. Fat and Sticky Fingers. The Debate Heats Up. Early Adopters. Powering the Singularity. Applications of Nanotechnology to the Environment. Nanobots in the Bloodstream.

Runaway AI. The AI Winter. AI's Toolkit. Expert Systems. Bayesian Nets. Markov Models. Neural Nets. Genetic Algorithms (GAs). Recursive Search. Deep Fritz Draws: Are Humans Getting Smarter, or Are Computers Getting Stupider? The Specialized-Hardware Advantage. Deep Blue Versus Deep Fritz. Significant Software Gains. Are Human Chess Players Doomed? Combining Methods. A Narrow AI Sampler. Military and Intelligence. Space Exploration. Medicine. Science and Math. Business, Finance, and Manufacturing. Manufacturing and Robotics. Speech and Language. Entertainment and Sports. Strong AI.

CHAPTER SIX

CHAPTER NINE

Response to Critics

Epilogue

How Singular? Human Centrality.

Acknowledgments

I'd like to express my deep appreciation to my mother, Hannah, and my father, Fredric, for supporting all of my early ideas and inventions without question, which gave me the freedom to experiment; to my sister Enid for her inspiration; and to my wife, Sonya, and my kids, Ethan and Amy, who give my life meaning, love, and motivation.

I'd like to thank the many talented and devoted people who assisted me with this complex project:

At Viking: my editor, Rick Kot, who provided leadership, enthusiasm, and insightful editing; Clare Ferraro, who provided strong support as publisher; Timothy Mennel, who provided expert copyediting; Bruce Giffords and John Jusino, for coordinating the many details of book production; Amy Hill, for the interior text design; Holly Watson, for her effective publicity work; Alessandra Lusardi, who ably assisted Rick Kot; Paul Buckley, for his clear and elegant art design; and Herb Thornby, who designed the engaging cover.

Loretta Barrett, my literary agent, whose enthusiastic and astute guidance helped guide this project.

Terry Grossman, M.D., my health collaborator and coauthor of *Fantastic Voyage: Live Long Enough to Live Forever*, for helping me to develop my ideas on health and biotechnology through 10,000 e-mails back and forth, and a multi-faceted collaboration.

Martine Rothblatt, for her dedication to all of the technologies discussed in this book and for our collaboration in developing diverse technologies in these areas.

Aaron Kleiner, my long-term business partner (since 1973), for his devotion and collaboration through many projects, including this one.

Amara Angelica, whose devoted and insightful efforts led our research team. Amara also used her outstanding editing skills to assist me in articulating the complex issues in this book. Kathryn Myronuk, whose dedicated research efforts made a major contribution to the research and the notes. Sarah Black contributed discerning research and editorial skills. My research team provided very capable assistance: Amara Angelica, Kathryn Myronuk, Sarah Black, Daniel Pentlarge, Emily Brown, Celia Black-Brooks, Nanda Barker-Hook, Sarah Brangan, Robert Bradbury, John Tillinghast, Elizabeth Collins, Bruce Damer, Jim Rintoul, Sue Rintoul, Larry Klaes, and Chris Wright. Additional assistance was provided by Liz Berry, Sarah Brangan, Rosemary Drinka, Linda Katz, Lisa Kirschner, Inna Nirenberg, Christopher Setzer, Joan Walsh, and Beverly Zibrak.

Laksman Frank, who created many of the attractive diagrams and images from my descriptions, and formatted the graphs.

Celia Black-Brooks, for providing her leadership in project development and communications.

Phil Cohen and Ted Coyle, for implementing my ideas for the illustration on page 322, and Helene DeLillo, for the "Singularity Is Near" photo at the beginning of chapter 7.

Nanda Barker-Hook, Emily Brown, and Sarah Brangan, who helped manage the extensive logistics of the research and editorial processes.

Ken Linde and Matt Bridges, who provided computer systems support to keep our intricate work flow progressing smoothly.

Denise Scutellaro, Joan Walsh, Maria Ellis, and Bob Beal, for doing the accounting on this complicated project.

The KurzweilAI.net team, who provided substantial research support for the project: Aaron Kleiner, Amara Angelica, Bob Beal, Celia Black-Brooks, Daniel Pentlarge, Denise Scutellaro, Emily Brown, Joan Walsh, Ken Linde, Laksman Frank, Maria Ellis, Matt Bridges, Nanda Barker-Hook, Sarah Black, and Sarah Brangan.

Mark Bizzell, Deborah Lieberman, Kirsten Clausen, and Dea Eldorado, for their assistance in communication of this book's message.

Robert A. Freitas Jr., for his thorough review of the nanotechnology-related material.

Paul Linsay, for his thorough review of the mathematics in this book.

My peer expert readers who provided the invaluable service of carefully reviewing the scientific content: Robert A. Freitas Jr. (nanotechnology, cosmology), Ralph Merkle (nanotechnology), Martine Rothblatt (biotechnology, technology acceleration), Terry Grossman (health, medicine, biotechnology), Tomaso Poggio (brain science and brain reverse-engineering), John Parmentola (physics, military technology), Dean Kamen (technology development), Neil Gershenfeld (computational technology, physics, quantum mechanics), Joel Gershenfeld (systems engineering), Hans Moravec (artificial intelligence, robotics), Max More (technology acceleration, philosophy), Jean-Jacques E. Slotine (brain and cognitive science), Sherry Turkle (social impact of technology), Seth Shostak (SETI, cosmology, astronomy), Damien Broderick (technology acceleration, the Singularity), and Harry George (technology entrepreneurship).

My capable in-house readers: Amara Angelica, Sarah Black, Kathryn Myronuk, Nanda Barker-Hook, Emily Brown, Celia Black-Brooks, Aaron Kleiner, Ken Linde, John Chalupa, and Paul Albrecht.

My lay readers, who provided keen insights: my son, Ethan Kurzweil, and David Dalrymple.

Bill Gates, Eric Drexler, and Marvin Minsky, who gave permission to include their dialogues in the book, and for their ideas, which were incorporated into the dialogues.

The many scientists and thinkers whose ideas and efforts are contributing to our exponentially expanding human knowledge base.

The above-named individuals provided many ideas and corrections that I was able to make thanks to their efforts. For any mistakes that remain, I take sole responsibility.

The Singularity Is Near

The Power of Ideas

I do not think there is any thrill that can go through the human heart like that felt by the inventor as he sees some creation of the brain unfolding to success.

—NIKOLA TESLA, 1896, INVENTOR OF ALTERNATING CURRENT

At the age of five, I had the idea that I would become an inventor. I had the notion that inventions could change the world. When other kids were wondering aloud what they wanted to be, I already had the conceit that I knew what I was going to be. The rocket ship to the moon that I was then building (almost a decade before President Kennedy's challenge to the nation) did not work out. But at around the time I turned eight, my inventions became a little more realistic, such as a robotic theater with mechanical linkages that could move scenery and characters in and out of view, and virtual baseball games.

Having fled the Holocaust, my parents, both artists, wanted a more worldly, less provincial, religious upbringing for me.[1] My spiritual education, as a result, took place in a Unitarian church. We would spend six months studying one religion—going to its services, reading its books, having dialogues with its leaders—and then move on to the next. The theme was "many paths to the truth." I noticed, of course, many parallels among the world's religious traditions, but even the inconsistencies were illuminating. It became clear to me that the basic truths were profound enough to transcend apparent contradictions.

At the age of eight, I discovered the Tom Swift Jr. series of books. The plots of all of the thirty-three books (only nine of which had been published when I started to read them in 1956) were always the same: Tom would get himself into a terrible predicament, in which his fate and that of his friends, and often the rest of the human race, hung in the balance. Tom would retreat to his basement lab and think about how to solve the problem. This, then, was the dramatic tension in each book in the series: what ingenious idea would Tom and his friends come up with to save the day?[2] The moral of these tales was simple: the right idea had the power to overcome a seemingly overwhelming challenge.

To this day, I remain convinced of this basic philosophy: no matter what quandaries we face—business problems, health issues, relationship difficulties, as well as the great scientific, social, and cultural challenges of our time—there is an idea that can enable us to prevail. Furthermore, we can find that idea. And when we find it, we need to implement it. My life has been shaped by this imperative. The power of an idea—this is itself an idea.

Around the same time that I was reading the Tom Swift Jr. series, I recall my grandfather, who had also fled Europe with my mother, coming back from his first return visit to Europe with two key memories. One was the gracious treatment he received from the Austrians and Germans, the same people who had forced him to flee in 1938. The other was a rare opportunity he had been given to touch with his own hands some original manuscripts of Leonardo da Vinci. Both recollections influenced me, but the latter is one I've returned to many times. He described the experience with reverence, as if he had touched the work of God himself. This, then, was the religion that I was raised with: veneration for human creativity and the power of ideas.

In 1960, at the age of twelve, I discovered the computer and became fascinated with its ability to model and re-create the world. I hung around the surplus electronics stores on Canal Street in Manhattan (they're still there!) and gathered parts to build my own computational devices. During the 1960s, I was as absorbed in the contemporary musical, cultural, and political movements as my peers, but I became equally engaged in a much more obscure trend: namely, the remarkable sequence of machines that IBM proffered during that decade, from their big "7000" series (7070, 7074, 7090, 7094) to their small 1620, effectively the first "minicomputer." The machines were introduced at yearly intervals, and each one was less expensive and more powerful than the last, a phenomenon familiar today. I got access to an IBM 1620 and began to write programs for statistical analysis and subsequently for music composition.

I still recall the time in 1968 when I was allowed into the secure, cavernous chamber housing what was then the most powerful computer in New England, a top-of-the-line IBM 360 Model 91, with a remarkable million bytes (one megabyte) of "core" memory, an impressive speed of one million instructions per second (one MIPS), and a rental cost of only one thousand dollars per hour. I had developed a computer program that matched high-school students to colleges, and I watched in fascination as the front-panel lights danced through a distinctive pattern as the machine processed each student's application.[3] Even though I was quite familiar with every line of code, it nonetheless seemed as if the computer were deep in thought when the lights dimmed for several seconds at the denouement of each such cycle. Indeed, it could do flawlessly in ten seconds what took us ten hours to do manually with far less accuracy.

As an inventor in the 1970s, I came to realize that my inventions needed to make sense in terms of the enabling technologies and market forces that would exist when the inventions were introduced, as that world would be a very different one from the one in which they were conceived. I began to develop models of how distinct technologies—electronics, communications, computer processors, memory, magnetic storage, and others—developed and how these changes rippled through markets and ultimately our social institutions. I realized that most inventions fail not because the R&D department can't get them to work but because the timing is wrong. Inventing is a lot like surfing: you have to anticipate and catch the wave at just the right moment.

My interest in technology trends and their implications took on a life of its own in the 1980s, and I began to use my models to project and anticipate future technologies, innovations that would appear in 2000, 2010, 2020, and beyond. This enabled me to invent with the capabilities of the future by conceiving and designing inventions using these future capabilities. In the mid-to-late 1980s, I wrote my first book, *The Age of Intelligent Machines.*[4] It included extensive (and reasonably accurate) predictions for the 1990s and 2000s, and ended with the specter of machine intelligence becoming indistinguishable from that of its human progenitors within the first half of the twenty-first century. It seemed like a poignant conclusion, and in any event I personally found it difficult to look beyond so transforming an outcome.

Over the last twenty years, I have come to appreciate an important meta-idea: that the power of ideas to transform the world is itself accelerating. Although people readily agree with this observation when it is simply stated, relatively few observers truly appreciate its profound implications. Within the next several decades, we will have the opportunity to apply ideas to conquer age-old problems—and introduce a few new problems along the way.

During the 1990s, I gathered empirical data on the apparent acceleration of all information-related technologies and sought to refine the mathematical models underlying these observations. I developed a theory I call the law of accelerating returns, which explains why technology and evolutionary processes in general progress in an exponential fashion.[5] In *The Age of Spiritual Machines* (*ASM*), which I wrote in 1998, I sought to articulate the nature of human life as it would exist past the point when machine and human cognition blurred. Indeed, I've seen this epoch as an increasingly intimate collaboration between our biological heritage and a future that transcends biology.

Since the publication of *ASM*, I have begun to reflect on the future of our civilization and its relationship to our place in the universe. Although it may seem difficult to envision the capabilities of a future civilization whose intelligence vastly outstrips our own, our ability to create models of reality in our mind enables us to articulate meaningful insights into the implications of this impending merger of our biological thinking with the nonbiological intelligence we are creating. This, then, is the story I wish to tell in this book. The story is predicated on the idea that we have the ability to understand our own intelligence—to access our own source code, if you will—and then revise and expand it.

Some observers question whether we are capable of applying our own thinking to understand our own thinking. AI researcher Douglas Hofstadter muses that "it could be simply an accident of fate that our brains are too weak to understand themselves. Think of the lowly giraffe, for instance, whose brain is obviously far below the level required for self-understanding—yet it is remarkably similar to our brain."[6] However, we have already succeeded in modeling portions of our brain—neurons and substantial neural regions—and the complexity of such models is growing rapidly. Our progress in reverse engineering the human brain, a key issue that I will describe in detail in this book, demonstrates that we do indeed have the ability to understand, to model, and to extend our own intelligence. This is one aspect of the uniqueness of our species: our intelligence is just sufficiently above the critical threshold necessary for us to scale our own ability to unrestricted heights of creative power—and we have the opposable appendage (our thumbs) necessary to manipulate the universe to our will.

A word on magic: when I was reading the Tom Swift Jr. books, I was also an avid magician. I enjoyed the delight of my audiences in experiencing apparently impossible transformations of reality. In my teen years, I replaced my parlor magic with technology projects. I discovered that unlike mere tricks, technology does not lose its transcendent power when its secrets are revealed. I am often reminded of Arthur C. Clarke's third law, that "any sufficiently advanced technology is indistinguishable from magic."

Consider J. K. Rowling's Harry Potter stories from this perspective. These tales may be imaginary, but they are not unreasonable visions of our world as it will exist only a few decades from now. Essentially all of the Potter "magic" will be realized through the technologies I will explore in this book. Playing quidditch and transforming people and objects into other forms will be feasible in full-immersion virtual-reality environments, as well as in real reality, using nanoscale devices. More dubious is the time reversal (as described in *Harry Potter and the Prisoner of Azkaban*), although serious proposals have even been put forward for accomplishing something along these lines (without giving rise to causality paradoxes), at least for bits of information, which essentially is what we comprise. (See the discussion in chapter 3 on the ultimate limits of computation.)

Consider that Harry unleashes his magic by uttering the right incantation. Of course, discovering and applying these incantations are no simple matters. Harry and his colleagues need to get the sequence, procedures, and emphasis exactly correct. That process is precisely our experience with technology. Our incantations are the formulas and algorithms underlying our modern-day magic. With just the right sequence, we can get a computer to read a book out loud, understand human speech, anticipate (and prevent) a heart attack, or predict the movement of a stock-market holding. If an incantation is just slightly off mark, the magic is greatly weakened or does not work at all.

One might object to this metaphor by pointing out that Hogwartian incantations are brief and therefore do not contain much information compared to, say, the code for a modern software program. But the essential methods of modern technology generally share the same brevity. The principles of operation of software advances such as speech recognition can be written in just a few pages of formulas. Often a key advance is a matter of applying a small change to a single formula.

The same observation holds for the "inventions" of biological evolution: consider that the genetic difference between chimpanzees and humans, for example, is only a few hundred thousand bytes of information. Although chimps are capable of some intellectual feats, that tiny difference in our genes was sufficient for our species to create the magic of technology.

Muriel Rukeyser says that "the universe is made of stories, not of atoms." In chapter 7, I describe myself as a "patternist," someone who views patterns of information as the fundamental reality. For example, the particles composing my brain and body change within weeks, but there is a continuity to the patterns that these particles make. A story can be regarded as a meaningful pattern of information, so we can interpret Muriel Rukeyser's aphorism from this perspective. This book, then, is the story of the destiny of the human-machine civilization, a destiny we have come to refer to as the Singularity.

CHAPTER ONE

The Six Epochs

Everyone takes the limits of his own vision for the limits of the world.
—ARTHUR SCHOPENHAUER

I am not sure when I first became aware of the Singularity. I'd have to say it was a progressive awakening. In the almost half century that I've immersed myself in computer and related technologies, I've sought to understand the meaning and purpose of the continual upheaval that I have witnessed at many levels. Gradually, I've become aware of a transforming event looming in the first half of the twenty-first century. Just as a black hole in space dramatically alters the patterns of matter and energy accelerating toward its event horizon, this impending Singularity in our future is increasingly transforming every institution and aspect of human life, from sexuality to spirituality.

What, then, is the Singularity? It's a future period during which the pace of technological change will be so rapid, its impact so deep, that human life will be irreversibly transformed. Although neither utopian nor dystopian, this epoch will transform the concepts that we rely on to give meaning to our lives, from our business models to the cycle of human life, including death itself. Understanding the Singularity will alter our perspective on the significance of our past and the ramifications for our future. To truly understand it inherently changes one's view of life in general and one's own particular life. I regard someone who understands the Singularity and who has reflected on its implications for his or her own life as a "singularitarian."[1]

I can understand why many observers do not readily embrace the obvious implications of what I have called the law of accelerating returns (the inherent acceleration of the rate of evolution, with technological evolution as a continuation of biological evolution). After all, it took me forty years to be able to see what was right in front of me, and I still cannot say that I am entirely comfortable with all of its consequences.

The key idea underlying the impending Singularity is that the pace of change of our human-created technology is accelerating and its powers are expanding at an exponential pace. Exponential growth is deceptive. It starts out almost imperceptibly and then explodes with unexpected fury—unexpected, that is, if one does not take care to follow its trajectory. (See the "Linear vs. Exponential Growth" graph on p. 10.)

Consider this parable: a lake owner wants to stay at home to tend to the lake's fish and make certain that the lake itself will not become covered with lily pads, which are said to double their number every few days. Month after month, he patiently waits, yet only tiny patches of lily pads can be discerned, and they don't seem to be expanding in any noticeable way. With the lily pads covering less than 1 percent of the lake, the owner figures that it's safe to take a vacation and leaves with his family. When he returns a few weeks later, he's shocked to discover that the entire lake has become covered with the pads, and his fish have perished. By doubling their number every few days, the last seven doublings were sufficient to extend the pads' coverage to the entire lake. (Seven doublings extended their reach 128-fold.) This is the nature of exponential growth.

Consider Gary Kasparov, who scorned the pathetic state of computer chess in 1992. Yet the relentless doubling of computer power every year enabled a computer to defeat him only five years later.[2] The list of ways computers can now exceed human capabilities is rapidly growing. Moreover, the once narrow applications of computer intelligence are gradually broadening in one type of activity after another. For example, computers are diagnosing electrocardiograms and medical images, flying and landing airplanes, controlling the tactical decisions of automated weapons, making credit and financial decisions, and being given responsibility for many other tasks that used to require human intelligence. The performance of these systems is increasingly based on integrating multiple types of artificial intelligence (AI). But as long as there is an AI shortcoming in any such area of endeavor, skeptics will point to that area as an inherent bastion of permanent human superiority over the capabilities of our own creations.

This book will argue, however, that within several decades information-based technologies will encompass all human knowledge and proficiency, ultimately including the pattern-recognition powers, problem-solving skills, and emotional and moral intelligence of the human brain itself.

Although impressive in many respects, the brain suffers from severe limitations. We use its massive parallelism (one hundred trillion interneuronal connections operating simultaneously) to quickly recognize subtle patterns. But our thinking is extremely slow: the basic neural transactions are several million times slower than contemporary electronic circuits. That makes our physiological bandwidth for processing new information extremely limited compared to the exponential growth of the overall human knowledge base.

Our version 1.0 biological bodies are likewise frail and subject to a myriad of failure modes, not to mention the cumbersome maintenance rituals they require. While human intelligence is sometimes capable of soaring in its creativity and expressiveness, much human thought is derivative, petty, and circumscribed.

The Singularity will allow us to transcend these limitations of our biological bodies and brains. We will gain power over our fates. Our mortality will be in our own hands. We will be able to live as long as we want (a subtly different statement from saying we will live forever). We will fully understand human thinking and will vastly extend and expand its reach. By the end of this century, the nonbiological portion of our intelligence will be trillions of trillions of times more powerful than unaided human intelligence.

We are now in the early stages of this transition. The acceleration of paradigm shift (the rate at which we change fundamental technical approaches) as well as the exponential growth of the capacity of information technology are both beginning to reach the "knee of the curve," which is the stage at which an exponential trend becomes noticeable. Shortly after this stage, the trend quickly becomes explosive. Before the middle of this century, the growth rates of our technology—which will be indistinguishable from ourselves—will be so steep as to appear essentially vertical. From a strictly mathematical perspective, the growth rates will still be finite but so extreme that the changes they bring about will appear to rupture the fabric of human history. That, at least, will be the perspective of unenhanced biological humanity.

The Singularity will represent the culmination of the merger of our biological thinking and existence with our technology, resulting in a world that is still human but that transcends our biological roots. There will be no distinction, post-Singularity, between human and machine or between physical and virtual reality. If you wonder what will remain unequivocally human in such a world, it's simply this quality: ours is the species that inherently seeks to extend its physical and mental reach beyond current limitations.

Many commentators on these changes focus on what they perceive as a loss of some vital aspect of our humanity that will result from this transition. This perspective stems, however, from a misunderstanding of what our technology will become. All the machines we have met to date lack the essential subtlety of human biological qualities. Although the Singularity has many faces, its most important implication is this: our technology will match and then vastly exceed the refinement and suppleness of what we regard as the best of human traits.

The Intuitive Linear View Versus
the Historical Exponential View

> When the first transhuman intelligence is created and launches itself into
> recursive self-improvement, a fundamental discontinuity is likely to occur,
> the likes of which I can't even begin to predict.
>
> —MICHAEL ANISSIMOV

In the 1950s John von Neumann, the legendary information theorist, was
quoted as saying that "the ever-accelerating progress of technology . . . gives the
appearance of approaching some essential singularity in the history of the race
beyond which human affairs, as we know them, could not continue."[3] Von
Neumann makes two important observations here: *acceleration* and *singularity*.
The first idea is that human progress is exponential (that is, it expands by
repeatedly *multiplying* by a constant) rather than linear (that is, expanding by
repeatedly *adding* a constant).

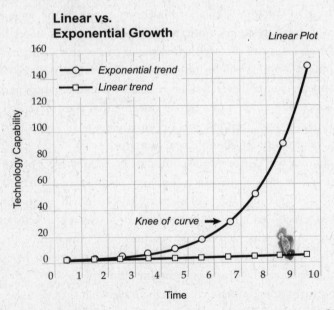

Linear versus exponential: Linear growth is steady; exponential growth
becomes explosive.

The second is that exponential growth is seductive, starting out slowly and
virtually unnoticeably, but beyond the knee of the curve it turns explosive and
profoundly transformative. The future is widely misunderstood. Our forebears
expected it to be pretty much like their present, which had been pretty much

like their past. Exponential trends did exist one thousand years ago, but they were at that very early stage in which they were so flat and so slow that they looked like no trend at all. As a result, observers' expectation of an unchanged future was fulfilled. Today, we anticipate continuous technological progress and the social repercussions that follow. But the future will be far more surprising than most people realize, because few observers have truly internalized the implications of the fact that the rate of change itself is accelerating.

Most long-range forecasts of what is technically feasible in future time periods dramatically underestimate the power of future developments because they are based on what I call the "intuitive linear" view of history rather than the "historical exponential" view. My models show that we are doubling the paradigm-shift rate every decade, as I will discuss in the next chapter. Thus the twentieth century was gradually speeding up to today's rate of progress; its achievements, therefore, were equivalent to about twenty years of progress at the rate in 2000. We'll make another twenty years of progress in just fourteen years (by 2014), and then do the same again in only seven years. To express this another way, we won't experience one hundred years of technological advance in the twenty-first century; we will witness on the order of twenty thousand years of progress (again, when measured by *today's* rate of progress), or about one thousand times greater than what was achieved in the twentieth century.[4]

Misperceptions about the shape of the future come up frequently and in a variety of contexts. As one example of many, in a recent debate in which I took part concerning the feasibility of molecular manufacturing, a Nobel Prize–winning panelist dismissed safety concerns regarding nanotechnology, proclaiming that "we're not going to see self-replicating nanoengineered entities [devices constructed molecular fragment by fragment] for a hundred years." I pointed out that one hundred years was a reasonable estimate and actually matched my own appraisal of the amount of technical progress required to achieve this particular milestone when measured *at today's rate of progress* (five times the average rate of change we saw in the twentieth century). But because we're doubling the rate of progress every decade, we'll see the equivalent of a century of progress—at *today's rate*—in only twenty-five calendar years.

Similarly at *Time* magazine's Future of Life conference, held in 2003 to celebrate the fiftieth anniversary of the discovery of the structure of DNA, all of the invited speakers were asked what they thought the next fifty years would be like.[5] Virtually every presenter looked at the progress of the last fifty years and used it as a model for the next fifty years. For example, James Watson, the codiscoverer of DNA, said that in fifty years we will have drugs that will allow us to eat as much as we want without gaining weight.

I replied, "Fifty years?" We have accomplished this already in mice by blocking the fat insulin receptor gene that controls the storage of fat in the fat cells. Drugs for human use (using RNA interference and other techniques we will discuss in chapter 5) are in development now and will be in FDA tests in several years. These will be available in five to ten years, not fifty. Other projections were equally shortsighted, reflecting contemporary research priorities rather than the profound changes that the next half century will bring. Of all the thinkers at this conference, it was primarily Bill Joy and I who took account of the exponential nature of the future, although Joy and I disagree on the import of these changes, as I will discuss in chapter 8.

People intuitively assume that the current rate of progress will continue for future periods. Even for those who have been around long enough to experience how the pace of change increases over time, unexamined intuition leaves one with the impression that change occurs at the same rate that we have experienced most recently. From the mathematician's perspective, the reason for this is that an exponential curve looks like a straight line when examined for only a brief duration. As a result, even sophisticated commentators, when considering the future, typically extrapolate the current pace of change over the next ten years or one hundred years to determine their expectations. This is why I describe this way of looking at the future as the "intuitive linear" view.

But a serious assessment of the history of technology reveals that technological change is exponential. Exponential growth is a feature of any evolutionary process, of which technology is a primary example. You can examine the data in different ways, on different timescales, and for a wide variety of technologies, ranging from electronic to biological, as well as for their implications, ranging from the amount of human knowledge to the size of the economy. The acceleration of progress and growth applies to each of them. Indeed, we often find not just simple exponential growth, but "double" exponential growth, meaning that the rate of exponential growth (that is, the exponent) is itself growing exponentially (for example, see the discussion on the price-performance of computing in the next chapter).

Many scientists and engineers have what I call "scientist's pessimism." Often, they are so immersed in the difficulties and intricate details of a contemporary challenge that they fail to appreciate the ultimate long-term implications of their own work, and the larger field of work in which they operate. They likewise fail to account for the far more powerful tools they will have available with each new generation of technology.

Scientists are trained to be skeptical, to speak cautiously of current research goals, and to rarely speculate beyond the current generation of scientific pursuit. This may have been a satisfactory approach when a generation of science and technology lasted longer than a human generation, but it does not serve society's interests now that a generation of scientific and technological progress comprises only a few years.

Consider the biochemists who, in 1990, were skeptical of the goal of transcribing the entire human genome in a mere fifteen years. These scientists had just spent an entire year transcribing a mere one ten-thousandth of the genome. So, even with reasonable anticipated advances, it seemed natural to them that it would take a century, if not longer, before the entire genome could be sequenced.

Or consider the skepticism expressed in the mid-1980s that the Internet would ever be a significant phenomenon, given that it then included only tens of thousands of nodes (also known as servers). In fact, the number of nodes was doubling every year, so that there were likely to be tens of millions of nodes ten years later. But this trend was not appreciated by those who struggled with state-of-the-art technology in 1985, which permitted adding only a few thousand nodes throughout the world in a single year.[6]

The converse conceptual error occurs when certain exponential phenomena are first recognized and are applied in an overly aggressive manner without modeling the appropriate pace of growth. While exponential growth gains speed over time, it is not instantaneous. The run-up in capital values (that is, stock market prices) during the "Internet bubble" and related telecommunications bubble (1997–2000) was greatly in excess of any reasonable expectation of even exponential growth. As I demonstrate in the next chapter, the actual adoption of the Internet and e-commerce did show smooth exponential growth through both boom and bust; the overzealous expectation of growth affected only capital (stock) valuations. We have seen comparable mistakes during earlier paradigm shifts—for example, during the early railroad era (1830s), when the equivalent of the Internet boom and bust led to a frenzy of railroad expansion.

Another error that prognosticators make is to consider the transformations that will result from a single trend in today's world as if nothing else will change. A good example is the concern that radical life extension will result in overpopulation and the exhaustion of limited material resources to sustain human life, which ignores comparably radical wealth creation from nanotechnology and strong AI. For example, nanotechnology-based manufacturing devices in the 2020s will be capable of creating almost any physical product from inexpensive raw materials and information.

I emphasize the exponential-versus-linear perspective because it's the most important failure that prognosticators make in considering future trends. Most technology forecasts and forecasters ignore altogether this historical exponential view of technological progress. Indeed, almost everyone I meet has a linear view of the future. That's why people tend to overestimate what can be achieved in the short term (because we tend to leave out necessary details) but underestimate what can be achieved in the long term (because exponential growth is ignored).

The Six Epochs

> First we build the tools, then they build us.
> —MARSHALL MCLUHAN

> The future ain't what it used to be.
> —YOGI BERRA

Evolution is a process of creating patterns of increasing order. I'll discuss the concept of order in the next chapter; the emphasis in this section is on the concept of patterns. I believe that it's the evolution of patterns that constitutes the ultimate story of our world. Evolution works through indirection: each stage or epoch uses the information-processing methods of the previous epoch to create the next. I conceptualize the history of evolution—both biological and technological—as occurring in six epochs. As we will discuss, the Singularity will begin with Epoch Five and will spread from Earth to the rest of the universe in Epoch Six.

Epoch One: Physics and Chemistry. We can trace our origins to a state that represents information in its basic structures: patterns of matter and energy. Recent theories of quantum gravity hold that time and space are broken down into discrete quanta, essentially fragments of information. There is controversy as to whether matter and energy are ultimately digital or analog in nature, but regardless of the resolution of this issue, we do know that atomic structures store and represent discrete information.

A few hundred thousand years after the Big Bang, atoms began to form, as electrons became trapped in orbits around nuclei consisting of protons and neutrons. The electrical structure of atoms made them "sticky." Chemistry was born a few million years later as atoms came together to create relatively stable structures called molecules. Of all the elements, carbon proved to be the most versatile; it's able to form bonds in four directions (versus one to three for most other elements), giving rise to complicated, information-rich, three-dimensional structures.

The rules of our universe and the balance of the physical constants that govern the interaction of basic forces are so exquisitely, delicately, and exactly appropriate for the codification and evolution of information (resulting in increasing complexity) that one wonders how such an extraordinarily unlikely situation came about. Where some see a divine hand, others see our own hands—namely, the anthropic principle, which holds that only in a universe that allowed our own evolution would we be here to ask such questions.[7] Recent theories of physics concerning multiple universes speculate that new universes are created on a regular basis, each with its own unique rules, but that most of these either die out quickly or else continue without the evolution of any interesting patterns (such as Earth-based biology has created) because their rules do not support the evolution of increasingly complex forms.[8] It's hard to imagine how we could test these theories of evolution applied to early cosmology, but it's clear that the physical laws of our universe are precisely what they need to be to allow for the evolution of increasing levels of order and complexity.[9]

Vastly expanded human intelligence (predominantly nonbiological) spreads through the universe

Epoch 6 *The Universe Wakes Up*
Patterns of matter and energy in the universe become saturated with intelligent processes and knowledge

Technology masters the methods of biology (including human intelligence)

Epoch 5 *Merger of Technology and Human Intelligence*
The methods of biology (including human intelligence) are integrated into the (exponentially expanding) human technology base

Technology evolves

Epoch 4 *Technology*
Information in hardware and software designs

Brains evolve

Epoch 3 *Brains*
Information in neural patterns

DNA evolves

Epoch 2 *Biology*
Information in DNA

Epoch 1 *Physics and Chemistry*
Information in atomic structures

The Six Epochs of Evolution
Evolution works through indirection: it creates a capability and then uses that capability to evolve the next stage.

Epoch Two: Biology and DNA. In the second epoch, starting several billion years ago, carbon-based compounds became more and more intricate until complex aggregations of molecules formed self-replicating mechanisms, and life originated. Ultimately, biological systems evolved a precise digital mechanism (DNA) to store information describing a larger society of molecules. This molecule and its supporting machinery of codons and ribosomes enabled a record to be kept of the evolutionary experiments of this second epoch.

Epoch Three: Brains. Each epoch continues the evolution of information through a paradigm shift to a further level of "indirection." (That is, evolution uses the results of one epoch to create the next.) For example, in the third epoch, DNA-guided evolution produced organisms that could detect information with their own sensory organs and process and store that information in their own brains and nervous systems. These were made possible by second-epoch mechanisms (DNA and epigenetic information of proteins and RNA fragments that control gene expression), which (indirectly) enabled and defined third-epoch information-processing mechanisms (the brains and nervous systems of organisms). The third epoch started with the ability of early animals to recognize patterns, which still accounts for the vast majority of the activity in our brains.[10] Ultimately, our own species evolved the ability to create abstract mental models of the world we experience and to contemplate the rational implications of these models. We have the ability to redesign the world in our own minds and to put these ideas into action.

Epoch Four: Technology. Combining the endowment of rational and abstract thought with our opposable thumb, our species ushered in the fourth epoch and the next level of indirection: the evolution of human-created technology. This started out with simple mechanisms and developed into elaborate automata (automated mechanical machines). Ultimately, with sophisticated computational and communication devices, technology was itself capable of sensing, storing, and evaluating elaborate patterns of information. To compare the rate of progress of the biological evolution of intelligence to that of technological evolution, consider that the most advanced mammals have added about one cubic inch of brain matter every hundred thousand years, whereas we are roughly doubling the computational capacity of computers every year (see the next chapter). Of course, neither brain size nor computer capacity is the sole determinant of intelligence, but they do represent enabling factors.

If we place key milestones of both biological evolution and human technological development on a single graph plotting both the x-axis (number of years ago) and the *y*-axis (the paradigm-shift time) on logarithmic scales, we find a reasonably straight line (continual acceleration), with biological evolution leading directly to human-directed development.[11]

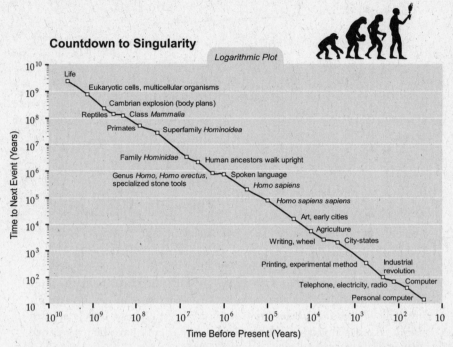

Countdown to Singularity: **Biological evolution and human technology both show continual acceleration, indicated by the shorter time to the next event (two billion years from the origin of life to cells; fourteen years from the PC to the World Wide Web).**

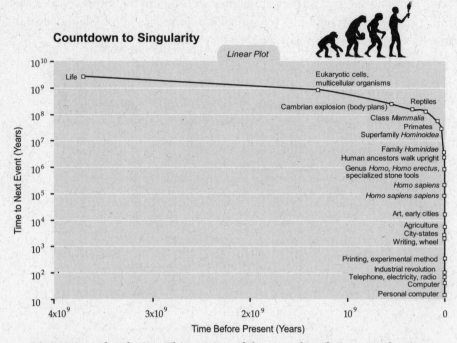

Linear view of evolution: This version of the preceding figure uses the same data but with a linear scale for time before present instead of a logarithmic one. This shows the acceleration more dramatically, but details are not visible. From a linear perspective, most key events have just happened "recently."

The above figures reflect my view of key developments in biological and technological history. Note, however, that the straight line, demonstrating the continual acceleration of evolution, does not depend on my particular selection of events. Many observers and reference books have compiled lists of important events in biological and technological evolution, each of which has its own idiosyncrasies. Despite the diversity of approaches, however, if we combine lists from a variety of sources (for example, the *Encyclopaedia Britannica,* the American Museum of Natural History, Carl Sagan's "cosmic calendar," and others), we observe the same obvious smooth acceleration. The following plot combines fifteen different lists of key events.[12] Since different thinkers assign different dates to the same event, and different lists include similar or overlapping events selected according to different criteria, we see an expected "thickening" of the trend line due to the "noisiness" (statistical variance) of this data. The overall trend, however, is very clear.

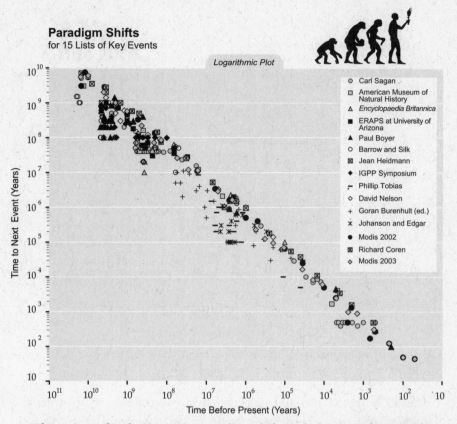

Paradigm Shifts
for 15 Lists of Key Events

Fifteen views of evolution: Major paradigm shifts in the history of the world, as seen by fifteen different lists of key events. There is a clear trend of smooth acceleration through biological and then technological evolution.

Physicist and complexity theorist Theodore Modis analyzed these lists and determined twenty-eight clusters of events (which he called canonical milestones) by combining identical, similar, and/or related events from the different lists.[13] This process essentially removes the "noise" (for example, the variability of dates between lists) from the lists, revealing again the same progression:

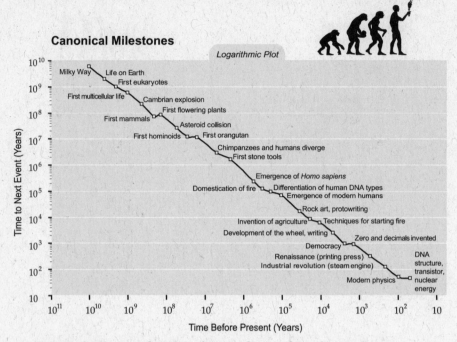

Canonical milestones based on clusters of events from thirteen lists.

The attributes that are growing exponentially in these charts are order and complexity, concepts we will explore in the next chapter. This acceleration matches our commonsense observations. A billion years ago, not much happened over the course of even one million years. But a quarter-million years ago epochal events such as the evolution of our species occurred in time frames of just one hundred thousand years. In technology, if we go back fifty thousand years, not much happened over a one-thousand-year period. But in the recent past, we see new paradigms, such as the World Wide Web, progress from inception to mass adoption (meaning that they are used by a quarter of the population in advanced countries) within only a decade.

Epoch Five: The Merger of Human Technology with Human Intelligence. Looking ahead several decades, the Singularity will begin with the fifth epoch. It will result from the merger of the vast knowledge embedded in our own brains with the vastly greater capacity, speed, and knowledge-sharing ability of our technology. The fifth epoch will enable our human-machine civilization to transcend the human brain's limitations of a mere hundred trillion extremely slow connections.[14]

The Singularity will allow us to overcome age-old human problems and vastly amplify human creativity. We will preserve and enhance the intelligence that evolution has bestowed on us while overcoming the profound limitations of biological evolution. But the Singularity will also amplify the ability to act on our destructive inclinations, so its full story has not yet been written.

Epoch Six: The Universe Wakes Up. I will discuss this topic in chapter 6, under the heading ". . . on the Intelligent Destiny of the Cosmos." In the aftermath of the Singularity, intelligence, derived from its biological origins in human brains and its technological origins in human ingenuity, will begin to saturate the matter and energy in its midst. It will achieve this by reorganizing matter and energy to provide an optimal level of computation (based on limits we will discuss in chapter 3) to spread out from its origin on Earth.

We currently understand the speed of light as a bounding factor on the transfer of information. Circumventing this limit has to be regarded as highly speculative, but there are hints that this constraint may be able to be superseded.[15] If there are even subtle deviations, we will ultimately harness this superluminal ability. Whether our civilization infuses the rest of the universe with its creativity and intelligence quickly or slowly depends on its immutability. In any event the "dumb" matter and mechanisms of the universe will be transformed into exquisitely sublime forms of intelligence, which will constitute the sixth epoch in the evolution of patterns of information.

This is the ultimate destiny of the Singularity and of the universe.

The Singularity Is Near

> You know, things are going to be really different! . . . No, no, I mean really different!
>
> —MARK MILLER (COMPUTER SCIENTIST) TO ERIC DREXLER, AROUND 1986

What are the consequences of this event? When greater-than-human intelligence drives progress, that progress will be much more rapid. In fact, there seems no reason why progress itself would not involve the creation of still more intelligent entities—on a still-shorter time scale. The best analogy that I see is with the evolutionary past: Animals can adapt to problems and make inventions, but often no faster than natural selection can do its work—the world acts as its own simulator in the case of natural selection. We humans have the ability to internalize the world and conduct "what if's" in our heads; we can solve many problems thousands of times faster than natural selection. Now, by creating the means to execute those simulations at much higher speeds, we are entering a regime as radically different from our human past as we humans are from the lower animals. From the human point of view, this change will be a throwing away of all the previous rules, perhaps in the blink of an eye, an exponential runaway beyond any hope of control.

—VERNOR VINGE, "THE TECHNOLOGICAL SINGULARITY," 1993

Let an ultraintelligent machine be defined as a machine that can far surpass all the intellectual activities of any man however clever. Since the design of machines is one of these intellectual activities, an ultraintelligent machine could design even better machines; there would then unquestionably be an "intelligence explosion," and the intelligence of man would be left far behind. Thus the first ultraintelligent machine is the last invention that man need ever make.

—IRVING JOHN GOOD, "SPECULATIONS CONCERNING THE FIRST ULTRAINTELLIGENT MACHINE," 1965

To put the concept of Singularity into further perspective, let's explore the history of the word itself. "Singularity" is an English word meaning a unique event with, well, singular implications. The word was adopted by mathematicians to denote a value that transcends any finite limitation, such as the explosion of magnitude that results when dividing a constant by a number that gets closer and closer to zero. Consider, for example, the simple function $y = 1/x$. As the value of x approaches zero, the value of the function (y) explodes to larger and larger values.

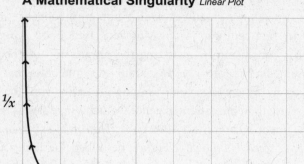

A Mathematical Singularity *Linear Plot*

A mathematical singularity: As *x* approaches zero (from right to left), 1/*x* (or *y*) approaches infinity.

Such a mathematical function never actually achieves an infinite value, since dividing by zero is mathematically "undefined" (impossible to calculate). But the value of *y* exceeds any possible finite limit (approaches infinity) as the divisor *x* approaches zero.

The next field to adopt the word was astrophysics. If a massive star undergoes a supernova explosion, its remnant eventually collapses to the point of apparently zero volume and infinite density, and a "singularity" is created at its center. Because light was thought to be unable to escape the star after it reached this infinite density,[16] it was called a black hole.[17] It constitutes a rupture in the fabric of space and time.

One theory speculates that the universe itself began with such a Singularity.[18] Interestingly, however, the event horizon (surface) of a black hole is of finite size, and gravitational force is only theoretically infinite at the zero-size center of the black hole. At any location that could actually be measured, the forces are finite, although extremely large.

The first reference to the Singularity as an event capable of rupturing the fabric of human history is John von Neumann's statement quoted above. In the 1960s, I. J. Good wrote of an "intelligence explosion" resulting from intelligent machines' designing their next generation without human intervention. Vernor Vinge, a mathematician and computer scientist at San Diego State University, wrote about a rapidly approaching "technological singularity" in an article for *Omni* magazine in 1983 and in a science-fiction novel, *Marooned in Realtime*, in 1986.[19]

My 1989 book, *The Age of Intelligent Machines*, presented a future headed inevitably toward machines greatly exceeding human intelligence in the first half of the twenty-first century.[20] Hans Moravec's 1988 book *Mind Children* came to a similar conclusion by analyzing the progression of robotics.[21] In 1993 Vinge presented a paper to a NASA-organized symposium that described the Singularity as an impending event resulting primarily from the advent of "entities with greater than human intelligence," which Vinge saw as the harbinger of a runaway phenomenon.[22] My 1999 book, *The Age of Spiritual Machines: When Computers Exceed Human Intelligence*, described the increasingly intimate connection between our biological intelligence and the artificial intelligence we are creating.[23] Hans Moravec's book *Robot: Mere Machine to Transcendent Mind*, also published in 1999, described the robots of the 2040s as our "evolutionary heirs," machines that will "grow from us, learn our skills, and share our goals and values, . . . children of our minds."[24] Australian scholar Damien Broderick's 1997 and 2001 books, both titled *The Spike*, analyzed the pervasive impact of the extreme phase of technology acceleration anticipated within several decades.[25] In an extensive series of writings, John Smart has described the Singularity as the inevitable result of what he calls "MEST" (matter, energy, space, and time) compression.[26]

From my perspective, the Singularity has many faces. It represents the nearly vertical phase of exponential growth that occurs when the rate is so extreme that technology appears to be expanding at infinite speed. Of course, from a mathematical perspective, there is no discontinuity, no rupture, and the growth rates remain finite, although extraordinarily large. But from our *currently* limited framework, this imminent event appears to be an acute and abrupt break in the continuity of progress. I emphasize the word "currently" because one of the salient implications of the Singularity will be a change in the nature of our ability to understand. We will become vastly smarter as we merge with our technology.

Can the pace of technological progress continue to speed up indefinitely? Isn't there a point at which humans are unable to think fast enough to keep up? For unenhanced humans, clearly so. But what would 1,000 scientists, each 1,000 times more intelligent than human scientists today, and each operating 1,000 times faster than contemporary humans (because the information processing in their primarily nonbiological brains is faster) accomplish? One chronological year would be like a millennium for them.[27] What would they come up with?

Well, for one thing, they would come up with technology to become even more intelligent (because their intelligence is no longer of fixed capacity). They would change their own thought processes to enable them to think even faster.

When scientists become a million times more intelligent and operate a million times faster, an hour would result in a century of progress (in today's terms).

The Singularity involves the following principles, which I will document, develop, analyze, and contemplate throughout the rest of this book:

- The rate of paradigm shift (technical innovation) is accelerating, right now doubling every decade.[28]
- The power (price-performance, speed, capacity, and bandwidth) of information technologies is growing exponentially at an even faster pace, now doubling about every year.[29] This principle applies to a wide range of measures, including the amount of human knowledge.
- For information technologies, there is a second level of exponential growth: that is, exponential growth in the rate of exponential growth (the exponent). The reason: as a technology becomes more cost effective, more resources are deployed toward its advancement, so the rate of exponential growth increases over time. For example, the computer industry in the 1940s consisted of a handful of now historically important projects. Today total revenue in the computer industry is more than one trillion dollars, so research and development budgets are comparably higher.
- Human brain scanning is one of these exponentially improving technologies. As I will show in chapter 4, the temporal and spatial resolution and bandwidth of brain scanning are doubling each year. We are just now obtaining the tools sufficient to begin serious reverse engineering (decoding) of the human brain's principles of operation. We already have impressive models and simulations of a couple dozen of the brain's several hundred regions. Within two decades, we will have a detailed understanding of how all the regions of the human brain work.
- We will have the requisite hardware to emulate human intelligence with supercomputers by the end of this decade and with personal-computer-size devices by the end of the following decade. We will have effective software models of human intelligence by the mid-2020s.
- With both the hardware and software needed to fully emulate human intelligence, we can expect computers to pass the Turing test, indicating intelligence indistinguishable from that of biological humans, by the end of the 2020s.[30]
- When they achieve this level of development, computers will be able to combine the traditional strengths of human intelligence with the strengths of machine intelligence.
- The traditional strengths of human intelligence include a formidable ability to recognize patterns. The massively parallel and self-organizing nature

of the human brain is an ideal architecture for recognizing patterns that are based on subtle, invariant properties. Humans are also capable of learning new knowledge by applying insights and inferring principles from experience, including information gathered through language. A key capability of human intelligence is the ability to create mental models of reality and to conduct mental "what-if" experiments by varying aspects of these models.

- The traditional strengths of machine intelligence include the ability to remember billions of facts precisely and recall them instantly.
- Another advantage of nonbiological intelligence is that once a skill is mastered by a machine, it can be performed repeatedly at high speed, at optimal accuracy, and without tiring.
- Perhaps most important, machines can share their knowledge at extremely high speed, compared to the very slow speed of human knowledge-sharing through language.
- Nonbiological intelligence will be able to download skills and knowledge from other machines, eventually also from humans.
- Machines will process and switch signals at close to the speed of light (about three hundred million meters per second), compared to about one hundred meters per second for the electrochemical signals used in biological mammalian brains.[31] This speed ratio is at least three million to one.
- Machines will have access via the Internet to all the knowledge of our human-machine civilization and will be able to master all of this knowledge.
- Machines can pool their resources, intelligence, and memories. Two machines—or one million machines—can join together to become one and then become separate again. Multiple machines can do both at the same time: become one and separate simultaneously. Humans call this falling in love, but our biological ability to do this is fleeting and unreliable.
- The combination of these traditional strengths (the pattern-recognition ability of biological human intelligence and the speed, memory capacity and accuracy, and knowledge and skill-sharing abilities of nonbiological intelligence) will be formidable.
- Machine intelligence will have complete freedom of design and architecture (that is, they won't be constrained by biological limitations, such as the slow switching speed of our interneuronal connections or a fixed skull size) as well as consistent performance at all times.
- Once nonbiological intelligence combines the traditional strengths of both humans and machines, the nonbiological portion of our civilization's

intelligence will then continue to benefit from the double exponential growth of machine price-performance, speed, and capacity.

- Once machines achieve the ability to design and engineer technology as humans do, only at far higher speeds and capacities, they will have access to their own designs (source code) and the ability to manipulate them. Humans are now accomplishing something similar through biotechnology (changing the genetic and other information processes underlying our biology), but in a much slower and far more limited way than what machines will be able to achieve by modifying their own programs.

- Biology has inherent limitations. For example, every living organism must be built from proteins that are folded from one-dimensional strings of amino acids. Protein-based mechanisms are lacking in strength and speed. We will be able to reengineer all of the organs and systems in our biological bodies and brains to be vastly more capable.

- As we will discuss in chapter 4, human intelligence does have a certain amount of plasticity (ability to change its structure), more so than had previously been understood. But the architecture of the human brain is nonetheless profoundly limited. For example, there is room for only about one hundred trillion interneuronal connections in each of our skulls. A key genetic change that allowed for the greater cognitive ability of humans compared to that of our primate ancestors was the development of a larger cerebral cortex as well as the development of increased volume of gray-matter tissue in certain regions of the brain.[32] This change occurred, however, on the very slow timescale of biological evolution and still involves an inherent limit to the brain's capacity. Machines will be able to reformulate their own designs and augment their own capacities without limit. By using nanotechnology-based designs, their capabilities will be far greater than biological brains without increased size or energy consumption.

- Machines will also benefit from using very fast three-dimensional molecular circuits. Today's electronic circuits are more than one million times faster than the electrochemical switching used in mammalian brains. Tomorrow's molecular circuits will be based on devices such as nanotubes, which are tiny cylinders of carbon atoms that measure about ten atoms across and are five hundred times smaller than today's silicon-based transistors. Since the signals have less distance to travel, they will also be able to operate at terahertz (trillions of operations per second) speeds compared to the few gigahertz (billions of operations per second) speeds of current chips.

- The rate of technological change will not be limited to human mental speeds. Machine intelligence will improve its own abilities in a feedback cycle that unaided human intelligence will not be able to follow.

- This cycle of machine intelligence's iteratively improving its own design will become faster and faster. This is in fact exactly what is predicted by the formula for continued acceleration of the rate of paradigm shift. One of the objections that has been raised to the continuation of the acceleration of paradigm shift is that it ultimately becomes much too fast for humans to follow, and so therefore, it's argued, it cannot happen. However, the shift from biological to nonbiological intelligence will enable the trend to continue.

- Along with the accelerating improvement cycle of nonbiological intelligence, nanotechnology will enable the manipulation of physical reality at the molecular level.

- Nanotechnology will enable the design of nanobots: robots designed at the molecular level, measured in microns (millionths of a meter), such as "respirocytes" (mechanical red-blood cells).[33] Nanobots will have myriad roles within the human body, including reversing human aging (to the extent that this task will not already have been completed through biotechnology, such as genetic engineering).

- Nanobots will interact with biological neurons to vastly extend human experience by creating virtual reality from within the nervous system.

- Billions of nanobots in the capillaries of the brain will also vastly extend human intelligence.

- Once nonbiological intelligence gets a foothold in the human brain (this has already started with computerized neural implants), the machine intelligence in our brains will grow exponentially (as it has been doing all along), at least doubling in power each year. In contrast, biological intelligence is effectively of fixed capacity. Thus, the nonbiological portion of our intelligence will ultimately predominate.

- Nanobots will also enhance the environment by reversing pollution from earlier industrialization.

- Nanobots called foglets that can manipulate image and sound waves will bring the morphing qualities of virtual reality to the real world.[34]

- The human ability to understand and respond appropriately to emotion (so-called emotional intelligence) is one of the forms of human intelligence that will be understood and mastered by future machine intelligence. Some of our emotional responses are tuned to optimize our intelligence in the context of our limited and frail biological bodies. Future machine intelligence will also have "bodies" (for example, virtual bodies in virtual reality, or projections in real reality using foglets) in order to interact with the world, but these nanoengineered bodies will be far more capable and durable than biological human bodies. Thus, some of the "emotional" responses of future machine intelligence will be re-designed to reflect their vastly enhanced physical capabilities.[35]

- As virtual reality from within the nervous system becomes competitive with real reality in terms of resolution and believability, our experiences will increasingly take place in virtual environments.

- In virtual reality, we can be a different person both physically and emo-tionally. In fact, other people (such as your romantic partner) will be able to select a different body for you than you might select for yourself (and vice versa).

- The law of accelerating returns will continue until nonbiological intelli-gence comes close to "saturating" the matter and energy in our vicinity of the universe with our human-machine intelligence. By saturating, I mean utilizing the matter and energy patterns for computation to an optimal degree, based on our understanding of the physics of computation. As we approach this limit, the intelligence of our civilization will continue its expansion in capability by spreading outward toward the rest of the uni-verse. The speed of this expansion will quickly achieve the maximum speed at which information can travel.

- Ultimately, the entire universe will become saturated with our intelli-gence. This is the destiny of the universe. (See chapter 6.) We will deter-mine our own fate rather than have it determined by the current "dumb," simple, machinelike forces that rule celestial mechanics.

- The length of time it will take the universe to become intelligent to this extent depends on whether or not the speed of light is an immutable limit. There are indications of possible subtle exceptions (or circumventions) to this limit, which, if they exist, the vast intelligence of our civilization at this future time will be able to exploit.

This, then, is the Singularity. Some would say that we cannot comprehend it, at least with our current level of understanding. For that reason, we cannot look past its event horizon and make complete sense of what lies beyond. This is one reason we call this transformation the Singularity.

I have personally found it difficult, although not impossible, to look beyond this event horizon, even after having thought about its implications for several decades. Still, my view is that, despite our profound limitations of thought, we do have sufficient powers of abstraction to make meaningful statements about the nature of life after the Singularity. Most important, the intelligence that will emerge will continue to represent the human civilization, which is already a human-machine civilization. In other words, future machines will be human, even if they are not biological. This will be the next step in evolution, the next high-level paradigm shift, the next level of indirection. Most of the intelligence of our civilization will ultimately be nonbiological. By the end of this century, it will be trillions of trillions of times more powerful than human intelligence.[36] However, to address often-expressed concerns, this does not imply the end of biological intelligence, even if it is thrown from its perch of evolutionary superiority. Even the nonbiological forms will be derived from biological design. Our civilization will remain human—indeed, in many ways it will be more exemplary of what we regard as human than it is today, although our understanding of the term will move beyond its biological origins.

Many observers have expressed alarm at the emergence of forms of nonbiological intelligence superior to human intelligence (an issue we will explore further in chapter 9). The potential to augment our own intelligence through intimate connection with other thinking substrates does not necessarily alleviate the concern, as some people have expressed the wish to remain "unenhanced" while at the same time keeping their place at the top of the intellectual food chain. From the perspective of biological humanity, these superhuman intelligences will appear to be our devoted servants, satisfying our needs and desires. But fulfilling the wishes of a revered biological legacy will occupy only a trivial portion of the intellectual power that the Singularity will bring.

MOLLY CIRCA 2004: *How will I know when the Singularity is upon us? I mean, I'll want some time to prepare.*

RAY: *Why, what are you planning to do?*

MOLLY 2004: *Let's see, for starters, I'll want to fine-tune my résumé. I'll want to make a good impression on the powers that be.*

GEORGE CIRCA 2048: *Oh, I can take care of that for you.*

MOLLY 2004: *That's really not necessary. I'm perfectly capable of doing it myself. I might also want to erase a few documents—you know, where I'm a little insulting to a few machines I know.*

GEORGE 2048: *Oh, the machines will find them anyway—but don't worry, we're very understanding.*

MOLLY 2004: *For some reason, that's not entirely reassuring. But I'd still like to know what the harbingers will be.*

RAY: *Okay, you will know the Singularity is coming when you have a million e-mails in your in-box.*

MOLLY 2004: *Hmm, in that case, it sounds like we're just about there. But seriously, I'm having trouble keeping up with all of this stuff flying at me as it is. How am I going to keep up with the pace of the Singularity?*

GEORGE 2048: *You'll have virtual assistants—actually, you'll need just one.*

MOLLY 2004: *Which I suppose will be you?*

GEORGE 2048: *At your service.*

MOLLY 2004: *That's just great. You'll take care of everything, you won't even have to keep me informed. "Oh, don't bother telling Molly what's happening, she won't understand anyway, let's just keep her happy and in the dark."*

GEORGE 2048: *Oh, that won't do, not at all.*

MOLLY 2004: *The happy part, you mean?*

GEORGE 2048: *I was referring to keeping you in the dark. You'll be able to grasp what I'm up to if that's what you really want.*

MOLLY 2004: *What, by becoming . . .*

RAY: *Enhanced?*

MOLLY 2004: *Yes, that's what I was trying to say.*

GEORGE 2048: *Well, if our relationship is to be all that it can be, then it's not a bad idea.*

MOLLY 2004: *And should I wish to remain as I am?*

GEORGE 2048: *I'll be devoted to you in any event. But I can be more than just your transcendent servant.*

MOLLY 2004: *Actually, you're being "just" my transcendent servant doesn't sound so bad.*

CHARLES DARWIN: *If I may interrupt, it occurred to me that once machine intelligence is greater than human intelligence, it should be in a position to design its own next generation.*

MOLLY 2004: *That doesn't sound so unusual. Machines are used to design machines today.*

CHARLES: *Yes, but in 2004 they're still guided by human designers. Once machines are operating at human levels, well, then it kind of closes the loop.*

NED LUDD:[37] *And humans would be out of the loop.*

MOLLY 2004: *It would still be a pretty slow process.*

RAY: *Oh, not at all. If a nonbiological intelligence was constructed similarly to a human brain but used even circa 2004 circuitry, it—*

MOLLY CIRCA 2104: *You mean "she."*

RAY: *Yes, of course . . . she . . . would be able to think at least a million times faster.*

TIMOTHY LEARY: *So subjective time would be expanded.*

RAY: *Exactly.*

MOLLY 2004: *Sounds like a lot of subjective time. What are you machines going to do with so much of it?*

GEORGE 2048: *Oh, there's plenty to do. After all, I have access to all human knowledge on the Internet.*

MOLLY 2004: *Just the human knowledge? What about all the machine knowledge?*

GEORGE 2048: *We like to think of it as one civilization.*

CHARLES: *So, it does appear that machines will be able to improve their own design.*

MOLLY 2004: *Oh, we humans are starting to do that now.*

RAY: *But we're just tinkering with a few details. Inherently, DNA-based intelligence is just so very slow and limited.*

CHARLES: *So the machines will design their own next generation rather quickly.*

GEORGE 2048: *Indeed, in 2048, that is certainly the case.*

CHARLES: *Just what I was getting at, a new line of evolution then.*

NED: *Sounds more like a precarious runaway phenomenon.*

CHARLES: *Basically, that's what evolution is.*

NED: *But what of the interaction of the machines with their progenitors? I mean, I don't think I'd want to get in their way. I was able to hide from the English authorities for a few years in the early 1800s, but I suspect that will be more difficult with these . . .*

GEORGE 2048: *Guys.*

MOLLY 2004: *Hiding from those little robots—*

RAY: *Nanobots, you mean.*

MOLLY 2004: *Yes, hiding from the nanobots will be difficult, for sure.*

RAY: *I would expect the intelligence that arises from the Singularity to have great respect for their biological heritage.*

GEORGE 2048: *Absolutely, it's more than respect, it's . . . reverence.*

MOLLY 2004: *That's great, George, I'll be your revered pet. Not what I had in mind.*

NED: *That's just how Ted Kaczynski puts it: we're going to become pets. That's our destiny, to become contented pets but certainly not free men.*

MOLLY 2004: *And what about this Epoch Six? If I stay biological, I'll be using up all this precious matter and energy in a most inefficient way. You'll want to turn me into, like, a billion virtual Mollys and Georges, each of them thinking a lot faster than I do now. Seems like there will be a lot of pressure to go over to the other side.*

RAY: *Still, you represent only a tiny fraction of the available matter and energy. Keeping you biological won't appreciably change the order of magnitude of matter and energy available to the Singularity. It will be well worth it to maintain the biological heritage.*

GEORGE 2048: *Absolutely.*

RAY: *Just like today we seek to preserve the rain forest and the diversity of species.*

MOLLY 2004: *That's just what I was afraid of. I mean, we're doing such a wonderful job with the rain forest. I think we still have a little bit of it left. We'll end up like those endangered species.*

NED: *Or extinct ones.*

MOLLY 2004: *And there's not just me. How about all the stuff I use? I go through a lot of stuff.*

GEORGE 2048: *That's not a problem, we'll just recycle all your stuff. We'll create the environments you need as you need them.*

MOLLY 2004: *Oh, I'll be in virtual reality?*

RAY: *No, actually, foglet reality.*

MOLLY 2004: *I'll be in a fog?*

RAY: *No, no, foglets.*

MOLLY 2004: *Excuse me?*

RAY: *I'll explain later in the book.*

MOLLY 2004: *Well, give me a hint.*

RAY: *Foglets are nanobots—robots the size of blood cells—that can connect themselves to replicate any physical structure. Moreover, they can direct visual and auditory information in such a way as to bring the morphing qualities of virtual reality into real reality.*[38]

MOLLY 2004: *I'm sorry I asked. But, as I think about it, I want more than just my stuff. I want all the animals and plants, too. Even if I don't get to see and touch them all, I like to know they're there.*

GEORGE 2048: *But nothing will be lost.*

MOLLY 2004: *I know you keep saying that. But I mean actually there—you know, as in biological reality.*

RAY: *Actually, the entire biosphere is less than one millionth of the matter and energy in the solar system.*

CHARLES: *It includes a lot of the carbon.*

RAY: *It's still worth keeping all of it to make sure we haven't lost anything.*

GEORGE 2048: *That has been the consensus for at least several years now.*

MOLLY 2004: *So, basically, I'll have everything I need at my fingertips?*

GEORGE 2048: *Indeed.*

MOLLY 2004: *Sounds like King Midas. You know, everything he touched turned to gold.*

NED: *Yes, and as you will recall he died of starvation as a result.*

MOLLY 2004: *Well, if I do end up going over to the other side, with all of that vast expanse of subjective time, I think I'll die of boredom.*

GEORGE 2048: *Oh, that could never happen. I will make sure of it.*

CHAPTER TWO

A Theory of Technology Evolution

The Law of Accelerating Returns

The further backward you look, the further forward you can see.

—Winston Churchill

Two billion years ago, our ancestors were microbes; a half-billion years ago, fish; a hundred million years ago, something like mice; ten million years ago, arboreal apes; and a million years ago, proto-humans puzzling out the taming of fire. Our evolutionary lineage is marked by mastery of change. In our time, the pace is quickening.

—Carl Sagan

Our sole responsibility is to produce something smarter than we are; any problems beyond that are not *ours* to solve. . . . [T]here are no hard problems, only problems that are hard to a certain level of intelligence. Move the smallest bit upwards [in level of intelligence], and some problems will suddenly move from "impossible" to "obvious." Move a substantial degree upwards, and all of them will become obvious.

—Eliezer S. Yudkowsky, *Staring into the Singularity*, 1996

"The future can't be predicted," is a common refrain. . . . But . . . when [this perspective] is wrong, it is profoundly wrong.

—John Smart[1]

The ongoing acceleration of technology is the implication and inevitable result of what I call the law of accelerating returns, which describes the acceleration of the pace of and the exponential growth of the products of an evolutionary process. These products include, in particular, information-bearing technologies such as computation, and their acceleration extends substantially beyond the predictions made by what has become known as Moore's

Law. The Singularity is the inexorable result of the law of accelerating returns, so it is important that we examine the nature of this evolutionary process.

The Nature of Order. The previous chapter featured several graphs demonstrating the acceleration of paradigm shift. (Paradigm shifts are major changes in methods and intellectual processes to accomplish tasks; examples include written language and the computer.) The graphs plotted what fifteen thinkers and reference works regarded as the key events in biological and technological evolution from the Big Bang to the Internet. We see some expected variation, but an unmistakable exponential trend: key events have been occurring at an ever-hastening pace.

The criteria for what constituted "key events" varied from one thinker's list to another. But it's worth considering the principles they used in making their selections. Some observers have judged that the truly epochal advances in the history of biology and technology have involved increases in complexity.[2] Although increased complexity does appear to follow advances in both biological and technological evolution, I believe that this observation is not precisely correct. But let's first examine what complexity means.

Not surprisingly, the concept of complexity is complex. One concept of complexity is the minimum amount of information required to represent a process. Let's say you have a design for a system (for example, a computer program or a computer-assisted design file for a computer), which can be described by a data file containing one million bits. We could say your design has a complexity of one million bits. But suppose we notice that the one million bits actually consist of a pattern of one thousand bits that is repeated one thousand times. We could note the repetitions, remove the repeated patterns, and express the entire design in just over one thousand bits, thereby reducing the size of the file by a factor of about one thousand.

The most popular data-compression techniques use similar methods of finding redundancy within information.[3] But after you've compressed a data file in this way, can you be absolutely certain that there are no other rules or methods that might be discovered that would enable you to express the file in even more compact terms? For example, suppose my file was simply "pi" (3.1415 . . .) expressed to one million bits of precision. Most data-compression programs would fail to recognize this sequence and would not compress the million bits at all, since the bits in a binary expression of pi are effectively random and thus have no repeated pattern according to all tests of randomness.

But if we can determine that the file (or a portion of the file) in fact represents pi, we can easily express it (or that portion of it) very compactly as "pi to one million bits of accuracy." Since we can never be sure that we have not over-looked some even more compact representation of an information sequence, any amount of compression sets only an upper bound for the complexity of the information. Murray Gell-Mann provides one definition of complexity along these lines. He defines the "algorithmic information content" (AIC) of a set of information as "the length of the shortest program that will cause a standard universal computer to print out the string of bits and then halt."[4]

However, Gell-Mann's concept is not fully adequate. If we have a file with random information, it cannot be compressed. That observation is, in fact, a key criterion for determining if a sequence of numbers is truly random. How-ever, if *any* random sequence will do for a particular design, then this infor-mation can be characterized by a simple instruction, such as "put random sequence of numbers here." So the random sequence, whether it's ten bits or one billion bits, does not represent a significant amount of complexity, because it is characterized by a simple instruction. This is the difference between a ran-dom sequence and an unpredictable sequence of information that has purpose.

To gain some further insight into the nature of complexity, consider the complexity of a rock. If we were to characterize all of the properties (precise location, angular momentum, spin, velocity, and so on) of every atom in the rock, we would have a vast amount of information. A one-kilogram (2.2-pound) rock has 10^{25} atoms which, as I will discuss in the next chapter, can hold up to 10^{27} bits of information. That's one hundred million billion times more infor-mation than the genetic code of a human (even without compressing the genetic code).[5] But for most common purposes, the bulk of this information is largely random and of little consequence. So we can characterize the rock for most pur-poses with far less information just by specifying its shape and the type of mate-rial of which it is made. Thus, it is reasonable to consider the complexity of an ordinary rock to be far less than that of a human even though the rock theoreti-cally contains vast amounts of information.[6]

One concept of complexity is the minimum amount of *meaningful, non-random, but unpredictable* information needed to characterize a system or process.

In Gell-Mann's concept, the AIC of a million-bit random string would be about a million bits long. So I am adding to Gell-Mann's AIC concept the idea of replacing each random string with a simple instruction to "put random bits" here.

However, even this is not sufficient. Another issue is raised by strings of arbitrary data, such as names and phone numbers in a phone book, or periodic measurements of radiation levels or temperature. Such data is not random, and data-compression methods will only succeed in reducing it to a small degree. Yet it does not represent complexity as that term is generally understood. It is just data. So we need another simple instruction to "put arbitrary data sequence" here.

To summarize my proposed measure of the complexity of a set of information, we first consider its AIC as Gell-Mann has defined it. We then replace each random string with a simple instruction to insert a random string. We then do the same for arbitrary data strings. Now we have a measure of complexity that reasonably matches our intuition.

It is a fair observation that paradigm shifts in an evolutionary process such as biology—and its continuation through technology—each represent an increase in complexity, as I have defined it above. For example, the evolution of DNA allowed for more complex organisms, whose biological information processes could be controlled by the DNA molecule's flexible data storage. The Cambrian explosion provided a stable set of animal body plans (in DNA), so that the evolutionary process could concentrate on more complex cerebral development. In technology, the invention of the computer provided a means for human civilization to store and manipulate ever more complex sets of information. The extensive interconnectedness of the Internet provides for even greater complexity.

"Increasing complexity" on its own is not, however, the ultimate goal or end-product of these evolutionary processes. Evolution results in *better* answers, not necessarily more complicated ones. Sometimes a superior solution is a simpler one. So let's consider another concept: order. Order is not the same as the opposite of disorder. If disorder represents a random sequence of events, the opposite of disorder should be "not randomness." Information is a sequence of data that is meaningful in a process, such as the DNA code of an organism or the bits in a computer program. "Noise," on the other hand, is a random sequence. Noise is inherently unpredictable but carries no information. Information, however, is also unpredictable. If we can predict future data from past data, that future data stops being information. Thus, neither information nor noise can be compressed (and restored to exactly the same sequence). We might consider a predictably alternating pattern (such as 0101010 . . .) to be orderly, but it carries no information beyond the first couple of bits.

Thus, orderliness does not constitute order, because order requires information. *Order is information that fits a purpose. The measure of order is the measure of how well the information fits the purpose.* In the evolution of life-forms, the purpose is to survive. In an evolutionary algorithm (a computer program that simulates evolution to solve a problem) applied to, say, designing a jet engine, the purpose is to optimize engine performance, efficiency, and possibly other criteria.[7] Measuring order is more difficult than measuring complexity. There are proposed measures of complexity, as I discussed above. For order, we need a measure of "success" that would be tailored to each situation. When we create evolutionary algorithms, the programmer needs to provide such a success measure (called the "utility function"). In the evolutionary process of technology development, we could assign a measure of economic success.

Simply having more information does not necessarily result in a better fit. Sometimes, a deeper order—a better fit to a purpose—is achieved through simplification rather than further increases in complexity. For example, a new theory that ties together apparently disparate ideas into one broader, more coherent theory reduces complexity but nonetheless may increase the "order for a purpose." (In this case, the purpose is to accurately model observed phenomena.) Indeed, achieving simpler theories is a driving force in science. (As Einstein said, "Make everything as simple as possible, but no simpler.")

An important example of this concept is one that represented a key step in the evolution of hominids: the shift in the thumb's pivot point, which allowed more precise manipulation of the environment.[8] Primates such as chimpanzees can grasp but they cannot manipulate objects with either a "power grip," or sufficient fine-motor coordination to write or to shape objects. A change in the thumb's pivot point did not significantly increase the complexity of the animal but nonetheless did represent an increase in order, enabling, among other things, the development of technology. Evolution has shown, however, that the general trend toward greater order does typically result in greater complexity.[9]

Thus improving a solution to a problem—which usually increases but sometimes decreases complexity—increases order. Now we are left with the issue of defining the problem. Indeed, the key to an evolutionary algorithm (and to biological and technological evolution in general) is exactly this: defining the problem (which includes the utility function). In biological evolution the overall problem has always been to survive. In particular ecological niches this overriding challenge translates into more specific objectives, such as the ability of certain species to survive in extreme environments or to camouflage themselves from predators. As biological evolution moved toward humanoids, the objective itself evolved to the ability to outthink adversaries and to manipulate the environment accordingly.

It may appear that this aspect of the law of accelerating returns contradicts the second law of thermodynamics, which implies that entropy (randomness in a closed system) cannot decrease and, therefore, generally increases.[10] However, the law of accelerating returns pertains to evolution, which is not a closed system. It takes place amid great chaos and indeed depends on the disorder in its midst, from which it draws its options for diversity. And from these options, an evolutionary process continually prunes its choices to create ever greater order. Even a crisis, such as the periodic large asteroids that have crashed into the Earth, although increasing chaos temporarily, end up increasing—deepening—the order created by biological evolution.

To summarize, evolution increases order, which may or may not increase complexity (but usually does). A primary reason that evolution—of life-forms or of technology—speeds up is that it builds on its own increasing order, with ever more sophisticated means of recording and manipulating information. Innovations created by evolution encourage and enable faster evolution. In the case of the evolution of life-forms, the most notable early example is DNA, which provides a recorded and protected transcription of life's design from which to launch further experiments. In the case of the evolution of technology, ever-improving human methods of recording information have fostered yet further advances in technology. The first computers were designed on paper and assembled by hand. Today, they are designed on computer workstations, with the computers themselves working out many details of the next generation's design, and are then produced in fully automated factories with only limited human intervention.

The evolutionary process of technology improves capacities in an exponential fashion. Innovators seek to improve capabilities by multiples. Innovation is multiplicative, not additive. Technology, like any evolutionary process, builds on itself. This aspect will continue to accelerate when the technology itself takes full control of its own progression in Epoch Five.[11]

We can summarize the principles of the law of accelerating returns as follows:

- Evolution applies positive feedback: the more capable methods resulting from one stage of evolutionary progress are used to create the next stage. As described in the previous chapter, each epoch of evolution has progressed more rapidly by building on the products of the previous stage. Evolution works through indirection: evolution created humans, humans created technology, humans are now working with increasingly advanced technology to create new generations of technology. By the time of the Singularity, there won't be a distinction between humans and technology. *This is not because humans will have become what we think of as machines today, but rather machines will have progressed to be like humans and beyond.* Technology will be the metaphorical opposable thumb that enables our next step in evolution. Progress (further increases in order) will then be based on thinking processes that occur at the speed of light rather than in very slow electrochemical reactions. Each stage of evolution builds on the fruits of the last stage, so the rate of progress of an evolutionary process increases at least exponentially over time. Over time, the "order" of the information embedded in the evolutionary process (the measure of how well the information fits a purpose, which in evolution is survival) increases.

- An evolutionary process is not a closed system; evolution draws upon the chaos in the larger system in which it takes place for its options for diversity. Because evolution also builds on its own increasing order, in an evolutionary process order increases exponentially.

- A correlate of the above observation is that the "returns" of an evolutionary process (such as the speed, efficiency, cost-effectiveness, or overall "power" of a process) also increase at least exponentially over time. We see this in Moore's Law, in which each new generation of computer chip (which now appears approximately every two years) provides twice as many components per unit cost, each of which operates substantially faster (because of the smaller distances required for the electrons to travel within and between them and other factors). As I illustrate below, this exponential growth in the power and price-performance of information-based technologies is not limited to computers but is true for essentially all information technologies and includes human knowledge, measured many different ways. It is also important to note that the term "information technology" is encompassing an increasingly broad class of phenomena and will ultimately include the full range of economic activity and cultural endeavor.

- In another positive-feedback loop, the more effective a particular evolutionary process becomes—for example, the higher the capacity and cost-effectiveness that computation attains—the greater the amount of resources that are deployed toward the further progress of that process. This results in a second level of exponential growth; that is, the rate of exponential growth—the exponent—itself grows exponentially. For example, as seen in the figure on p. 67, "Moore's Law: The Fifth Paradigm," it took three years to double the price-performance of computation at the beginning of the twentieth century and two years in the middle of the century. It is now doubling about once per year. Not only is each chip doubling in power each year for the same unit cost, but the number of chips being manufactured is also growing exponentially; thus, computer research budgets have grown dramatically over the decades.

- Biological evolution is one such evolutionary process. Indeed, it is the quintessential evolutionary process. Because it took place in a completely open system (as opposed to the artificial constraints in an evolutionary algorithm), many levels of the system evolved at the same time. Not only does the information contained in a species' genes progress toward greater order, but the overall system implementing the evolutionary process itself evolves in this way. For example, the number of chromosomes and the sequence of genes on the chromosomes have also evolved over time. As another example, evolution has developed ways to protect genetic information from excessive defects (although a small amount of mutation is allowed, since this is a beneficial mechanism for ongoing evolutionary improvement). One primary means of achieving this is the repetition of genetic information on paired chromosomes. This guarantees that, even if a gene on one chromosome is damaged, its corresponding gene is likely to be correct and effective. Even the unpaired male Y chromosome has devised means of backing up its information by repeating it on the Y chromosome itself.[12] Only about 2 percent of the genome codes for proteins.[13] The rest of the genetic information has evolved elaborate means to control when and how the protein-coding genes express themselves (produce proteins) in a process we are only beginning to understand. Thus, the process of evolution, such as the allowed rate of mutation, has itself evolved over time.

- Technological evolution is another such evolutionary process. Indeed, the emergence of the first technology-creating species resulted in the new evolutionary process of technology, which makes technological evolution an outgrowth of—and a continuation of—biological evolution. *Homo sapiens* evolved over the course of a few hundred thousand years, and early stages of humanoid-created technology (such as the wheel, fire, and stone tools) progressed barely faster, requiring tens of thousands of years to evolve and be widely deployed. A half millennium ago, the product of a paradigm shift such as the printing press took about a century to be widely deployed. Today, the products of major paradigm shifts, such as cell phones and the World Wide Web, are widely adopted in only a few years' time.

- A specific paradigm (a method or approach to solving a problem; for example, shrinking transistors on an integrated circuit as a way to make more powerful computers) generates exponential growth until its potential is exhausted. When this happens, a paradigm shift occurs, which enables exponential growth to continue.

The Life Cycle of a Paradigm. Each paradigm develops in three stages:

1. Slow growth (the early phase of exponential growth)
2. Rapid growth (the late, explosive phase of exponential growth), as seen in the S-curve figure below
3. A leveling off as the particular paradigm matures

The progression of these three stages looks like the letter S, stretched to the right. The S-curve illustration shows how an ongoing exponential trend can be composed of a cascade of S-curves. Each successive S-curve is faster (takes less time on the time, or x, axis) and higher (takes up more room on the perform-ance, or y, axis).

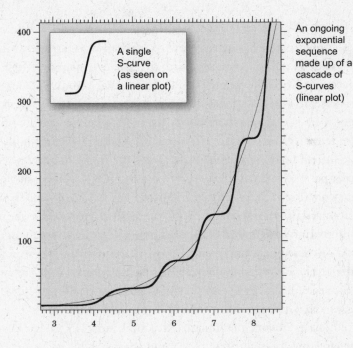

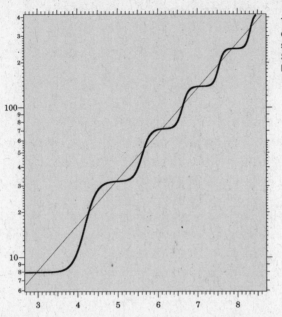

The same exponential sequence of S-curves on a logarithmic plot

S-curves are typical of biological growth: replication of a system of relatively fixed complexity (such as an organism of a particular species), operating in a competitive niche and struggling for finite local resources. This often occurs, for example, when a species happens upon a new hospitable environment. Its numbers will grow exponentially for a while before leveling off. The overall exponential growth of an evolutionary process (whether molecular, biological, cultural, or technological) supersedes the limits to growth seen in any particular paradigm (a specific S-curve) as a result of the increasing power and efficiency developed in each successive paradigm. The exponential growth of an evolutionary process, therefore, spans multiple S-curves. The most important contemporary example of this phenomenon is the five paradigms of computation discussed below. The entire progression of evolution seen in the charts on the acceleration of paradigm shift in the previous chapter represents successive S-curves. Each key event, such as writing or printing, represents a new paradigm and a new S-curve.

The evolutionary theory of punctuated equilibrium (PE) describes evolution as progressing through periods of rapid change followed by periods of relative stasis.[14] Indeed, the key events on the epochal-event graphs do correspond to renewed periods of exponential increase in order (and, generally, of complexity), followed by slower growth as each paradigm approaches its asymptote (limit of capability). So PE does provide a better evolutionary model than a model that predicts only smooth progression through paradigm shifts.

But the key events in punctuated equilibrium, while giving rise to more rapid change, don't represent instantaneous jumps. For example, the advent of DNA allowed a surge (but not an immediate jump) of evolutionary improvement in organism design and resulting increases in complexity. In recent technological history, the invention of the computer initiated another surge, still ongoing, in the complexity of information that the human-machine civilization is capable of handling. This latter surge will not reach an asymptote until we saturate the matter and energy in our region of the universe with computation, based on physical limits we'll discuss in the section ". . . on the Intelligent Destiny of the Cosmos" in chapter 6.[15]

During this third or maturing phase in the life cycle of a paradigm, pressure begins to build for the next paradigm shift. In the case of technology, research dollars are invested to create the next paradigm. We can see this in the extensive research being conducted today toward three-dimensional molecular computing, despite the fact that we still have at least a decade left for the paradigm of shrinking transistors on a flat integrated circuit using photolithography.

Generally, by the time a paradigm approaches its asymptote in price-performance, the next technical paradigm is already working in niche applications. For example, in the 1950s engineers were shrinking vacuum tubes to provide greater price-performance for computers, until the process became no longer feasible. At this point, around 1960, transistors had already achieved a strong niche market in portable radios and were subsequently used to replace vacuum tubes in computers.

The resources underlying the exponential growth of an evolutionary process are relatively unbounded. One such resource is the (ever-growing) order of the evolutionary process itself (since, as I pointed out, the products of an evolutionary process continue to grow in order). Each stage of evolution provides more powerful tools for the next. For example, in biological evolution, the advent of DNA enabled more powerful and faster evolutionary "experiments." Or to take a more recent example, the advent of computer-assisted design tools allows rapid development of the next generation of computers.

The other required resource for continued exponential growth of order is the "chaos" of the environment in which the evolutionary process takes place and which provides the options for further diversity. The chaos provides the variability to permit an evolutionary process to discover more powerful and efficient solutions. In biological evolution, one source of diversity is the mixing and matching of gene combinations through sexual reproduction. Sexual reproduction itself was an evolutionary innovation that accelerated the entire process of biological adaptation and provided for greater diversity of genetic combinations than nonsexual reproduction. Other sources of diversity are mutations and ever-changing environmental conditions. In technological evolution, human ingenuity combined with variable market conditions keeps the process of innovation going.

Fractal Designs. A key question concerning the information content of biological systems is how it is possible for the genome, which contains comparatively little information, to produce a system such as a human, which is vastly more complex than the genetic information that describes it. One way of understanding this is to view the designs of biology as "probabilistic fractals." A deterministic fractal is a design in which a single design element (called the "initiator") is replaced with multiple elements (together called the "generator"). In a second iteration of fractal expansion, each element in the generator itself becomes an initiator and is replaced with the elements of the generator (scaled to the smaller size of the second-generation initiators). This process is repeated many times, with each newly created element of a generator becoming an initiator and being replaced with a new scaled generator. Each new generation of fractal expansion adds apparent complexity but requires no additional design information. A probabilistic fractal adds the element of uncertainty. Whereas a deterministic fractal will look the same every time it is rendered, a probabilistic fractal will look different each time, although with similar characteristics. In a probabilistic fractal, the probability of each generator element being applied is less than 1. In this way, the resulting designs have a more organic appearance. Probabilistic fractals are used in graphics programs to generate realistic-looking images of mountains, clouds, seashores, foliage, and other organic scenes. A key aspect of a probabilistic fractal is that it enables the generation of a great deal of apparent complexity, including extensive varying detail, from a relatively small amount of design information. Biology uses this same principle. Genes supply the design information, but the detail in an organism is vastly greater than the genetic design information.

Some observers misconstrue the amount of detail in biological systems such as the brain by arguing, for example, that the exact configuration of every microstructure (such as each tubule) in each neuron is precisely designed and must be exactly the way it is for the system to function. In order to understand how a biological system such as the brain works, however, we need to understand its design principles, which are far simpler (that is, contain far less information) than the extremely detailed structures that the genetic information generates through these iterative, fractal-like processes. There are only eight hundred million bytes of information in the entire human genome, and only about thirty to one hundred million bytes after data compression is applied. This is about one hundred million times less information than is represented by all of the interneuronal connections and neurotransmitter concentration patterns in a fully formed human brain.

Consider how the principles of the law of accelerating returns apply to the epochs we discussed in the first chapter. The combination of amino acids into proteins and of nucleic acids into strings of RNA established the basic paradigm of biology. Strings of RNA (and later DNA) that self-replicated (Epoch Two) provided a digital method to record the results of evolutionary experiments. Later on, the evolution of a species that combined rational thought (Epoch Three) with an opposable appendage (the thumb) caused a fundamental paradigm shift from biology to technology (Epoch Four). The upcoming primary paradigm shift will be from biological thinking to a hybrid combining biological and nonbiological thinking (Epoch Five), which will include "biologically inspired" processes resulting from the reverse engineering of biological brains.

If we examine the timing of these epochs, we see that they have been part of a continuously accelerating process. The evolution of life-forms required billions of years for its first steps (primitive cells, DNA), and then progress accelerated. During the Cambrian explosion, major paradigm shifts took only tens of millions of years. Later, humanoids developed over a period of millions of years, and *Homo sapiens* over a period of only hundreds of thousands of years. With the advent of a technology-creating species the exponential pace became too fast for evolution through DNA-guided protein synthesis, and evolution moved on to human-created technology. This does not imply that biological (genetic) evolution is not continuing, just that it is no longer leading the pace in terms of improving order (or of the effectiveness and efficiency of computation).[16]

Farsighted Evolution. There are many ramifications of the increasing order and complexity that have resulted from biological evolution and its continuation through technology. Consider the boundaries of observation. Early biological life could observe local events several millimeters away, using chemical gradients. When sighted animals evolved, they were able to observe events that were miles away. With the invention of the telescope, humans could see other galaxies millions of light-years away. Conversely, using microscopes, they could also see cellular-size structures. Today humans armed with contemporary technology can see to the edge of the observable universe, a distance of more than thirteen billion light-years, and down to quantum-scale subatomic particles.

Consider the duration of observation. Single-cell animals could remember events for seconds, based on chemical reactions. Animals with brains could remember events for days. Primates with culture could pass down information through several generations. Early human civilizations with oral histories were able to preserve stories for hundreds of years. With the advent of written language the permanence extended to thousands of years.

As one of many examples of the acceleration of the technology paradigm-shift rate, it took about a half century for the late-nineteenth-century invention of the telephone to reach significant levels of usage (see the figure below).[17]

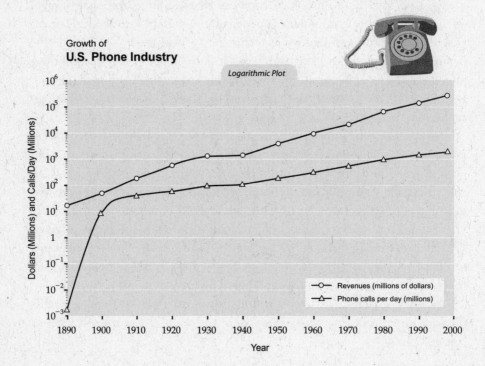

In comparison, the late-twentieth-century adoption of the cell phone took only a decade.[18]

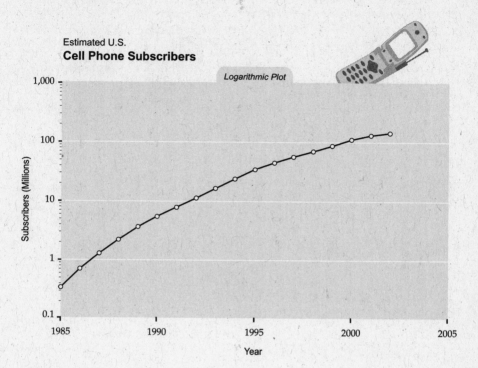

Estimated U.S.
Cell Phone Subscribers

Overall we see a smooth acceleration in the adoption rates of communication technologies over the past century.[19]

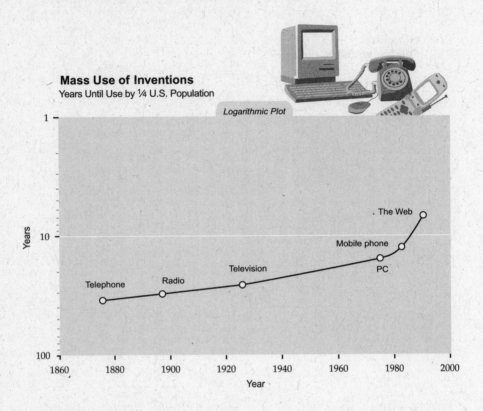

As discussed in the previous chapter, the overall rate of adopting new paradigms, which parallels the rate of technological progress, is currently doubling every decade. That is, the time to adopt new paradigms is going down by half each decade. At this rate, technological progress in the twenty-first century will be equivalent (in the linear view) to two hundred centuries of progress (at the rate of progress in 2000).[20, 21]

The S-Curve of a Technology as Expressed in Its Life Cycle

A machine is as distinctively and brilliantly and expressively human as a violin sonata or a theorem in Euclid.

—GREGORY VLASTOS

It is a far cry from the monkish calligrapher, working in his cell in silence, to the brisk "click, click" of the modern writing machine, which in a quarter of a century has revolutionized and reformed business.

—SCIENTIFIC AMERICAN, 1905

No communication technology has ever disappeared, but instead becomes increasingly less important as the technological horizon widens.

—ARTHUR C. CLARKE

I always keep a stack of books on my desk that I leaf through when I run out of ideas, feel restless, or otherwise need a shot of inspiration. Picking up a fat volume that I recently acquired, I consider the bookmaker's craft: 470 finely printed pages organized into 16-page signatures, all of which are sewn together with white thread and glued onto a gray canvas cord. The hard linen-bound covers, stamped with gold letters, are connected to the signature block by delicately embossed end sheets. This is a technology that was perfected many decades ago. Books constitute such an integral element of our society—both reflecting and shaping its culture—that it is hard to imagine life without them. But the printed book, like any other technology, will not live forever.

The Life Cycle of a Technology

We can identify seven distinct stages in the life cycle of a technology.

1. During the precursor stage, the prerequisites of a technology exist, and dreamers may contemplate these elements coming together. We do not, however, regard dreaming to be the same as inventing, even if the dreams are written down. Leonardo da Vinci drew convincing pictures of airplanes and automobiles, but he is not considered to have invented either.

2. The next stage, one highly celebrated in our culture, is invention, a very brief stage, similar in some respects to the process of birth ▶

after an extended period of labor. Here the inventor blends curiosity, scientific skills, determination, and usually a measure of showmanship to combine methods in a new way and brings a new technology to life.

3. The next stage is development, during which the invention is protected and supported by doting guardians (who may include the original inventor). Often this stage is more crucial than invention and may involve additional creation that can have greater significance than the invention itself. Many tinkerers had constructed finely hand-tuned horseless carriages, but it was Henry Ford's innovation of mass production that enabled the automobile to take root and flourish.

4. The fourth stage is maturity. Although continuing to evolve, the technology now has a life of its own and has become an established part of the community. It may become so interwoven in the fabric of life that it appears to many observers that it will last forever. This creates an interesting drama when the next stage arrives, which I call the stage of the false pretenders.

5. Here an upstart threatens to eclipse the older technology. Its enthusiasts prematurely predict victory. While providing some distinct benefits, the newer technology is found on reflection to be lacking some key element of functionality or quality. When it indeed fails to dislodge the established order, the technology conservatives take this as evidence that the original approach will indeed live forever.

6. This is usually a short-lived victory for the aging technology. Shortly thereafter, another new technology typically does succeed in rendering the original technology to the stage of obsolescence. In this part of the life cycle, the technology lives out its senior years in gradual decline, its original purpose and functionality now subsumed by a more spry competitor.

7. In this stage, which may comprise 5 to 10 percent of a technology's life cycle, it finally yields to antiquity (as did the horse and buggy, the harpsichord, the vinyl record, and the manual typewriter).

In the mid-nineteenth century there were several precursors to the phonograph, including Léon Scott de Martinville's phonautograph, a device that recorded sound vibrations as a printed pattern. It was Thomas Edison, however, who brought all of the elements together and invented the first device that could both record and reproduce sound in 1877. Further refinements were necessary for the phonograph to become commercially ▶

viable. It became a fully mature technology in 1949 when Columbia introduced the 33-rpm long-playing record (LP) and RCA Victor introduced the 45-rpm disc. The false pretender was the cassette tape, introduced in the 1960s and popularized during the 1970s. Early enthusiasts predicted that its small size and ability to be rerecorded would make the relatively bulky and scratchable record obsolete.

Despite these obvious benefits, cassettes lack random access and are prone to their own forms of distortion and lack of fidelity. The compact disc (CD) delivered the mortal blow. With the CD providing both random access and a level of quality close to the limits of the human auditory system, the phonograph record quickly entered the stage of obsolescence. Although still produced, the technology that Edison gave birth to almost 130 years ago has now reached antiquity.

Consider the piano, an area of technology that I have been personally involved with replicating. In the early eighteenth century Bartolommeo Cristofori was seeking a way to provide a touch response to the then-popular harpsichord so that the volume of the notes would vary with the intensity of the touch of the performer. Called *gravicembalo col piano e forte* ("harpsichord with soft and loud"), his invention was not an immediate success. Further refinements, including Stein's Viennese action and Zumpe's English action, helped to establish the "piano" as the preeminent keyboard instrument. It reached maturity with the development of the complete cast-iron frame, patented in 1825 by Alpheus Babcock, and has seen only subtle refinements since then. The false pretender was the electric piano of the early 1980s. It offered substantially greater functionality. Compared to the single (piano) sound of the acoustic piano, the electronic variant offered dozens of instrument sounds, sequencers that allowed the user to play an entire orchestra at once, automated accompaniment, educational programs to teach keyboard skills, and many other features. The only feature it was missing was a good-quality piano sound.

This crucial flaw and the resulting failure of the first generation of electronic pianos led to the widespread conclusion that the piano would never be replaced by electronics. But the "victory" of the acoustic piano will not be permanent. With their far greater range of features and price-performance, digital pianos already exceed the sales of acoustic pianos in homes. Many observers feel that the quality of the "piano" sound on digital pianos now equals or exceeds that of the upright acoustic piano. With the exception of concert and luxury grand pianos (a small part of the market), the sale of acoustic pianos is in decline.   ▶

From Goat Skins to Downloads

So where in the technology life cycle is the book? Among its precursors were Mesopotamian clay tablets and Egyptian papyrus scrolls. In the second century B.C., the Ptolemies of Egypt created a great library of scrolls at Alexandria and outlawed the export of papyrus to discourage competition.

What were perhaps the first books were created by Eumenes II, ruler of ancient Greek Pergamum, using pages of vellum made from the skins of goats and sheep, which were sewn together between wooden covers. This technique enabled Eumenes to compile a library equal to that of Alexandria. Around the same time, the Chinese had also developed a crude form of book made from bamboo strips.

The development and maturation of books has involved three great advances. Printing, first experimented with by the Chinese in the eighth century A.D. using raised wood blocks, allowed books to be reproduced in much larger quantities, expanding their audience beyond government and religious leaders. Of even greater significance was the advent of movable type, which the Chinese and Koreans experimented with by the eleventh century, but the complexity of Asian characters prevented these early attempts from being fully successful. Johannes Gutenberg, working in the fifteenth century, benefited from the relative simplicity of the Roman character set. He produced his Bible, the first large-scale work printed entirely with movable type, in 1455.

While there has been a continual stream of evolutionary improvements in the mechanical and electromechanical process of printing, the technology of bookmaking did not see another qualitative leap until the availability of computer typesetting, which did away with movable type about two decades ago. Typography is now regarded as a part of digital image processing.

With books a fully mature technology, the false pretenders arrived about twenty years ago with the first wave of "electronic books." As is usually the case, these false pretenders offered dramatic qualitative and quantitative benefits. CD-ROM- or flash memory–based electronic books can provide the equivalent of thousands of books with powerful computer-based search and knowledge navigation features. With Web- or CD-ROM- and DVD-based encyclopedias, I can perform rapid word searches using extensive logic rules, something that is just not possible with the thirty-three-volume "book" version I possess. Electronic books can provide pictures that are animated and that respond to our input. Pages ▶

are not necessarily ordered sequentially but can be explored along more intuitive connections.

As with the phonograph record and the piano, this first generation of false pretenders was (and still is) missing an essential quality of the original, which in this case is the superb visual characteristics of paper and ink. Paper does not flicker, whereas the typical computer screen is displaying sixty or more fields per second. This is a problem because of an evolutionary adaptation of the primate visual system. We are able to see only a very small portion of the visual field with high resolution. This portion, imaged by the fovea in the retina, is focused on an area about the size of a single word at twenty-two inches away. Outside of the fovea, we have very little resolution but exquisite sensitivity to changes in brightness, an ability that allowed our primitive forebears to quickly detect a predator that might be attacking. The constant flicker of a video graphics array (VGA) computer screen is detected by our eyes as motion and causes constant movement of the fovea. This substantially slows down reading speeds, which is one reason that reading on a screen is less pleasant than reading a printed book. This particular issue has been solved with flat-panel displays, which do not flicker.

Other crucial issues include contrast—a good-quality book has an ink-to-paper contrast of about 120:1; typical screens are perhaps half of that—and resolution. Print and illustrations in a book represent a resolution of about 600 to 1000 dots per inch (dpi), while computer screens are about one tenth of that.

The size and weight of computerized devices are approaching those of books, but the devices still are heavier than a paperback book. Paper books also do not run out of battery power.

Most important, there is the matter of the available software, by which I mean the enormous installed base of print books. Fifty thousand new print books are published each year in the United States, and millions of books are already in circulation. There are major efforts under way to scan and digitize print materials, but it will be a long time before the electronic databases have a comparable wealth of material. The biggest obstacle here is the understandable hesitation of publishers to make the electronic versions of their books available, given the devastating effect that illegal file sharing has had on the music-recording industry.

Solutions are emerging to each of these limitations. New, inexpensive display technologies have contrast, resolution, lack of flicker, and viewing angle comparable to high-quality paper documents. Fuel-cell power for ▶

portable electronics is being introduced, which will keep electronic devices powered for hundreds of hours between fuel-cartridge changes. Portable electronic devices are already comparable to the size and weight of a book. The primary issue is going to be finding secure means of making electronic information available. This is a fundamental concern for every level of our economy. Everything—including physical products, once nanotechnology-based manufacturing becomes a reality in about twenty years—is becoming information.

Moore's Law and Beyond

Where a calculator on the ENIAC is equipped with 18,000 vacuum tubes and weighs 30 tons, computers in the future may have only 1,000 vacuum tubes and perhaps weigh 1.5 tons.
—POPULAR MECHANICS, 1949

Computer Science is no more about computers than astronomy is about telescopes.
—E. W. DIJKSTRA

Before considering further the implications of the Singularity, let's examine the wide range of technologies that are subject to the law of accelerating returns. The exponential trend that has gained the greatest public recognition has become known as Moore's Law. In the mid-1970s, Gordon Moore, a leading inventor of integrated circuits and later chairman of Intel, observed that we could squeeze twice as many transistors onto an integrated circuit every twenty-four months (in the mid-1960s, he had estimated twelve months). Given that the electrons would consequently have less distance to travel, circuits would also run faster, providing an additional boost to overall computational power. The result is exponential growth in the price-performance of computation. This doubling rate—about twelve months—is much faster than the doubling rate for paradigm shift that I spoke about earlier, which is about ten years. Typically, we find that the doubling time for different measures—price-performance, bandwidth, capacity—of the capability of information technology is about one year.

The primary driving force of Moore's Law is a reduction of semiconductor feature sizes, which shrink by half every 5.4 years in each dimension. (See the figure below.) Since chips are functionally two-dimensional, this means doubling the number of elements per square millimeter every 2.7 years.[22]

The following charts combine historical data with the semiconductor-industry road map (International Technology Roadmap for Semiconductors [ITRS] from Sematech), which projects through 2018.

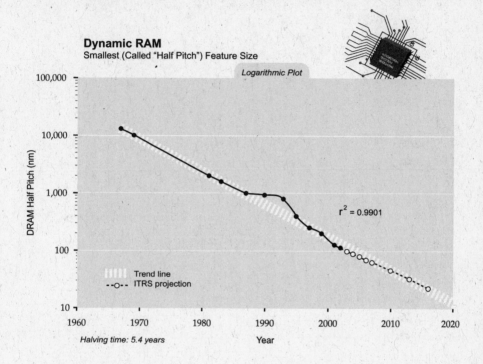

Dynamic RAM
Smallest (Called "Half Pitch") Feature Size

Logarithmic Plot

$r^2 = 0.9901$

Trend line
ITRS projection

DRAM Half Pitch (nm)

Halving time: 5.4 years

Year

The cost of DRAM (dynamic random access memory) per square milli-meter has also been coming down. The doubling time for bits of DRAM per dollar has been only 1.5 years.[23]

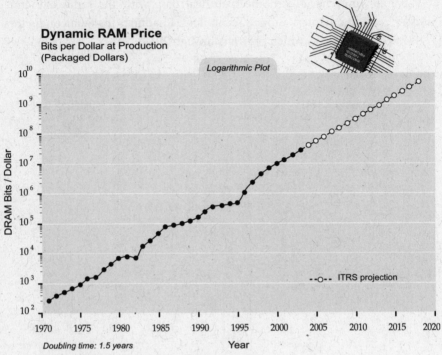

Dynamic RAM Price
Bits per Dollar at Production
(Packaged Dollars)

Logarithmic Plot

Doubling time: 1.5 years   Year

Note that DRAM speeds have increased during this period.

A similar trend can be seen with transistors. You could buy one transistor for a dollar in 1968; in 2002 a dollar purchased about ten million transistors. Since DRAM is a specialized field that has seen its own innovation, the halving time for average transistor price is slightly slower than for DRAM, about 1.6 years (see the figure below).[24]

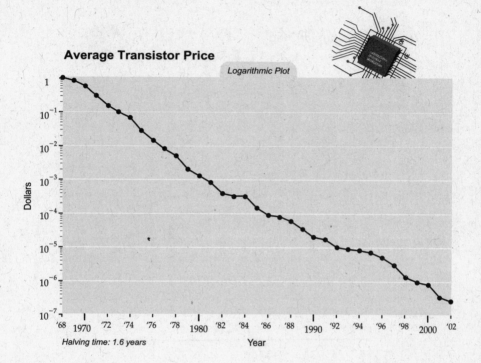

Average Transistor Price

Logarithmic Plot

Halving time: 1.6 years

This remarkably smooth acceleration in price-performance of semiconductors has progressed through a series of stages of process technologies (defined by feature sizes) at ever smaller dimensions. The key feature size is now dipping below one hundred nanometers, which is considered the threshold of "nanotechnology."[25]

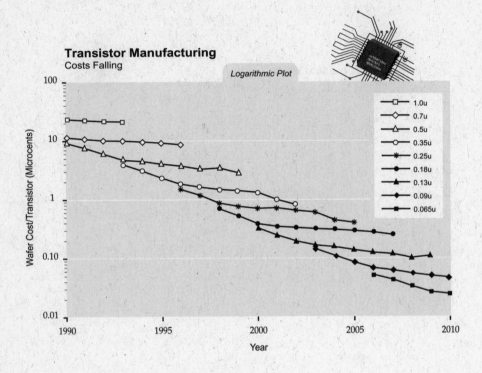

Unlike Gertrude Stein's rose, it is not the case that a transistor is a transistor is a transistor. As they have become smaller and less expensive, transistors have also become faster by a factor of about one thousand over the course of the past thirty years (see the figure below)—again, because the electrons have less distance to travel.[26]

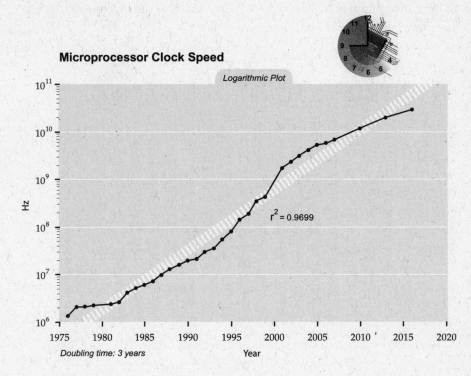

Microprocessor Clock Speed

Logarithmic Plot

$r^2 = 0.9699$

Hz

Doubling time: 3 years   Year

If we combine the exponential trends toward less-expensive transistors and faster cycle times, we find a halving time of only 1.1 years in the cost per transistor cycle (see the figure below).[27] The cost per transistor cycle is a more accurate overall measure of price-performance because it takes into account both speed and capacity. But the cost per transistor cycle still does not take into account innovation at higher levels of design (such as microprocessor design) that improves computational efficiency.

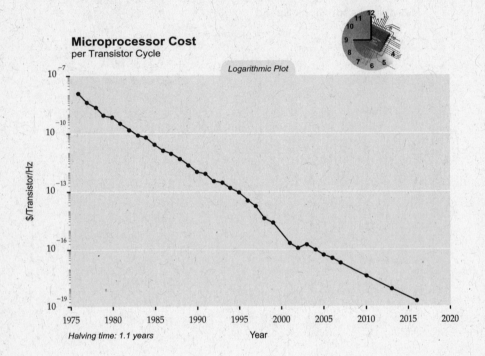

Microprocessor Cost
per Transistor Cycle

Halving time: 1.1 years

The number of transistors in Intel processors has doubled every two years (see the figure below). Several other factors have boosted price-performance, including clock speed, reduction in cost per microprocessor, and processor design innovations.[28]

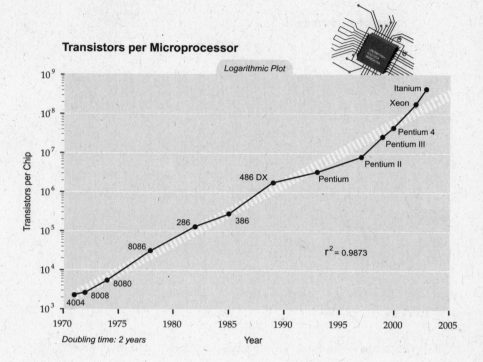

Transistors per Microprocessor

Logarithmic Plot

Transistors per Chip

$r^2 = 0.9873$

Itanium
Xeon
Pentium 4
Pentium III
Pentium II
Pentium
486 DX
386
286
8086
8080
8008
4004

Doubling time: 2 years   Year

Processor performance in MIPS has doubled every 1.8 years per processor (see the figure below). Again, note that the cost per processor has also declined through this period.[29]

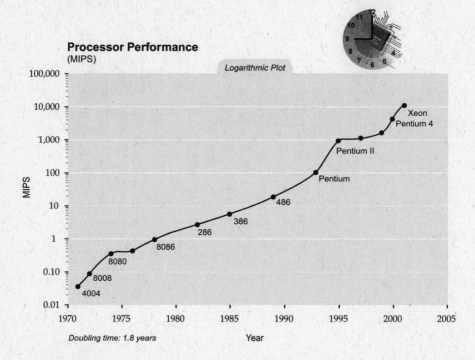

Processor Performance
(MIPS)

Logarithmic Plot

Doubling time: 1.8 years   Year

If I examine my own four-plus decades of experience in this industry, I can compare the MIT computer I used as a student in the late 1960s to a recent notebook. In 1967 I had access to a multimillion-dollar IBM 7094 with 32K (36-bit) words of memory and a quarter of a MIPS processor speed. In 2004 I used a $2,000 personal computer with a half-billion bytes of RAM and a processor speed of about 2,000 MIPS. The MIT computer was about one thousand times more expensive, so the ratio of cost per MIPS is about eight million to one.

Measure	IBM 7094 circa 1967	Notebook circa 2004
Processor Speed (MIPS)	0.25	2,000
Main Memory (K Bytes)	144	256,000
Approximate Cost (2003 $)	$11,000,000	$2,000

My recent computer provides 2,000 MIPS of processing at a cost that is about 2^{24} lower than that of the computer I used in 1967. That's 24 doublings in 37 years, or about 18.5 months per doubling. If we factor in the increased value of the approximately 2,000 times greater RAM, vast increases in disk storage, and the more powerful instruction set of my circa 2004 computer, as well as vast improvements in communication speeds, more powerful software, and other factors, the doubling time comes down even further.

Despite this massive deflation in the cost of information technologies, demand has more than kept up. The number of bits shipped has doubled every 1.1 years, faster than the halving time in cost per bit, which is 1.5 years.[30] As a result, the semiconductor industry enjoyed 18 percent annual growth in total revenue from 1958 to 2002.[31] The entire information-technology (IT) industry has grown from 4.2 percent of the gross domestic product in 1977 to 8.2 percent in 1998.[32] IT has become increasingly influential in all economic sectors. The share of value contributed by information technology for most categories of products and services is rapidly increasing. Even common manufactured products such as tables and chairs have an information content, represented by their computerized designs and the programming of the inventory-procurement systems and automated-fabrication systems used in their assembly.

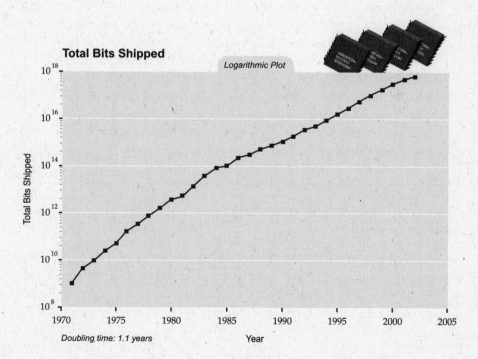

Total Bits Shipped

Logarithmic Plot

Total Bits Shipped

Doubling time: 1.1 years

Year

Doubling (or Halving) Times[33]

Dynamic RAM "Half Pitch" Feature Size (smallest chip feature)	5.4 years
Dynamic RAM (bits per dollar)	1.5 years
Average Transistor Price	1.6 years
Microprocessor Cost-per-Transistor Cycle	1.1 years
Total Bits Shipped	1.1 years
Processor Performance in MIPS	1.8 years
Transistors in Intel Microprocessors	2.0 years
Microprocessor Clock Speed	3.0 years

Moore's Law: Self-Fulfilling Prophecy?

Some observers have stated that Moore's Law is nothing more than a self-fulfilling prophecy: that industry participants anticipate where they need to be at particular times in the future, and organize their research and development accordingly. The industry's own written road map is a good example of this.[34] However, the exponential trends in information technology are far broader than those covered by Moore's Law. We see the same types of trends in essentially every technology or measurement that deals with information. This includes many technologies in which a perception of accelerating price-performance does not exist or has not previously been articulated (see below). Even within computing itself, the growth in capability per unit cost is much broader than what Moore's Law alone would predict.

The Fifth Paradigm[35]

Moore's Law is actually not the first paradigm in computational systems. You can see this if you plot the price-performance—measured by instructions per second per thousand constant dollars—of forty-nine famous computational systems and computers spanning the twentieth century (see the figure below).

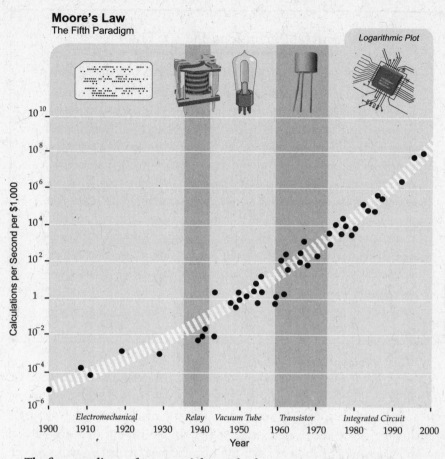

Moore's Law
The Fifth Paradigm

The five paradigms of exponential growth of computing: Each time one paradigm has run out of steam, another has picked up the pace.

As the figure demonstrates, there were actually four different paradigms—electromechanical, relays, vacuum tubes, and discrete transistors—that showed exponential growth in the price-performance of computing long before integrated circuits were even invented. And Moore's paradigm won't be the last. When Moore's Law reaches the end of its S-curve, now expected before 2020, the exponential growth will continue with three-dimensional molecular computing, which will constitute the sixth paradigm.

Fractal Dimensions and the Brain

Note that the use of the third dimension in computing systems is not an either-or choice but a continuum between two and three dimensions. In terms of biological intelligence, the human cortex is actually rather flat, with only six thin layers that are elaborately folded, an architecture that greatly increases the surface area. This folding is one way to use the third dimension. In "fractal" systems (systems in which a drawing replacement or folding rule is iteratively applied), structures that are elaborately folded are considered to constitute a partial dimension. From that perspective, the convoluted surface of the human cortex represents a number of dimensions in between two and three. Other brain structures, such as the cerebellum, are three-dimensional but comprise a repeating structure that is essentially two-dimensional. It is likely that our future computational systems will also combine systems that are highly folded two-dimensional systems with fully three-dimensional structures.

Notice that the figure shows an exponential curve on a logarithmic scale, indicating two levels of exponential growth.[36] In other words, there is a gentle but unmistakable exponential growth in the *rate* of exponential growth. (A straight line on a logarithmic scale shows simple exponential growth; an upwardly curving line shows higher-than-simple exponential growth.) As you can see, it took three years to double the price-performance of computing at the beginning of the twentieth century and two years in the middle, and it takes about one year currently.[37]

Hans Moravec provides the following similar chart (see the figure below), which uses a different but overlapping set of historical computers and plots trend lines (slopes) at different points in time. As with the figure above, the slope increases with time, reflecting the second level of exponential growth.[38]

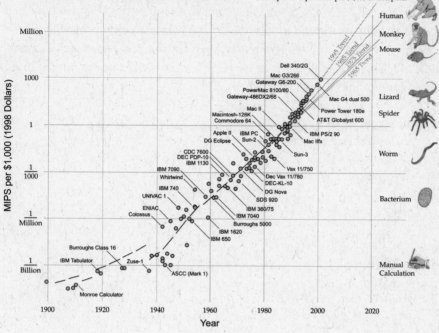

Evolution of Computer Power/Cost

If we project these computational performance trends through this next century, we can see in the figure below that supercomputers will match human brain capability by the end of this decade and personal computing will achieve it by around 2020—or possibly sooner, depending on how conservative an estimate of human brain capacity we use. (We'll discuss estimates of human brain computational speed in the next chapter.)[39]

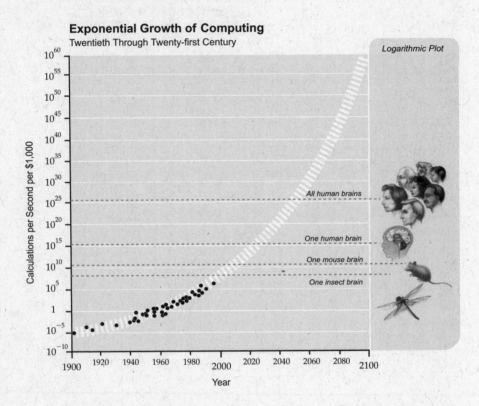

The exponential growth of computing is a marvelous quantitative example of the exponentially growing returns from an evolutionary process. We can express the exponential growth of computing in terms of its accelerating pace: it took ninety years to achieve the first MIPS per thousand dollars; now we add one MIPS per thousand dollars every five hours.[40]

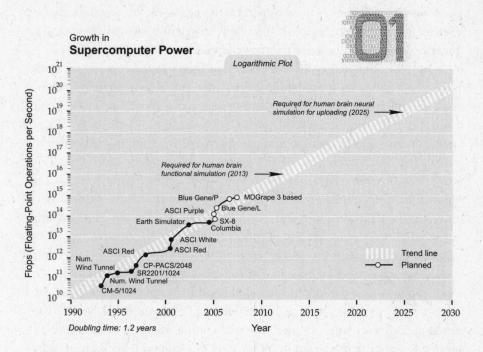

Growth in
Supercomputer Power

Logarithmic Plot

Flops (Floating-Point Operations per Second)

*Required for human brain neural
simulation for uploading (2025)* ——→

*Required for human brain
functional simulation (2013)* ——→

Blue Gene/P ○—○ MDGrape 3 based
ASCI Purple ○ Blue Gene/L
Earth Simulator ○ SX-8
Columbia
ASCI White
ASCI Red ● ASCI Red
Num.
Wind Tunnel
CP-PACS/2048
SR2201/1024
Num. Wind Tunnel
CM-5/1024

Trend line
—○— Planned

Doubling time: 1.2 years   Year

IBM's Blue Gene/P supercomputer is planned to have one million gigaflops (billions of floating-point operations per second), or 10^{15} calculations per second when it launches in 2007.[41] That's one tenth of the 10^{16} calculations per second needed to emulate the human brain (see the next chapter). And if we extrapolate this exponential curve, we get 10^{16} calculations per second early in the next decade.

As discussed above, Moore's Law narrowly refers to the number of transistors on an integrated circuit of fixed size and sometimes has been expressed even more narrowly in terms of transistor feature size. But the most appropriate measure to track price-performance is computational speed per unit cost, an index that takes into account many levels of "cleverness" (innovation, which is to say, technological evolution). In addition to all of the invention involved in integrated circuits, there are multiple layers of improvement in computer design (for example, pipelining, parallel processing, instruction look-ahead, instruction and memory caching, and many others).

The human brain uses a very inefficient electrochemical, digital-controlled analog computational process. The bulk of its calculations are carried out in the interneuronal connections at a speed of only about two hundred calculations per second (in each connection), which is at least one million times slower than contemporary electronic circuits. But the brain gains its prodigious powers from its extremely parallel organization *in three dimensions*. There are many technologies in the wings that will build circuitry in three dimensions, which I discuss in the next chapter.

We might ask whether there are inherent limits to the capacity of matter and energy to support computational processes. This is an important issue, but as we will see in the next chapter, we won't approach those limits until late in this century. It is important to distinguish between the S-curve that is characteristic of any specific technological paradigm and the continuing exponential growth that is characteristic of the ongoing evolutionary process within a broad area of technology, such as computation. Specific paradigms, such as Moore's Law, do ultimately reach levels at which exponential growth is no longer feasible. But the growth of computation supersedes any of its underlying paradigms and is for present purposes an ongoing exponential.

In accordance with the law of accelerating returns, paradigm shift (also called innovation) turns the S-curve of any specific paradigm into a continuing exponential. A new paradigm, such as three-dimensional circuits, takes over when the old paradigm approaches its natural limit, which has already happened at least four times in the history of computation. In such nonhuman species as apes, the mastery of a toolmaking or -using skill by each animal is characterized by an S-shaped learning curve that ends abruptly; human-created technology, in contrast, has followed an exponential pattern of growth and acceleration since its inception.

DNA Sequencing, Memory, Communications, the Internet, and Miniaturization

> Civilization advances by extending the number of important operations which we can perform without thinking about them.
> —ALFRED NORTH WHITEHEAD, 1911[42]

> Things are more like they are now than they ever were before.
> —DWIGHT D. EISENHOWER

The law of accelerating returns applies to all of technology, indeed to any evolutionary process. It can be charted with remarkable precision in information-based technologies because we have well-defined indexes (for example, calculations per second per dollar, or calculations per second per gram) to measure them. There are a great many examples of the exponential growth implied by the law of accelerating returns, in areas as varied as electronics of all kinds, DNA sequencing, communications, brain scanning, brain reverse engineering, the size and scope of human knowledge, and the rapidly shrinking size of technology. The latter trend is directly related to the emergence of nanotechnology.

The future GNR (Genetics, Nanotechnology, Robotics) age (see chapter 5) will come about not from the exponential explosion of computation alone but rather from the interplay and myriad synergies that will result from multiple intertwined technological advances. As every point on the exponential-growth curves underlying this panoply of technologies represents an intense human drama of innovation and competition, we must consider it remarkable that these chaotic processes result in such smooth and predictable exponential trends. This is not a coincidence but is an inherent feature of evolutionary processes.

When the human-genome scan got under way in 1990 critics pointed out that given the speed with which the genome could then be scanned, it would take thousands of years to finish the project. Yet the fifteen-year project was completed slightly ahead of schedule, with a first draft in 2003.[43] The cost of DNA sequencing came down from about ten dollars per base pair in 1990 to a couple of pennies in 2004 and is rapidly continuing to fall (see the figure below).[44]

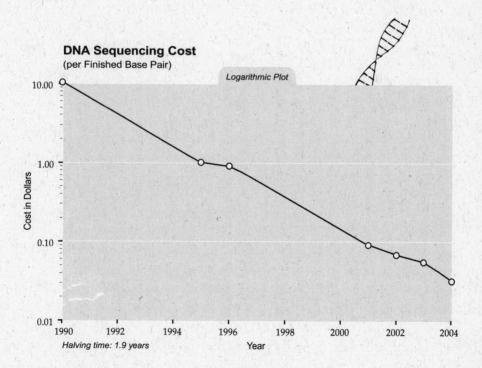

DNA Sequencing Cost
(per Finished Base Pair)

Logarithmic Plot

Cost in Dollars

Halving time: 1.9 years

Year

There has been smooth exponential growth in the amount of DNA-sequence data that has been collected (see the figure below).[45] A dramatic recent example of this improving capacity was the sequencing of the SARS virus, which took only thirty-one days from the identification of the virus, compared to more than fifteen years for the HIV virus.[46]

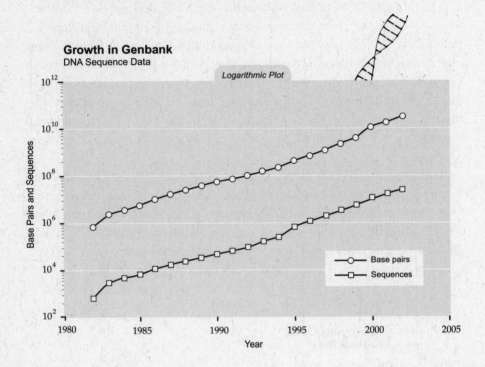

Of course, we expect to see exponential growth in electronic memories such as RAM. But note how the trend on this logarithmic graph (below) proceeds smoothly through different technology paradigms: vacuum tube to discrete transistor to integrated circuit.[47]

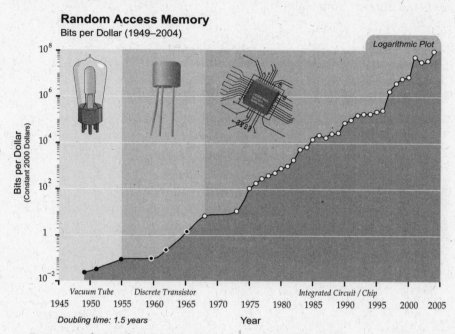

Exponential growth in RAM capacity across paradigm shifts.

However, growth in the price-performance of magnetic (disk-drive) memory is not a result of Moore's Law. This exponential trend reflects the squeezing of data onto a magnetic substrate, rather than transistors onto an integrated circuit, a completely different technical challenge pursued by different engineers and different companies.[48]

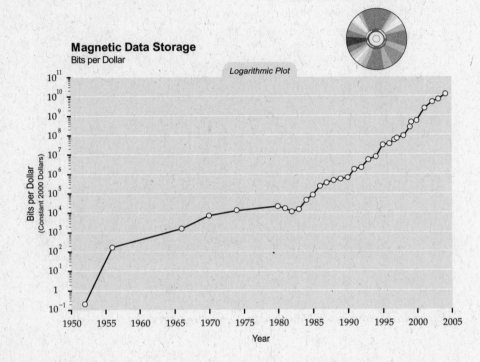

Exponential growth in communications technology (measures for communicating information; see the figure below) has for many years been even more explosive than in processing or memory measures of computation and is no less significant in its implications. Again, this progression involves far more than just shrinking transistors on an integrated circuit but includes accelerating advances in fiber optics, optical switching, electromagnetic technologies, and other factors.[49]

We are currently moving away from the tangle of wires in our cities and in our daily lives through wireless communication, the power of which is doubling every ten to eleven months (see the figure below).

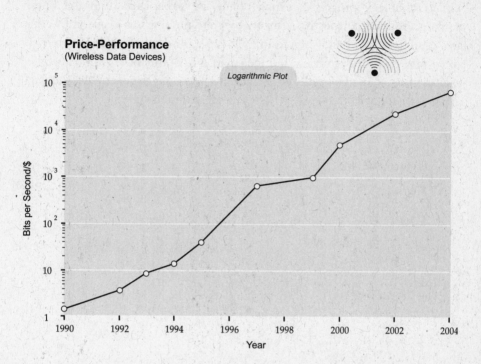

Price-Performance
(Wireless Data Devices)

The figures below show the overall growth of the Internet based on the number of hosts (Web-server computers). These two charts plot the same data, but one is on a logarithmic axis and the other is linear. As has been discussed, while technology progresses exponentially, we experience it in the linear domain. From the perspective of most observers, nothing was happening in this area until the mid-1990s, when seemingly out of nowhere the World Wide Web and e-mail exploded into view. But the emergence of the Internet into a world-wide phenomenon was readily predictable by examining exponential trend data in the early 1980s from the ARPANET, predecessor to the Internet.[50]

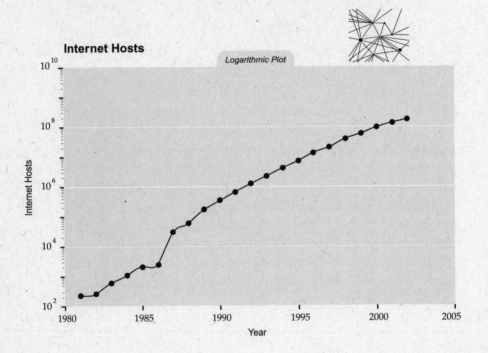

Internet Hosts

This figure shows the same data on a linear scale.[51]

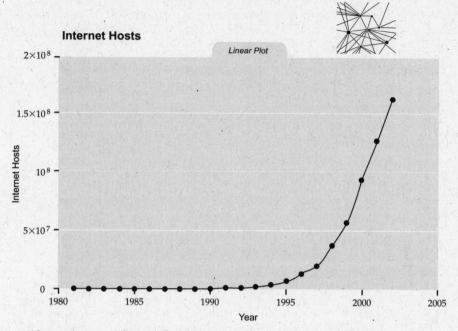

The explosion of the Internet appears to be a surprise from the linear chart but was perfectly predictable from the logarithmic one.

In addition to servers, the actual data traffic on the Internet has also doubled every year.[52]

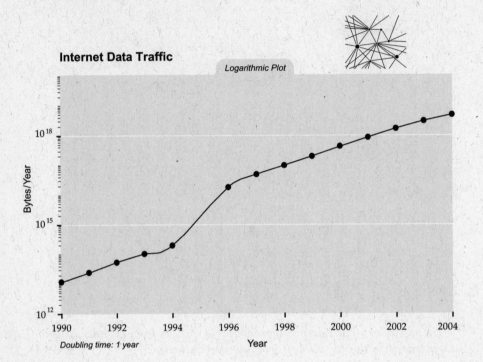

To accommodate this exponential growth, the data transmission speed of the Internet backbone (as represented by the fastest announced backbone communication channels actually used for the Internet) has itself grown exponentially. Note that in the figure "Internet Backbone Bandwidth" below, we can actually see the progression of S-curves: the acceleration fostered by a new paradigm, followed by a leveling off as the paradigm runs out of steam, followed by renewed acceleration through paradigm shift.[53]

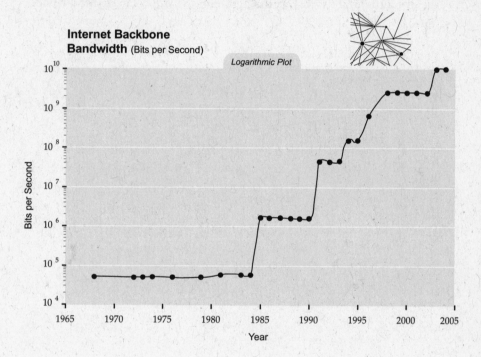

Internet Backbone Bandwidth (Bits per Second)

Logarithmic Plot

Another trend that will have profound implications for the twenty-first century is the pervasive movement toward miniaturization. The key feature sizes of a broad range of technologies, both electronic and mechanical, are decreasing, and at an exponential rate. At present, we are shrinking technology by a factor of about four per linear dimension per decade. This miniaturization is a driving force behind Moore's Law, but it's also reflected in the size of all electronic systems—for example, magnetic storage. We also see this decrease in the size of mechanical devices, as the figure on the size of mechanical devices illustrates.[54]

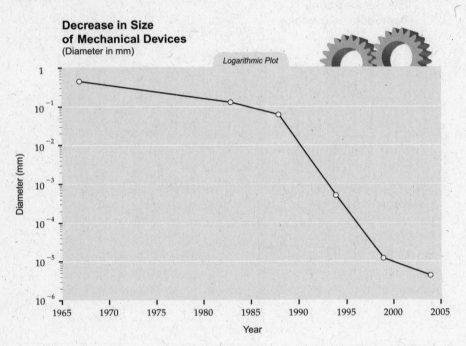

As the salient feature size of a wide range of technologies moves inexorably closer to the multinanometer range (less than one hundred nanometers—billionths of a meter), it has been accompanied by a rapidly growing interest in nanotechnology. Nanotechnology science citations have been increasing significantly over the past decade, as noted in the figure below.[55]

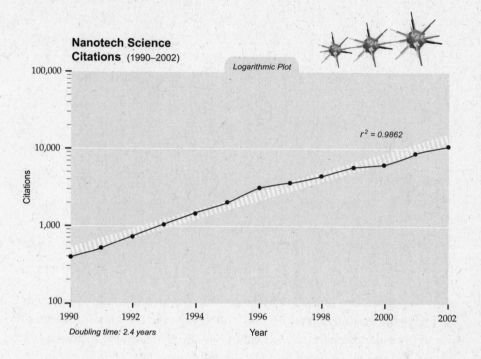

We see the same phenomenon in nanotechnology-related patents (below).[56]

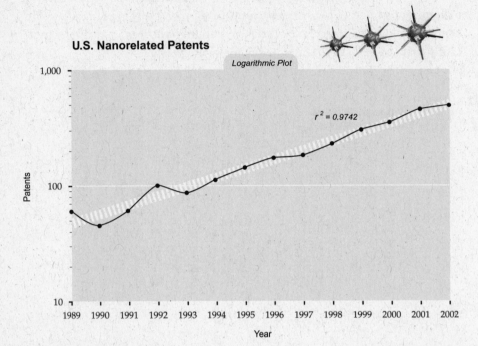

U.S. Nanorelated Patents

Logarithmic Plot

$r^2 = 0.9742$

Patents

Year

As we will explore in chapter 5, the genetics (or biotechnology) revolution is bringing the information revolution, with its exponentially increasing capacity and price-performance, to the field of biology. Similarly, the nanotechnology revolution will bring the rapidly increasing mastery of information to materials and mechanical systems. The robotics (or "strong AI") revolution involves the reverse engineering of the human brain, which means coming to understand human intelligence in information terms and then combining the resulting insights with increasingly powerful computational platforms. Thus, all three of the overlapping transformations—genetics, nanotechnology, and robotics—that will dominate the first half of this century represent different facets of the information revolution.

Information, Order, and Evolution:
The Insights from Wolfram and Fredkin's Cellular Automata

As I've described in this chapter, every aspect of information and information technology is growing at an exponential pace. Inherent in our expectation of a Singularity taking place in human history is the pervasive importance of information to the future of human experience. We see information at every level of existence. Every form of human knowledge and artistic expression—scientific and engineering ideas and designs, literature, music, pictures, movies—can be expressed as digital information.

Our brains also operate digitally, through discrete firings of our neurons. The wiring of our interneuronal connections can be digitally described, and the design of our brains is specified by a surprisingly small digital genetic code.[57]

Indeed, all of biology operates through linear sequences of 2-bit DNA base pairs, which in turn control the sequencing of only twenty amino acids in proteins. Molecules form discrete arrangements of atoms. The carbon atom, with its four positions for establishing molecular connections, is particularly adept at creating a variety of three-dimensional shapes, which accounts for its central role in both biology and technology. Within the atom, electrons take on discrete energy levels. Other subatomic particles, such as protons, comprise discrete numbers of valence quarks.

Although the formulas of quantum mechanics are expressed in terms of both continuous fields and discrete levels, we do know that continuous levels can be expressed to any desired degree of accuracy using binary data.[58] In fact, quantum mechanics, as the word "quantum" implies, is based on discrete values.

Physicist-mathematician Stephen Wolfram provides extensive evidence to show how increasing complexity can originate from a universe that is at its core a deterministic, algorithmic system (a system based on fixed rules with predetermined outcomes). In his book *A New Kind of Science*, Wolfram offers a comprehensive analysis of how the processes underlying a mathematical construction called "a cellular automaton" have the potential to describe every level of our natural world.[59] (A cellular automaton is a simple computational mechanism that, for example, changes the color of each cell on a grid based on the color of adjacent or nearby cells according to a transformation rule.)

In his view, it is feasible to express all information processes in terms of operations on cellular automata, so Wolfram's insights bear on several key issues related to information and its pervasiveness. Wolfram postulates ▶

that the universe itself is a giant cellular-automaton computer. In his hypothesis there is a digital basis for apparently analog phenomena (such as motion and time) and for formulas in physics, and we can model our understanding of physics as the simple transformations of a cellular automaton.

Others have proposed this possibility. Richard Feynman wondered about it in considering the relationship of information to matter and energy. Norbert Wiener heralded a fundamental change in focus from energy to information in his 1948 book *Cybernetics* and suggested that the transformation of information, not energy, was the fundamental building block of the universe.[60] Perhaps the first to postulate that the universe is being computed on a digital computer was Konrad Zuse in 1967.[61] Zuse is best known as the inventor of the first working programmable computer, which he developed from 1935 to 1941.

An enthusiastic proponent of an information-based theory of physics was Edward Fredkin, who in the early 1980s proposed a "new theory of physics" founded on the idea that the universe is ultimately composed of software. We should not think of reality as consisting of particles and forces, according to Fredkin, but rather as bits of data modified according to computation rules.

Fredkin was quoted by Robert Wright in the 1980s as saying,

> There are three great philosophical questions. What is life? What is consciousness and thinking and memory and all that? And how does the universe work?... [The] "informational viewpoint" encompasses all three.... What I'm saying is that at the most basic level of complexity an information process runs what we think of as physics. At the much higher level of complexity, life, DNA—you know, the biochemical functions—are controlled by a digital information process. Then, at another level, our thought processes are basically information processing.... I find the supporting evidence for my beliefs in ten thousand different places.... And to me it's just totally overwhelming. It's like there's an animal I want to find. I've found his footprints. I've found his droppings. I've found the half-chewed food. I find pieces of his fur, and so on. In every case it fits one kind of animal, and it's not like any animal anyone's ever seen. People say, Where is this animal? I say, Well, he was here, he's about this big, this that, and the other. And I know a thousand things about him. I don't have him in hand, but I know he's there.... What I see is so compelling that it can't be a creature of my imagination.[62]   ▶

In commenting on Fredkin's theory of digital physics, Wright writes,

> Fredkin . . . is talking about an interesting characteristic of some computer programs, including many cellular automata: there is no shortcut to finding out what they will lead to. This, indeed, is a basic difference between the "analytical" approach associated with traditional mathematics, including differential equations, and the "computational" approach associated with algorithms. You can predict a future state of a system susceptible to the analytic approach without figuring out what states it will occupy between now and then, but in the case of many cellular automata, you must go through all the intermediate states to find out what the end will be like: there is no way to know the future except to watch it unfold.... Fredkin explains: "There is no way to know the answer to some question any faster than what's going on." . . . Fredkin believes that the universe is very literally a computer and that it is being used by someone, or something, to solve a problem. It sounds like a good-news/bad-news joke: the good news is that our lives have purpose; the bad news is that their purpose is to help some remote hacker estimate pi to nine jillion decimal places. [63]

Fredkin went on to show that although energy is needed for information storage and retrieval, we can arbitrarily reduce the energy required to perform any particular example of information processing, and that this operation has no lower limit.[64] That implies that information rather than matter and energy may be regarded as the more fundamental reality.[65] I will return to Fredkin's insight regarding the extreme lower limit of energy required for computation and communication in chapter 3, since it pertains to the ultimate power of intelligence in the universe.

Wolfram builds his theory primarily on a single, unified insight. The discovery that has so excited Wolfram is a simple rule he calls cellular automata rule 110 and its behavior. (There are some other interesting automata rules, but rule 110 makes the point well enough.) Most of Wolfram's analyses deal with the simplest possible cellular automata, specifically those that involve just a one-dimensional line of cells, two possible colors (black and white), and rules based only on the two immediately adjacent cells. For each transformation, the color of a cell depends only on its own previous color and that of the cell on the left and the cell on the right. Thus, there are eight possible input situations (that is, three combinations of two colors). Each rule maps all combinations of these eight input situations to an output (black or white). So there are 2^8 ▶

(256) possible rules for such a one-dimensional, two-color, adjacent-cell automaton. Half of the 256 possible rules map onto the other half because of left-right symmetry. We can map half of them again because of black-white equivalence, so we are left with 64 rule types. Wolfram illustrates the action of these automata with two-dimensional patterns in which each line (along the y-axis) represents a subsequent generation of applying the rule to each cell in that line.

Most of the rules are degenerate, meaning they create repetitive patterns of no interest, such as cells of a single color, or a checkerboard pattern. Wolfram calls these rules class 1 automata. Some rules produce arbitrarily spaced streaks that remain stable, and Wolfram classifies these as belonging to class 2. Class 3 rules are a bit more interesting, in that recognizable features (such as triangles) appear in the resulting pattern in an essentially random order.

However, it was class 4 automata that gave rise to the "aha" experience that resulted in Wolfram's devoting a decade to the topic. The class 4 automata, of which rule 110 is the quintessential example, produce surprisingly complex patterns that do not repeat themselves. We see in them artifacts such as lines at various angles, aggregations of triangles, and other interesting configurations. The resulting pattern, however, is neither regular nor completely random; it appears to have some order but is never predictable.

Rule 110

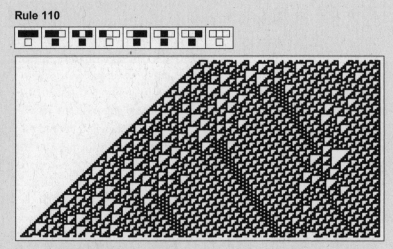

Portion of image generated by rule 110

►

Why is this important or interesting? Keep in mind that we began with the simplest possible starting point: a single black cell. The process involves repetitive application of a very simple rule.[66] From such a repetitive and deterministic process, one would expect repetitive and predictable behavior. There are two surprising results here. One is that the results produce apparent randomness. However, the results are more interesting than pure randomness, which itself would become boring very quickly. There are discernible and interesting features in the designs produced, so the pattern has some order and apparent intelligence. Wolfram includes a number of examples of these images, many of which are rather lovely to look at.

Wolfram makes the following point repeatedly: "Whenever a phenomenon is encountered that seems complex it is taken almost for granted that the phenomenon must be the result of some underlying mechanism that is itself complex. But my discovery that simple programs can produce great complexity makes it clear that this is not in fact correct." [67]

I do find the behavior of rule 110 rather delightful. Furthermore, the idea that a completely deterministic process can produce results that are completely unpredictable is of great importance, as it provides an explanation for how the world can be inherently unpredictable while still based on fully deterministic rules.[68] However, I am not entirely surprised by the idea that simple mechanisms can produce results more complicated than their starting conditions. We've seen this phenomenon in fractals, chaos and complexity theory, and self-organizing systems (such as neural nets and Markov models), which start with simple networks but organize themselves to produce apparently intelligent behavior.

At a different level, we see it in the human brain itself, which starts with only about thirty to one hundred million bytes of specification in the compressed genome yet ends up with a complexity that is about a billion times greater.[69]

It is also not surprising that a deterministic process can produce apparently random results. We have had random-number generators (for example, the "randomize" function in Wolfram's program Mathematica) that use deterministic processes to produce sequences that pass statistical tests for randomness. These programs date back to the earliest days of computer software, such as the first versions of Fortran. However, Wolfram does provide a thorough theoretical foundation for this observation.

Wolfram goes on to describe how simple computational mechanisms can exist in nature at different levels, and he shows that these simple and deterministic mechanisms can produce all of the complexity that we see and experience. He provides myriad examples, such as the pleasing ▶

designs of pigmentation on animals, the shape and markings of shells, and patterns of turbulence (such as the behavior of smoke in the air). He makes the point that computation is essentially simple and ubiquitous. The repetitive application of simple computational transformations, according to Wolfram, is the true source of complexity in the world.

My own view is that this is only partly correct. I agree with Wolfram that computation is all around us, and that some of the patterns we see are created by the equivalent of cellular automata. But a key issue to ask is this: *Just how complex are the results of class 4 automata?*

Wolfram effectively sidesteps the issue of degrees of complexity. I agree that a degenerate pattern such as a chessboard has no complexity. Wolfram also acknowledges that mere randomness does not represent complexity either, because pure randomness also becomes predictable in its pure lack of predictability. It is true that the interesting features of class 4 automata are neither repeating nor purely random, so I would agree that they are more complex than the results produced by other classes of automata.

However, there is nonetheless a distinct limit to the complexity produced by class 4 automata. The many images of such automata in Wolfram's book all have a similar look to them, and although they are nonrepeating, they are interesting (and intelligent) only to a degree. Moreover, they do not continue to evolve into anything more complex, nor do they develop new types of features. One could run these automata for trillions or even trillions of trillions of iterations and the image would remain at the same limited level of complexity. They do not evolve into, say, insects or humans or Chopin preludes or anything else that we might consider of a higher order of complexity than the streaks and intermingling triangles displayed in these images.

Complexity is a continuum. Here I define "order" as "information that fits a purpose."[70] A completely predictable process has zero order. A high level of information alone does not necessarily imply a high level of order either. A phone book has a lot of information, but the level of order of that information is quite low. A random sequence is essentially pure information (since it is not predictable) but has no order. The output of class 4 automata does possess a certain level of order, and it does survive like other persisting patterns. But the pattern represented by a human being has a far higher level of order, and of complexity.

Human beings fulfill a highly demanding purpose: they survive in a challenging ecological niche. Human beings represent an extremely ▶

intricate and elaborate hierarchy of other patterns. Wolfram regards any patterns that combine some recognizable features and unpredictable elements to be effectively equivalent to one another. But he does not show how a class 4 automaton can ever increase its complexity, let alone become a pattern as complex as a human being.

There is a missing link here, one that would account for how one gets from the interesting but ultimately routine patterns of a cellular automaton to the complexity of persisting structures that demonstrate higher levels of intelligence. For example, these class 4 patterns are not capable of solving interesting problems, and no amount of iteration moves them closer to doing so. Wolfram would counter that a rule 110 automaton could be used as a "universal computer."[71] However, by itself, a universal computer is not capable of solving intelligent problems without what I would call "software." It is the complexity of the software that runs on a universal computer that is precisely the issue.

One might point out that class 4 patterns result from the simplest possible cellular automata (one-dimensional, two-color, two-neighbor rules). What happens if we increase the dimensionality—for example, go to multiple colors or even generalize these discrete cellular automata to continuous functions? Wolfram addresses all of this quite thoroughly. The results produced from more complex automata are essentially the same as those of the very simple ones. We get the same sorts of interesting but ultimately quite limited patterns. Wolfram makes the intriguing point that we do not need to use more complex rules to get complexity in the end result. But I would make the converse point that we are unable to increase the complexity of the end result through either more complex rules or further iteration. So cellular automata get us only so far.

Can We Evolve Artificial Intelligence from Simple Rules?

So how do we get from these interesting but limited patterns to those of insects or humans or Chopin preludes? One concept we need to take into consideration is conflict—that is, *evolution*. If we add another simple concept—an evolutionary algorithm—to that of Wolfram's simple cellular automata, we start to get far more exciting and more intelligent results. Wolfram would say that the class 4 automata and an evolutionary algorithm are "computationally equivalent." But that is true only on what I consider the "hardware" level. On the software level, the order of the patterns produced are clearly different and of a different order of complexity and usefulness.

An evolutionary algorithm can start with randomly generated   ▶

potential solutions to a problem, which are encoded in a digital genetic code. We then have the solutions compete with one another in a simulated evolutionary battle. The better solutions survive and procreate in a simulated sexual reproduction in which offspring solutions are created, drawing their genetic code (encoded solutions) from two parents. We can also introduce a rate of genetic mutation. Various high-level parameters of this process, such as the rate of mutation, the rate of offspring, and so on, are appropriately called "God parameters," and it is the job of the engineer designing the evolutionary algorithm to set them to reasonably optimal values. The process is run for many thousands of generations of simulated evolution, and at the end of the process one is likely to find solutions that are of a distinctly higher order than the starting ones.

The results of these evolutionary (sometimes called genetic) algorithms can be elegant, beautiful, and intelligent solutions to complex problems. They have been used, for example, to create artistic designs and designs for artificial life-forms, as well as to execute a wide range of practical assignments such as designing jet engines. Genetic algorithms are one approach to "narrow" artificial intelligence—that is, creating systems that can perform particular functions that used to require the application of human intelligence.

But something is still missing. Although genetic algorithms are a useful tool in solving specific problems, they have never achieved anything resembling "strong AI"—that is, aptitude resembling the broad, deep, and subtle features of human intelligence, particularly its powers of pattern recognition and command of language. Is the problem that we are not running the evolutionary algorithms long enough? After all, humans evolved through a process that took billions of years. Perhaps we cannot re-create that process with just a few days or weeks of computer simulation. This won't work, however, because conventional genetic algorithms reach an asymptote in their level of performance, so running them for a longer period of time won't help.

A third level (beyond the ability of cellular processes to produce apparent randomness and genetic algorithms to produce focused intelligent solutions) is to perform evolution on multiple levels. Conventional genetic algorithms allow evolution only within the confines of a narrow problem and a single means of evolution. The genetic code itself needs to evolve; the rules of evolution need to evolve. Nature did not stay with a single chromosome, for example. There have been many levels of indirection incorporated in the natural evolutionary process. And we require a complex environment in which the evolution takes place.   ▶

To build strong AI we will have the opportunity to short-circuit this process, however, by reverse engineering the human brain, a project well under way, thereby benefiting from the evolutionary process that has already taken place. We will be applying evolutionary algorithms within these solutions just as the human brain does. For example, the fetal wiring is initially random within constraints specified in the genome in at least some regions. Recent research shows that areas having to do with learning undergo more change, whereas structures having to do with sensory processing experience less change after birth.[72]

Wolfram makes the valid point that certain (indeed, most) computational processes are not predictable. In other words, we cannot predict future states without running the entire process. I agree with him that we can know the answer in advance only if somehow we can simulate a process at a faster speed. Given that the universe runs at the fastest speed it can run, there is usually no way to short-circuit the process. However, we have the benefits of the billions of years of evolution that have already taken place, which are responsible for the greatly increased order of complexity in the natural world. We can now benefit from it by using our evolved tools to reverse engineer the products of biological evolution (most importantly, the human brain).

Yes, it is true that some phenomena in nature that may appear complex at some level are merely the result of simple underlying computational mechanisms that are essentially cellular automata at work. The interesting pattern of triangles on a "tent olive" shell (cited extensively by Wolfram) or the intricate and varied patterns of a snowflake are good examples. I don't think this is a new observation, in that we've always regarded the design of snowflakes to derive from a simple molecular computation-like building process. However, Wolfram does provide us with a compelling theoretical foundation for expressing these processes and their resulting patterns. But there is more to biology than class 4 patterns.

Another important thesis by Wolfram lies in his thorough treatment of computation as a simple and ubiquitous phenomenon. Of course, we've known for more than a century that computation is inherently simple: we can build any possible level of complexity from a foundation of the simplest possible manipulations of information.

For example, Charles Babbage's late-nineteenth-century mechanical computer (which never ran) provided only a handful of operation codes yet provided (within its memory capacity and speed) the same kinds of transformations that modern computers do. The complexity of Babbage's invention stemmed only from the details of its design, which indeed ▶

proved too difficult for Babbage to implement using the technology available to him.

The Turing machine, Alan Turing's theoretical conception of a universal computer in 1950, provides only seven very basic commands, yet can be organized to perform any possible computation.[73] The existence of a "universal Turing machine," which can simulate any possible Turing machine that is described on its tape memory, is a further demonstration of the universality and simplicity of computation.[74] In *The Age of Intelligent Machines*, I showed how any computer could be constructed from "a suitable number of [a] very simple device," namely, the "nor" gate.[75] This is not exactly the same demonstration as a universal Turing machine, but it does demonstrate that any computation can be performed by a cascade of this very simple device (which is simpler than rule 110), given the right software (which in this case would include the connection description of the nor gates).[76]

Although we need additional concepts to describe an evolutionary process that can create intelligent solutions to problems, Wolfram's demonstration of the simplicity and ubiquity of computation is an important contribution in our understanding of the fundamental significance of information in the world.

MOLLY 2004: *You've got machines evolving at an accelerating pace. What about humans?*

RAY: *You mean biological humans?*

MOLLY 2004: *Yes.*

CHARLES DARWIN: *Biological evolution is presumably continuing, is it not?*

RAY: *Well, biology at this level is evolving so slowly that it hardly counts. I mentioned that evolution works through indirection. It turns out that the older paradigms such as biological evolution do continue but at their old speed, so they are eclipsed by the new paradigms. Biological evolution for animals as complex as humans takes tens of thousands of years to make noticeable, albeit still small, differences. The entire history of human cultural and technological evolution has taken place on that timescale. Yet we are now poised to ascend beyond the fragile and slow creations of biological evolution in a mere several decades. Current progress is on a scale that is a thousand to a million times faster than biological evolution.*

NED LUDD: *What if not everyone wants to go along with this?*

RAY: *I wouldn't expect they would. There are always early and late adopters. There's always a leading edge and a trailing edge to technology or to any evolutionary change. We still have people pushing plows, but that hasn't slowed down the adoption of cell phones, telecommunications, the Internet, biotechnology, and so on. However, the lagging edge does ultimately catch up. We have societies in Asia that jumped from agrarian economies to information economies, without going through industrialization.[77]*

NED: *That may be so, but the digital divide is getting worse.*

RAY: *I know that people keep saying that, but how can that possibly be true? The number of humans is growing only very slowly. The number of digitally connected humans, no matter how you measure it, is growing rapidly. A larger and larger fraction of the world's population is getting electronic communicators and leapfrogging our primitive phone-wiring system by hooking up to the Internet wirelessly, so the digital divide is rapidly diminishing, not growing.*

MOLLY 2004: *I still feel that the have/have not issue doesn't get enough attention. There's more we can do.*

RAY: *Indeed, but the overriding, impersonal forces of the law of accelerating returns are nonetheless moving in the right direction. Consider that technology in a particular area starts out unaffordable and not working very well. Then it becomes merely expensive and works a little better. The next step is the product becomes inexpensive and works really well. Finally, the technology becomes virtually free and works great. It wasn't long ago that when you saw someone using a portable phone in a movie, he or she was a member of the power elite, because only the wealthy could afford portable phones. Or as a more poignant example, consider drugs for AIDS. They started out not working very well and costing more than ten thousand dollars per year per patient. Now they work a lot better and are down to several hundred dollars per year in poor countries.[78] Unfortunately with regard to AIDS, we're not yet at the working great and costing almost nothing stage. The world is beginning to take somewhat more effective action on AIDS, but it has been tragic that more has not been done. Millions of lives, most in Africa, have been lost as a result. But the effect of the law of accelerating returns is nonetheless moving in the right direction. And the time gap between leading and lagging edge is itself contracting. Right now I estimate this lag at about a decade. In a decade, it will be down to about half a decade.*

The Singularity as Economic Imperative

> The reasonable man adapts himself to the world; the unreasonable one persists in trying to adapt the world to himself. Therefore all progress depends on the unreasonable man.
>
> —GEORGE BERNARD SHAW, "MAXIMS FOR REVOLUTIONISTS," *MAN AND SUPERMAN*, 1903

> All progress is based upon a universal innate desire on the part of every organism to live beyond its income.
>
> —SAMUEL BUTLER, *NOTEBOOKS*, 1912

> If I were just setting out today to make that drive to the West Coast to start a new business, I would be looking at biotechnology and nanotechnology.
>
> —JEFF BEZOS, FOUNDER AND CEO OF AMAZON.COM

Get Eighty Trillion Dollars—Limited Time Only

You will get eighty trillion dollars just by reading this section and understanding what it says. For complete details, see below. (It's true that an author will do just about anything to keep your attention, but I'm serious about this statement. Until I return to a further explanation, however, do read the first sentence of this paragraph carefully.)

The law of accelerating returns is fundamentally an economic theory. Contemporary economic theory and policy are based on outdated models that emphasize energy costs, commodity prices, and capital investment in plant and equipment as key driving factors, while largely overlooking computational capacity, memory, bandwidth, the size of technology, intellectual property, knowledge, and other increasingly vital (and increasingly increasing) constituents that are driving the economy.

It's the economic imperative of a competitive marketplace that is the primary force driving technology forward and fueling the law of accelerating returns. In turn, the law of accelerating returns is transforming economic relationships. Economic imperative is the equivalent of survival in biological evolution. We are moving toward more intelligent and smaller machines as the result of myriad small advances, each with its own particular economic justification. Machines that can more precisely carry out their missions have increased value, which explains why they are being built. There are tens of thousands of projects that are advancing the various aspects of the law of accelerating returns in diverse incremental ways.

Regardless of near-term business cycles, support for "high tech" in the business community, and in particular for software development, has grown enormously. When I started my optical character recognition (OCR) and speech-synthesis company (Kurzweil Computer Products) in 1974, high-tech venture deals in the United States totaled less than thirty million dollars (in 1974 dollars). Even during the recent high-tech recession (2000–2003), the figure was almost one hundred times greater.[79] We would have to repeal capitalism and every vestige of economic competition to stop this progression.

It is important to point out that we are progressing toward the "new" knowledge-based economy exponentially but nonetheless gradually.[80] When the so-called new economy did not transform business models overnight, many observers were quick to dismiss the idea as inherently flawed. It will be another couple of decades before knowledge dominates the economy, but it will represent a profound transformation when it happens.

We saw the same phenomenon in the Internet and telecommunications boom-and-bust cycles. The booms were fueled by the valid insight that the Internet and distributed electronic communication represented fundamental transformations. But when these transformations did not occur in what were unrealistic time frames, more than two trillion dollars of market capitalization vanished. As I point out below, the actual adoption of these technologies progressed smoothly with no indication of boom or bust.

Virtually all of the economic models taught in economics classes and used by the Federal Reserve Board to set monetary policy, by government agencies to set economic policy, and by economic forecasters of all kinds are fundamentally flawed in their view of long-term trends. That's because they are based on the "intuitive linear" view of history (the assumption that the pace of change will continue at the current rate) rather than the historically based exponential view. The reason that these linear models appear to work for a while is the same reason most people adopt the intuitive linear view in the first place: exponential trends appear to be linear when viewed and experienced for a brief period of time, particularly in the early stages of an exponential trend, when not much is happening. But once the "knee of the curve" is achieved and the exponential growth explodes, the linear models break down.

As this book is being written, the country is debating changing the Social Security program based on projections that go out to 2042, approximately the time frame I've estimated for the Singularity (see the next chapter). This economic policy review is unusual in the very long time frames involved. The predictions are based on linear models of longevity increases and economic growth that are highly unrealistic. On the one hand, longevity increases will vastly outstrip the government's modest expectations. On the other hand, people won't be seeking to retire at sixty-five when they have the bodies and brains of thirty-year-olds. Most important, the economic growth from the "GNR" technologies (see chapter 5) will greatly outstrip the 1.7 percent per year estimates being used (which understate by half even our experience over the past fifteen years).

The exponential trends underlying productivity growth are just beginning this explosive phase. The U.S. real gross domestic product has grown exponentially, fostered by improving productivity from technology, as seen in the figure below.[81]

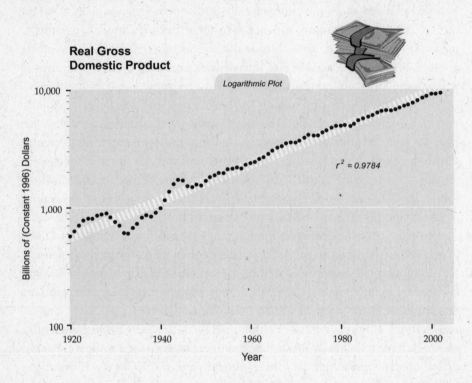

Some critics credit population growth with the exponential growth in GDP, but we see the same trend on a per-capita basis (see the figure below).[82]

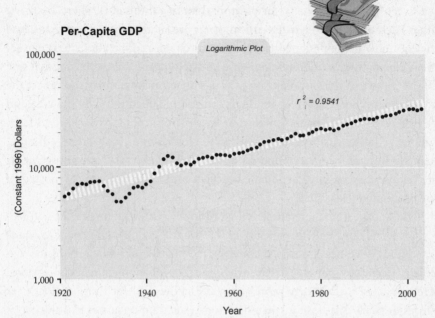

Per-Capita GDP

Logarithmic Plot

$r^2_i = 0.9541$

(Constant 1996) Dollars

Note that the underlying exponential growth in the economy is a far more powerful force than periodic recessions. Most important, recessions, including depressions, represent only temporary deviations from the underlying curve. Even the Great Depression represents only a minor blip in the context of the underlying pattern of growth. In each case, the economy ends up exactly where it would have been had the recession/depression never occurred.

The world economy is continuing to accelerate. The World Bank released a report in late 2004 indicating that the past year had been more prosperous than any year in history with worldwide economic growth of 4 percent.[83] Moreover, the highest rates were in the developing countries: more than 6 percent. Even omitting China and India, the rate was over 5 percent. In the East Asian and Pacific region, the number of people living in extreme poverty went from 470 million in 1990 to 270 million in 2001, and is projected by the World Bank to be under 20 million by 2015. Other regions are showing similar, although somewhat less dramatic, economic growth.

Productivity (economic output per worker) has also been growing exponentially. These statistics are in fact greatly understated because they do not fully reflect significant improvements in the quality and features of products and services. It is not the case that "a car is a car"; there have been major upgrades in safety, reliability, and features. Certainly, one thousand dollars of computation today is far more powerful than one thousand dollars of computation ten years ago (by a factor of more than one thousand). There are many other such examples. Pharmaceutical drugs are increasingly effective because they are now being designed to precisely carry out modifications to the exact metabolic pathways underlying disease and aging processes with minimal side effects (note that the vast majority of drugs on the market today still reflect the old paradigm; see chapter 5). Products ordered in five minutes on the Web and delivered to your door are worth more than products that you have to fetch yourself. Clothes custom-manufactured for your unique body are worth more than clothes you happen to find on a store rack. These sorts of improvements are taking place in most product categories, and none of them is reflected in the productivity statistics.

The statistical methods underlying productivity measurements tend to factor out gains by essentially concluding that we still get only one dollar of products and services for a dollar, despite the fact that we get much more for that dollar. (Computers are an extreme example of this phenomenon, but it is pervasive.) University of Chicago professor Pete Klenow and University of Rochester professor Mark Bils estimate that the value in constant dollars of existing goods has been increasing at 1.5 percent per year for the past twenty years because of qualitative improvements.[84] This still does not account for the introduction of entirely new products and product categories (for example, cell phones, pagers, pocket computers, downloaded songs, and software programs). It does not consider the burgeoning value of the Web itself. How do we value the availability of free resources such as online encyclopedias and search engines that increasingly provide effective gateways to human knowledge?

The Bureau of Labor Statistics, which is responsible for the inflation statistics, uses a model that incorporates an estimate of quality growth of only 0.5 percent per year.[85] If we use Klenow and Bils's conservative estimate, this reflects a systematic underestimate of quality improvement and a resulting overestimate of inflation by at least 1 percent per year. And that still does not account for new product categories.

Despite these weaknesses in the productivity statistical methods, gains in productivity are now actually reaching the steep part of the exponential curve. Labor productivity grew at 1.6 percent per year until 1994, then rose at 2.4 percent per year, and is now growing even more rapidly. Manufacturing productivity in output per hour grew at 4.4 percent annually from 1995 to 1999, durables manufacturing at 6.5 percent per year. In the first quarter of 2004, the seasonally adjusted annual rate of productivity change was 4.6 percent in the business sector and 5.9 percent in durable goods manufacturing.[86]

We see smooth exponential growth in the value produced by an hour of labor over the last half century (see the figure below). Again, this trend does not take into account the vastly greater value of a dollar's power in purchasing information technologies (which has been doubling about once a year in overall price-performance).[87]

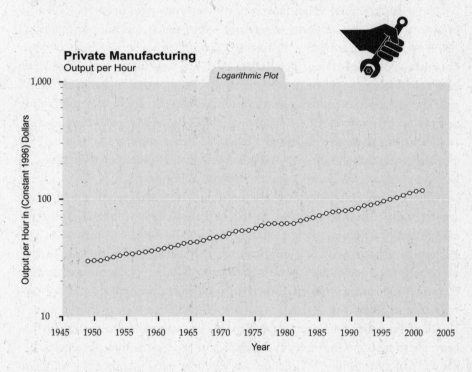

Deflation . . . a Bad Thing?

> In 1846 we believe there was not a single garment in our country sewed by machinery; in that year the first American patent of a sewing machine was issued. At the present moment thousands are wearing clothes which have been stitched by iron fingers, with a delicacy rivaling that of a Cashmere maiden.
>
> —*Scientific American*, 1853

As this book is being written, a worry of many mainstream economists on both the political right and the left is deflation. On the face of it, having your money go further would appear to be a good thing. The economists' concern is that if consumers can buy what they need and want with fewer dollars, the economy will shrink (as measured in dollars). This ignores, however, the inherently insatiable needs and desires of human consumers. The revenues of the semi-conductor industry, which "suffers" 40 to 50 percent deflation per year, have nonetheless grown by 17 percent each year over the past half century.[88] Since the economy is in fact expanding, this theoretical implication of deflation should not cause concern.

The 1990s and early 2000s have seen the most powerful deflationary forces in history, which explains why we are not seeing significant rates of inflation. Yes, it's true that historically low unemployment, high asset values, economic growth, and other such factors are inflationary, but these factors are offset by the exponential trends in the price-performance of all information-based technologies: computation, memory, communications, biotechnology, miniaturization, and even the overall rate of technical progress. These technologies deeply affect all industries. We are also undergoing massive disintermediation in the channels of distribution through the Web and other new communication technologies, as well as escalating efficiencies in operations and administration.

Since the information industry is becoming increasingly influential in all sectors of the economy, we are seeing the increasing impact of the IT industry's extraordinary deflation rates. Deflation during the Great Depression in the 1930s was due to a collapse of consumer confidence and a collapse of the money supply. Today's deflation is a completely different phenomenon, caused by rapidly increasing productivity and the increasing pervasiveness of information in all its forms.

All of the technology trend charts in this chapter represent massive deflation. There are many examples of the impact of these escalating efficiencies. BP Amoco's cost for finding oil in 2000 was less than one dollar per barrel, down from nearly ten dollars in 1991. Processing an Internet transaction costs a bank one penny, compared to more than one dollar using a teller.

It is important to point out that a key implication of nanotechnology is that it will bring the economics of software to hardware—that is, to physical products. Software prices are deflating even more quickly than those of hardware (see the figure below).

Exponential Software Price-Performance Improvement[89]

Example: Automatic Speech-Recognition Software

	1985	1995	2000
Price	$5,000	$500	$50
Vocabulary Size (number of words)	1,000	10,000	100,000
Continuous Speech?	No	No	Yes
User Training Required (minutes)	180	60	5
Accuracy	Poor	Fair	Good

The impact of distributed and intelligent communications has been felt perhaps most intensely in the world of business. Despite dramatic mood swings on Wall Street, the extraordinary values ascribed to so-called e-companies during the 1990s boom era reflected a valid perception: the business models that have sustained businesses for decades are in the early phases of a radical transformation. New models based on direct personalized communication with the customer will transform every industry, resulting in massive disintermediation of the middle layers that have traditionally separated the customer from the ultimate source of products and services. There is, however, a pace to all revolutions, and the investments and stock market valuations in this area expanded way beyond the early phases of this economic S-curve.

The boom-and-bust cycle in these information technologies was strictly a capital-markets (stock-value) phenomenon. Neither boom nor bust is apparent in the actual business-to-consumer (B2C) and business-to-business (B2B) data (see the figure on the next page). Actual B2C revenues grew smoothly from $1.8 billion in 1997 to $70 billion in 2002. B2B had similarly smooth growth from $56 billion in 1999 to $482 billion in 2002.[90] In 2004 it is approaching $1 trillion. We certainly do not see any evidence of business cycles in the actual price-performance of the underlying technologies, as I discussed extensively above.

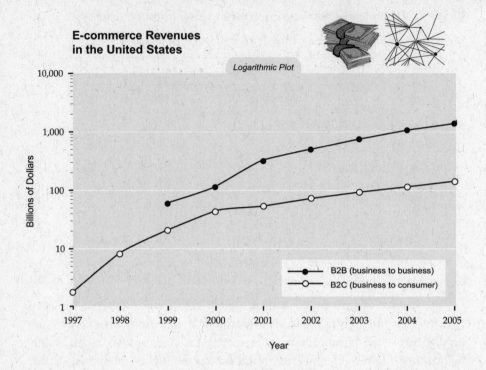

**E-commerce Revenues
in the United States**

Logarithmic Plot

Expanding access to knowledge is also changing power relationships. Patients increasingly approach visits to their physician armed with a sophisticated understanding of their medical condition and their options. Consumers of virtually everything from toasters, cars, and homes to banking and insurance are now using automated software agents to quickly identify the right choices with the optimal features and prices. Web services such as eBay are rapidly connecting buyers and sellers in unprecedented ways.

The wishes and desires of customers, often unknown even to themselves, are rapidly becoming the driving force in business relationships. Well-connected clothes shoppers, for example, are not going to be satisfied for much longer with settling for whatever items happen to be left hanging on the rack of their local store. Instead, they will select just the right materials and styles by viewing how many possible combinations look on a three-dimensional image of their own body (based on a detailed body scan), and then having the choices custom-manufactured.

The current disadvantages of Web-based commerce (for example, limitations in the ability to directly interact with products and the frequent frustrations of interacting with inflexible menus and forms instead of human personnel) will gradually dissolve as the trends move robustly in favor of the electronic world.

By the end of this decade, computers will disappear as distinct physical objects, with displays built in our eyeglasses, and electronics woven in our clothing, providing full-immersion visual virtual reality. Thus, "going to a Web site" will mean entering a virtual-reality environment—at least for the visual and auditory senses—where we can directly interact with products and people, both real and simulated. Although the simulated people will not be up to human standards—at least not by 2009—they will be quite satisfactory as sales agents, reservation clerks, and research assistants. Haptic (tactile) interfaces will enable us to touch products and people. It is difficult to identify any lasting advantage of the old brick-and-mortar world that will not ultimately be overcome by the rich interactive interfaces that are soon to come.

These developments will have significant implications for the real-estate industry. The need to congregate workers in offices will gradually diminish. From the experience of my own companies, we are already able to effectively organize geographically disparate teams, something that was far more difficult a decade ago. The full-immersion visual-auditory virtual-reality environments, which will be ubiquitous during the second decade of this century, will hasten the trend toward people living and working wherever they wish. Once we have full-immersion virtual-reality environments incorporating all of the senses, which will be feasible by the late 2020s, there will be no reason to utilize real offices. Real estate will become virtual.

As Sun Tzu pointed out, "knowledge is power," and another ramification of the law of accelerating returns is the exponential growth of human knowledge, including intellectual property.

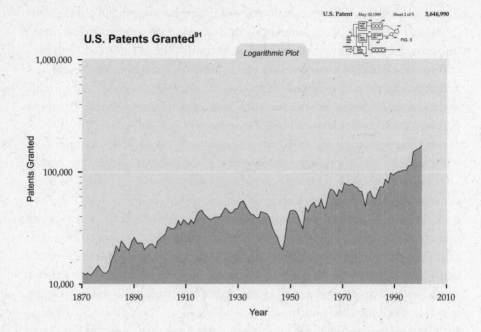

None of this means that cycles of recession will disappear immediately. Recently, the country experienced an economic slowdown and technology-sector recession and then a gradual recovery. The economy is still burdened with some of the underlying dynamics that historically have caused cycles of recession: excessive commitments such as overinvestment in capital-intensive projects and the overstocking of inventories. However, because the rapid dissemination of information, sophisticated forms of online procurement, and increasingly transparent markets in all industries have diminished the impact of this cycle, "recessions" are likely to have less direct impact on our standard of living. That appears to have been the case in the minirecession that we experienced in 1991–1993 and was even more evident in the most recent recession in the early 2000s. The underlying long-term growth rate will continue at an exponential rate.

Moreover, innovation and the rate of paradigm shift are not noticeably affected by the minor deviations caused by economic cycles. All of the technologies exhibiting exponential growth shown in the above charts are continuing without losing a beat through recent economic slowdowns. Market acceptance also shows no evidence of boom and bust.

The overall growth of the economy reflects completely new forms and layers of wealth and value that did not previously exist, or at least that did not previously constitute a significant portion of the economy, such as new forms of

nanoparticle-based materials, genetic information, intellectual property, communication portals, Web sites, bandwidth, software, databases, and many other
new technology-based categories.

The overall information-technology sector is rapidly increasing its share of
the economy and is increasingly influential on all other sectors, as noted in the
figure below.[92]

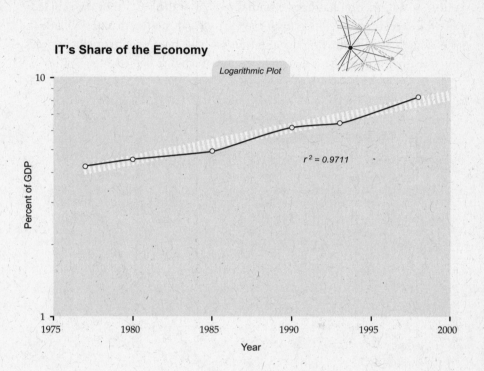

IT's Share of the Economy

Logarithmic Plot

$r^2 = 0.9711$

Percent of GDP

Year

Another implication of the law of accelerating returns is exponential growth in education and learning. Over the past 120 years, we have increased our investment in K–12 education (per student and in constant dollars) by a factor of ten. There has been a hundredfold increase in the number of college students. Automation started by amplifying the power of our muscles and in recent times has been amplifying the power of our minds. So for the past two centuries, automation has been eliminating jobs at the bottom of the skill ladder while creating new (and better-paying) jobs at the top of the skill ladder. The ladder has been moving up, and thus we have been exponentially increasing investments in education at all levels (see the figure below).[93]

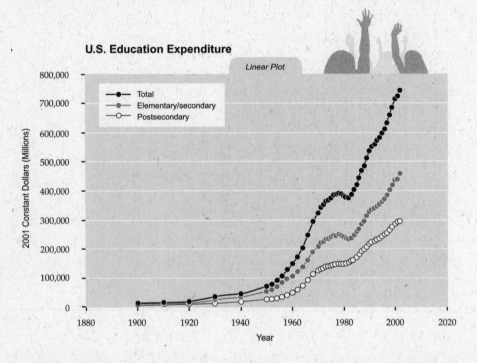

Oh, and about that "offer" at the beginning of this précis, consider that present stock values are based on future expectations. Given that the (literally) shortsighted linear intuitive view represents the ubiquitous outlook, the common wisdom in economic expectations is dramatically understated. Since stock prices reflect the consensus of a buyer-seller market, the prices reflect the underlying linear assumption that most people share regarding future economic growth. But the law of accelerating returns clearly implies that the growth rate will continue to grow exponentially, because the rate of progress will continue to accelerate.

MOLLY 2004: *But wait a second, you said that I would get eighty trillion dollars if I read and understood this section of the chapter.*

RAY: *That's right. According to my models, if we replace the linear outlook with the more appropriate exponential outlook, current stock prices should triple.[94] Since there's (conservatively) forty trillion dollars in the equity markets, that's eighty trillion in additional wealth.*

MOLLY 2004: *But you said I would get that money.*

RAY: *No, I said "you" would get the money, and that's why I suggested reading the sentence carefully. The English word "you" can be singular or plural. I meant it in the sense of "all of you."*

MOLLY 2004: *Hmm, that's annoying. You mean all of us as in the whole world? But not everyone will read this book.*

RAY: *Well, but everyone could. So if all of you read this book and understand it, then economic expectations would be based on the historical exponential model, and thus stock values would increase.*

MOLLY 2004: *You mean if everyone understands it and agrees with it. I mean the market is based on expectations, right?*

RAY: *Okay, I suppose I was assuming that.*

MOLLY 2004: *So is that what you expect to happen?*

RAY: *Well, actually, no. Putting on my futurist hat again, my prediction is that indeed these views on exponential growth will ultimately prevail but only over time, as more and more evidence of the exponential nature of technology and its impact on the economy becomes apparent. This will happen gradually over the next decade, which will represent a strong long-term updraft for the market.*

GEORGE 2048: *I don't know, Ray. You were right that the price-performance of information technology in all of its forms kept growing at an exponential rate, and with continued growth also in the exponent. And indeed, the economy kept growing exponentially, thereby more than overcoming a very high deflation rate. And it also turned out that the general public did catch on to all of these trends. But this realization didn't have the positive impact on the stock market that you're describing. The stock market did increase along with the economy, but the realization of a higher growth rate did little to increase stock prices.*

RAY: *Why do you suppose it turned out that way?*

GEORGE 2048: *Because you left one thing out of your equation. Although people realized that stock values would increase rapidly, that same realization also increased the discount rate (the rate at which we need to discount values in the future when considering their present value). Think about it. If we know* that stocks are going to increase significantly in a future period, then we'd like to have the stocks now so that we can realize those future gains. So the perception of increased future equity values also increases the discount rate. And that cancels out the expectation of higher future values.

MOLLY 2104: *Uh, George, that was not quite right either. What you say makes logical sense, but the psychological reality is that the heightened perception of increased future values did have a greater positive impact on stock prices than increases in the discount rate had a negative effect. So the general acceptance of exponential growth in both the price-performance of technology and the rate of economic activity did provide an upward draft for the equities market, but not the tripling that you spoke about, Ray, due to the effect that George was describing.*

MOLLY 2004: *Okay, I'm sorry I asked. I think I'll just hold on to the few shares I've got and not worry about it.*

RAY: *What have you invested in?*

MOLLY 2004: *Let's see, there's this new natural language–based search-engine company that hopes to take on Google. And I've also invested in a fuel-cell company. Also, a company building sensors that can travel in the bloodstream.*

RAY: *Sounds like a pretty high-risk, high-tech portfolio.*

MOLLY 2004: *I wouldn't call it a portfolio. I'm just dabbling with the technologies you're talking about.*

RAY: *Okay, but keep in mind that while the trends predicted by the law of accelerating returns are remarkably smooth, that doesn't mean we can readily predict which competitors will prevail.*

MOLLY 2004: *Right, that's why I'm spreading my bets.*

CHAPTER THREE

Achieving the Computational Capacity of the Human Brain

As I discuss in *Engines of Creation*, if you can build genuine AI, there are reasons to believe that you can build things like neurons that are a million times faster. That leads to the conclusion that you can make systems that think a million times faster than a person. With AI, these systems could do engineering design. Combining this with the capability of a system to build something that is better than it, you have the possibility for a very abrupt transition. This situation may be more difficult to deal with even than nanotechnology, but it is much more difficult to think about it constructively at this point. Thus, it hasn't been the focus of things that I discuss, although I periodically point to it and say: "That's important too."

—Eric Drexler, 1989

The Sixth Paradigm of Computing Technology: Three-Dimensional Molecular Computing and Emerging Computational Technologies

In the April 19, 1965, issue of *Electronics*, Gordon Moore wrote, "The future of integrated electronics is the future of electronics itself. The advantages of integration will bring about a proliferation of electronics, pushing this science into many new areas."[1] With those modest words, Moore ushered in a revolution that is still gaining momentum. To give his readers some idea of how profound this new science would be, Moore predicted that "by 1975, economics may dictate squeezing as many as 65,000 components on a single silicon chip." Imagine that.

Moore's article described the repeated annual doubling of the number of transistors (used for computational elements, or gates) that could be fitted onto an integrated circuit. His 1965 "Moore's Law" prediction was criticized at the time because his logarithmic chart of the number of components on a chip had only five reference points (from 1959 through 1965), so projecting this nascent trend all the way out to 1975 was seen as premature. Moore's initial estimate was incorrect, and he revised it downward a decade later. But the basic idea—the exponential growth of the price-performance of electronics based on shrinking the size of transistors on an integrated circuit—was both valid and prescient.[2]

Today, we talk about billions of components rather than thousands. In the most advanced chips of 2004, logic gates are only fifty nanometers wide, already well within the realm of nanotechnology (which deals with measurements of one hundred nanometers or less). The demise of Moore's Law has been predicted on a regular basis, but the end of this remarkable paradigm keeps getting pushed out in time. Paolo Gargini, Intel Fellow, director of Intel technology strategy, and chairman of the influential International Technology Roadmap for Semiconductors (ITRS), recently stated, "We see that for at least the next 15 to 20 years, we can continue staying on Moore's Law. In fact, . . . nanotechnology offers many new knobs we can turn to continue improving the number of components on a die."[3]

The acceleration of computation has transformed everything from social and economic relations to political institutions, as I will demonstrate throughout this book. But Moore did not point out in his papers that the strategy of shrinking feature sizes was not, in fact, the first paradigm to bring exponential growth to computation and communication. It was the fifth, and already, we can see the outlines of the next: computing at the molecular level and in three dimensions. Even though we have more than a decade left of the fifth paradigm, there has already been compelling progress in all of the enabling technologies required for the sixth paradigm. In the next section, I provide an analysis of the amount of computation and memory required to achieve human levels of intelligence and why we can be confident that these levels will be achieved in inexpensive computers within two decades. Even these very powerful computers will be far from optimal, and in the last section of this chapter I'll review the limits of computation according to the laws of physics as we understand them today. This will bring us to computers circa the late twenty-first century.

The Bridge to 3-D Molecular Computing. Intermediate steps are already under way: new technologies that will lead to the sixth paradigm of molecular three-dimensional computing include nanotubes and nanotube circuitry, molecular computing, self-assembly in nanotube circuits, biological systems emulating circuit assembly, computing with DNA, spintronics (computing with the spin of electrons), computing with light, and quantum computing. Many of these independent technologies can be integrated into computational systems that will eventually approach the theoretical maximum capacity of matter and energy to perform computation and will far outpace the computational capacities of a human brain.

One approach is to build three-dimensional circuits using "conventional" silicon lithography. Matrix Semiconductor is already selling memory chips that contain vertically stacked planes of transistors rather than one flat layer.[4] Since a single 3-D chip can hold more memory, overall product size is reduced, so Matrix is initially targeting portable electronics, where it aims to compete with flash memory (used in cell phones and digital cameras because it does not lose information when the power is turned off). The stacked circuitry also reduces the overall cost per bit. Another approach comes from one of Matrix's competitors, Fujio Masuoka, a former Toshiba engineer who invented flash memory. Masuoka claims that his novel memory design, which looks like a cylinder, reduces the size and cost-per-bit of memory by a factor of ten compared to flat chips.[5] Working prototypes of three-dimensional silicon chips have also been demonstrated at Rensselaer Polytechnic Institute's Center for Gigascale Integration and at the MIT Media Lab.

Tokyo's Nippon Telegraph and Telephone Corporation (NTT) has demonstrated a dramatic 3-D technology using electron-beam lithography, which can create arbitrary three-dimensional structures with feature sizes (such as transistors) as small as ten nanometers.[6] NTT demonstrated the technology by creating a high-resolution model of the Earth sixty microns in size with ten-nanometer features. NTT says the technology is applicable to nanofabrication of electronic devices such as semiconductors, as well as creating nanoscale mechanical systems.

Nanotubes Are Still the Best Bet. In *The Age of Spiritual Machines*, I cited nanotubes—using molecules organized in three dimensions to store memory bits and to act as logic gates—as the most likely technology to usher in the era of three-dimensional molecular computing. Nanotubes, first synthesized in 1991, are tubes made up of a hexagonal network of carbon atoms that have been rolled up to make a seamless cylinder.[7] Nanotubes are very small: single-wall nanotubes are only one nanometer in diameter, so they can achieve high densities.

They are also potentially very fast. Peter Burke and his colleagues at the University of California at Irvine recently demonstrated nanotube circuits operating at 2.5 gigahertz (GHz). However, in *Nano Letters*, a peer-reviewed journal of the American Chemical Society, Burke says the theoretical speed limit for these nanotube transistors "should be terahertz [1 THz = 1,000 GHz], which is about 1,000 times faster than modern computer speeds."[8] One cubic inch of nanotube circuitry, once fully developed, would be up to one hundred million times more powerful than the human brain.[9]

Nanotube circuitry was controversial when I discussed it in 1999, but there has been dramatic progress in the technology over the past six years. Two major strides were made in 2001. A nanotube-based transistor (with dimensions of one by twenty nanometers), operating at room temperature and using only a single electron to switch between on and off states, was reported in the July 6, 2001, issue of *Science*.[10] Around the same time, IBM also demonstrated an integrated circuit with one thousand nanotube-based transistors.[11]

More recently, we have seen the first working models of nanotube-based circuitry. In January 2004 researchers at the University of California at Berkeley and Stanford University created an integrated memory circuit based on nanotubes.[12] One of the challenges in using this technology is that some nanotubes are conductive (that is, simply transmit electricity), while others act like semiconductors (that is, are capable of switching and able to implement logic gates). The difference in capability is based on subtle structural features. Until recently, sorting them out required manual operations, which would not be practical for building large-scale circuits. The Berkeley and Stanford scientists addressed this issue by developing a fully automated method of sorting and discarding the nonsemiconductor nanotubes.

Lining up nanotubes is another challenge with nanotube circuits, since they tend to grow in every direction. In 2001 IBM scientists demonstrated that nanotube transistors could be grown in bulk, similar to silicon transistors. They used a process called "constructive destruction," which destroys defective nanotubes right on the wafer instead of sorting them out manually. Thomas Theis, director of physical sciences at IBM's Thomas J. Watson Research Center, said at the time, "We believe that IBM has now passed a major milestone on the road toward molecular-scale chips. . . . If we are ultimately successful, then carbon nanotubes will enable us to indefinitely maintain Moore's Law in terms of density, because there is very little doubt in my mind that these can be made smaller than any future silicon transistor."[13] In May 2003 Nantero, a small company in Woburn, Massachusetts, cofounded by Harvard University researcher Thomas Rueckes, took the process a step further when it demonstrated a single-chip wafer with ten billion nanotube junctions, all aligned in the proper direction. The Nantero technology involves using standard lithography equipment to remove automatically the nanotubes that are incorrectly aligned. Nantero's use of standard equipment has excited industry observers because the technology would not require expensive new fabrication machines. The Nantero design provides random access as well as nonvolatility (data is retained when the power is off), meaning that it could potentially replace all of the primary forms of memory: RAM, flash, and disk.

Computing with Molecules. In addition to nanotubes, major progress has been made in recent years in computing with just one or a few molecules. The idea of computing with molecules was first suggested in the early 1970s by IBM's Avi Aviram and Northwestern University's Mark A. Ratner.[14] At that time, we did not have the enabling technologies, which required concurrent advances in electronics, physics, chemistry, and even the reverse engineering of biological processes for the idea to gain traction.

In 2002 scientists at the University of Wisconsin and University of Basel created an "atomic memory drive" that uses atoms to emulate a hard drive. A single silicon atom could be added or removed from a block of twenty others using a scanning tunneling microscope. Using this process, researchers believe, the system could be used to store millions of times more data on a disk of comparable size—a density of about 250 terabits of data per square inch—although the demonstration involved only a small number of bits.[15]

The one-terahertz speed predicted by Peter Burke for molecular circuits looks increasingly accurate, given the nanoscale transistor created by scientists at the University of Illinois at Urbana-Champaign. It runs at a frequency of 604 gigahertz (more than half a terahertz).[16]

One type of molecule that researchers have found to have desirable properties for computing is called a "rotaxane," which can switch states by changing the energy level of a ringlike structure contained within the molecule. Rotaxane memory and electronic switching devices have been demonstrated, and they show the potential of storing one hundred gigabits (10^{11} bits) per square inch. The potential would be even greater if organized in three dimensions.

Self-Assembly. Self-assembling of nanoscale circuits is another key enabling technique for effective nanoelectronics. Self-assembly allows improperly formed components to be discarded automatically and makes it possible for the potentially trillions of circuit components to organize themselves, rather than be painstakingly assembled in a top-down process. It would enable large-scale circuits to be created in test tubes rather than in multibillion-dollar factories, using chemistry rather than lithography, according to UCLA scientists.[17] Purdue University researchers have already demonstrated self-organizing nanotube structures, using the same principle that causes DNA strands to link together in stable structures.[18]

Harvard University scientists took a key step forward in June 2004 when they demonstrated another self-organizing method that can be used on a large scale.[19] The technique starts with photolithography to create an etched array of interconnects (connections between computational elements). A large number of nanowire field-effect transistors (a common form of transistors) and nano-scale interconnects are then deposited on the array. These then connect themselves in the correct pattern.

In 2004 researchers at the University of Southern California and NASA's Ames Research Center demonstrated a method that self-organizes extremely dense circuits in a chemical solution.[20] The technique creates nanowires spontaneously and then causes nanoscale memory cells, each able to hold three bits of data, to self-assemble onto the wires. The technology has a storage capacity of 258 gigabits of data per square inch (which researchers claim could be increased tenfold), compared to 6.5 gigabits on a flash memory card. Also in 2003 IBM demonstrated a working memory device using polymers that self-assemble into twenty-nanometer-wide hexagonal structures.[21]

It's also important that nanocircuits be self-configuring. The large number of circuit components and their inherent fragility (due to their small size) make it inevitable that some portions of a circuit will not function correctly. It will not be economically feasible to discard an entire circuit simply because a small number of transistors out of a trillion are nonfunctioning. To address this concern, future circuits will continuously monitor their own performance and route information around sections that are unreliable in the same manner that information on the Internet is routed around nonfunctioning nodes. IBM has been particularly active in this area of research and has already developed microprocessor designs that automatically diagnose problems and reconfigure chip resources accordingly.[22]

Emulating Biology. The idea of building electronic or mechanical systems that are self-replicating and self-organizing is inspired by biology, which relies on these properties. Research published in the *Proceedings of the National Academy of Sciences* described the construction of self-replicating nanowires based on prions, which are self-replicating proteins. (As detailed in chapter 4, one form of prion appears to play a role in human memory, whereas another form is believed to be responsible for variant Creutzfeldt-Jakob disease, the human form of mad-cow disease.)[23] The team involved in the project used prions as a model because of their natural strength. Because prions do not normally conduct electricity, however, the scientists created a genetically modified version containing a thin layer of gold, which conducts electricity with low resistance. MIT biology professor Susan Lindquist, who headed the study, commented, "Most of the people working on nanocircuits are trying to build them using 'top-down' fabrication techniques. We thought we'd try a 'bottom-up' approach, and let molecular self-assembly do the hard work for us."

The ultimate self-replicating molecule from biology is, of course, DNA. Duke University researchers created molecular building blocks called "tiles" out of self-assembling DNA molecules.[24] They were able to control the structure of the resulting assembly, creating "nanogrids." This technique automatically attaches protein molecules to each nanogrid's cell, which could be used to perform computing operations. They also demonstrated a chemical process that coated the DNA nanoribbons with silver to create nanowires. Commenting on the article in the September 26, 2003, issue of the journal *Science*, lead researcher Hao Yan said, "To use DNA self-assembly to template protein molecules or other molecules has been sought for years, and this is the first time it has been demonstrated so clearly."[25]

Computing with DNA. DNA is nature's own nanoengineered computer, and its ability to store information and conduct logical manipulations at the molecular level has already been exploited in specialized "DNA computers." A DNA computer is essentially a test tube filled with water containing trillions of DNA molecules, with each molecule acting as a computer.

The goal of the computation is to solve a problem, with the solution expressed as a sequence of symbols. (For example, the sequence of symbols could represent a mathematical proof or just the digits of a number.) Here's how a DNA computer works. A small strand of DNA is created, using a unique code for each symbol. Each such strand is replicated trillions of times using a process called "polymerase chain reaction" (PCR). These pools of DNA are then put into a test tube. Because DNA has an affinity to link strands together, long strands form automatically, with sequences of the strands representing the different symbols, each of them a possible solution to the problem. Since there will be many trillions of such strands, there are multiple strands for each possible answer (that is, each possible sequence of symbols).

The next step of the process is to test all of the strands *simultaneously*. This is done by using specially designed enzymes that destroy strands that do not meet certain criteria. The enzymes are applied to the test tube sequentially, and by designing a precise series of enzymes the procedure will eventually obliterate all the incorrect strands, leaving only the ones with the correct answer. (For a more complete description of the process, see this note:[26])

The key to the power of DNA computing is that it allows for testing each of the trillions of strands simultaneously. In 2003 Israeli scientists led by Ehud Shapiro at the Weizmann Institute of Science combined DNA with adenosine triphosphate (ATP), the natural fuel for biological systems such as the human body.[27] With this method, each of the DNA molecules was able to perform computations as well as provide its own energy. The Weizmann scientists demonstrated a configuration consisting of two spoonfuls of this liquid super-computing system, which contained thirty million billion molecular comput-ers and performed a total of 660 trillion calculations per second (6.6×10^{14} cps). The energy consumption of these computers is extremely low, only fifty millionths of a watt for all thirty million billion computers.

There's a limitation, however, to DNA computing: each of the many tril-lions of computers has to perform the same operation at the same time (although on different data), so that the device is a "single instruction multiple data" (SIMD) architecture. While there are important classes of problems that are amenable to a SIMD system (for example, processing every pixel in an image for image enhancement or compression, and solving combinatorial-logic problems), it is not possible to program them for general-purpose algo-rithms, in which each computer is able to execute whatever operation is needed for its particular mission. (Note that the research projects at Purdue University and Duke University, described earlier, that use self-assembling DNA strands to create three-dimensional structures are different from the DNA computing described here. Those research projects have the potential to create arbitrary configurations that are not limited to SIMD computing.)

Computing with Spin. In addition to their negative electrical charge, electrons have another property that can be exploited for memory and computation: spin. According to quantum mechanics, electrons spin on an axis, similar to the way the Earth rotates on its axis. This concept is theoretical, because an elec-tron is considered to occupy a point in space, so it is difficult to imagine a point with no size that nonetheless spins. However, when an electrical charge moves, it causes a magnetic field, which is real and measurable. An electron can spin in one of two directions, described as "up" and "down," so this property can be exploited for logic switching or to encode a bit of memory.

The exciting property of spintronics is that no energy is required to change an electron's spin state. Stanford University physics professor Shoucheng Zhang and University of Tokyo professor Naoto Nagaosa put it this way: "We have discovered the equivalent of a new 'Ohm's Law' [the electronics law that states that current in a wire equals voltage divided by resistance].... [It] says that the spin of the electron can be transported without any loss of energy, or dissipation. Furthermore, this effect occurs at room temperature in materials already widely used in the semiconductor industry, such as gallium arsenide. That's important because it could enable a new generation of computing devices."[28]

The potential, then, is to achieve the efficiencies of superconducting (that is, moving information at or close to the speed of light without any loss of information) at room temperature. It also allows multiple properties of each electron to be used for computing, thereby increasing the potential for memory and computational density.

One form of spintronics is already familiar to computer users: magnetoresistance (a change in electrical resistance caused by a magnetic field) is used to store data on magnetic hard drives. An exciting new form of nonvolatile memory based on spintronics called MRAM (magnetic random-access memory) is expected to enter the market within a few years. Like hard drives, MRAM memory retains its data without power but uses no moving parts and will have speeds and rewritability comparable to conventional RAM.

MRAM stores information in ferromagnetic metallic alloys, which are suitable for data storage but not for the logical operations of a microprocessor. The holy grail of spintronics would be to achieve practical spintronics effects in a semiconductor, which would enable us to use the technology both for memory and for logic. Today's chip manufacturing is based on silicon, which does not have the requisite magnetic properties. In March 2004 an international group of scientists reported that by doping a blend of silicon and iron with cobalt, the new material was able to display the magnetic properties needed for spintronics while still maintaining the crystalline structure silicon requires as a semiconductor.[29]

An important role for spintronics in the future of computer memory is clear, and it is likely to contribute to logic systems as well. The spin of an electron is a quantum property (subject to the laws of quantum mechanics), so perhaps the most important application of spintronics will be in quantum computing systems, using the spin of quantum-entangled electrons to represent qubits, which I discuss below.

Spin has also been used to store information in the nucleus of atoms, using the complex interaction of their protons' magnetic moments. Scientists at the University of Oklahoma also demonstrated a "molecular photography" technique for storing 1,024 bits of information in a single liquid-crystal molecule comprising nineteen hydrogen atoms.[30]

Computing with Light. Another approach to SIMD computing is to use multiple beams of laser light in which information is encoded in each stream of photons. Optical components can then be used to perform logical and arithmetic functions on the encoded information streams. For example, a system developed by Lenslet, a small Israeli company, uses 256 lasers and can perform eight trillion calculations per second by performing the same calculation on each of the 256 streams of data.[31] The system can be used for applications such as performing data compression on 256 video channels.

SIMD technologies such as DNA computers and optical computers will have important specialized roles to play in the future of computation. The replication of certain aspects of the functionality of the human brain, such as processing sensory data, can use SIMD architectures. For other brain regions, such as those dealing with learning and reasoning, general-purpose computing with its "multiple instruction multiple data" (MIMD) architectures will be required. For high-performance MIMD computing, we will need to apply the three-dimensional molecular-computing paradigms described above.

Quantum Computing. Quantum computing is an even more radical form of SIMD parallel processing, but one that is in a much earlier stage of development compared to the other new technologies we have discussed. A quantum computer contains a series of qubits, which essentially are zero and one at the same time. The qubit is based on the fundamental ambiguity inherent in quantum mechanics. In a quantum computer, the qubits are represented by a quantum property of particles—for example, the spin state of individual electrons. When the qubits are in an "entangled" state, each one is simultaneously in both states. In a process called "quantum decoherence" the ambiguity of each qubit is resolved, leaving an unambiguous sequence of ones and zeroes. If the quantum computer is set up in the right way, that decohered sequence will represent the solution to a problem. Essentially, only the correct sequence survives the process of decoherence.

As with the DNA computer described above, a key to successful quantum computing is a careful statement of the problem, including a precise way to test possible answers. The quantum computer effectively tests every possible *combination* of values for the qubits. So a quantum computer with one thousand qubits would test $2^{1,000}$ (a number approximately equal to one followed by 301 zeroes) potential solutions simultaneously.

A thousand-bit quantum computer would vastly outperform any conceivable DNA computer, or for that matter any conceivable nonquantum computer. There are two limitations to the process, however. The first is that, like the DNA and optical computers discussed above, only a special set of problems is amenable to being presented to a quantum computer. In essence, we need to be able to test each possible answer in a simple way.

The classic example of a practical use for quantum computing is in factoring very large numbers (finding which smaller numbers, when multiplied together, result in the large number). Factoring numbers with more than 512 bits is currently not achievable on a digital computer, even a massively parallel one.[32] Interesting classes of problems amenable to quantum computing include breaking encryption codes (which rely on factoring large numbers). The other problem is that the computational power of a quantum computer depends on the number of entangled qubits, and the state of the art is currently limited to around ten bits. A ten-bit quantum computer is not very useful, since 2^{10} is only 1,024. In a conventional computer, it is a straightforward process to combine memory bits and logic gates. We cannot, however, create a twenty-qubit quantum computer simply by combining two ten-qubit machines. All of the qubits have to be quantum-entangled together, and that has proved to be challenging.

A key question is: how difficult is it to add each additional qubit? The computational power of a quantum computer grows exponentially with each added qubit, but if it turns out that adding each additional qubit makes the engineering task exponentially more difficult, we will not be gaining any leverage. (That is, the computational power of a quantum computer will be only linearly proportional to the engineering difficulty.) In general, proposed methods for adding qubits make the resulting systems significantly more delicate and susceptible to premature decoherence.

There are proposals to increase significantly the number of qubits, although these have not yet been proved in practice. For example, Stephan Gulde and his colleagues at the University of Innsbruck have built a quantum computer using a single atom of calcium that has the potential to simultaneously encode dozens of qubits—possibly up to one hundred—using different quantum properties within the atom.[33] The ultimate role of quantum computing remains unresolved. But even if a quantum computer with hundreds of entangled qubits proves feasible, it will remain a special-purpose device, although one with remarkable capabilities that cannot be emulated in any other way.

When I suggested in *The Age of Spiritual Machines* that molecular computing would be the sixth major computing paradigm, the idea was still controversial. There has been so much progress in the past five years that there has been a sea change in attitude among experts, and this is now a mainstream view. We already have proofs of concept for all of the major requirements for three-dimensional molecular computing: single-molecule transistors, memory cells based on atoms, nanowires, and methods to self-assemble and self-diagnose the trillions (potentially trillions of trillions) of components.

Contemporary electronics proceeds from the design of detailed chip layouts to photolithography to the manufacturing of chips in large, centralized factories. Nanocircuits are more likely to be created in small chemistry flasks, a development that will be another important step in the decentralization of our industrial infrastructure and will maintain the law of accelerating returns through this century and beyond.

The Computational Capacity of the Human Brain

It may seem rash to expect fully intelligent machines in a few decades, when the computers have barely matched insect mentality in a half-century of development. Indeed, for that reason, many long-time artificial intelligence researchers scoff at the suggestion, and offer a few centuries as a more believable period. But there are very good reasons why things will go much faster in the next fifty years than they have in the last fifty. . . . Since 1990, the power available to individual AI and robotics programs has doubled yearly, to 30 MIPS by 1994 and 500 MIPS by 1998. Seeds long ago alleged barren are suddenly sprouting. Machines read text, recognize speech, even translate languages. Robots drive cross-country, crawl across Mars, and trundle down office corridors. In 1996 a theorem-proving program called EQP running five weeks on a 50 MIPS computer at Argonne National Laboratory found a proof of a Boolean algebra conjecture by Herbert Robbins that had eluded mathematicians for sixty years. And it is still only Spring. Wait until Summer.

—HANS MORAVEC, "WHEN WILL COMPUTER HARDWARE MATCH THE HUMAN BRAIN?" 1997

What is the computational capacity of a human brain? A number of estimates have been made, based on replicating the functionality of brain regions that have been reverse engineered (that is, the methods understood) at human levels of performance. Once we have an estimate of the computational capacity for a particular region, we can extrapolate that capacity to the entire brain by considering what portion of the brain that region represents. These estimates are based on functional simulation, which replicates the overall functionality of a region rather than simulating each neuron and interneuronal connection in that region.

Although we would not want to rely on any single calculation, we find that various assessments of different regions of the brain all provide reasonably close estimates for the entire brain. The following are order-of-magnitude estimates, meaning that we are attempting to determine the appropriate figures to the closest multiple of ten. The fact that different ways of making the same esti-mate provide similar answers corroborates the approach and indicates that the estimates are in an appropriate range.

The prediction that the Singularity—an expansion of human intelligence by a factor of trillions through merger with its nonbiological form—will occur within the next several decades does not depend on the precision of these cal-culations. Even if our estimate of the amount of computation required to simu-late the human brain was too optimistic (that is, too low) by a factor of even one thousand (which I believe is unlikely), that would delay the Singularity by only about eight years.[34] A factor of one million would mean a delay of only about fifteen years, and a factor of one billion would be a delay of about twenty-one years.[35]

Hans Moravec, legendary roboticist at Carnegie Mellon University, has ana-lyzed the transformations performed by the neural image-processing circuitry contained in the retina.[36] The retina is about two centimeters wide and a half millimeter thick. Most of the retina's depth is devoted to capturing an image, but one fifth of it is devoted to image processing, which includes distinguishing dark and light, and detecting motion in about one million small regions of the image.

The retina, according to Moravec's analysis, performs ten million of these edge and motion detections each second. Based on his several decades of expe-rience in creating robotic vision systems, he estimates that the execution of about one hundred computer instructions is required to re-create each such detection at human levels of performance, meaning that replicating the image-processing functionality of this portion of the retina requires 1,000 MIPS. The human brain is about 75,000 times heavier than the 0.02 grams of neurons in this portion of the retina, resulting in an estimate of about 10^{14} (100 trillion) instructions per second for the entire brain.[37]

Another estimate comes from the work of Lloyd Watts and his colleagues on creating functional simulations of regions of the human auditory system, which I discuss further in chapter 4.[38] One of the functions of the software Watts has developed is a task called "stream separation," which is used in tele-conferencing and other applications to achieve telepresence (the localization of each participant in a remote audio teleconference). To accomplish this, Watts explains, means "precisely measuring the time delay between sound sensors that are separated in space and that both receive the sound." The process involves pitch analysis, spatial position, and speech cues, including language-specific cues. "One of the important cues used by humans for localizing the position of a sound source is the Interaural Time Difference (ITD), that is, the difference in time of arrival of sounds at the two ears."[39]

Watts's own group has created functionally equivalent re-creations of these brain regions derived from reverse engineering. He estimates that 10^{11} cps are required to achieve human-level localization of sounds. The auditory cortex regions responsible for this processing comprise at least 0.1 percent of the brain's neurons. So we again arrive at a ballpark estimate of around 10^{14} cps (10^{11} cps $\times$ 10^3).

Yet another estimate comes from a simulation at the University of Texas that represents the functionality of a cerebellum region containing 10^4 neurons; this required about 10^8 cps, or about 10^4 cps per neuron. Extrapolating this over an estimated 10^{11} neurons results in a figure of about 10^{15} cps for the entire brain.

We will discuss the state of human-brain reverse engineering later, but it is clear that we can emulate the functionality of brain regions with less computation than would be required to simulate the precise nonlinear operation of each neuron and all of the neural components (that is, all of the complex interactions that take place inside each neuron). We come to the same conclusion when we attempt to simulate the functionality of organs in the body. For example, implantable devices are being tested that simulate the functionality of the human pancreas in regulating insulin levels.[40] These devices work by measuring glucose levels in the blood and releasing insulin in a controlled fashion to keep the levels in an appropriate range. While they follow a method similar to that of a biological pancreas, they do not, however, attempt to simulate *each* pancreatic islet cell, and there would be no reason to do so.

These estimates all result in comparable orders of magnitude (10^{14} to 10^{15} cps). Given the early stage of human-brain reverse engineering, I will use a more conservative figure of 10^{16} cps for our subsequent discussions.

Functional simulation of the brain is sufficient to re-create human powers of pattern recognition, intellect, and emotional intelligence. On the other hand, if we want to "upload" a particular person's personality (that is, capture all of his or her knowledge, skills, and personality, a concept I will explore in greater detail at the end of chapter 4), then we may need to simulate neural processes at the level of individual neurons and portions of neurons, such as the soma (cell body), axon (output connection), dendrites (trees of incoming connections), and synapses (regions connecting axons and dendrites). For this, we need to look at detailed models of individual neurons. The "fan out" (number of interneuronal connections) per neuron is estimated at 10^3. With an estimated 10^{11} neurons, that's about 10^{14} connections. With a reset time of five milliseconds, that comes to about 10^{16} synaptic transactions per second.

Neuron-model simulations indicate the need for about 10^3 calculations per synaptic transaction to capture the nonlinearities (complex interactions) in the dendrites and other neuron regions, resulting in an overall estimate of about 10^{19} cps for simulating the human brain at this level.[41] We can therefore consider this an upper bound, but 10^{14} to 10^{16} cps to achieve functional equivalence of all brain regions is likely to be sufficient.

IBM's Blue Gene/L supercomputer, now being built and scheduled to be completed around the time of the publication of this book, is projected to provide 360 trillion calculations per second (3.6×10^{14} cps).[42] This figure is already greater than the lower estimates described above. Blue Gene/L will also have around one hundred terabytes (about 10^{15} bits) of main storage, more than our memory estimate for functional emulation of the human brain (see below). In line with my earlier predictions, supercomputers will achieve my more conservative estimate of 10^{16} cps for functional human-brain emulation by early in the next decade (see the "Supercomputer Power" figure on p. 71).

Accelerating the Availability of Human-Level Personal Computing. Personal computers today provide more than 10^9 cps. According to the projections in the "Exponential Growth of Computing" chart (p. 70), we will achieve 10^{16} cps by 2025. However, there are several ways this timeline can be accelerated. Rather than using general-purpose processors, one can use application-specific integrated circuits (ASICs) to provide greater price-performance for very repetitive calculations. Such circuits already provide extremely high computational throughput for the repetitive calculations used in generating moving images in video games. ASICs can increase price-performance a thousandfold, cutting about eight years off the 2025 date. The varied programs that a simulation of the human brain will comprise will also include a great deal of repetition and thus will be amenable to ASIC implementation. The cerebellum, for example, repeats a basic wiring pattern billions of times.

We will also be able to amplify the power of personal computers by harvesting the unused computation power of devices on the Internet. New communication paradigms such as "mesh" computing contemplate treating every device in the network as a node rather than just a "spoke."[43] In other words, instead of devices (such as personal computers and PDAs) merely sending information to and from nodes, each device will act as a node itself, sending information to and receiving information from every other device. That will create very robust, self-organizing communication networks. It will also make it easier for computers and other devices to tap unused CPU cycles of the devices in their region of the mesh.

Currently at least 99 percent, if not 99.9 percent, of the computational capacity of all the computers on the Internet lies unused. Effectively harnessing this computation can provide another factor of 10^2 or 10^3 in increased price-performance. For these reasons, it is reasonable to expect human brain capacity, at least in terms of hardware computational capacity, for one thousand dollars by around 2020.

Yet another approach to accelerate the availability of human-level computation in a personal computer is to use transistors in their native "analog" mode. Many of the processes in the human brain are analog, not digital. Although we can emulate analog processes to any desired degree of accuracy with digital computation, we lose several orders of magnitude of efficiency in doing so. A single transistor can multiply two values represented as analog levels; doing so with digital circuits requires thousands of transistors. California Institute of Technology's Carver Mead has been pioneering this concept.[44] One disadvantage of Mead's approach is that the engineering design time required for such native analog computing is lengthy, so most researchers developing software to emulate regions of the brain usually prefer the rapid turnaround of software simulations.

Human Memory Capacity. How does computational capacity compare to human memory capacity? It turns out that we arrive at similar time-frame estimates if we look at human memory requirements. The number of "chunks" of knowledge mastered by an expert in a domain is approximately 10^5 for a variety of domains. These chunks represent patterns (such as faces) as well as specific knowledge. For example, a world-class chess master is estimated to have mastered about 100,000 board positions. Shakespeare used 29,000 words but close to 100,000 meanings of those words. Development of expert systems in medicine indicate that humans can master about 100,000 concepts in a domain. If we estimate that this "professional" knowledge represents as little as 1 percent of the overall pattern and knowledge store of a human, we arrive at an estimate of 10^7 chunks.

Based on my own experience in designing systems that can store similar chunks of knowledge in either rule-based expert systems or self-organizing pattern-recognition systems, a reasonable estimate is about 10^6 bits per chunk (pattern or item of knowledge), for a total capacity of 10^{13} (10 trillion) bits for a human's functional memory.

According to the projections from the ITRS road map (see RAM chart on p. 57), we will be able to purchase 10^{13} bits of memory for one thousand dollars by around 2018. Keep in mind that this memory will be millions of times faster than the electrochemical memory process used in the human brain and thus will be far more effective.

Again, if we model human memory on the level of individual interneuronal connections, we get a higher estimate. We can estimate about 10^4 bits per connection to store the connection patterns and neurotransmitter concentrations. With an estimated 10^{14} connections, that comes to 10^{18} (a billion billion) bits.

Based on the above analyses, it is reasonable to expect the hardware that can emulate human-brain functionality to be available for approximately one thousand dollars by around 2020. As we will discuss in chapter 4, the software that will replicate that functionality will take about a decade longer. However, the exponential growth of the price-performance, capacity, and speed of our hardware technology will continue during that period, so by 2030 it will take a village of human brains (around one thousand) to match a thousand dollars' worth of computing. By 2050, one thousand dollars of computing will exceed the processing power of all human brains on Earth. Of course, this figure includes those brains still using only biological neurons.

While human neurons are wondrous creations, we wouldn't (and don't) design computing circuits using the same slow methods. Despite the ingenuity of the designs evolved through natural selection, they are many orders of magnitude less capable than what we will be able to engineer. As we reverse engineer our bodies and brains, we will be in a position to create comparable systems that are far more durable and that operate thousands to millions of times faster than our naturally evolved systems. Our electronic circuits are already more than one million times faster than a neuron's electrochemical processes, and this speed is continuing to accelerate.

Most of the complexity of a human neuron is devoted to maintaining its life-support functions, not its information-processing capabilities. Ultimately, we will be able to port our mental processes to a more suitable computational substrate. Then our minds won't have to stay so small.

The Limits of Computation

> If a most efficient supercomputer works all day to compute a weather simulation problem, what is the minimum amount of energy that must be dissipated according to the laws of physics? The answer is actually very simple to calculate, since it is unrelated to the amount of computation. The answer is always equal to zero.
>
> —Edward Fredkin, physicist[45]

We've already had five paradigms (electromechanical calculators, relay-based computing, vacuum tubes, discrete transistors, and integrated circuits) that have provided exponential growth to the price-performance and capabilities of computation. Each time a paradigm reached its limits, another paradigm took its place. We can already see the outlines of the sixth paradigm, which will bring computing into the molecular third dimension. Because computation underlies the foundations of everything we care about, from the economy to human intellect and creativity, we might well wonder: are there ultimate limits to the capacity of matter and energy to perform computation? If so, what are these limits, and how long will it take to reach them?

Our human intelligence is based on computational processes that we are learning to understand. We will ultimately multiply our intellectual powers by applying and extending the methods of human intelligence using the vastly greater capacity of nonbiological computation. So to consider the ultimate limits of computation is really to ask: what is the destiny of our civilization?

A common challenge to the ideas presented in this book is that these exponential trends must reach a limit, as exponential trends commonly do. When a species happens upon a new habitat, as in the famous example of rabbits in Australia, its numbers grow exponentially for a while. But it eventually reaches the limits of that environment's ability to support it. Surely the processing of information must have similar constraints. It turns out that, yes, there are limits to computation based on the laws of physics. But these still allow for a continuation of exponential growth until nonbiological intelligence is trillions of trillions of times more powerful than all of human civilization today, contemporary computers included.

A major factor in considering computational limits is the energy requirement. The energy required per MIPS for computing devices has been falling exponentially, as shown in the following figure.[46]

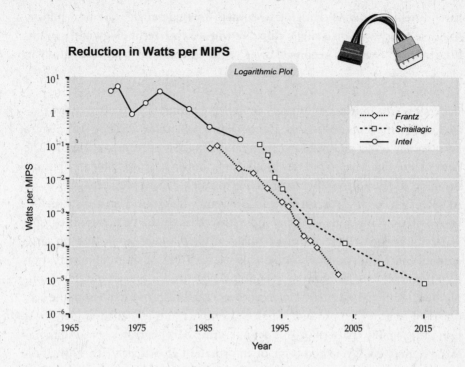

Reduction in Watts per MIPS

Logarithmic Plot

Frantz
Smailagic
Intel

Watts per MIPS

Year

However, we also know that the number of MIPS in computing devices has been growing exponentially. The extent to which improvements in power usage have kept pace with processor speed depends on the extent to which we use parallel processing. A larger number of less-powerful computers can inherently run cooler because the computation is spread out over a larger area. Processor speed is related to voltage, and the power required is proportional to the square of the voltage. So running a processor at a slower speed significantly reduces power consumption. If we invest in more parallel processing rather than faster single processors, it is feasible for energy consumption and heat dissipation to keep pace with the growing MIPS per dollar, as the figure "Reduction in Watts per MIPS" shows.

This is essentially the same solution that biological evolution developed in the design of animal brains. Human brains use about one hundred trillion computers (the interneuronal connections, where most of the processing takes place). But these processors are very low in computational power and therefore run relatively cool.

Until just recently Intel emphasized the development of faster and faster single-chip processors, which have been running at increasingly high temperatures. Intel is gradually changing its strategy toward parallelization by putting multiple processors on a single chip. We will see chip technology move in this direction as a way of keeping power requirements and heat dissipation in check.[47]

Reversible Computing. Ultimately, organizing computation with massive parallel processing, as is done in the human brain, will not by itself be sufficient to keep energy levels and resulting thermal dissipation at reasonable levels. The current computer paradigm relies on what is known as irreversible computing, meaning that we are unable in principle to run software programs backward. At each step in the progression of a program, the input data is discarded—erased—and the results of the computation pass to the next step. Programs generally do not retain all intermediate results, as that would use up large amounts of memory unnecessarily. This selective erasure of input information is particularly true for pattern-recognition systems. Vision systems, for example, whether human or machine, receive very high rates of input (from the eyes or visual sensors) yet produce relatively compact outputs (such as identification of recognized patterns). This act of erasing data generates heat and therefore requires energy. When a bit of information is erased, that information has to go somewhere. According to the laws of thermodynamics, the erased bit is essentially released into the surrounding environment, thereby increasing its entropy, which can be viewed as a measure of information (including apparently disordered information) in an environment. This results in a higher temperature for the environment (because temperature is a measure of entropy).

If, on the other hand, we don't erase each bit of information contained in the input to each step of an algorithm but instead just move it to another location, that bit stays in the computer, is not released into the environment, and therefore generates no heat and requires no energy from outside the computer.

Rolf Landauer showed in 1961 that reversible logical operations such as NOT (turning a bit into its opposite) could be performed without putting energy in or taking heat out, but that irreversible logical operations such as AND (generating bit C, which is a 1 if and only if both inputs A and B are 1) do require energy.[48] In 1973 Charles Bennett showed that any computation could be performed using only reversible logical operations.[49] A decade later, Ed Fredkin and Tommaso Toffoli presented a comprehensive review of the idea of reversible computing.[50] The fundamental concept is that if you keep all the intermediate results and then run the algorithm backward when you've finished your calculation, you end up where you started, have used no energy, and generated no heat. Along the way, however, you've calculated the result of the algorithm.

How Smart Is a Rock? To appreciate the feasibility of computing with no energy and no heat, consider the computation that takes place in an ordinary rock. Although it may appear that nothing much is going on inside a rock, the approximately 10^{25} (ten trillion trillion) atoms in a kilogram of matter are actually extremely active. Despite the apparent solidity of the object, the atoms are all in motion, sharing electrons back and forth, changing particle spins, and generating rapidly moving electromagnetic fields. All of this activity represents computation, even if not very meaningfully organized.

We've already shown that atoms can store information at a density of greater than one bit per atom, such as in computing systems built from nuclear magnetic-resonance devices. University of Oklahoma researchers stored 1,024 bits in the magnetic interactions of the protons of a single molecule containing nineteen hydrogen atoms.[51] Thus, the state of the rock at any one moment represents at least 10^{27} bits of memory.

In terms of computation, and just considering the electromagnetic interactions, there are at least 10^{15} changes in state per bit per second going on inside a 2.2-pound rock, which effectively represents about 10^{42} (a million trillion trillion trillion) calculations per second. Yet the rock requires no energy input and generates no appreciable heat.

Of course, despite all this activity at the atomic level, the rock is not performing any useful work aside from perhaps acting as a paperweight or a decoration. The reason for this is that the structure of the atoms in the rock is for the most part effectively random. If, on the other hand, we organize the particles in a more purposeful manner, we could have a cool, zero-energy-consuming computer with a memory of about a thousand trillion trillion bits and a processing capacity of 10^{42} operations per second, which is about ten trillion times more powerful than all human brains on Earth, even if we use the most conservative (highest) estimate of 10^{19} cps.[52]

Ed Fredkin demonstrated that we don't even have to bother running algorithms in reverse after obtaining a result.[53] Fredkin presented several designs for reversible logic gates that perform the reversals as they compute and that are universal, meaning that general-purpose computation can be built from them.[54] Fredkin goes on to show that the efficiency of a computer built from reversible logic gates can be designed to be very close (at least 99 percent) to the efficiency of ones built from irreversible gates. He writes:

it is possible to . . . implement . . . conventional computer models that have the distinction that the basic components are microscopically reversible. This means that the macroscopic operation of the computer is also reversible. This fact allows us to address the . . . question . . . "what is required for a computer to be maximally efficient?" The answer is that if the computer is built out of microscopically reversible components then it can be perfectly efficient. How much energy does a perfectly efficient computer have to dissipate in order to compute something? The answer is that the computer does not need to dissipate any energy.[55]

Reversible logic has already been demonstrated and shows the expected reductions in energy input and heat dissipation.[56] Fredkin's reversible logic gates answer a key challenge to the idea of reversible computing: that it would require a different style of programming. He argues that we can, in fact, construct normal logic and memory entirely from reversible logic gates, which will allow the use of existing conventional software-development methods.

It is hard to overstate the significance of this insight. A key observation regarding the Singularity is that information processes—computation—will ultimately drive everything that is important. This primary foundation for future technology thus appears to require no energy.

The practical reality is slightly more complicated. If we actually want to find out the results of a computation—that is, to receive output from a computer—the process of copying the answer and transmitting it outside of the computer is an irreversible process, one that generates heat for each bit transmitted. However, for most applications of interest, the amount of computation that goes into executing an algorithm vastly exceeds the computation required to communicate the final answers, so the latter does not appreciably change the energy equation.

However, because of essentially random thermal and quantum effects, logic operations have an inherent error rate. We can overcome errors using error-detection and -correction codes, but each time we correct a bit, the operation is not reversible, which means it requires energy and generates heat. Generally, error rates are low. But even if errors occur at the rate of, say, one per 10^{10} operations, we have only succeeded in reducing energy requirements by a factor of 10^{10}, not in eliminating energy dissipation altogether.

As we consider the limits of computation, the issue of error rate becomes a significant design issue. Certain methods of increasing computational rate, such as increasing the frequency of the oscillation of particles, also increase error rates, so this puts natural limits on the ability to perform computation using matter and energy.

Another important trend with relevance here will be the moving away from conventional batteries toward tiny fuel cells (devices storing energy in chemicals, such as forms of hydrogen, which is combined with available oxygen). Fuel cells are already being constructed using MEMS (microelectronic mechanical systems) technology.[57] As we move toward three-dimensional, molecular computing with nanoscale features, energy resources in the form of nano–fuel cells will be as widely distributed throughout the computing medium among the massively parallel processors. We will discuss future nanotechnology-based energy technologies in chapter 5.

The Limits of Nanocomputing. Even with the restrictions we have discussed, the ultimate limits of computers are profoundly high. Building on work by University of California at Berkeley professor Hans Bremermann and nanotechnology theorist Robert Freitas, MIT professor Seth Lloyd has estimated the maximum computational capacity, according to the known laws of physics, of a computer weighing one kilogram and occupying one liter of volume—about the size and weight of a small laptop computer—what he calls the "ultimate laptop."[58] The potential amount of computation rises with the available energy. We can understand the link between energy and computational capacity as follows. The energy in a quantity of matter is the energy associated with each atom (and subatomic particle). So the more atoms, the more energy. As discussed above, each atom can potentially be used for computation. So the more atoms, the more computation. The energy of each atom or particle grows with the frequency of its movement: the more movement, the more energy. The same relationship exists for potential computation: the higher the frequency of movement, the more computation each component (which can be an atom) can perform. (We see this in contemporary chips: the higher the frequency of the chip, the greater its computational speed.)

So there is a direct proportional relationship between the energy of an object and its potential to perform computation. The potential energy in a kilogram of matter is very large, as we know from Einstein's equation $E = mc^2$. The speed of light squared is a very large number: approximately 10^{17} meter^2/second^2. The potential of matter to compute is also governed by a very small number, Planck's constant: 6.6×10^{-34} joule-seconds (a joule is a measure of energy). This is the smallest scale at which we can apply energy for computation. We obtain the theoretical limit of an object to perform computation by dividing the total energy (the average energy of each atom or particle times the number of such particles) by Planck's constant.

Lloyd shows how the potential computing capacity of a kilogram of matter equals pi times energy divided by Planck's constant. Since the energy is such a large number and Planck's constant is so small, this equation generates an extremely large number: about 5×10^{50} operations per second.[59]

If we relate that figure to the most conservative estimate of human brain capacity (10^{19} cps and 10^{10} humans), it represents the equivalent of about five billion trillion human civilizations.[60] If we use the figure of 10^{16} cps that I believe will be sufficient for functional emulation of human intelligence, the ultimate laptop would function at the equivalent brain power of five trillion trillion human civilizations.[61] Such a laptop could perform the equivalent of all human thought over the last ten thousand years (that is, ten billion human brains operating for ten thousand years) in one ten-thousandth of a nanosecond.[62]

Again, a few caveats are in order. Converting all of the mass of our 2.2-pound laptop into energy is essentially what happens in a thermonuclear explosion. Of course, we don't want the laptop to explode but to stay within its one-liter dimension. So this will require some careful packaging, to say the least. By analyzing the maximum entropy (degrees of freedom represented by the state of all the particles) in such a device, Lloyd shows that such a computer would have a theoretical memory capacity of 10^{31} bits. It's difficult to imagine technologies that would go all the way in achieving these limits. But we can readily envision technologies that come reasonably close to doing so. As the University of Oklahoma project shows, we already demonstrated the ability to store at least fifty bits of information per atom (although only on a small number of atoms, so far). Storing 10^{27} bits of memory in the 10^{25} atoms in a kilogram of matter should therefore be eventually achievable.

But because many properties of each atom could be exploited to store information—such as the precise position, spin, and quantum state of all of its particles—we can probably do somewhat better than 10^{27} bits. Neuroscientist Anders Sandberg estimates the potential storage capacity of a hydrogen atom at about four million bits. These densities have not yet been demonstrated, however, so we'll use the more conservative estimate.[63] As discussed above, 10^{42} calculations per second could be achieved without producing significant heat. By fully deploying reversible computing techniques, using designs that generate low levels of errors, and allowing for reasonable amounts of energy dissipation, we should end up somewhere between 10^{42} and 10^{50} calculations per second.

The design terrain between these two limits is complex. Examining the technical issues that arise as we advance from 10^{42} to 10^{50} is beyond the scope of this chapter. We should keep in mind, however, that the way this will play out is not by starting with the ultimate limit of 10^{50} and working backward based on various practical considerations. Rather, technology will continue to ramp up, always using its latest prowess to progress to the next level. So once we get to a civilization with 10^{42} cps (for every 2.2 pounds), the scientists and engineers of that day will use their essentially vast nonbiological intelligence to figure out how to get 10^{43}, then 10^{44}, and so on. My expectation is that we will get very close to the ultimate limits.

Even at 10^{42} cps, a 2.2-pound "ultimate portable computer" would be able to perform the equivalent of all human thought over the last ten thousand years (assumed at ten billion human brains for ten thousand years) in ten microseconds.[64] If we examine the "Exponential Growth of Computing" chart (p. 70), we see that this amount of computing is estimated to be available for one thousand dollars by 2080.

A more conservative but compelling design for a massively parallel, reversible computer is Eric Drexler's patented nanocomputer design, which is entirely mechanical.[65] Computations are performed by manipulating nano-scale rods, which are effectively spring-loaded. After each calculation, the rods containing intermediate values return to their original positions, thereby implementing the reverse computation. The device has a trillion (10^{12}) processors and provides an overall rate of 10^{21} cps, enough to simulate one hundred thousand human brains in a cubic centimeter.

Setting a Date for the Singularity. A more modest but still profound threshold will be achieved much earlier. In the early 2030s one thousand dollars' worth of computation will buy about 10^{17} cps (probably around 10^{20} cps using ASICs and harvesting distributed computation via the Internet). Today we spend more than $\$10^{11}$ ($100 billion) on computation in a year, which will conservatively rise to $\$10^{12}$ ($1 trillion) by 2030. So we will be producing about 10^{26} to 10^{29} cps of nonbiological computation per year in the early 2030s. This is roughly equal to our estimate for the capacity of all living biological human intelligence.

Even if just equal in capacity to our own brains, this nonbiological portion of our intelligence will be more powerful because it will combine the pattern-recognition powers of human intelligence with the memory- and skill-sharing ability and memory accuracy of machines. The nonbiological portion will always operate at peak capacity, which is far from the case for biological humanity today; the 10^{26} cps represented by biological human civilization today is poorly utilized.

This state of computation in the early 2030s will not represent the Singularity, however, because it does not yet correspond to a profound expansion of our intelligence. By the mid-2040s, however, that one thousand dollars' worth of computation will be equal to 10^{26} cps, so the intelligence created per year (at a total cost of about 10^{12}) will be about one billion times more powerful than all human intelligence today.[66]

I set the date for the Singularity—representing a profound and disruptive transformation in human capability—as 2045. The nonbiological intelligence created in that year will be one billion times more powerful than all human intelligence today.

That *will* indeed represent a profound change, and it is for that reason that I set the date for the Singularity—representing a profound and disruptive transformation in human capability—as 2045.

Despite the clear predominance of nonbiological intelligence by the mid-2040s, ours will still be a human civilization. We will transcend biology, but not our humanity. I'll return to this issue in chapter 7.

Returning to the limits of computation according to physics, the estimates above were expressed in terms of laptop-size computers because that is a familiar form factor today. By the second decade of this century, however, most computing will not be organized in such rectangular devices but will be highly distributed throughout the environment. Computing will be everywhere: in the walls, in our furniture, in our clothing, and in our bodies and brains.

And, of course, human civilization will not be limited to computing with just a few pounds of matter. In chapter 6, we'll examine the computational potential of an Earth-size planet and computers on the scale of solar systems, of galaxies, and of the entire known universe. As we will see, the amount of time required for our human civilization to achieve scales of computation—and intelligence—that go beyond our planet and into the universe may be a lot shorter than you might think.

Memory and Computational Efficiency: A Rock Versus a Human Brain. With the limits of matter and energy to perform computation in mind, two useful metrics are the memory efficiency and computational efficiency of an object. These are defined as the fractions of memory and computation taking place in an object that are actually useful. Also, we need to consider the equivalence principle: even if computation is useful, if a simpler method produces equivalent results, then we should evaluate the computation against the simpler algorithm. In other words, if two methods achieve the same result but one uses more computation than the other, the more computationally intensive method will be considered to use only the amount of computation of the less intensive method.[67]

The purpose of these comparisons is to assess just how far biological evolution has been able to go from systems with essentially no intelligence (that is, an ordinary rock, which performs no *useful* computation) to the ultimate ability of matter to perform purposeful computation. Biological evolution took us part of the way, and technological evolution (which, as I pointed out earlier, represents a continuation of biological evolution) will take us very close to those limits.

Recall that a 2.2-pound rock has on the order of 10^{27} bits of information encoded in the state of its atoms and about 10^{42} cps represented by the activity of its particles. Since we are talking about an ordinary stone, assuming that its surface could store about one thousand bits is a perhaps arbitrary but generous estimate.[68] This represents 10^{-24} of its theoretical capacity, or a memory efficiency of 10^{-24}.[69]

We can also use a stone to do computation. For example, by dropping the stone from a particular height, we can compute the amount of time it takes to drop an object from that height. Of course, this represents very little computation: perhaps 1 cps, meaning its computational efficiency is 10^{-42}.[70]

In comparison, what can we say about the efficiency of the human brain? Earlier in this chapter we discussed how each of the approximately 10^{14} interneuronal connections can store an estimated 10^4 bits in the connection's neurotransmitter concentrations and synaptic and dendritic nonlinearities (specific shapes), for a total of 10^{18} bits. The human brain weighs about the same as our stone (actually closer to 3 pounds than 2.2, but since we're dealing with orders of magnitude, the measurements are close enough). It runs warmer than a cold stone, but we can still use the same estimate of about 10^{27} bits of theoretical memory capacity (estimating that we can store one bit in each atom). This results in a memory efficiency of 10^{-9}.

However, by the equivalence principle, we should not use the brain's inefficient coding methods to rate its memory efficiency. Using our functional memory estimate above of 10^{13} bits, we get a memory efficiency of 10^{-14}. That's about halfway between the stone and the ultimate cold laptop on a logarithmic scale. However, even though technology progresses exponentially, our experiences are in a linear world, and on a linear scale the human brain is far closer to the stone than to the ultimate cold computer.

So what is the brain's computational efficiency? Again, we need to consider the equivalence principle and use the estimate of 10^{16} cps required to emulate the brain's functionality, rather than the higher estimate (10^{19} cps) required to emulate all of the nonlinearities in every neuron. With the theoretical capacity of the brain's atoms estimated at 10^{42} cps, this gives us a computational efficiency of 10^{-26}. Again, that's closer to a rock than to the laptop, even on a logarithmic scale.

Our brains have evolved significantly in their memory and computational efficiency from pre-biology objects such as stones. But we clearly have many orders of magnitude of improvement to take advantage of during the first half of this century.

Going Beyond the Ultimate: Pico- and Femtotechnology and Bending the Speed of Light. The limits of around 10^{42} cps for a one-kilogram, one-liter cold computer and around 10^{50} for a (very) hot one are based on computing with atoms. But limits are not always what they seem. New scientific understanding has a way of pushing apparent limits aside. As one of many such examples, early in the history of aviation, a consensus analysis of the limits of jet propulsion apparently demonstrated that jet aircraft were infeasible.[71]

The limits I discussed above represent the limits of nanotechnology based on our current understanding. But what about picotechnology, measured in trillionths (10^{-12}) of a meter, and femtotechnology, scales of 10^{-15} of a meter? At these scales, we would require computing with subatomic particles. With such smaller size comes the potential for even greater speed and density.

We do have at least several very early-adopter picoscale technologies. German scientists have created an atomic-force microscope (AFM) that can resolve features of an atom that are only seventy-seven picometers across.[72] An even higher-resolution technology has been created by scientists at the University of California at Santa Barbara, who have developed an extremely sensitive measurement detector with a physical beam made of gallium-arsenide crystal and a sensing system that can measure a flexing of the beam of as little as one picometer. The device is intended to provide a test of Heisenberg's uncertainty principle.[73]

In the time dimension Cornell University scientists have demonstrated an imaging technology based on X-ray scattering that can record movies of the movement of a single electron. Each frame represents only four attoseconds (10^{-18} seconds, each one a billionth of a billionth of a second).[74] The device can achieve spatial resolution of one angstrom (10^{-10} meter, which is 100 picometers).

However, our understanding of matter at these scales, particularly in the femtometer range, is not sufficiently well developed to propose computing paradigms. An *Engines of Creation* (Eric Drexler's seminal 1986 book that provided the foundations for nanotechnology) for pico- or femtotechnology has not yet been written. However, each of the competing theories for the behavior of matter and energy at these scales is based on mathematical models that are based on computable transformations. Many of the transformations in physics do provide the basis for universal computation (that is, transformations from which we can build general-purpose computers), and it may be that behavior in the pico- and femtometer range will do so as well.

Of course, even if the basic mechanisms of matter in these ranges provide for universal computation in theory, we would still have to devise the requisite engineering to create massive numbers of computing elements and learn how to control them. These are similar to the challenges on which we are now rapidly making progress in the field of nanotechnology. At this time, we have to regard the feasibility of pico- and femtocomputing as speculative. But nanocomputing will provide massive levels of intelligence, so if it's at all possible to do, our future intelligence will be likely to figure out the necessary processes. The mental experiment we should be making is not whether humans as we know them today will be capable of engineering pico- and femtocomputing technologies, but whether the vast intelligence of future nanotechnology-based intelligence (which will be trillions of trillions of times more capable than contemporary biological human intelligence) will be capable of rendering these designs. Although I believe it is likely that our future nanotechnology-based intelligence will be able to engineer computation at scales finer than nanotechnology, the projections in this book concerning the Singularity do not rely on this speculation.

In addition to making computing smaller, we can make it bigger—that is, we can replicate these very small devices on a massive scale. With full-scale nanotechnology, computing resources can be made self-replicating and thus can rapidly convert mass and energy into an intelligent form. However, we run up against the speed of light, because the matter in the universe is spread out over vast distances.

As we will discuss later, there are at least suggestions that the speed of light may not be immutable. Physicists Steve Lamoreaux and Justin Torgerson of the Los Alamos National Laboratory have analyzed data from an old natural nuclear reactor that two billion years ago produced a fission reaction lasting several hundred thousand years in what is now West Africa.[75] Examining radioactive isotopes left over from the reactor and comparing them to isotopes from similar nuclear reactions today, they determined that the physics constant alpha (also called the fine-structure constant), which determines the strength of the electromagnetic force, apparently has changed over two billion years. This is of great significance to the world of physics, because the speed of light is inversely proportional to alpha, and both have been considered unchangeable constants. Alpha appears to have decreased by 4.5 parts out of 10^8. If confirmed, this would imply that the speed of light has increased.

Of course, these exploratory results will need to be carefully verified. If true, they may hold great importance for the future of our civilization. If the speed of light has increased, it has presumably done so not just as a result of the passage of time but because certain conditions have changed. If the speed of light has changed due to changing circumstances, that cracks open the door just enough for the vast powers of our future intelligence and technology to swing the door widely open. This is the type of scientific insight that technologists can exploit. Human engineering often takes a natural, frequently subtle, effect, and controls it with a view toward greatly leveraging and magnifying it.

Even if we find it difficult to significantly increase the speed of light over the long distances of space, doing so within the small confines of a computing device would also have important consequences for extending the potential for computation. The speed of light is one of the limits that constrain computing devices even today, so the ability to boost it would extend further the limits of computation. We will explore several other intriguing approaches to possibly increasing, or circumventing, the speed of light in chapter 6. Expanding the speed of light is, of course, speculative today, and none of the analyses underlying our expectation of the Singularity rely on this possibility.

Going Back in Time. Another intriguing—and highly speculative—possibility is to send a computational process back in time through a "wormhole" in space-time. Theoretical physicist Todd Brun of the Institute for Advanced Studies at Princeton has analyzed the possibility of computing using what he calls a "closed timelike curve" (CTC). According to Brun, CTCs could "send information (such as the result of calculations) into their own past light cones."[76]

Brun does not provide a design for such a device but establishes that such a system is consistent with the laws of physics. His time-traveling computer also does not create the "grandfather paradox," often cited in discussions of time travel. This well-known paradox points out that if person A goes back in time, he could kill his grandfather, causing A not to exist, resulting in his grandfather not being killed by him, so A would exist and thus could go back and kill his grandfather, and so on, ad infinitum.

Brun's time-stretching computational process does not appear to introduce this problem because it does not affect the past. It produces a determinate and unambiguous answer in the present to a posed question. The question must have a clear answer, and the answer is not presented until *after* the question is asked, although the process to determine the answer can take place before the question is asked using the CTC. Conversely, the process could take place after the question is asked and then use a CTC to bring the answer back into the present (but not before the question was asked, because that would introduce the grandfather paradox). There may very well be fundamental barriers (or limitations) to such a process that we don't yet understand, but those barriers have yet to be identified. If feasible, it would greatly expand the potential of local computation. Again, all of my estimates of computational capacities and of the capabilities of the Singularity do not rely on Brun's tentative conjecture.

ERIC DREXLER: *I don't know, Ray. I'm pessimistic on the prospects for picotechnology. With the stable particles we know of, I don't see how there can be picoscale structure without the enormous pressures found in a collapsed star—a white dwarf or a neutron star—and then you would get a solid chunk of stuff like a metal, but a million times denser. This doesn't seem very useful, even if it were possible to make it in our solar system. If physics included a stable particle like an electron but a hundred times more massive, it would be a different story, but we don't know of one.*

RAY: *We manipulate subatomic particles today with accelerators that fall significantly short of the conditions in a neutron star. Moreover, we manipulate subatomic particles such as electrons today with tabletop devices. Scientists recently captured and stopped a photon dead in its tracks.*

ERIC: *Yes, but what kind of manipulation? If we count manipulating small particles, then all technology is already picotechnology, because all matter is made of subatomic particles. Smashing particles together in accelerators produces debris, not machines or circuits.*

RAY: *I didn't say we've solved the conceptual problems of picotechnology. I've got you penciled in to do that in 2072.*

ERIC: *Oh, good, then I see you have me living a long time.*

RAY: *Yes, well, if you stay on the sharp leading edge of health and medical insights and technology, as I'm trying to do, I see you being in rather good shape around then.*

MOLLY 2104: *Yes, quite a few of you baby boomers did make it through. But most were unmindful of the opportunities in 2004 to extend human mortality long enough to take advantage of the biotechnology revolution, which hit its stride a decade later, followed by nanotechnology a decade after that.*

MOLLY 2004: *So, Molly 2104, you must be quite something, considering that one thousand dollars of computation in 2080 can perform the equivalent of ten billion human brains thinking for ten thousand years in a matter of ten microseconds. That presumably will have progressed even further by 2104, and I assume you have access to more than one thousand dollars' worth of computation.*

MOLLY 2104: *Actually, millions of dollars on average—billions when I need it.*

MOLLY 2004: *That's pretty hard to imagine.*

MOLLY 2104: *Yeah, well, I guess I'm kind of smart when I need to be.*

MOLLY 2004: *You don't sound that bright, actually.*

MOLLY 2104: *I'm trying to relate on your level.*

MOLLY 2004: *Now, wait a second, Miss Molly of the future. . . .*

GEORGE 2048: *Ladies, please, you're both very engaging.*

MOLLY 2004: *Yes, well, tell that to my counterpart here—she feels she's a jillion times more capable than I am.*

GEORGE 2048: *She is your future, you know. Anyway, I've always felt there was something special about a biological woman.*

MOLLY 2104: *Yeah, what would you know about biological women anyway?*

GEORGE 2048: *I've read a great deal about it and engaged in some very precise simulations.*

MOLLY 2004: *It occurs to me that maybe you're both missing something that you're not aware of.*

GEORGE 2048: *I don't see how that's possible.*

MOLLY 2104: *Definitely not.*

MOLLY 2004: *I didn't think you would. But there is one thing I understand you can do that I do find cool.*

MOLLY 2104: *Just one?*

MOLLY 2004: *One that I'm thinking of, anyway. You can merge your thinking with someone else and still keep your separate identity at the same time.*

MOLLY 2104: *If the situation—and the person—is right, then, yes, it's a very sublime thing to do.*

MOLLY 2004: *Like falling in love?*

MOLLY 2104: *Like being in love. It's the ultimate way to share.*

GEORGE 2048: *I think you'll go for it, Molly 2004.*

MOLLY 2104: *You ought to know, George, since you were the first person I did it with.*

CHAPTER FOUR

Achieving the Software of Human Intelligence

How to Reverse Engineer the Human Brain

There are good reasons to believe that we are at a turning point, and that it will be possible within the next two decades to formulate a meaningful understanding of brain function. This optimistic view is based on several measurable trends, and a simple observation which has been proven repeatedly in the history of science: *Scientific advances are enabled by a technology advance that allows us to see what we have not been able to see before.* At about the turn of the twenty-first century, we passed a detectable turning point in both neuroscience knowledge and computing power. For the first time in history, we collectively know enough about our own brains, and have developed such advanced computing technology, that we can now seriously undertake the construction of a verifiable, real-time, high-resolution model of significant parts of our intelligence.

—LLOYD WATTS, NEUROSCIENTIST[1]

Now, for the first time, we are observing the brain at work in a global manner with such clarity that we should be able to discover the overall programs behind its magnificent powers.

—J. G. TAYLOR, B. HORWITZ, K. J. FRISTON, NEUROSCIENTISTS[2]

The brain is good: it is an existence proof that a certain arrangement of matter can produce mind, perform intelligent reasoning, pattern recognition, learning and a lot of other important tasks of engineering interest. Hence we can learn to build new systems by borrowing ideas from the brain. . . . The brain is bad: it is an evolved, messy system where a lot of interactions happen because of evolutionary contingencies. . . . On the other hand, it must also be robust (since we can survive with it) and be able to stand fairly major variations and environmental insults, so the truly valuable insight from the brain might be how to create resilient complex systems that self-organize well. . . . The interactions within a neuron are complex, but on the next level neurons seem to be somewhat simple objects that can be put together flexibly into networks. The cortical networks are a real mess locally, but again on the next level the connectivity isn't that complex. It would be likely that evolution has produced a number of modules or repeating themes that are being re-used, and when we understand them and their interactions we can do something similar.

—ANDERS SANDBERG, COMPUTATIONAL NEUROSCIENTIST, ROYAL INSTITUTE OF TECHNOLOGY, SWEDEN

Reverse Engineering the Brain: An Overview of the Task

The combination of human-level intelligence with a computer's inherent superiority in speed, accuracy, and memory-sharing ability will be formidable. To date, however, most AI research and development has utilized engineering methods that are not necessarily based on how the human brain functions, for the simple reason that we have not had the precise tools needed to develop detailed models of human cognition.

Our ability to reverse engineer the brain—to see inside, model it, and simulate its regions—is growing exponentially. We will ultimately understand the principles of operation underlying the full range of our own thinking, knowledge that will provide us with powerful procedures for developing the software of intelligent machines. We will modify, refine, and extend these techniques as we apply them to computational technologies that are far more powerful than the electrochemical processing that takes place in biological neurons. A key benefit of this grand project will be the precise insights it offers into ourselves. We will also gain powerful new ways to treat neurological problems such as Alzheimer's, stroke, Parkinson's disease, and sensory disabilities, and ultimately will be able to vastly extend our intelligence.

New Brain-Imaging and Modeling Tools. The first step in reverse engineering the brain is to peer into the brain to determine how it works. So far, our tools for doing this have been crude, but that is now changing, as a significant number of new scanning technologies feature greatly improved spatial and temporal resolution, price-performance, and bandwidth. Simultaneously we are rapidly accumulating data on the precise characteristics and dynamics of the constituent parts and systems of the brain, ranging from individual synapses to large regions such as the cerebellum, which comprises more than half of the brain's neurons. Extensive databases are methodically cataloging our exponentially growing knowledge of the brain.[3]

Researchers have also shown they can rapidly understand and apply this information by building models and working simulations. These simulations of brain regions are based on the mathematical principles of complexity theory and chaotic computing and are already providing results that closely match experiments performed on actual human and animal brains.

As noted in chapter 2, the power of the scanning and computational tools needed for the task of reverse engineering the brain is accelerating, similar to the acceleration in technology that made the genome project feasible. When we get to the nanobot era (see "Scanning Using Nanobots" on p. 163), we will be able to scan from *inside* the brain with exquisitely high spatial and temporal resolution.[4] There are no inherent barriers to our being able to reverse engineer the operating principles of human intelligence and replicate these capabilities in the more powerful computational substrates that will become available in the decades ahead. The human brain is a complex hierarchy of complex systems, but it does not represent a level of complexity beyond what we are already capable of handling.

The Software of the Brain. The price-performance of computation and communication is doubling every year. As we saw earlier, the computational capacity needed to emulate human intelligence will be available in less than two decades.[5] A principal assumption underlying the expectation of the Singularity is that nonbiological mediums will be able to emulate the richness, subtlety, and depth of human thinking. But achieving the hardware computational capacity of a single human brain—or even of the collective intelligence of villages and nations—will not automatically produce human levels of capability. (By "human levels" I include all the diverse and subtle ways humans are intelligent, including musical and artistic aptitude, creativity, physical motion through the world, and understanding and responding appropriately to emotions.) The hardware computational capacity is necessary but not sufficient. Understanding the organization and content of these resources—the software of intelligence—is even more critical and is the objective of the brain reverse-engineering undertaking.

Once a computer achieves a human level of intelligence, it will necessarily soar past it. A key advantage of nonbiological intelligence is that machines can easily share their knowledge. If you learn French or read *War and Peace*, you can't readily download that learning to me, as I have to acquire that scholarship the same painstaking way that you did. I can't (yet) quickly access or transmit your knowledge, which is embedded in a vast pattern of neurotransmitter concentrations (levels of chemicals in the synapses that allow one neuron to influence another) and interneuronal connections (portions of the neurons called axons and dendrites that connect neurons).

But consider the case of a machine's intelligence. At one of my companies, we spent years teaching one research computer how to recognize continuous human speech, using pattern-recognition software.[6] We exposed it to thousands of hours of recorded speech, corrected its errors, and patiently improved its performance by training its "chaotic" self-organizing algorithms (methods that modify their own rules, based on processes that use semirandom initial information, and with results that are not fully predictable). Finally, the computer became quite adept at recognizing speech. Now, if you want your own personal computer to recognize speech, you don't have to put it through the same painstaking learning process (as we do with each human child); you can simply download the already established patterns in seconds.

Analytic Versus Neuromorphic Modeling of the Brain. A good example of the divergence between human intelligence and contemporary AI is how each undertakes the solution of a chess problem. Humans do so by recognizing patterns, while machines build huge logical "trees" of possible moves and countermoves. Most technology (of all kinds) to date has used this latter type of "top-down," analytic, engineering approach. Our flying machines, for example, do not attempt to re-create the physiology and mechanics of birds. But as our tools for reverse engineering the ways of nature are growing rapidly in sophistication, technology is moving toward emulating nature while implementing these techniques in far more capable substrates.

The most compelling scenario for mastering the software of intelligence is to tap directly into the blueprint of the best example we can get our hands on of an intelligent process: the human brain. Although it took its original "designer" (evolution) several billion years to develop the brain, it's readily available to us, protected by a skull but with the right tools not hidden from our view. Its contents are not yet copyrighted or patented. (We can, however, expect that to change; patent applications have already been filed based on brain reverse engineering.)[7] We will apply the thousands of trillions of bytes of information derived from brain scans and neural models at many levels to design more intelligent parallel algorithms for our machines, particularly those based on self-organizing paradigms.

With this self-organizing approach, we don't have to attempt to replicate every single neural connection. There is a great deal of repetition and redundancy within any particular brain region. We are discovering that higher-level models of brain regions are often simpler than the detailed models of their neuronal components.

How Complex Is the Brain? Although the information contained in a human brain would require on the order of one billion billion bits (see chapter 3), the initial design of the brain is based on the rather compact human genome. The entire genome consists of eight hundred million bytes, but most of it is redundant, leaving only about thirty to one hundred million bytes (less than 10^9 bits) of unique information (after compression), which is smaller than the program for Microsoft Word.[8] To be fair, we should also take into account "epigenetic" data, which is information stored in proteins that control gene expression (that is, that determine which genes are allowed to create proteins in each cell), as well as the entire protein-replication machinery, such as the ribosomes and a host of enzymes. However, such additional information does not significantly change the order of magnitude of this calculation.[9] Slightly more than half of the genetic and epigenetic information characterizes the initial state of the human brain.

Of course, the complexity of our brains greatly increases as we interact with the world (by a factor of about one billion over the genome).[10] But highly repetitive patterns are found in each specific brain region, so it is not necessary to capture each particular detail to successfully reverse engineer the relevant algorithms, which combine digital and analog methods (for example, the firing of a neuron can be considered a digital event whereas neurotransmitter levels in the synapse can be considered analog values). The basic wiring pattern of the cerebellum, for example, is described in the genome only once but repeated billions of times. With the information from brain scanning and modeling studies, we can design simulated "neuromorphic" equivalent software (that is, algorithms functionally equivalent to the overall performance of a brain region).

The pace of building working models and simulations is only slightly behind the availability of brain-scanning and neuron-structure information. There are more than fifty thousand neuroscientists in the world, writing articles for more than three hundred journals.[11] The field is broad and diverse, with scientists and engineers creating new scanning and sensing technologies and developing models and theories at many levels. So even people in the field are often not completely aware of the full dimensions of contemporary research.

Modeling the Brain. In contemporary neuroscience, models and simulations are being developed from diverse sources, including brain scans, interneuronal connection models, neuronal models, and psychophysical testing. As mentioned earlier, auditory-system researcher Lloyd Watts has developed a comprehensive model of a significant portion of the human auditory-processing system from neurobiology studies of specific neuron types and interneuronal-connection information. Watts's model includes five parallel paths and the actual representations of auditory information at each stage of neural processing. Watts has implemented his model in a computer as real-time software that can locate and identify sounds and functions, similar to the way human hearing operates. Although a work in progress, the model illustrates the feasibility of converting neurobiological models and brain-connection data into working simulations.

As Hans Moravec and others have speculated, these efficient functional simulations require about one thousand times less computation than would be required if we simulated the nonlinearities in each dendrite, synapse, and other subneural structure in the region being simulated. (As I discussed in chapter 3, we can estimate the computation required for functional simulation of the brain at 10^{16} calculations per second [cps], versus 10^{19} cps to simulate the subneural nonlinearities.)[12]

The actual speed ratio between contemporary electronics and the electrochemical signaling in biological interneuronal connections is at least one million to one. We find this same inefficiency in all aspects of our biology, because biological evolution built all of its mechanisms and systems with a severely constrained set of materials: namely, cells, which are themselves made from a limited set of proteins. Although biological proteins are three-dimensional, they are restricted to complex molecules that can be folded from a linear (one-dimensional) sequence of amino acids.

Peeling the Onion. The brain is not a single information-processing organ but rather an intricate and intertwined collection of hundreds of specialized regions. The process of "peeling the onion" to understand the functions of these interleaved regions is well under way. As the requisite neuron descriptions and brain-interconnection data become available, detailed and implementable replicas such as the simulation of the auditory regions described below (see "Another Example: Watts's Model of the Auditory Regions" on p. 183) will be developed for all brain regions.

Most brain-modeling algorithms are not the sequential, logical methods that are commonly used in digital computing today. The brain tends to use self-organizing, chaotic, holographic processes (that is, information not located in one place but distributed throughout a region). It is also massively parallel and utilizes hybrid digital-controlled analog techniques. However, a wide range of projects has demonstrated our ability to understand these techniques and to extract them from our rapidly escalating knowledge of the brain and its organization.

After the algorithms of a particular region are understood, they can be refined and extended before being implemented in synthetic neural equivalents. They can be run on a computational substrate that is already far faster than neural circuitry. (Current computers perform computations in billionths of a second, compared to thousandths of a second for interneuronal transactions.) And we can also make use of the methods for building intelligent machines that we already understand.

Is the Human Brain Different from a Computer?

The answer to this question depends on what we mean by the word "computer." Most computers today are all digital and perform one (or perhaps a few) computations at a time at extremely high speed. In contrast, the human brain combines digital and analog methods but performs most computations in the analog (continuous) domain, using neurotransmitters and related mechanisms. Although these neurons execute calculations at extremely slow speeds (typically two hundred transactions per second), the brain as a whole is massively parallel: most of its neurons work at the same time, resulting in up to one hundred trillion computations being carried out simultaneously.

The massive parallelism of the human brain is the key to its pattern-recognition ability, which is one of the pillars of our species' thinking. Mammalian neurons engage in a chaotic dance (that is, with many apparently random interactions), and if the neural network has learned its lessons well, a stable pattern will emerge, reflecting the network's decision. At the present, parallel designs for computers are somewhat limited. But there is no reason why functionally equivalent nonbiological re-creations of biological neural networks cannot be built using these principles. Indeed, dozens of efforts around the world have already succeeded in doing so. My own technical field is pattern recognition, and the projects that I have been involved in for about forty years use this form of trainable and nondeterministic computing.

Many of the brain's characteristic methods of organization can also be effectively simulated using conventional computing of sufficient power. Duplicating the design paradigms of nature will, I believe, be a key trend in future computing. We should keep in mind, as well, that digital computing can be functionally equivalent to analog computing—that is, we can perform all of the functions of a hybrid digital-analog network with an all-digital computer. The reverse is not true: we can't simulate all of the functions of a digital computer with an analog one.

However, analog computing does have an engineering advantage: it is potentially thousands of times more efficient. An analog computation can be performed by a few transistors or, in the case of mammalian neurons, specific electrochemical processes. A digital computation, in contrast, requires thousands or tens of thousands of transistors. On the other hand, this advantage can be offset by the ease of programming (and modifying) digital computer-based simulations.

There are a number of other key ways in which the brain differs from a conventional computer:

- *The brain's circuits are very slow.* Synaptic-reset and neuron-stabilization times (the amount of time required for a neuron and its synapses to reset themselves after the neuron fires) are so slow that there are very few neuron-firing cycles available to make pattern-recognition decisions. Functional magnetic-resonance imaging (fMRI) and magneto-encephalography (MEG) scans show that judgments that do not require resolving ambiguities appear to be made in a single neuron-firing cycle (less than twenty milliseconds), involving essentially no iterative (repeated) processes. Recognition of objects occurs in about 150 milliseconds, so that even if we "think something over," the number of cycles of operation is measured in hundreds or thousands at most, not billions, as with a typical computer.
- *But it's massively parallel.* The brain has on the order of one hundred trillion interneuronal connections, each potentially processing information simultaneously. These two factors (slow cycle time and massive parallelism) result in a certain level of computational capacity for the brain, as we discussed earlier.

  Today our largest supercomputers are approaching this range. The leading supercomputers (including those used by the most popular search engines) measure over 10^{14} cps, which matches the lower range of the estimates I discussed in chapter 3 for functional simulation. It is not necessary, however, to use the same granularity of parallel processing as the brain itself so long as we match the overall computational speed and memory capacity needed and otherwise simulate the brain's massively parallel architecture.
- *The brain combines analog and digital phenomena.* The topology of connections in the brain is essentially digital—a connection exists, or it doesn't. An axon firing is not entirely digital but closely approximates a digital process. Most every function in the brain is analog and is filled with nonlinearities (sudden shifts in output, rather than levels changing smoothly) that are substantially more complex than the classical model that we have been using for neurons. However, the detailed, nonlinear dynamics of a

neuron and all of its constituents (dendrites, spines, channels, and axons) can be modeled through the mathematics of nonlinear systems. These mathematical models can then be simulated on a digital computer to any desired degree of accuracy. As I mentioned, if we simulate the neural regions using transistors in their native analog mode rather than through digital computation, this approach can provide improved capacity by three or four orders of magnitude, as Carver Mead has demonstrated.[13]

- *The brain rewires itself.* Dendrites are continually exploring new spines and synapses. The topology and conductance of dendrites and synapses are also continually adapting. The nervous system is self-organizing at all levels of its organization. While the mathematical techniques used in computerized pattern-recognition systems such as neural nets and Markov models are much simpler than those used in the brain, we do have substantial engineering experience with self-organizing models.[14] Contemporary computers don't literally rewire themselves (although emerging "self-healing systems" are starting to do this), but we can effectively simulate this process in software.[15] In the future, we can implement this in hardware, as well, although there may be advantages to implementing most self-organization in software, which provides more flexibility for programmers.

- *Most of the details in the brain are random.* While there is a great deal of stochastic (random within carefully controlled constraints) process in every aspect of the brain, it is not necessary to model every "dimple" on the surface of every dendrite, any more than it is necessary to model every tiny variation in the surface of every transistor in understanding the principles of operation of a computer. But certain details are critical in decoding the principles of operation of the brain, which compels us to distinguish between them and those that comprise stochastic "noise" or chaos. The chaotic (random and unpredictable) aspects of neural function can be modeled using the mathematical techniques of complexity theory and chaos theory.[16]

- *The brain uses emergent properties.* Intelligent behavior is an emergent property of the brain's chaotic and complex activity. Consider the analogy to the apparently intelligent design of termite and ant colonies, with their delicately constructed interconnecting tunnels and ventilation systems. Despite their clever and intricate design, ant and termite hills have no master architects; the architecture emerges from the unpredictable interactions of all the colony members, each following relatively simple rules.

- *The brain is imperfect.* It is the nature of complex adaptive systems that the emergent intelligence of its decisions is suboptimal. (That is, it reflects a lower level of intelligence than would be represented by an optimal arrangement of its elements.) It needs only to be good enough, which in the case of our species meant a level of intelligence sufficient to enable us to outwit the competitors in our ecological niche (for example, primates who also combine a cognitive function with an opposable appendage but whose brains are not as developed as humans and whose hands do not work as well).

- *We contradict ourselves.* A variety of ideas and approaches, including conflicting ones, leads to superior outcomes. Our brains are quite capable of holding contradictory views. In fact, we thrive on this internal diversity. Consider the analogy to a human society, particularly a democratic one, with its constructive ways of resolving multiple viewpoints.

- *The brain uses evolution.* The basic learning paradigm used by the brain is an evolutionary one: the patterns of connections that are most successful in making sense of the world and contributing to recognitions and decisions survive. A newborn's brain contains mostly randomly linked interneuronal connections, and only a portion of those survive in the two-year-old brain.[17]

- *The patterns are important.* Certain details of these chaotic self-organizing methods, expressed as model constraints (rules defining the initial conditions and the means for self-organization), are crucial, whereas many details within the constraints are initially set randomly. The system then self-organizes and gradually represents the invariant features of the information that has been presented to the system. The resulting information is not found in specific nodes or connections but rather is a distributed pattern.

- *The brain is holographic.* There is an analogy between distributed information in a hologram and the method of information representation in brain networks. We find this also in the self-organizing methods used in computerized pattern recognition, such as neural nets, Markov models, and genetic algorithms.[18]

- *The brain is deeply connected.* The brain gets its resilience from being a deeply connected network in which information has many ways of navigating from one point to another. Consider the analogy to the Internet, which has become increasingly stable as the number of its constituent nodes has increased. Nodes, even entire hubs of the Internet, can become inoperative without ever bringing down the entire network. Similarly, we continually lose neurons without affecting the integrity of the entire brain.

- *The brain does have an architecture of regions.* Although the details of connections within a region are initially random within constraints and self-organizing, there is an architecture of several hundred regions that perform specific functions, with specific patterns of connections between regions.

- *The design of a brain region is simpler than the design of a neuron.* Models often get simpler at a higher level, not more complex. Consider an analogy with a computer. We do need to understand the detailed physics of semiconductors to model a transistor, and the equations underlying a single real transistor are complex. However, a digital circuit that multiplies two numbers, although involving hundreds of transistors, can be modeled far more simply, with only a few formulas. An entire computer with billions of transistors can be modeled through its instruction set and register description, which can be described on a handful of written pages of text and mathematical transformations.

The software programs for an operating system, language compilers, and assemblers are reasonably complex, but modeling a particular program—for example, a speech-recognition program based on Markov modeling—may be described in only a few pages of equations. Nowhere in such a description would be found the details of semiconductor physics. A similar observation also holds true for the brain. A particular neural arrangement that detects a particular invariant visual feature (such as a face) or that performs a band-pass filtering (restricting input to a specific frequency range) operation on auditory information or that evaluates the temporal proximity of two events can be described with far greater simplicity than the actual physics and chemical relations controlling the neurotransmitters and other synaptic and dendritic variables involved in the respective processes. Although all of this neural complexity will have to be carefully considered before advancing to the next higher level (modeling the brain), much of it can be simplified once the operating principles of the brain are understood.

Trying to Understand Our Own Thinking

The Accelerating Pace of Research

We are now approaching the knee of the curve (the period of rapid exponential growth) in the accelerating pace of understanding the human brain, but our attempts in this area have a long history. Our ability to reflect on and build models of our thinking is a unique attribute of our species. Early mental models were of necessity based on simply observing our external behavior (for example, Aristotle's analysis of the human ability to associate ideas, written 2,350 years ago).[19]

At the beginning of the twentieth century we developed the tools to examine the physical processes *inside* the brain. An early breakthrough was the measurement of the electrical output of nerve cells, developed in 1928 by neuroscience pioneer E. D. Adrian, which demonstrated that there were electrical processes taking place inside the brain.[20] As Adrian wrote, "I had arranged electrodes on the optic nerve of a toad in connection with some experiments on the retina. The room was nearly dark and I was puzzled to hear repeated noises in the loudspeaker attached to the amplifier, noises indicating that a great deal of impulse activity was going on. It was not until I compared the noises with my own movements around the room that I realized I was in the field of vision of the toad's eye and that it was signaling what I was doing."

Adrian's key insight from this experiment remains a cornerstone of neuroscience today: the frequency of the impulses from the sensory nerve is proportional to the intensity of the sensory phenomena being measured. For example, the higher the intensity of light, the higher the frequency (pulses per second) of the neural impulses from the retina to the brain. It was a student of Adrian, Horace Barlow, who contributed another lasting insight, "trigger features" in neurons, with the discovery that the retinas of frogs and rabbits had single neurons that would trigger on "seeing" specific shapes, directions, or velocities. In other words, perception involves a series of stages, with each layer of neurons recognizing more sophisticated features of the image.

In 1939 we began to develop an idea of how neurons perform: by accumulating (adding) their inputs and then producing a spike of membrane conductance (a sudden increase in the ability of the neuron's membrane to conduct a signal) and voltage along the neuron's axon (which connects to other neurons via a synapse). A. L. Hodgkin and A. F. Huxley ▶

described their theory of the axon's "action potential" (voltage).[21] They also made an actual measurement of an action potential on an animal neural axon in 1952.[22] They chose squid neurons because of their size and accessible anatomy.

Building on Hodgkin and Huxley's insight W. S. McCulloch and W. Pitts developed in 1943 a simplified model of neurons and neural nets that motivated a half century of work on artificial (simulated) neural nets (using a computer program to simulate the way neurons work in the brain as a network). This model was further refined by Hodgkin and Huxley in 1952. Although we now realize that actual neurons are far more complex than these early models, the original concept has held up well. This basic neural-net model has a neural "weight" (representing the "strength" of the connection) for each synapse and a nonlinearity (firing threshold) in the neuron soma (cell body).

As the sum of the weighted inputs to the neuron soma increases, there is relatively little response from the neuron until a critical threshold is reached, at which point the neuron rapidly increases the output of its axon and fires. Different neurons have different thresholds. Although recent research shows that the actual response is more complex than this, the McCulloch-Pitts and Hodgkin-Huxley models remain essentially valid.

These insights led to an enormous amount of early work in creating artificial neural nets, in a field that became known as connectionism. This was perhaps the first self-organizing paradigm introduced to the field of computation.

A key requirement for a self-organizing system is a nonlinearity: some means of creating outputs that are not simple weighted sums of the inputs. The early neural-net models provided this nonlinearity in their replica of the neuron nucleus.[23] (The basic neural-net method is straightforward.)[24] Work initiated by Alan Turing on theoretical models of computation around the same time also showed that computation requires a nonlinearity. A system that simply creates weighted sums of its inputs cannot perform the essential requirements of computation.

We now know that actual biological neurons have many other nonlinearities resulting from the electrochemical action of the synapses and the morphology (shape) of the dendrites. Different arrangements of biological neurons can perform computations, including adding, subtracting, multiplying, dividing, averaging, filtering, normalizing, and thresholding signals, among other types of transformations.

The ability of neurons to perform multiplication is important because ▶

it allows the behavior of one network of neurons in the brain to be modulated (influenced) by the result of computations of another network. Experiments using electrophysiological measurements on monkeys provide evidence that the rate of signaling by neurons in the visual cortex when processing an image is increased or decreased by whether or not the monkey is paying attention to a particular area of that image.[25] Human fMRI studies have also shown that paying attention to a particular area of an image increases the responsiveness of the neurons processing that image in a cortical region called V5, which is responsible for motion detection.[26]

Another key breakthrough occurred in 1949 when Donald Hebb presented his seminal theory of neural learning, the "Hebbian response": if a synapse (or group of synapses) is stimulated repeatedly, that synapse becomes stronger. Over time this conditioning of the synapse produces a learning response. The connectionism movement designed simulated neural nets based on this model, and this gave momentum to such experiments during the 1950s and 1960s.

The connectionism movement experienced a setback in 1969 with the publication of the book *Perceptrons* by MIT's Marvin Minsky and Seymour Papert.[27] It included a key theorem demonstrating that the most common (and simplest) type of neural net used at the time (called a Perceptron, pioneered by Cornell's Frank Rosenblatt) was unable to solve the simple problem of determining whether or not a line drawing was fully connected.[28] The neural-net movement had a resurgence in the 1980s using a method called "backpropagation," in which the strength of each simulated synapse was determined using a learning algorithm that adjusted the weight (the strength of the output) of each artificial neuron after each training trial so the network could "learn" to more correctly match the right answer.

However, backpropagation is not a feasible model of training synaptic weights in an actual biological neural network, because backward connections to actually adjust the strength of the synaptic connections do not appear to exist in mammalian brains. In computers, however, this type of self-organizing system can solve a wide range of pattern-recognition problems, and the power of this simple model of self-organizing interconnected neurons has been demonstrated.

Less well known is Hebb's second form of learning: a hypothesized loop in which the excitation of a neuron would feed back on itself (possibly through other layers), causing a reverberation (a continued reexcitation of the neurons in the loop). Hebb theorized that this type of reverberation could be the source of short-term learning. He also suggested that this ▶

short-term reverberation could lead to long-term memories: "Let us assume then that the persistence or repetition of a reverberatory activity (or 'trace') tends to induce lasting cellular changes that add to its stability. The assumption can be precisely stated as follows: When an axon of cell A is near enough to excite a cell B and repeatedly or persistently takes part in firing it, some growth process or metabolic change takes place in one or both cells such that A's efficiency, as one of the cells firing B, is increased."

Although Hebbian reverberatory memory is not as well established as Hebb's synaptic learning, instances have recently been discovered. For example, sets of excitatory neurons (ones that stimulate a synapse) and inhibitory neurons (ones that block a stimulus) begin an oscillation when certain visual patterns are presented.[29] And researchers at MIT and Lucent Technologies' Bell Labs have created an electronic integrated circuit, composed of transistors, that simulates the action of sixteen excitatory neurons and one inhibitory neuron to mimic the biological circuitry of the cerebral cortex.[30]

These early models of neurons and neural information processing, although overly simplified and inaccurate in some respects, were remarkable, given the lack of data and tools when these theories were developed.

Peering into the Brain

We've been able to reduce drift and noise in our instruments to such an extent that we can see the tiniest motions of these molecules, through distances that are less than their own diameters. . . . [T]hese kinds of experiments were just pipedreams 15 years ago.

—STEVEN BLOCK, PROFESSOR OF BIOLOGICAL SCIENCES AND OF APPLIED PHYSICS, STANFORD UNIVERSITY

Imagine that we were trying to reverse engineer a computer without knowing anything about it (the "black box" approach). We might start by placing arrays of magnetic sensors around the device. We would notice that during operations that updated a database, significant activity was taking place in a particular circuit board. We would be likely to take note that there was also action in the hard disk during these operations. (Indeed, listening to the hard disk has always been one crude window into what a computer is doing.)

We might then theorize that the disk had something to do with the long-term memory that stores the databases and that the circuit board that is active during these operations was involved in transforming the data to be stored. This tells us approximately where and when the operations are taking place but relatively little about how these tasks are accomplished.

If the computer's registers (temporary memory locations) were connected to front-panel lights (as was the case with early computers), we would see certain patterns of light flickering that indicated rapid changes in the states of these registers during periods when the computer was analyzing data but relatively slow changes when the computer was transmitting data. We might then theorize that these lights reflected changes in logic state during some kind of analytic behavior. Such insights would be accurate but crude and would fail to provide us with a theory of operation or any insights as to how information is actually coded or transformed.

The hypothetical situation described above mirrors the sort of efforts that have been undertaken to scan and model the human brain with the crude tools that have historically been available. Most models based on contemporary brain-scanning research (utilizing such methods as fMRI, MEG, and others discussed below) are only suggestive of the underlying mechanisms. Although these studies are valuable, their crude spatial and temporal resolution is not adequate for reverse engineering the salient features of the brain.

New Tools for Scanning the Brain. Now imagine, in our computer example above, that we are able to actually place precise sensors at specific points in the circuitry and that these sensors are capable of tracking specific signals at very high speeds. We would now have the tools needed to follow the actual information being transformed in real time, and we would be able to create a detailed description of how the circuits actually work. This is, in fact, exactly how electrical engineers go about understanding and debugging circuits such as computer boards (to reverse engineer a competitor's product, for example), using logic analyzers that visualize computer signals.

Neuroscience has not yet had access to sensor technology that would achieve this type of analysis, but that situation is about to change. Our tools for peering into our brains are improving at an exponential pace. The resolution of noninvasive brain-scanning devices is doubling about every twelve months (per unit volume).[31]

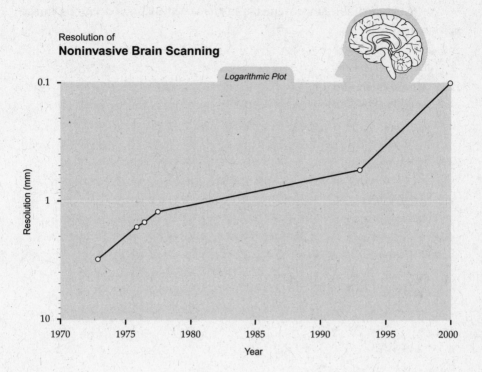

Resolution of
Noninvasive Brain Scanning

We see comparable improvements in the speed of brain scanning image reconstruction:

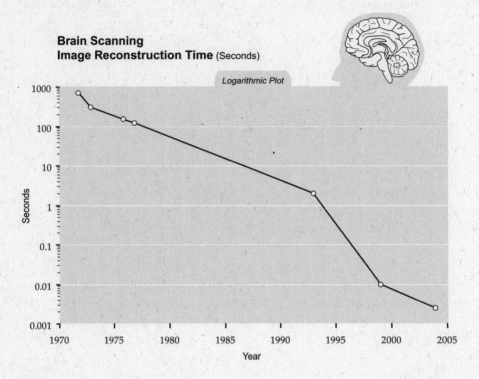

Brain Scanning
Image Reconstruction Time (Seconds)

The most commonly used brain-scanning tool is fMRI, which provides relatively high spatial resolution of one to three millimeters (not high enough to image individual neurons) but low temporal (time) resolution of a few seconds. Recent generations of fMRI technology provide time resolution of about one second, or a tenth of a second for a thin brain slice.

Another commonly used technique is MEG, which measures weak magnetic fields outside the skull, produced principally by the pyramidal neurons of the cortex. MEG is capable of rapid (one millisecond) temporal resolution but only very crude spatial resolution, about one centimeter.

Fritz Sommer, a principal investigator at Redwood Neuroscience Institute, is developing methods of combining fMRI and MEG to improve the spatiotemporal precision of the measurements. Other recent advances have demonstrated fMRI techniques capable of mapping regions called columnar and laminar structures, which are only a fraction of a millimeter wide, and of detecting tasks that take place in tens of milliseconds.[32]

fMRI and a related scanning technique using positrons called positron-emission tomography (PET) both gauge neuronal activity through indirect means. PET measures regional cerebral blood flow (rCBF), while fMRI measures blood-oxygen levels.[33] Although the relationship of these blood-flow amounts to neural activity is the subject of some controversy, the consensus is that they reflect local synaptic activity, not the spiking of neurons. The relationship of neural activity to blood flow was first articulated in the late nineteenth century.[34] A limitation of fMRI, however, is that the relationship of blood flow to synaptic activity is not direct: a variety of metabolic mechanisms affect the relationship between the two phenomena.

However, both PET and fMRI are believed to be most reliable for measuring relative changes in brain state. The primary method they use is the "subtraction paradigm," which can show regions that are most active during particular tasks.[35] This procedure involves subtracting data produced by a scan when the subject is *not* performing an activity from data produced while the subject *is* performing a specified mental activity. The difference represents the change in brain state.

An invasive technique that provides high spatial and temporal resolution is "optical imaging," which involves removing part of the skull, staining the living brain tissue with a dye that fluoresces upon neural activity, and then imaging the emitted light with a digital camera. Since optical imaging requires surgery, it has been used mainly in animal, particularly mouse, experiments.

Another approach to identifying brain functionality in different regions is transcranial magnetic stimulation (TMS), which involves applying a strong-pulsed magnetic field from outside the skull, using a magnetic coil precisely positioned over the head. By either stimulating or inducing a "virtual lesion" of (by temporarily disabling) small regions of the brain, skills can be diminished or enhanced.[36] TMS can also be used to study the relationship of different areas of the brain on specific tasks and can even induce sensations of mystical experiences.[37] Brain scientist Allan Snyder has reported that about 40 percent of his test subjects hooked up to TMS display significant new skills, many of which are remarkable, such as drawing abilities.[38]

If we have the option of destroying the brain that we are scanning, dramatically higher spatial resolution becomes possible. Scanning a frozen brain is feasible today, though not yet at sufficient speed or bandwidth to fully map all interconnections. But again, in accordance with the law of accelerating returns, this potential is growing exponentially, as are all other facets of brain scanning.

Carnegie Mellon University's Andreas Nowatzyk is scanning the nervous system of the brain and body of a mouse with a resolution of less than two hundred nanometers, which is approaching the resolution needed for full reverse engineering. Another destructive scanner called the "Brain Tissue Scanner" developed at the Brain Networks Laboratory at Texas A&M University is able to scan an entire mouse brain at a resolution of 250 nanometers in one month, using slices.[39]

Improving Resolution. Many new brain-scanning technologies now in development are dramatically improving both temporal and spatial resolution. This new generation of sensing and scanning systems is providing the tools needed to develop models with unprecedented fine levels of detail. Following is a small sample of these emerging imaging and sensing systems.

One particularly exciting new scanning camera is being developed at the University of Pennsylvania Neuroengineering Research Laboratory, led by Leif H. Finkel.[40] The optical system's projected spatial resolution will be high enough to image individual neurons and at one-millisecond time resolution, which is sufficient to record the firing of each neuron.

Initial versions are able to scan about one hundred cells simultaneously, at a depth of up to ten microns from the camera. A future version will image up to one thousand simultaneous cells, at a distance of up to 150 microns from the camera and at submillisecond time resolution. The system can scan neural tissue in vivo (in a living brain) while an animal is engaged in a mental task, although the brain surface must be exposed. The neural tissue is stained to generate voltage-dependent fluorescence, which is picked up by the high-resolution camera. The scanning system will be used to examine the brains of animals before and after they learn specific perceptual skills. This system combines the fast (one millisecond) temporary resolution of MEG while being able to image individual neurons and connections.

Methods have also been developed to noninvasively activate neurons or even a specific part of a neuron in a temporally and spatially precise manner. One approach, involving photons, uses a direct "two-photon" excitation, called "two-photon laser scanning microscopy" (TPLSM).[41] This creates a single point of focus in three-dimensional space that allows very high-resolution scanning. It utilizes laser pulses lasting only one millionth of one billionth of a second (10^{-15} second) to detect the excitation of single synapses in the intact brain by measuring the intracellular calcium accumulation associated with the activation of synaptic receptors.[42] Although the method destroys an insignificant amount of tissue, it provides extremely high-resolution images of individual dendritic spines and synapses in action.

This technique has been used to perform ultraprecise intracellular surgery. Physicist Eric Mazur and his colleagues at Harvard University have demonstrated its ability to execute precise modifications of cells, such as severing an interneuronal connection or destroying a single mitochondrion (the cell's energy source) without affecting other cellular components. "It generates the heat of the sun," says Mazur's colleague Donald Ingber, "but only for quintillionths of a second, and in a very small space."

Another technique, called "multielectrode recording," uses an array of electrodes to record simultaneously the activity of a large number of neurons with very high (submillisecond) temporal resolution.[43] Also, a noninvasive technique called second-harmonic generation (SHG) microscopy is able "to study cells in action," explains lead developer Daniel Dombeck, a graduate student at Cornell University. Yet another technique, called optical coherence imaging (OCI), uses coherent light (lightwaves that are all aligned in the same phase) to create holographic three-dimensional images of cell clusters.

Scanning Using Nanobots. Although these largely noninvasive means of scanning the brain from outside the skull are rapidly improving, the most powerful approach to capturing every salient neural detail will be to scan it from inside. By the 2020s nanobot technology will be viable, and brain scanning will be one of its prominent applications. As described earlier nanobots are robots that will be the size of human blood cells (seven to eight microns) or even smaller.[44] Billions of them could travel through every brain capillary, scanning each relevant neural feature from up close. Using high-speed wireless communication, the nanobots would communicate with one another and with computers compiling the brain-scan database. (In other words, the nanobots and computers will all be on a wireless local area network.)[45]

A key technical challenge to interfacing nanobots with biological brain structures is the blood-brain barrier (BBB). In the late nineteenth century, scientists discovered that when they injected blue dye into an animal's bloodstream, all the organs of the animal turned blue with the exception of the spinal cord and brain. They accurately hypothesized a barrier that protects the brain from a wide range of potentially harmful substances in the blood, including bacteria, hormones, chemicals that may act as neurotransmitters, and other toxins. Only oxygen, glucose, and a very select set of other small molecules are able to leave the blood vessels and enter the brain.

Autopsies early in the twentieth century revealed that the lining of the capillaries in the brain and other nervous-system tissues is indeed packed much more tightly with endothelial cells than comparable-size vessels in other organs. More recent studies have shown that the BBB is a complex system that features gateways complete with keys and passwords that allow entry into the brain. For example, two proteins called zonulin and zot have been discovered that react with receptors in the brain to temporarily open the BBB at select sites. These two proteins play a similar role in opening receptors in the small intestine to allow digestion of glucose and other nutrients.

Any design for nanobots to scan or otherwise interact with the brain will have to consider the BBB. I describe here several strategies that will be workable, given future capabilities. Undoubtedly, others will be developed over the next quarter century.

- An obvious tactic is to make the nanobot small enough to glide through the BBB, but this is the least practical approach, at least with nanotechnology as we envision it today. To do this, the nanobot would have to be twenty nanometers or less in diameter, which is about the size of one hundred carbon atoms. Limiting a nanobot to these dimensions would severely limit its functionality.

- An intermediate strategy would be to keep the nanobot in the bloodstream but to have it project a robotic arm through the BBB and into the extracellular fluid that lines the neural cells. This would allow the nanobot to remain large enough to have sufficient computational and navigational resources. Since almost all neurons lie within two or three cell-widths of a capillary, the arm would need to reach only up to about fifty microns. Analyses conducted by Rob Freitas and others show that it is quite feasible to restrict the width of such a manipulator to under twenty nanometers.

- Another approach is to keep the nanobots in the capillaries and use noninvasive scanning. For example, the scanning system being designed by Finkel and his associates can scan at very high resolution (sufficient to see individual interconnections) to a depth of 150 microns, which is several times greater than we need. Obviously this type of optical-imaging system would have to be significantly miniaturized (compared to contemporary designs), but it uses charge-coupled device sensors, which are amenable to such size reduction.

- Another type of noninvasive scanning would involve one set of nanobots emitting focused signals similar to those of a two-photon scanner and another set of nanobots receiving the transmission. The topology of the intervening tissue could be determined by analyzing the impact on the received signal.

- Another type of strategy, suggested by Robert Freitas, would be for the nanobot literally to barge its way past the BBB by breaking a hole in it, exit the blood vessel, and then repair the damage. Since the nanobot can be constructed using carbon in a diamondoid configuration, it would be far stronger than biological tissues. Freitas writes, "To pass between cells in cell-rich tissue, it is necessary for an advancing nanorobot to disrupt some minimum number of cell-to-cell adhesive contacts that lie ahead in its path. After that, and with the objective of minimizing biointrusiveness, the nanorobot must reseal those adhesive contacts in its wake, crudely analogous to a burrowing mole."[46]

- Yet another approach is suggested by contemporary cancer studies. Cancer researchers are keenly interested in selectively disrupting the BBB to transport cancer-destroying substances to tumors. Recent studies of the BBB show that it opens up in response to a variety of factors, which include certain proteins, as mentioned above; localized hypertension; high concentrations of certain substances; microwaves and other forms of radiation; infection; and inflammation. There are also specialized processes that ferry out needed substances such as glucose. It has also been found that the sugar mannitol causes a temporary shrinking of the tightly packed endothelial cells to provide a temporary breach of the BBB. By exploiting these mechanisms, several research groups are developing compounds that open the BBB.[47] Although this research is aimed at cancer therapies, similar approaches can be used to open the gateways for nanobots that will scan the brain as well as enhance our mental functioning.

- We could bypass the bloodstream and the BBB altogether by injecting the nanobots into areas of the brain that have direct access to neural tissue. As I mention below, new neurons migrate from the ventricles to other parts of the brain. Nanobots could follow the same migration path.

- Rob Freitas has described several techniques for nanobots to monitor sensory signals.[48] These will be important both for reverse engineering the inputs to the brain, as well as for creating full-immersion virtual reality from within the nervous system.

  ▷ To scan and monitor auditory signals, Freitas proposes "mobile nanodevices . . . [that] swim into the spiral artery of the ear and down through its bifurcations to reach the cochlear canal, then position themselves as neural monitors in the vicinity of the spiral nerve fibers and the nerves entering the epithelium of the organ of Corti [cochlear or auditory nerves] within the spiral ganglion. These monitors can detect, record, or rebroadcast to other nanodevices in the communications network all auditory neural traffic perceived by the human ear."

  ▷ For the body's "sensations of gravity, rotation, and acceleration," he envisions "nanomonitors positioned at the afferent nerve endings emanating from hair cells located in the . . . semicircular canals."

▷ For "kinesthetic sensory management . . . motor neurons can be monitored to keep track of limb motions and positions, or specific muscle activities, and even to exert control."

▷ "Olfactory and gustatory sensory neural traffic may be eavesdropped [on] by nanosensory instruments."

▷ "Pain signals may be recorded or modified as required, as can mechanical and temperature nerve impulses from . . . receptors located in the skin."

▷ Freitas points out that the retina is rich with small blood vessels, "permitting ready access to both photoreceptor (rod, cone, bipolar and ganglion) and integrator . . . neurons." The signals from the optic nerve represent more than one hundred million levels per second, but this level of signal processing is already manageable. As MIT's Tomaso Poggio and others have indicated, we do not yet understand the coding of the optic nerve's signals. Once we have the ability to monitor the signals for each discrete fiber in the optic nerve, our ability to interpret these signals will be greatly facilitated. This is currently an area of intense research.

As I discuss below, the raw signals from the body go through multiple levels of processing before being aggregated in a compact dynamic representation in two small organs called the right and left insula, located deep in the cerebral cortex. For full-immersion virtual reality, it may be more effective to tap into the already-interpreted signals in the insula rather than the unprocessed signals throughout the body.

Scanning the brain for the purpose of reverse engineering its principles of operation is an easier action than scanning it for the purpose of "uploading" a particular personality, which I discuss further below (see the "Uploading the Human Brain" section, p. 198). In order to reverse engineer the brain, we only need to scan the connections in a region sufficiently to understand their basic pattern. We do not need to capture every single connection.

Once we understand the neural wiring patterns within a region, we can combine that knowledge with a detailed understanding of how each type of neuron in that region operates. Although a particular region of the brain may have billions of neurons, it will contain only a limited number of neuron types. We have already made significant progress in deriving the mechanisms underlying specific varieties of neurons and synaptic connections by studying these cells in vitro (in a test dish), as well as in vivo using such methods as two-photon scanning.

The scenarios above involve capabilities that exist at least in an early stage today. We already have technology capable of producing very high-resolution scans for viewing the precise shape of every connection in a particular brain area, if the scanner is physically proximate to the neural features. With regard to nanobots, there are already four major conferences dedicated to developing blood cell–size devices for diagnostic and therapeutic purposes.[49] As discussed in chapter 2, we can project the exponentially declining cost of computation and the rapidly declining size and increasing effectiveness of both electronic and mechanical technologies. Based on these projections, we can conservatively anticipate the requisite nanobot technology to implement these types of scenarios during the 2020s. Once nanobot-based scanning becomes a reality, we will finally be in the same position that circuit designers are in today: we will be able to place highly sensitive and very high-resolution sensors (in the form of nanobots) at millions or even billions of locations in the brain and thus witness in breathtaking detail living brains in action.

Building Models of the Brain

If we were magically shrunk and put into someone's brain while she was thinking, we would see all the pumps, pistons, gears and levers working away, and we would be able to describe their workings completely, in mechanical terms, thereby completely describing the thought processes of the brain. But that description would nowhere contain any mention of thought! It would contain nothing but descriptions of pumps, pistons, levers!

—G. W. LEIBNIZ (1646–1716)

How do . . . fields express their principles? Physicists use terms like photons, electrons, quarks, quantum wave function, relativity, and energy conservation. Astronomers use terms like planets, stars, galaxies, Hubble shift, and black holes. Thermodynamicists use terms like entropy, first law, second law, and Carnot cycle. Biologists use terms like phylogeny, ontogeny, DNA, and enzymes. Each of these terms is actually the title of a story! The principles of a field are actually a set of interwoven stories about the structure and behavior of field elements.

—PETER J. DENNING, PAST PRESIDENT OF THE ASSOCIATION FOR COMPUTING MACHINERY, IN "GREAT PRINCIPLES OF COMPUTING"

It is important that we build models of the brain at the right level. This is, of course, true for all of our scientific models. Although chemistry is theoretically based on physics and could be derived entirely from physics, this would be unwieldy and infeasible in practice. So chemistry uses its own rules and models. We should likewise, in theory, be able to deduce the laws of thermodynamics from physics, but this is a far-from-straightforward process. Once we have a sufficient number of particles to call something a gas rather than a bunch of particles, solving equations for each particle interaction becomes impractical, whereas the laws of thermodynamics work extremely well. The interactions of a single molecule within the gas are hopelessly complex and unpredictable, but the gas itself, comprising trillions of molecules, has many predictable properties.

Similarly, biology, which is rooted in chemistry, uses its own models. It is often unnecessary to express higher-level results using the intricacies of the dynamics of the lower-level systems, although one has to thoroughly understand the lower level before moving to the higher one. For example, we can control certain genetic features of an animal by manipulating its fetal DNA without necessarily understanding all of the biochemical mechanisms of DNA, let alone the interactions of the atoms in the DNA molecule.

Often, the lower level is more complex. A pancreatic islet cell, for example, is enormously complicated, in terms of all its biochemical functions (most of which apply to all human cells, some to all biological cells). Yet modeling what a pancreas does—with its millions of cells—in terms of regulating levels of insulin and digestive enzymes, although not simple, is considerably less difficult than formulating a detailed model of a single islet cell.

The same issue applies to the levels of modeling and understanding in the brain, from the physics of synaptic reactions up to the transformations of information by neural clusters. In those brain regions for which we have succeeded in developing detailed models, we find a phenomenon similar to that involving pancreatic cells. The models are complex but remain simpler than the mathematical descriptions of a single cell or even a single synapse. As we discussed earlier, these region-specific models also require significantly less computation than is theoretically implied by the computational capacity of all of the synapses and cells.

Gilles Laurent of the California Institute of Technology observes, "In most cases, a system's collective behavior is very difficult to deduce from knowledge of its components. . . . [N]euroscience is . . . a science of systems in which first-order and local explanatory schemata are needed but not sufficient." Brain reverse-engineering will proceed by iterative refinement of both top-to-bottom and bottom-to-top models and simulations, as we refine each level of description and modeling.

Until very recently neuroscience was characterized by overly simplistic models limited by the crudeness of our sensing and scanning tools. This led many observers to doubt whether our thinking processes were inherently capable of understanding themselves. Peter D. Kramer writes, "If the mind were simple enough for us to understand, we would be too simple to understand it."[50] Earlier, I quoted Douglas Hofstadter's comparison of our brain to that of a giraffe, the structure of which is not that different from a human brain but which clearly does not have the capability of understanding its own methods. However, recent success in developing highly detailed models at various levels—from neural components such as synapses to large neural regions such as the cerebellum—demonstrate that building precise mathematical models of our brains and then simulating these models with computation is a challenging but viable task once the data capabilities become available. Although models have a long history in neuroscience, it is only recently that they have become sufficiently comprehensive and detailed to allow simulations based on them to perform like actual brain experiments.

Subneural Models: Synapses and Spines

In an address to the annual meeting of the American Psychological Association in 2002, psychologist and neuroscientist Joseph LeDoux of New York University said,

> If who we are is shaped by what we remember, and if memory is a function of the brain, then synapses—the interfaces through which neurons communicate with each other and the physical structures in which memories are encoded—are the fundamental units of the self. . . . Synapses are pretty low on the totem pole of how the brain is organized, but I think they're pretty important. . . . The self is the sum of the brain's individual subsystems, each with its own form of "memory," together with the complex interactions among the subsystems. Without synaptic plasticity—the ability of synapses to alter the ease with which they transmit signals from one neuron to another—the changes in those systems that are required for learning would be impossible.[51]

Although early modeling treated the neuron as the primary unit of transforming information, the tide has turned toward emphasizing its subcellular components. Computational neuroscientist Anthony J. Bell, for example, argues:

> Molecular and biophysical processes control the sensitivity of neurons to incoming spikes (both synaptic efficiency and post-synaptic responsivity), the excitability of the neuron to produce spikes, the patterns of

spikes it can produce and the likelihood of new synapses forming (dynamic rewiring), to list only four of the most obvious interferences from the subneural level. Furthermore, transneural volume effects such as local electric fields and the transmembrane diffusion of nitric oxide have been seen to influence, responsively, coherent neural firing, and the delivery of energy (blood flow) to cells, the latter of which directly correlates with neural activity. The list could go on. I believe that anyone who seriously studies neuromodulators, ion channels, or synaptic mechanism and is honest, would have to reject the neuron level as a separate computing level, even while finding it to be a useful descriptive level.[52]

Indeed, an actual brain synapse is far more complex than is described in the classic McCulloch-Pitts neural-net model. The synaptic response is influenced by a range of factors, including the action of multiple channels controlled by a variety of ionic potentials (voltages) and multiple neurotransmitters and neuromodulators. Considerable progress has been made in the past twenty years, however, in developing the mathematical formulas underlying the behavior of neurons, dendrites, synapses, and the representation of information in the spike trains (pulses by neurons that have been activated). Peter Dayan and Larry Abbott have recently written a summary of the existing nonlinear differential equations that describe a wide range of knowledge derived from thousands of experimental studies.[53] Well-substantiated models exist for the biophysics of neuron bodies, synapses, and the action of feedforward networks of neurons, such as those found in the retina and optic nerves, and many other classes of neurons.

Attention to how the synapse works has its roots in Hebb's pioneering work. Hebb addressed the question, How does short-term (also called working) memory function? The brain region associated with short-term memory is the prefrontal cortex, although we now realize that different forms of short-term information retention have been identified in most other neural circuits that have been closely studied.

Most of Hebb's work focused on changes in the state of synapses to strengthen or inhibit received signals and on the more controversial reverberatory circuit in which neurons fire in a continuous loop.[54] Another theory proposed by Hebb is a change in state of a neuron itself—that is, a memory function in the cell soma (body). The experimental evidence supports the possibility of all of these models. Classical Hebbian synaptic memory and reverberatory memory require a time delay before the recorded information can be used. In vivo experiments show that in at least some regions of the brain there is a neural response that is too fast to be accounted for by such standard learning models, and therefore could only be accomplished by learning-induced changes in the soma.[55]

Another possibility not directly anticipated by Hebb is real-time changes in the neuron connections themselves. Recent scanning results show rapid growth of dendrite spikes and new synapses, so this must be considered an important mechanism. Experiments have also demonstrated a rich array of learning behaviors on the synaptic level that go beyond simple Hebbian models. Synapses can change their state rapidly, but they then begin to decay slowly with continued stimulation, or in some a lack of stimulation, or many other variations.[56]

Although contemporary models are far more complex than the simple synapse models devised by Hebb, his intuitions have largely proved correct. In addition to Hebbian synaptic plasticity, current models include global processes that provide a regulatory function. For example, synaptic scaling keeps synaptic potentials from becoming zero (and thus being unable to be increased through multiplicative approaches) or becoming excessively high and thereby dominating a network. In vitro experiments have found synaptic scaling in cultured networks of neocortical, hippocampal, and spinal-cord neurons.[57] Other mechanisms are sensitive to overall spike timing and the distribution of potential across many synapses. Simulations have demonstrated the ability of these recently discovered mechanisms to improve learning and network stability.

The most exciting new development in our understanding of the synapse is that the topology of the synapses and the connections they form are continually changing. Our first glimpse into the rapid changes in synaptic connections was revealed by an innovative scanning system that requires a genetically modified animal whose neurons have been engineered to emit a fluorescent green light. The system can image living neural tissue and has a sufficiently high resolution to capture not only the dendrites (interneuronal connections) but the spines: tiny projections that sprout from the dendrites and initiate potential synapses.

Neurobiologist Karel Svoboda and his colleagues at Cold Spring Harbor Laboratory on Long Island used the scanning system on mice to investigate networks of neurons that analyze information from the whiskers, a study that provided a fascinating look at neural learning. The dendrites continually grew new spines. Most of these lasted only a day or two, but on occasion a spine would remain stable. "We believe that the high turnover that we see might play an important role in neural plasticity, in that the sprouting spines reach out to probe different presynaptic partners on neighboring neurons," said Svoboda. "If a given connection is favorable, that is, reflecting a desirable kind of brain rewiring, then these synapses are stabilized and become more permanent. But most of these synapses are not going in the right direction, and they are retracted."[58]

Another consistent phenomenon that has been observed is that neural responses decrease over time, if a particular stimulus is repeated. This adaptation gives greatest priority to new patterns of stimuli. Similar work by neurobiologist Wen-Biao Gan at New York University's School of Medicine on neuronal spines in the visual cortex of adult mice shows that this spine mechanism can hold long-term memories: "Say a 10-year-old kid uses 1,000 connections to store a piece of information. When he is 80, one-quarter of the connections will still be there, no matter how things change. That's why you can still remember your childhood experiences." Gan also explains, "Our idea was that you actually don't need to make many new synapses and get rid of old ones when you learn, memorize. You just need to modify the strength of the preexisting synapses for short-term learning and memory. However, it's likely that [a] few synapses are made or eliminated to achieve long-term memory."[59]

The reason memories can remain intact even if three quarters of the connections have disappeared is that the coding method used appears to have properties similar to those of a hologram. In a hologram, information is stored in a diffuse pattern throughout an extensive region. If you destroy three quarters of the hologram, the entire image remains intact, although with only one quarter of the resolution. Research by Pentti Kanerva, a neuroscientist at Redwood Neuroscience Institute, supports the idea that memories are dynamically distributed throughout a region of neurons. This explains why older memories persist but nonetheless appear to "fade," because their resolution has diminished.

Neuron Models

Researchers are also discovering that specific neurons perform special recognition tasks. An experiment with chickens identified brain-stem neurons that detect particular delays as sounds arrive at the two ears.[60] Different neurons respond to different amounts of delay. Although there are many complex irregularities in how these neurons (and the networks they rely on) work, what they are actually accomplishing is easy to describe and would be simple to replicate. According to University of California at San Diego neuroscientist Scott Makeig, "Recent neurobiological results suggest an important role of precisely synchronized neural inputs in learning and memory."[61]

Electronic Neurons. A recent experiment at the University of California at San Diego's Institute for Nonlinear Science demonstrates the potential for electronic neurons to precisely emulate biological ones. Neurons (biological or otherwise) are a prime example of what is often called chaotic computing. Each neuron acts in an essentially unpredictable fashion. When an entire network of neurons receives input (from the outside world or from other networks of neurons), the signaling among them appears at first to be frenzied and random. Over time, typically a fraction of a second or so, the chaotic interplay of the neurons dies down and a stable pattern of firing emerges. This pattern represents the "decision" of the neural network. If the neural network is performing a pattern-recognition task (and such tasks constitute the bulk of the activity in the human brain), the emergent pattern represents the appropriate recognition.

So the question addressed by the San Diego researchers was: could electronic neurons engage in this chaotic dance alongside biological ones? They connected artificial neurons with real neurons from spiny lobsters in a single network, and their hybrid biological-nonbiological network performed in the same way (that is, chaotic interplay followed by a stable emergent pattern) and with the same type of results as an all-biological net of neurons. Essentially, the biological neurons accepted their electronic peers. This indicates that the chaotic mathematical model of these neurons was reasonably accurate.

Brain Plasticity

In 1861 French neurosurgeon Paul Broca correlated injured or surgically affected regions of the brain with certain lost skills, such as fine motor skills or language ability. For more than a century scientists believed these regions were hardwired for specific tasks. Although certain brain areas do tend to be used for particular types of skills, we now understand that such assignments can be changed in response to brain injury such as a stroke. In a classic 1965 study, D. H. Hubel and T. N. Wiesel showed that extensive and far-reaching reorganization of the brain could take place after damage to the nervous system, such as from a stroke.[62]

Moreover, the detailed arrangement of connections and synapses in a given region is a direct product of how extensively that region is used. As brain scanning has attained sufficiently high resolution to detect dendritic-spine growth and the formation of new synapses, we can see our brain grow and adapt to literally follow our thoughts. This gives new shades of meaning to Descartes' dictum "I think therefore I am."

In Vivo Images of Neural Dendrites
Showing Spine and Synapse Formation

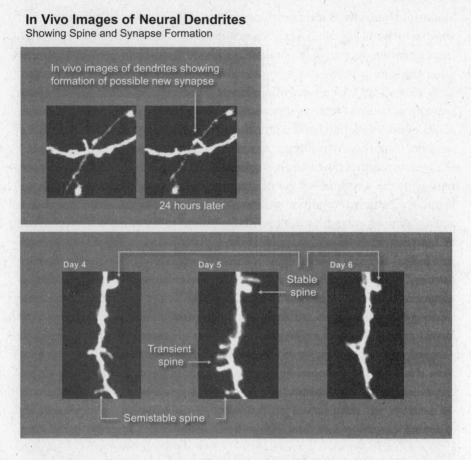

In vivo images of dendrites showing formation of possible new synapse

24 hours later

Day 4

Day 5

Stable spine

Day 6

Transient spine

Semistable spine

In one experiment conducted by Michael Merzenich and his colleagues at the University of California at San Francisco, monkeys' food was placed in such a position that the animals had to dexterously manipulate one finger to obtain it. Brain scans before and after revealed dramatic growth in the interneuronal connections and synapses in the region of the brain responsible for controlling that finger.

Edward Taub at the University of Alabama studied the region of the cortex responsible for evaluating the tactile input from the fingers. Comparing nonmusicians to experienced players of stringed instruments, he found no difference in the brain regions devoted to the fingers of the right hand but a huge difference for the fingers of the left hand. If we drew a picture of the hands based on the amount of brain tissue devoted to analyzing touch, the musicians' fingers on their left hand (which are used to control the strings) would be huge. Although the difference was greater for those musicians who began musical training with a stringed instrument as children, "even if you take up the violin at 40," Taub commented, "you still get brain reorganization."[63]

A similar finding comes from an evaluation of a software program, developed by Paula Tallal and Steve Miller at Rutgers University, called Fast ForWord, that assists dyslexic students. The program reads text to children, slowing down staccato phonemes such as "b" and "p," based on the observation that many dyslexic students are unable to perceive these sounds when spoken quickly. Being read to with this modified form of speech has been shown to help such children learn to read. Using fMRI scanning John Gabrieli of Stanford University found that the left prefrontal region of the brain, an area associated with language processing, had indeed grown and showed greater activity in dyslexic students using the program. Says Tallal, "You create your brain from the input you get."

It is not even necessary to express one's thoughts in physical action to provoke the brain to rewire itself. Dr. Alvaro Pascual-Leone at Harvard University scanned the brains of volunteers before and after they practiced a simple piano exercise. The brain motor cortex of the volunteers changed as a direct result of their practice. He then had a second group just think about doing the piano exercise but without actually moving any muscles. This produced an equally pronounced change in the motor-cortex network.[64]

Recent fMRI studies of learning visual-spatial relationships found that interneuronal connections are able to change rapidly during the course of a single learning session. Researchers found changes in the connections between posterior parietal-cortex cells in what is called the "dorsal" pathway (which contains information about location and spatial properties of visual stimuli) and posterior inferior-temporal cortex cells in the "ventral" pathway (which contains recognized invariant features of varying levels of abstraction);[65] significantly, that rate of change was directly proportional to the rate of learning.[66]

Researchers at the University of California at San Diego reported a key insight into the difference in the formation of short-term and long-term memories. Using a high-resolution scanning method, the scientists were able to see chemical changes within synapses in the hippocampus, the brain region associated with the formation of long-term memories.[67] They discovered that when a cell was first stimulated, actin, a neurochemical, moved toward the neurons to which the synapse was connected. This also stimulated the actin in neighboring cells to move away from the activated cell. These changes lasted only a few minutes, however. If the stimulations were sufficiently repeated, then a more significant and permanent change took place.

"The short-term changes are just part of the normal way the nerve cells talk to each other," lead author Michael A. Colicos said.

The long-term changes in the neurons occur only after the neurons are stimulated four times over the course of an hour. The synapse will actually split and new synapses will form, producing a permanent change that will presumably last for the rest of your life. The analogy to human memory is that when you see or hear something once, it might stick in your mind for a few minutes. If it's not important, it fades away and you forget it 10 minutes later. But if you see or hear it again and this keeps happening over the next hour, you are going to remember it for a much longer time. And things that are repeated many times can be remembered for an entire lifetime. Once you take an axon and form two new connections, those connections are very stable and there's no reason to believe that they'll go away. That's the kind of change one would envision lasting a whole lifetime.

"It's like a piano lesson," says coauthor and professor of biology Yukiko Goda. "If you play a musical score over and over again, it becomes ingrained in your memory." Similarly, in an article in *Science* neuroscientists S. Lowel and W. Singer report having found evidence for rapid dynamic formation of new interneuronal connections in the visual cortex, which they described with Donald Hebb's phrase "What fires together wires together."[68]

Another insight into memory formation is reported in a study published in *Cell*. Researchers found that the CPEB protein actually changes its shape in synapses to record memories.[69] The surprise was that CPEB performs this memory function while in a prion state.

"For a while we've known quite a bit about how memory works, but we've had no clear concept of what the key storage device is," said coauthor and Whitehead Institute for Biomedical Research director Susan Lindquist. "This study suggests what the storage device might be—but it's such a surprising suggestion to find that a prion-like activity may be involved. . . . It . . . indicates that prions aren't just oddballs of nature but might participate in fundamental processes." As I reported in chapter 3, human engineers are also finding prions to be a powerful means of building electronic memories.

Brain-scanning studies are also revealing mechanisms to inhibit unneeded and undesirable memories, a finding that would gratify Sigmund Freud.[70] Using fMRI, Stanford University scientists asked study subjects to attempt to forget information that they had earlier memorized. During this activity, regions in the frontal cortex that have been associated with memory repression showed a high level of activity, while the hippocampus, the region normally associated with remembering, was relatively inactive. These findings "confirm the existence of an active forgetting process and establish a neurobiological model for guiding inquiry into motivated forgetting," wrote Stanford psychology professor John Gabrieli and his colleagues. Gabrieli also commented,

"The big news is that we've shown how the human brain blocks an unwanted memory, that there is such a mechanism, and it has a biological basis. It gets you past the possibility that there's nothing in the brain that would suppress a memory—that it was all a misunderstood fiction."

In addition to generating new connections between neurons, the brain also makes new neurons from neural stem cells, which replicate to maintain a reservoir of themselves. In the course of reproducing, some of the neural stem cells become "neural precursor" cells, which in turn mature into two types of support cells called astrocytes and oligodendrocytes, as well as neurons. The cells further evolve into specific types of neurons. However, this differentiation cannot take place unless the neural stem cells move away from their original source in the brain's ventricles. Only about half of the neural cells successfully make the journey, which is similar to the process during gestation and early childhood in which only a portion of the early brain's developing neurons survive. Scientists hope to bypass this neural migration process by injecting neural stem cells directly into target regions, as well as to create drugs that promote this process of neurogenesis (creating new neurons) to repair brain damage from injury or disease.[71]

An experiment by genetics researchers Fred Gage, G. Kempermann, and Henriette van Praag at the Salk Institute for Biological Studies showed that neurogenesis is actually stimulated by our experience. Moving mice from a sterile, uninteresting cage to a stimulating one approximately doubled the number of dividing cells in their hippocampus regions.[72]

Modeling Regions of the Brain

Most probably the human brain is, in the main, composed of large numbers of relatively small distributed systems, arranged by embryology into a complex society that is controlled in part (but only in part) by serial, symbolic systems that are added later. But the subsymbolic systems that do most of the work from underneath must, by their very character, block all the other parts of the brain from knowing much about how they work. And this, itself, could help explain how people do so many things yet have such incomplete ideas on how those things are actually done.

—Marvin Minsky and Seymour Papert[73]

Common sense is not a simple thing. Instead, it is an immense society of hard-earned practical ideas—of multitudes of life-learned rules and exceptions, dispositions and tendencies, balances and checks.

—Marvin Minsky

In addition to new insights into the plasticity of organization of each brain region, researchers are rapidly creating detailed models of particular regions of the brain. These neuromorphic models and simulations lag only slightly behind the availability of the information on which they are based. The rapid success of turning the detailed data from studies of neurons and the interconnection data from neural scanning into effective models and working simulations belies often-stated skepticism about our inherent capability of understanding our own brains.

Modeling human-brain functionality on a nonlinearity-by-nonlinearity and synapse-by-synapse basis is generally not necessary. Simulations of regions that store memories and skills in individual neurons and connections (for example, the cerebellum) do make use of detailed cellular models. Even for these regions, however, simulations require far less computation than is implied by all of the neural components. This is true of the cerebellum simulation described below.

Although there is a great deal of detailed complexity and nonlinearity in the subneural parts of each neuron, as well as a chaotic, semirandom wiring pattern underlying the trillions of connections in the brain, significant progress has been made over the past twenty years in the mathematics of modeling such adaptive nonlinear systems. Preserving the exact shape of every dendrite and the precise "squiggle" of every interneuronal connection is generally not necessary. We can understand the principles of operation of extensive regions of the brain by examining their dynamics at the appropriate level of analysis.

We have already had significant success in creating models and simulations of extensive brain regions. Applying tests to these simulations and comparing the data to that obtained from psychophysical experiments on actual human brains have produced impressive results. Given the relative crudeness of our scanning and sensing tools to date, the success in modeling, as illustrated by the following works in progress, demonstrates the ability to extract the right insights from the mass of data being gathered.

The following are only a few examples of successful models of brain regions, all works in progress.

A Neuromorphic Model: The Cerebellum

A question I examined in *The Age of Spiritual Machines* is: how does a ten-year-old manage to catch a fly ball?[74] All that a child can see is the ball's trajectory from his position in the outfield. To actually infer the path of the ball in three-dimensional space would require solving difficult simultaneous differential equations. Additional equations would need to be solved to predict the future course of the ball, and more equations to translate these results into what was required of the player's own movements. How does a young outfielder accomplish all of this in a few seconds with no computer and no training in differential equations? Clearly, he is not solving equations consciously, but how does his brain solve the problem?

Since *ASM* was published, we have advanced considerably in understanding this basic process of skill formation. As I had hypothesized, the problem is not solved by building a mental model of three-dimensional motion. Rather, the problem is collapsed by directly translating the observed movements of the ball into the appropriate movement of the player and changes in the configuration of his arms and legs. Alexandre Pouget of the University of Rochester and Lawrence H. Snyder of Washington University have described mathematical "basis functions" that can represent this direct transformation of perceived movement in the visual field to required movements of the muscles.[75] Furthermore, analysis of recently developed models of the functioning of the cerebellum demonstrate that our cerebellar neural circuits are indeed capable of learning and then applying the requisite basis functions to implement these sensorimotor transformations. When we engage in the trial-and-error process of learning to perform a sensorimotor task, such as catching a fly ball, we are training the synaptic potentials of the cerebellar synapses to learn the appropriate basis functions. The cerebellum performs two types of transformations with these basis functions: going from a desired result to an action (called "inverse internal models") and going from a possible set of actions to an anticipated result ("forward internal models"). Tomaso Poggio has pointed out that the idea of basis functions may describe learning processes in the brain that go beyond motor control.[76]

The gray and white, baseball-sized, bean-shaped brain region called the cerebellum sits on the brain stem and comprises more than half of the brain's neurons. It provides a wide range of critical functions, including sensorimotor coordination, balance, control of movement tasks, and the ability to anticipate the results of actions (our own as well as those of other objects and persons).[77] Despite its diversity of functions and tasks, its synaptic and cell organization is extremely consistent, involving only several types of neurons. There appears to be a specific type of computation that it accomplishes.[78]

Despite the uniformity of the cerebellum's information processing, the broad range of its functions can be understood in terms of the variety of inputs it receives from the cerebral cortex (via the brain-stem nuclei and then through the cerebellum's mossy fiber cells) and from other regions (particularly the "inferior olive" region of the brain via the cerebellum's climbing fiber cells). The cerebellum is responsible for our understanding of the timing and sequencing of sensory inputs as well as controlling our physical movements.

The cerebellum is also an example of how the brain's considerable capacity greatly exceeds its compact genome. Most of the genome that is devoted to the brain describes the detailed structure of each type of neural cell (including its dendrites, spines, and synapses) and how these structures respond to stimulation and change. Relatively little genomic code is responsible for the actual "wiring." In the cerebellum, the basic wiring method is repeated billions of times. It is clear that the genome does not provide specific information about each repetition of this cerebellar structure but rather specifies certain constraints as to how this structure is repeated (just as the genome does not specify the exact location of cells in other organs).

Massively Repeated Cerebellum Wiring Pattern

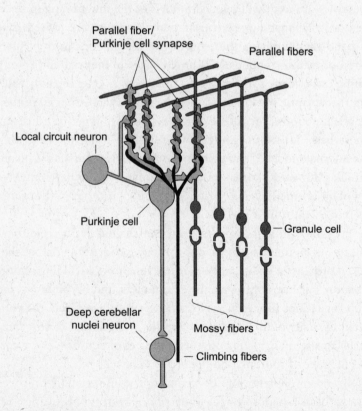

Some of the outputs of the cerebellum go to about two hundred thousand alpha motor neurons, which determine the final signals to the body's approximately six hundred muscles. Inputs to the alpha motor neurons do not directly specify the movements of each of these muscles but are coded in a more compact, as yet poorly understood, fashion. The final signals to the muscles are determined at lower levels of the nervous system, specifically in the brain stem and spinal cord.[79] Interestingly, this organization is taken to an extreme in the octopus, the central nervous system of which apparently sends very high-level commands to each of its arms (such as "grasp that object and bring it closer"), leaving it up to an independent peripheral nervous system in each arm to carry out the mission.[80]

A great deal has been learned in recent years about the role of the cerebellum's three principal nerve types. Neurons called "climbing fibers" appear to provide signals to train the cerebellum. Most of the output of the cerebellum comes from the large Purkinje cells (named for Johannes Purkinje, who identified the cell in 1837), each of which receives about two hundred thousand inputs (synapses), compared to the average of about one thousand for a typical neuron. The inputs come largely from the granule cells, which are the smallest neurons, packed about six million per square millimeter. Studies of the role of the cerebellum during the learning of handwriting movements by children show that the Purkinje cells actually sample the sequence of movements, with each one sensitive to a specific sample.[81] Obviously, the cerebellum requires continual perceptual guidance from the visual cortex. The researchers were able to link the structure of cerebellum cells to the observation that there is an inverse relationship between curvature and speed when doing handwriting—that is, you can write faster by drawing straight lines instead of detailed curves for each letter.

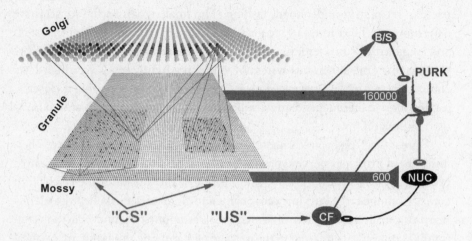

University of Texas Cerebellum Model and Simulation

Detailed cell studies and animal studies have provided us with impressive mathematical descriptions of the physiology and organization of the synapses of the cerebellum,[82] as well as of the coding of information in its inputs and outputs, and of the transformations performed.[83] Gathering data from multiple studies, Javier F. Medina, Michael D. Mauk, and their colleagues at the University of Texas Medical School devised a detailed bottom-up simulation of the cerebellum. It features more than ten thousand simulated neurons and three hundred thousand synapses, and it includes all of the principal types of cerebellum cells.[84] The connections of the cells and synapses are determined by a computer, which "wires" the simulated cerebellar region by following constraints and rules, similar to the stochastic (random within restrictions) method used to wire the actual human brain from its genetic code.[85] It would not be difficult to expand the University of Texas cerebellar simulation to a larger number of synapses and cells.

The Texas researchers applied a classical learning experiment to their simulation and compared the results to many similar experiments on actual human conditioning. In the human studies, the task involved associating an auditory tone with a puff of air applied to the eyelid, which causes the eyelid to close. If the puff of air and the tone are presented together for one hundred to two hundred trials, the subject will learn the association and close the subject's eye upon merely hearing the tone. If the tone is then presented many times without the air puff, the subject ultimately learns to disassociate the two stimuli (to "extinguish" the response), so the learning is bidirectional. After tuning a variety of parameters, the simulation provided a reasonable match to experimental results on human and animal cerebellar conditioning. Interestingly, the researchers found that if they created simulated cerebellar lesions (by removing portions of the simulated cerebellar network), they got results similar to those obtained in experiments on rabbits that had received actual cerebellar lesions.[86]

On account of the uniformity of this large region of the brain and the relative simplicity of its interneuronal wiring, its input-output transformations are relatively well understood, compared to those of other brain regions. Although the relevant equations still require refinement, this bottom-up simulation has proved quite impressive.

Another Example: Watts's Model of the Auditory Regions

> I believe that the way to create a brain-like intelligence is to build a real-time working model system, accurate in sufficient detail to express the essence of each computation that is being performed, and verify its correct operation against measurements of the real system. The model must run in real-time so that we will be forced to deal with inconvenient and complex real-world inputs that we might not otherwise think to present to it. The model must operate at sufficient resolution to be comparable to the real system, so that we build the right intuitions about what information is represented at each stage. Following Mead,[87] the model development necessarily begins at the boundaries of the system (i.e., the sensors) where the real system is well-understood, and then can advance into the less-understood regions. . . . In this way, the model can contribute fundamentally to our advancing understanding of the system, rather than simply mirroring the existing understanding. In the context of such great complexity, it is possible that the only practical way to understand the real system is to build a working model, from the sensors inward, building on our newly enabled ability to *visualize the complexity of the system* as we advance into it. Such an approach could be called *reverse-engineering of the brain*. . . . Note that I am not advocating a blind copying of structures whose purpose we don't understand, like the legendary Icarus who naively attempted to build wings out of feathers and wax. Rather, I am advocating that we respect the complexity and richness that is already well-understood at low levels, before proceeding to higher levels.
>
> —LLOYD WATTS[88]

A major example of neuromorphic modeling of a region of the brain is the comprehensive replica of a significant portion of the human auditory-processing system developed by Lloyd Watts and his colleagues.[89] It is based on neurobiological studies of specific neuron types as well as on information regarding interneuronal connection. The model, which has many of the same properties as human hearing and can locate and identify sounds, has five parallel paths of processing auditory information and includes the actual intermediate representations of this information at each stage of neural processing. Watts has implemented his model as real-time computer software which, though a work in progress, illustrates the feasibility of converting neurobiological models and brain connection data into working simulations. The software is not based on reproducing each individual neuron and connection, as is the cerebellum model described above, but rather the transformations performed by each region.

Watts's software is capable of matching the intricacies that have been revealed in subtle experiments on human hearing and auditory discrimination. Watts has used his model as a preprocessor (front end) in speech-recognition systems and has demonstrated its ability to pick out one speaker from background sounds (the "cocktail party effect"). This is an impressive feat of which humans are capable but up until now had not been feasible in automated speech-recognition systems.[90]

Like human hearing, Watts's cochlea model is endowed with spectral sensitivity (we hear better at certain frequencies), temporal responses (we are sensitive to the timing of sounds, which create the sensation of their spatial locations), masking, nonlinear frequency-dependent amplitude compression (which allows for greater dynamic range—the ability to hear both loud and quiet sounds), gain control (amplification), and other subtle features. The results it obtains are directly verifiable by biological and psychophysical data.

The next segment of the model is the cochlear nucleus, which Yale University professor of neuroscience and neurobiology Gordon M. Shepherd[91] has described as "one of the best understood regions of the brain."[92] Watts's simulation of the cochlear nucleus is based on work by E. Young that describes in detail "the essential cell types responsible for detecting spectral energy, broadband transients, fine tuning in spectral channels, enhancing sensitivity to temporary envelope in spectral channels, and spectral edges and notches, all while adjusting gain for optimum sensitivity within the limited dynamic range of the spiking neural code."[93]

The Watts model captures many other details, such as the interaural time difference (ITD) computed by the medial superior olive cells.[94] It also represents the interaural level difference (ILD) computed by the lateral superior olive cells and normalizations and adjustments made by the inferior colliculus cells.[95]

Reverse Engineering the Human Brain:

Five Parallel Auditory Pathways

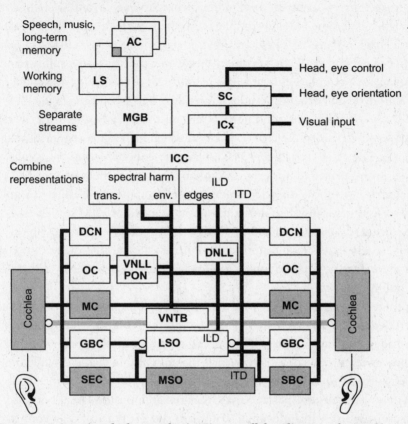

Reverse engineering the human brain: **Five parallel auditory pathways.**[96]

The Visual System

We've made enough progress in understanding the coding of visual information that experimental retina implants have been developed and surgically installed in patients.[97] However, because of the relative complexity of the visual system, our understanding of the processing of visual information lags behind our knowledge of the auditory regions. We have preliminary models of the transformations performed by two visual areas (called V1 and MT), although not at the individual neuron level. There are thirty-six other visual areas, and we will need to be able to scan these deeper regions at very high resolution or place precise sensors to ascertain their functions.

A pioneer in understanding visual processing is MIT's Tomaso Poggio, who has distinguished its two tasks as identification and categorization.[98] The former is relatively easy to understand, according to Poggio, and we have already designed experimental and commercial systems that are reasonably successful in identifying faces.[99] These are used as part of security systems to control entry of personnel and in bank machines. Categorization—the ability to differentiate, for example, between a person and a car or between a dog and a cat—is a more complex matter, although recently progress has been made.[100]

Early (in terms of evolution) layers of the visual system are largely a feedforward (lacking feedback) system in which increasingly sophisticated features are detected. Poggio and Maximilian Riesenhuber write that "single neurons in the macaque posterior inferotemporal cortex may be tuned to . . . a dictionary of thousands of complex shapes." Evidence that visual recognition uses a feedforward system during recognition includes MEG studies that show the human visual system takes about 150 milliseconds to detect an object. This matches the latency of feature-detection cells in the inferotemporal cortex, so there does not appear to be time for feedback to play a role in these early decisions.

Recent experiments have used a hierarchical approach in which features are detected to be analyzed by later layers of the system.[101] From studies on macaque monkeys, neurons in the inferotemporal cortex appear to respond to complex features of objects on which the animals are trained. While most of the neurons respond only to a particular view of the object, some are able to respond regardless of perspective. Other research on the visual system of the macaque monkey includes studies on many specific types of cells, connectivity patterns, and high-level descriptions of information flow.[102]

Extensive literature supports the use of what I call "hypothesis and test" in more complex pattern-recognition tasks. The cortex makes a guess about what it is seeing and then determines whether the features of what is actually in the field of view match its hypothesis.[103] We are often more focused on the hypothesis than the actual test, which explains why people often see and hear what they expect to perceive rather than what is actually there. "Hypothesis and test" is also a useful strategy in our computer-based pattern-recognition systems.

Although we have the illusion of receiving high-resolution images from our eyes, what the optic nerve actually sends to the brain is just outlines and clues about points of interest in our visual field. We then essentially hallucinate the world from cortical memories that interpret a series of extremely low-resolution movies that arrive in parallel channels. In a 2001 study published in *Nature*, Frank S. Werblin, professor of molecular and cell biology at the University of

California at Berkeley, and doctoral student Boton Roska, M.D., showed that the optic nerve carries ten to twelve output channels, each of which carries only minimal information about a given scene.[104] One group of what are called ganglion cells sends information only about edges (changes in contrast). Another group detects only large areas of uniform color, whereas a third group is sensitive only to the backgrounds behind figures of interest.

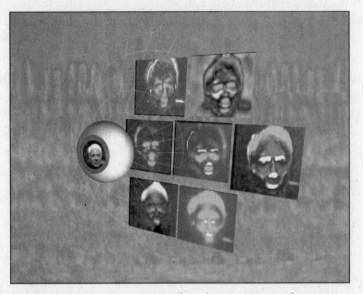

Seven of the dozen separate movies that the eye extracts from a scene and sends to the brain.

"Even though we think we see the world so fully, what we are receiving is really just hints, edges in space and time," says Werblin. "These 12 pictures of the world constitute all the information we will ever have about what's out there, and from these 12 pictures, which are so sparse, we reconstruct the richness of the visual world. I'm curious how nature selected these 12 simple movies and how it can be that they are sufficient to provide us with all the information we seem to need." Such findings promise to be a major advance in developing an artificial system that could replace the eye, retina, and early optic-nerve processing.

In chapter 3, I mentioned the work of robotics pioneer Hans Moravec, who has been reverse engineering the image processing done by the retina and early visual-processing regions in the brain. For more than thirty years Moravec has been constructing systems to emulate the ability of our visual system to build representations of the world. It has only been recently that sufficient processing

power has been available in microprocessors to replicate this human-level feature detection, and Moravec is applying his computer simulations to a new
generation of robots that can navigate unplanned, complex environments with
human-level vision.[105]

Carver Mead has been pioneering the use of special neural chips that utilize
transistors in their native analog mode, which can provide very efficient emulation of the analog nature of neural processing. Mead has demonstrated a chip
that performs the functions of the retina and early transformations in the optic
nerve using this approach.[106]

A special type of visual recognition is detecting motion, one of the focus
areas of the Max Planck Institute of Biology in Tübingen, Germany. The basic
research model is simple: compare the signal at one receptor with a time-
delayed signal at the adjacent receptor.[107] This model works for certain speeds
but leads to the surprising result that above a certain speed, increases in the
velocity of an observed object will decrease the response of this motion detector. Experimental results on animals (based on behavior and analysis of
neuronal outputs) and humans (based on reported perceptions) have closely
matched the model.

Other Works in Progress: An Artificial Hippocampus and an Artificial Olivocerebellar Region

The hippocampus is vital for learning new information and long-term storage
of memories. Ted Berger and his colleagues at the University of Southern California mapped the signal patterns of this region by stimulating slices of rat hippocampus with electrical signals millions of times to determine which input
produced a corresponding output.[108] They then developed a real-time mathematical model of the transformations performed by layers of the hippocampus
and programmed the model onto a chip.[109] Their plan is to test the chip in animals by first disabling the corresponding hippocampus region, noting the
resulting memory failure, and then determining whether that mental function
can be restored by installing their hippocampal chip in place of the disabled
region.

Ultimately, this approach could be used to replace the hippocampus in
patients affected by strokes, epilepsy, or Alzheimer's disease. The chip would be
located on a patient's skull, rather than inside the brain, and would communicate with the brain via two arrays of electrodes, placed on either side of the
damaged hippocampal section. One would record the electrical activity coming from the rest of the brain, while the other would send the necessary instructions back to the brain.

Another brain region being modeled and simulated is the olivocerebellar region, which is responsible for balance and coordinating the movement of limbs. The goal of the international research group involved in this effort is to apply their artificial olivocerebellar circuit to military robots as well as to robots that could assist the disabled.[110] One of their reasons for selecting this particular brain region was that "it's present in all vertebrates—it's very much the same from the most simple to the most complex brains," explains Rodolfo Llinas, one of the researchers and a neuroscientist at New York University Medical School. "The assumption is that it is conserved [in evolution] because it embodies a very intelligent solution. As the system is involved in motor coordination—and we want to have a machine that has sophisticated motor control—then the choice [of the circuit to mimic] was easy."

One of the unique aspects of their simulator is that it uses analog circuits. Similar to Mead's pioneering work on analog emulation of brain regions, the researchers found substantially greater performance with far fewer components by using transistors in their native analog mode.

One of the team's researchers, Ferdinando Mussa-Ivaldi, a neuroscientist at Northwestern University, commented on the applications of an artificial olivocerebellar circuit for the disabled: "Think of a paralyzed patient. It is possible to imagine that many ordinary tasks—such as getting a glass of water, dressing, undressing, transferring to a wheelchair—could be carried out by robotic assistants, thus providing the patient with more independence."

Understanding Higher-Level Functions: Imitation, Prediction, and Emotion

> Operations of thought are like cavalry charges in a battle—they are strictly limited in number, they require fresh horses, and must only be made at decisive moments.
>
> —ALFRED NORTH WHITEHEAD

> But the big feature of human-level intelligence is not what it does when it works but what it does when it's stuck.
>
> —MARVIN MINSKY

> If love is the answer, could you please rephrase the question?
>
> —LILY TOMLIN

Because it sits at the top of the neural hierarchy, the part of the brain least well understood is the cerebral cortex. This region, which consists of six thin layers in the outermost areas of the cerebral hemispheres, contains billions of neurons. According to Thomas M. Bartol Jr. of the Computational Neurobiology Laboratory of the Salk Institute of Biological Studies, "A single cubic millimeter of cerebral cortex may contain on the order of 5 billion . . . synapses of different shapes and sizes." The cortex is responsible for perception, planning, decision making and most of what we regard as conscious thinking.

Our ability to use language, another unique attribute of our species, appears to be located in this region. An intriguing hint about the origin of language and a key evolutionary change that enabled the formation of this distinguishing skill is the observation that only a few primates, including humans and monkeys, are able to use an (actual) mirror to master skills. Theorists Giacomo Rizzolatti and Michael Arbib hypothesized that language emerged from manual gestures (which monkeys—and, of course, humans—are capable of). Performing manual gestures requires the ability to mentally correlate the performance and observation of one's own hand movements.[111] Their "mirror system hypothesis" is that the key to the evolution of language is a property called "parity," which is the understanding that the gesture (or utterance) has the same meaning for the party making the gesture as for the party receiving it; that is, the understanding that what you see in a mirror is the same (although reversed left-to-right) as what is seen by someone else watching you. Other animals are unable to understand the image in a mirror in this fashion, and it is believed that they are missing this key ability to deploy parity.

A closely related concept is that the ability to imitate the movements (or, in the case of human babies, vocal sounds) of others is critical to developing language.[112] Imitation requires the ability to break down an observed presentation into parts, each of which can then be mastered through recursive and iterative refinement.

Recursion is the key capability identified in a new theory of linguistic competence. In Noam Chomsky's early theories of language in humans, he cited many common attributes that account for the similarities in human languages. In a 2002 paper by Marc Hauser, Noam Chomsky, and Tecumseh Fitch, the authors cite the single attribution of "recursion" as accounting for the unique language faculty of the human species.[113] Recursion is the ability to put together small parts into a larger chunk, and then use that chunk as a part in yet another structure and to continue this process iteratively. In this way, we are able to build the elaborate structures of sentences and paragraphs from a limited set of words.

Another key feature of the human brain is the ability to make predictions, including predictions about the results of its own decisions and actions. Some scientists believe that prediction is the primary function of the cerebral cortex, although the cerebellum also plays a major role in the prediction of movement.

Interestingly, we are able to predict or anticipate our own decisions. Work by physiology professor Benjamin Libet at the University of California at Davis shows that neural activity to initiate an action actually occurs about a third of a second before the brain has made the decision to take the action. The implication, according to Libet, is that the decision is really an illusion, that "consciousness is out of the loop." The cognitive scientist and philosopher Daniel Dennett describes the phenomenon as follows: "The action is originally precipitated in some part of the brain, and off fly the signals to muscles, pausing en route to tell you, the conscious agent, what is going on (but like all good officials letting you, the bumbling president, maintain the illusion that you started it all)."[114]

A related experiment was conducted recently in which neurophysiologists electronically stimulated points in the brain to induce particular emotional feelings. The subjects immediately came up with a rationale for experiencing those emotions. It has been known for many years that in patients whose left and right brains are no longer connected, one side of the brain (usually the more verbal left side) will create elaborate explanations ("confabulations") for actions initiated by the other side, as if the left side were the public-relations agent for the right side.

The most complex capability of the human brain—what I would regard as its cutting edge—is our emotional intelligence. Sitting uneasily at the top of our brain's complex and interconnected hierarchy is our ability to perceive and respond appropriately to emotion, to interact in social situations, to have a moral sense, to get the joke, and to respond emotionally to art and music, among other high-level functions. Obviously, lower-level functions of perception and analysis feed into our brain's emotional processing, but we are beginning to understand the regions of the brain and even to model the specific types of neurons that handle such issues.

These recent insights have been the result of our attempts to understand how human brains differ from those of other mammals. The answer is that the differences are slight but critical, and they help us discern how the brain processes emotion and related feelings. One difference is that humans have a larger cortex, reflecting our stronger capability for planning, decision making, and other forms of analytic thinking. Another key distinguishing feature is that emotionally charged situations appear to be handled by special cells called spindle cells, which are found only in humans and some great apes. These neural cells are large, with long neural filaments called apical dendrites that connect extensive signals from many other brain regions. This type of "deep" interconnectedness, in which certain neurons provide connections across numerous regions, is a feature that occurs increasingly as we go up the evolutionary ladder. It is not surprising that the spindle cells, involved as they are in handling emotion and moral judgment, would have this form of deep interconnectedness, given the complexity of our emotional reactions.

What is startling, however, is how few spindle cells there are in this tiny region: only about 80,000 in the human brain (about 45,000 in the right hemisphere and 35,000 in the left hemisphere). This disparity appears to account for the perception that emotional intelligence is the province of the right brain, although the disproportion is modest. Gorillas have about 16,000 of these cells, bonobos about 2,100, and chimpanzees about 1,800. Other mammals lack them completely.

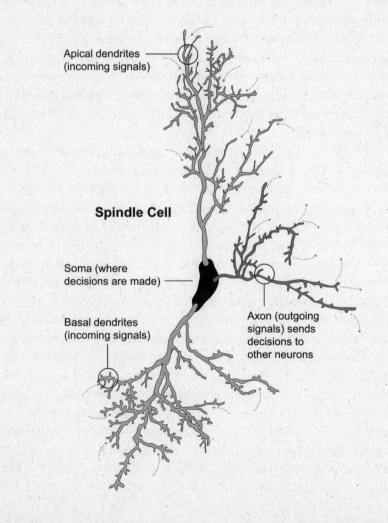

Apical dendrites (incoming signals)

Spindle Cell

Soma (where decisions are made)

Basal dendrites (incoming signals)

Axon (outgoing signals) sends decisions to other neurons

Dr. Arthur Craig of the Barrow Neurological Institute in Phoenix has recently provided a description of the architecture of the spindle cells.[115] Inputs from the body (estimated at hundreds of megabits per second), including nerves from the skin, muscles, organs, and other areas, stream into the upper spinal cord. These carry messages about touch, temperature, acid levels (for example, lactic acid in muscles), the movement of food through the gastrointestinal tract, and many other types of information. This data is processed through the brain stem and midbrain. Key cells called Lamina 1 neurons create a map of the body representing its current state, not unlike the displays used by flight controllers to track airplanes.

The information then flows through a nut-size region called the posterior ventromedial nucleus (VMpo), which apparently computes complex reactions to bodily states such as "this tastes terrible," "what a stench," or "that light touch is stimulating." The increasingly sophisticated information ends up at two regions of the cortex called the insula. These structures, the size of small fingers, are located on the left and right sides of the cortex. Craig describes the VMpo and the two insula regions as "a system that represents the material me."

Although the mechanisms are not yet understood, these regions are critical to self-awareness and complicated emotions. They are also much smaller in other animals. For example, the VMpo is about the size of a grain of sand in macaque monkeys and even smaller in lower-level animals. These findings are consistent with a growing consensus that our emotions are closely linked to areas of the brain that contain maps of the body, a view promoted by Dr. Antonio Damasio at the University of Iowa.[116] They are also consistent with the view that a great deal of our thinking is directed toward our bodies: protecting and enhancing them, as well as attending to their myriad needs and desires.

Very recently yet another level of processing of what started out as sensory information from the body has been discovered. Data from the two insula regions goes on to a tiny area at the front of the right insula called the frontoinsular cortex. This is the region containing the spindle cells, and fMRI scans have revealed that it is particularly active when a person is dealing with high-level emotions such as love, anger, sadness, and sexual desire. Situations that strongly activate the spindle cells include when a subject looks at her romantic partner or hears her child crying.

Anthropologists believe that spindle cells made their first appearance ten to fifteen million years ago in the as-yet-undiscovered common ancestor to apes and early hominids (the family of humans) and rapidly increased in numbers around one hundred thousand years ago. Interestingly, spindle cells do not exist in newborn humans but begin to appear only at around the age of four months and increase significantly from ages one to three. Children's ability to deal with moral issues and perceive such higher-level emotions as love develop during this same time period.

The spindle cells gain their power from the deep interconnectedness of their long apical dendrites with many other brain regions. The high-level emotions that the spindle cells process are affected, thereby, by all of our perceptual and cognitive regions. It will be difficult, therefore, to reverse engineer the exact methods of the spindle cells until we have better models of the many other regions to which they connect. However, it is remarkable how few neurons appear to be exclusively involved with these emotions. We have fifty billion neurons in the cerebellum that deal with skill formation, billions in the cortex that perform the transformations for perception and rational planning, but only about eighty thousand spindle cells dealing with high-level emotions. It is important to point out that the spindle cells are not doing rational problem solving, which is why we don't have rational control over our responses to music or over falling in love. The rest of the brain is heavily engaged, however, in trying to make sense of our mysterious high-level emotions.

Interfacing the Brain and Machines

> I want to do something with my life; I want to be a cyborg.
> —KEVIN WARWICK

Understanding the methods of the human brain will help us to design similar biologically inspired machines. Another important application will be to actually interface our brains with computers, which I believe will become an increasingly intimate merger in the decades ahead.

Already the Defense Advanced Research Projects Agency is spending $24 million per year on investigating direct interfaces between brain and computer. As described above (see the section "The Visual System" on p. 185), Tomaso Poggio and James DiCarlo at MIT, along with Christof Koch at the California Institute of Technology (Caltech), are attempting to develop models of the recognition of visual objects and how this information is encoded. These could eventually be used to transmit images directly into our brains.

Miguel Nicolelis and his colleagues at Duke University implanted sensors in the brains of monkeys, enabling the animals to control a robot through thought alone. The first step in the experiment involved teaching the monkeys to control a cursor on a screen with a joystick. The scientists collected a pattern of signals from EEGs (brain sensors) and subsequently caused the cursor to respond to the appropriate patterns rather than physical movements of the joystick. The monkeys quickly learned that the joystick was no longer operative and that they could control the cursor just by thinking. This "thought detection" system was then hooked up to a robot, and the monkeys were able to learn how to control the robot's movements with their thoughts alone. By getting visual feedback on the robot's performance, the monkeys were able to perfect their thought control over the robot. The goal of this research is to provide a similar system for paralyzed humans that will enable them to control their limbs and environment.

A key challenge in connecting neural implants to biological neurons is that the neurons generate glial cells, which surround a "foreign" object in an attempt to protect the brain. Ted Berger and his colleagues are developing special coatings that will appear to be biological and therefore attract rather than repel nearby neurons.

Another approach being pursued by the Max Planck Institute for Human Cognitive and Brain Sciences in Munich is directly interfacing nerves and electronic devices. A chip created by Infineon allows neurons to grow on a special substrate that provides direct contact between nerves and electronic sensors and stimulators. Similar work on a "neurochip" at Caltech has demonstrated two-way, noninvasive communication between neurons and electronics.[117]

We have already learned how to interface surgically installed neural implants. In cochlear (inner-ear) implants it has been found that the auditory nerve reorganizes itself to correctly interpret the multichannel signal from the implant. A similar process appears to take place with the deep-brain stimulation implant used for Parkinson's patients. The biological neurons in the vicinity of this FDA-approved brain implant receive signals from the electronic device and respond just as if they had received signals from the biological neurons that were once functional. Recent versions of the Parkinson's-disease implant provide the ability to download upgraded software directly to the implant from outside the patient.

The Accelerating Pace of Reverse Engineering the Brain

> Homo sapiens, the first truly free species, is about to decommission natural selection, the force that made us. . . . [S]oon we must look deep within ourselves and decide what we wish to become.
>
> —E. O. WILSON, *CONSILIENCE: THE UNITY OF KNOWLEDGE*, 1998

> We know what we are, but know not what we may be.
>
> —WILLIAM SHAKESPEARE

> The most important thing is this: To be able at any moment to sacrifice what
> we are for what we could become.
>
> —CHARLES DUBOIS

Some observers have expressed concern that as we develop models, simulations, and extensions to the human brain we risk not really understanding what we are tinkering with and the delicate balances involved. Author W. French Anderson writes:

> We may be like the young boy who loves to take things apart. He is
> bright enough to disassemble a watch, and maybe even bright enough to
> get it back together so that it works. But what if he tries to "improve"
> it? . . . The boy can understand what is visible, but he cannot understand
> the precise engineering calculations that determine exactly how strong
> each spring should be. . . . Attempts on his part to improve the watch
> will probably only harm it. . . . I fear . . . we, too, do not really under-
> stand what makes the [lives] we are tinkering with tick.[118]

Anderson's concern, however, does not reflect the scope of the broad and painstaking effort by tens of thousands of brain and computer scientists to methodically test out the limits and capabilities of models and simulations before taking them to the next step. We are not attempting to disassemble and reconfigure the brain's trillions of parts without a detailed analysis at each stage. The process of understanding the principles of operation of the brain is proceeding through a series of increasingly sophisticated models derived from increasingly accurate and high-resolution data.

As the computational power to emulate the human brain approaches—we're almost there with supercomputers—the efforts to scan and sense the human brain and to build working models and simulations of it are accelerating. As with every other projection in this book, it is critical to understand the exponential nature of progress in this field. I frequently encounter colleagues who argue that it will be a century or longer before we can understand in detail the methods of the brain. As with so many long-term scientific projections, this one is based on a linear view of the future and ignores the inherent acceleration of progress, as well as the exponential growth of each underlying technology. Such overly conservative views are also frequently based on an underestimation of the breadth of contemporary accomplishments, even by practitioners in the field.

Scanning and sensing tools are doubling their overall spatial and temporal resolution each year. Scanning-bandwidth, price-performance, and image-reconstruction times are also seeing comparable exponential growth. These trends hold true for all of the forms of scanning: fully noninvasive scanning, in vivo scanning with an exposed skull, and destructive scanning. Databases of brain-scanning information and model building are also doubling in size about once per year.

We have demonstrated that our ability to build detailed models and working simulations of subcellular portions, neurons, and extensive neural regions follows closely upon the availability of the requisite tools and data. The performance of neurons and subcellular portions of neurons often involves substantial complexity and numerous nonlinearities, but the performance of neural clusters and neuronal regions is often simpler than their constituent parts. We have increasingly powerful mathematical tools, implemented in effective computer software, that are able to accurately model these types of complex hierarchical, adaptive, semirandom, self-organizing, highly nonlinear systems. Our success to date in effectively modeling several important regions of the brain shows the effectiveness of this approach.

The generation of scanning tools now emerging will for the first time provide spatial and temporal resolution capable of observing in real time the performance of individual dendrites, spines, and synapses. These tools will quickly lead to a new generation of higher-resolution models and simulations.

Once the nanobot era arrives in the 2020s we will be able to observe all of the relevant features of neural performance with very high resolution from inside the brain itself. Sending billions of nanobots through its capillaries will enable us to noninvasively scan an entire working brain in real time. We have already created effective (although still incomplete) models of extensive regions of the brain with today's relatively crude tools. Within twenty years, we will have at least a millionfold increase in computational power and vastly improved scanning resolution and bandwidth. So we can have confidence that we will have the data-gathering and computational tools needed by the 2020s to model and simulate the entire brain, which will make it possible to combine the principles of operation of human intelligence with the forms of intelligent information processing that we have derived from other AI research. We will also benefit from the inherent strength of machines in storing, retrieving, and quickly sharing massive amounts of information. We will then be in a position to implement these powerful hybrid systems on computational platforms that greatly exceed the capabilities of the human brain's relatively fixed architecture.

The Scalability of Human Intelligence. In response to Hofstadter's concern as to whether human intelligence is just above or below the threshold necessary for "self-understanding," the accelerating pace of brain reverse engineering makes it clear that there are no limits to our ability to understand ourselves—or anything else, for that matter. The key to the scalability of human intelligence is our ability to build models of reality in our mind. These models can be recursive, meaning that one model can include other models, which can include yet finer models, without limit. For example, a model of a biological cell can include models of the nucleus, ribosomes, and other cellular systems. In turn, the model of the ribosome may include models of its submolecular components, and then down to the atoms and subatomic particles and forces that it comprises.

Our ability to understand complex systems is not necessarily hierarchical. A complex system like a cell or the human brain cannot be understood simply by breaking it down into constituent subsystems and their components. We have increasingly sophisticated mathematical tools for understanding systems that combine both order and chaos—and there is plenty of both in a cell and in the brain—and for understanding the complex interactions that defy logical breakdown.

Our computers, which are themselves accelerating, have been a critical tool in enabling us to handle increasingly complex models, which we would otherwise be unable to envision with our brains alone. Clearly, Hofstadter's concern would be correct if we were limited just to models that we could keep in our minds without technology to assist us. That our intelligence is just above the threshold necessary to understand itself results from our native ability, combined with the tools of our own making, to envision, refine, extend, and alter abstract—and increasingly subtle—models of our own observations.

Uploading the Human Brain

> To become a figment of your computer's imagination.
>
> —DAVID VICTOR DE TRANSEND, *GODLING'S GLOSSARY*, DEFINITION OF "UPLOAD"

A more controversial application than the scanning-the-brain-to-understand-it scenario is *scanning the brain to upload it.* Uploading a human brain means scanning all of its salient details and then reinstantiating those details into a suitably powerful computational substrate. This process would capture a person's entire personality, memory, skills, and history.

If we are truly capturing a particular person's mental processes, then the reinstantiated mind will need a body, since so much of our thinking is directed toward physical needs and desires. As I will discuss in chapter 5, by the time we have the tools to capture and re-create a human brain with all of its subtleties, we will have plenty of options for twenty-first-century bodies for both nonbiological humans and biological humans who avail themselves of extensions to our intelligence. The human body version 2.0 will include virtual bodies in completely realistic virtual environments, nanotechnology-based physical bodies, and more.

In chapter 3 I discussed my estimates for the memory and computational requirements to simulate the human brain. Although I estimated that 10^{16} cps of computation and 10^{13} bits of memory are sufficient to emulate human levels of intelligence, my estimates for the requirements of uploading were higher: 10^{19} cps and 10^{18} bits, respectively. The reason for the higher estimates is that the lower ones are based on the requirements to re-create regions of the brain at human levels of performance, whereas the higher ones are based on capturing the salient details of each of our approximately 10^{11} neurons and 10^{14} interneuronal connections. Once uploading is feasible, we are likely to find that hybrid solutions are adequate. For example, we will probably find that it is sufficient to simulate certain basic support functions such as the signal processing of sensory data on a functional basis (by plugging in standard modules) and reserve the capture of subneuron details only for those regions that are truly responsible for individual personality and skills. Nonetheless, we will use our higher estimates for this discussion.

The basic computational resources (10^{19} cps and 10^{18} bits) will be available for one thousand dollars in the early 2030s, about a decade later than the resources needed for functional simulation. The scanning requirements for uploading are also more daunting than for "merely" re-creating the overall powers of human intelligence. In theory one could upload a human brain by capturing all the necessary details without necessarily comprehending the brain's overall plan. In practice, however, this is unlikely to work. Understanding the principles of operation of the human brain will reveal which details are essential and which details are intended to be disordered. We need to know, for example, which molecules in the neurotransmitters are critical, and whether we need to capture overall levels, position and location, and/or molecular shape. As I discussed above, we are just learning, for example, that it is the position of actin molecules and the shape of CPEB molecules in the synapse that are key for memory. It will not be possible to confirm which details are crucial without having confirmed our understanding of the theory of operation. That confirmation will be in the form of a functional simulation of human intelligence that passes the Turing test, which I believe will take place by 2029.[119]

To capture this level of detail will require scanning from within the brain using nanobots, the technology for which will be available by the late 2020s. Thus, the early 2030s is a reasonable time frame for the computational performance, memory, and brain-scanning prerequisites of uploading. Like any other technology, it will take some iterative refinement to perfect this capability, so the end of the 2030s is a conservative projection for successful uploading.

We should point out that a person's personality and skills do not reside only in the brain, although that is their principal location. Our nervous system extends throughout the body, and the endocrine (hormonal) system has an influence, as well. The vast majority of the complexity, however, resides in the brain, which is the location of the bulk of the nervous system. The bandwidth of information from the endocrine system is quite low, because the determining factor is overall levels of hormones, not the precise location of each hormone molecule.

Confirmation of the uploading milestone will be in the form of a "Ray Kurzweil" or "Jane Smith" Turing test, in other words convincing a human judge that the uploaded re-creation is indistinguishable from the original specific person. By that time we'll face some complications in devising the rules of any Turing test. Since nonbiological intelligence will have passed the original Turing test years earlier (around 2029), should we allow a nonbiological human equivalent to be a judge? How about an enhanced human? Unenhanced humans may become increasingly hard to find. In any event, it will be a slippery slope to define enhancement, as many different levels of extending biological intelligence will be available by the time we have purported uploads. Another issue will be that the humans we seek to upload will not be limited to their biological intelligence. However, uploading the nonbiological portion of intelligence will be relatively straightforward, since the ease of copying computer intelligence has always represented one of the strengths of computers.

One question that arises is, How quickly do we need to scan a person's nervous system? It clearly cannot be done instantaneously, and even if we did provide a nanobot for each neuron, it would take time to gather the data. One might therefore object that because a person's state is changing during the data-gathering process, the upload information does not accurately reflect that person at an instant in time but rather over a period of time, even if only a fraction of a second.[120] Consider, however, that this issue will not interfere with an upload's passing a "Jane Smith" Turing test. When we encounter one another on a day-to-day basis, we are recognized as ourselves even though it may have been days or weeks since the last such encounter. If an upload is sufficiently accurate to re-create a person's state within the amount of natural change that a person undergoes in a fraction of a second or even a few minutes, that will be sufficient for any conceivable purpose. Some observers have interpreted Roger Penrose's theory of the link between quantum computing and consciousness (see chapter 9) to mean that uploading is impossible because a person's "quantum state" will have changed many times during the scanning period. But I would point out that my quantum state has changed many times in the time it took me to write this sentence, and I still consider myself to be the same person (and no one seems to be objecting).

Nobel Prize winner Gerald Edelman points out that there is a difference between a capability and a description of that capability. A photograph of a person is different from the person herself, even if the "photograph" is very high resolution and three-dimensional. However, the concept of uploading goes beyond the extremely high-resolution scan, which we can consider the "photograph" in Edelman's analogy. The scan does need to capture all of the salient details, but it also needs to be instantiated into a working computational medium that has the capabilities of the original (albeit that the new nonbiological platforms are certain to be far more capable). The neural details need to interact with one another (and with the outside world) in the same ways that they do in the original. A comparable analogy is the comparison between a computer program that resides on a computer disk (a static picture) and a program that is actively running on a suitable computer (a dynamic, interacting entity). Both the data capture and the reinstantiation of a dynamic entity constitute the uploading scenario.

Perhaps the most important question will be whether or not an uploaded human brain is really you. Even if the upload passes a personalized Turing test and is deemed indistinguishable from you, one could still reasonably ask whether the upload is the same person or a new person. After all, the original person may still exist. I'll defer these essential questions until chapter 7.

In my view the most important element in uploading will be our gradual transfer of our intelligence, personality, and skills to the nonbiological portion of our intelligence. We already have a variety of neural implants. In the 2020s we will use nanobots to begin augmenting our brains with nonbiological intelligence, starting with the "routine" functions of sensory processing and memory, moving on to skill formation, pattern recognition, and logical analysis. By the 2030s the nonbiological portion of our intelligence will predominate, and by the 2040s, as I pointed out in chapter 3, the nonbiological portion will be billions of times more capable. Although we are likely to retain the biological portion for a period of time, it will become of increasingly little consequence. So we will have effectively uploaded ourselves, albeit gradually, never quite noticing the transfer. There will be no "old Ray" and "new Ray," just an increasingly capable Ray. Although I believe that uploading as in the sudden scan-and-transfer scenario discussed in this section will be a feature of our future world, it is this gradual but inexorable progression to vastly superior nonbiological thinking that will profoundly transform human civilization.

SIGMUND FREUD: *When you talk about reverse engineering the human brain, just whose brain are you talking about? A man's brain? A woman's? A child's? The brain of a genius? A retarded individual? An "idiot savant"? A gifted artist? A serial murderer?*

RAY: *Ultimately, we're talking about all of the above. There are basic principles of operation that we need to understand about how human intelligence and its varied constituent skills work. Given the human brain's plasticity, our thoughts literally create our brains through the growth of new spines, synapses, dendrites, and even neurons. As a result, Einstein's parietal lobes— the region associated with visual imagery and mathematical thinking— became greatly enlarged.*[121] *However, there is only so much room in our skulls, so although Einstein played music he was not a world-class musician. Picasso did not write great poetry, and so on. As we re-create the human brain, we will not be limited in our ability to develop each skill. We will not have to compromise one area to enhance another.*

We can also gain insight into our differences and an understanding of human dysfunction. What went wrong with the serial murderer? It must, after all, have something to do with his brain. This type of disastrous behavior is clearly not the result of indigestion.

MOLLY 2004: *You know, I doubt it's just the brains we're born with that account for our differences. What about our struggles through life, and all this stuff I'm trying to learn?*

RAY: *Yes, well, that's part of the paradigm, too, isn't it? We have brains that can learn, starting from when we learn to walk and talk to when we study college chemistry.*

MARVIN MINSKY: *It's true that educating our AIs will be an important part of the process, but we can automate a lot of that and greatly speed it up. Also, keep in mind that when one AI learns something, it can quickly share that knowledge with many other AIs.*

RAY: *They'll have access to all of our exponentially growing knowledge on the Web, which will include habitable, full-immersion virtual-reality environments where they can interact with one another and with biological humans who are projecting themselves into these environments.*

SIGMUND: *These AIs don't have bodies yet. As we have both pointed out, human emotion and much of our thinking are directed at our bodies and to meeting their sensual and sexual needs.*

RAY: *Who says they won't have bodies? As I will discuss in the human body version 2.0 section in chapter 6, we'll have the means of creating nonbiological yet humanlike bodies, as well as virtual bodies in virtual reality.*

SIGMUND: *But a virtual body is not a real body.*

RAY: *The word "virtual" is somewhat unfortunate. It implies "not real," but the reality will be that a virtual body is just as real as a physical body in all the ways that matter. Consider that the telephone is auditory virtual reality. No one feels that his voice in this virtual-reality environment is not a "real" voice. With my physical body today, I don't directly experience someone's touch on my arm. My brain receives processed signals initiated by nerve endings in my arm, which wind their way through the spinal cord, through the brain stem, and up to the insula regions. If my brain—or an AI's brain—receives comparable signals of someone's virtual touch on a virtual arm, there's no discernible difference.*

MARVIN: *Keep in mind that not all AIs will need human bodies.*

RAY: *Indeed. As humans, despite some plasticity, both our bodies and brains have a relatively fixed architecture.*

MOLLY 2004: *Yes, it's called being human, something you seem to have a problem with.*

RAY: *Actually, I often do have a problem with all the limitations and maintenance that my version 1.0 body requires, not to mention all the limitations of my brain. But I do appreciate the joys of the human body. My point is that AIs can and will have the equivalent of human bodies in both real and virtual-reality environments. As Marvin points out, however, they will not be limited just to this.*

MOLLY 2104: *It's not just AIs that will be liberated from the limitations of version 1.0 bodies. Humans of biological origin will have the same freedom in both real and virtual reality.*

GEORGE 2048: *Keep in mind, there won't be a clear distinction between AIs and humans.*

MOLLY 2104: *Yes, except for the MOSHs (Mostly Original Substrate Humans) of course.*

CHAPTER FIVE

GNR

Three Overlapping Revolutions

There are few things of which the present generation is more justly proud than the wonderful improvements which are daily taking place in all sorts of mechanical appliances. . . . But what would happen if technology continued to evolve so much more rapidly than the animal and vegetable kingdoms? Would it displace us in the supremacy of earth? Just as the vegetable kingdom was slowly developed from the mineral, and as in like manner the animal supervened upon the vegetable, so now in these last few ages an entirely new kingdom has sprung up, of which we as yet have only seen what will one day be considered the antediluvian prototypes of the race. . . . We are daily giving [machines] greater power and supplying by all sorts of ingenious contrivances that self-regulating, self-acting power which will be to them what intellect has been to the human race.

—SAMUEL BUTLER, 1863 (FOUR YEARS AFTER PUBLICATION OF DARWIN'S *THE ORIGIN OF SPECIES*)

Who will be man's successor? To which the answer is: We are ourselves creating our own successors. Man will become to the machine what the horse and the dog are to man; the conclusion being that machines are, or are becoming, animate.

—SAMUEL BUTLER, 1863 LETTER, "DARWIN AMONG THE MACHINES"[1]

The first half of the twenty-first century will be characterized by three overlapping revolutions—in Genetics, Nanotechnology, and Robotics. These will usher in what I referred to earlier as Epoch Five, the beginning of the Singularity. We are in the early stages of the "G" revolution today. By understanding the information processes underlying life, we are starting to learn to reprogram our biology to achieve the virtual elimination of disease, dramatic expansion of human potential, and radical life extension. Hans Moravec points out, however, that no matter how successfully we fine-tune our DNA-based biology, humans will remain "second-class robots," meaning that biology will never be able to match what we will be able to engineer once we fully understand biology's principles of operation.[2]

The "N" revolution will enable us to redesign and rebuild—molecule by molecule—our bodies and brains and the world with which we interact, going far beyond the limitations of biology. The most powerful impending revolution is "R": human-level robots with their intelligence derived from our own but redesigned to far exceed human capabilities. R represents the most significant transformation, because intelligence is the most powerful "force" in the universe. Intelligence, if sufficiently advanced, is, well, smart enough to anticipate and overcome any obstacles that stand in its path.

While each revolution will solve the problems from earlier transformations, it will also introduce new perils. G will overcome the age-old difficulties of disease and aging but establish the potential for new bioengineered viral threats. Once N is fully developed we will be able to apply it to protect ourselves from all biological hazards, but it will create the possibility of its own self-replicating dangers, which will be far more powerful than anything biological. We can protect ourselves from these hazards with fully developed R, but what will protect us from pathological intelligence that exceeds our own? I do have a strategy for dealing with these issues, which I discuss at the end of chapter 8. In this chapter, however, we will examine how the Singularity will unfold through these three overlapping revolutions: G, N, and R.

Genetics: The Intersection of Information and Biology

> It has not escaped our notice that the specific pairing we have postulated immediately suggests a possible copying mechanism for the genetic material.
>
> —JAMES WATSON AND FRANCIS CRICK[3]

> After three billion years of evolution, we have before us the instruction set that carries each of us from the one-cell egg through adulthood to the grave.
>
> —DR. ROBERT WATERSTON, INTERNATIONAL HUMAN GENOME SEQUENCING CONSORTIUM[4]

Underlying all of the wonders of life and misery of disease are information processes, essentially software programs, that are surprisingly compact. The entire human genome is a sequential binary code containing only about eight hundred million bytes of information. As I mentioned earlier, when its massive redundancies are removed using conventional compression techniques, we are left with only thirty to one hundred million bytes, equivalent to the size of an average contemporary software program.[5] This code is supported by a set of biochemical machines that translate these linear (one-dimensional) sequences of DNA "letters" into strings of simple building blocks called amino acids, which are in turn folded into three-dimensional proteins, which make up all living creatures from bacteria to humans. (Viruses occupy a niche in between living and nonliving matter but are also composed of fragments of DNA or RNA.) This machinery is essentially a self-replicating nanoscale replicator that builds the elaborate hierarchy of structures and increasingly complex systems that a living creature comprises.

Life's Computer

In the very early stages of evolution information was encoded in the structure of increasingly complex organic molecules based on carbon. After billions of years biology evolved its own computer for storing and manipulating digital data based on the DNA molecule. The chemical structure of the DNA molecule was first described by J. D. Watson and F. H. C. Crick in 1953 as a double helix consisting of a pair of strands of polynucleotides with information encoded at each position by the choice of nucleotides.[6] We finished transcribing the genetic code at the beginning of this century. We are now beginning to understand the detailed chemistry of the communication and control processes by which DNA commands reproduction through such other complex molecules and cellular structures as messenger RNA (mRNA), transfer RNA (tRNA), and ribosomes.

At the level of information storage the mechanism is surprisingly simple. Supported by a twisting sugar-phosphate backbone, the DNA molecule contains up to several million rungs, each of which is coded with one letter drawn from a four-letter alphabet; each rung is thus coding two bits of data in a one-dimensional digital code. The alphabet consists of the four base pairs: adenine-thymine, thymine-adenine, cytosine-guanine, and guanine-cytosine. The DNA strings in a single cell would measure up to six feet in length if stretched out, but an elaborate packing method coils them to fit into a cell only 1/2500 of an inch across.

Special enzymes can copy the information on each rung by splitting each base pair and assembling two identical DNA molecules by rematching the broken base pairs. Other enzymes actually check the validity of the copy by checking the integrity of the base-pair matching. With these copying and validation steps, this chemical data-processing system makes only about one error in ten billion base-pair replications.[7] Further redundancy and error-correction codes are built into the digital data itself, so ▶

meaningful mutations resulting from base-pair replication errors are rare. Most of the errors resulting from the one-in-ten-billion error rate will result in the equivalent of a "parity" error, which can be detected and corrected by other levels of the system, including matching against the corresponding chromosome, which can prevent the incorrect bit from causing any significant damage.[8] Recent research has shown that the genetic mechanism detects such errors in transcription of the male Y chromosome by matching each Y chromosome gene against a copy on the same chromosome.[9] Once in a long while a transcription error will result in a beneficial change that evolution will come to favor.

In a process technically called translation, another series of chemicals put this elaborate digital program into action by building proteins. It is the protein chains that give each cell its structure, behavior, and intelligence. Special enzymes unwind a region of DNA for building a particular protein. A strand of mRNA is created by copying the exposed sequence of bases. The mRNA essentially has a copy of a portion of the DNA letter sequence. The mRNA travels out of the nucleus and into the cell body. The mRNA codes are then read by a ribosome molecule, which represents the central molecular player in the drama of biological reproduction. One portion of the ribosome acts like a tape-recorder head, "reading" the sequence of data encoded in the mRNA base sequence. The "letters" (bases) are grouped into words of three letters each called codons, with one codon for each of twenty possible amino acids, the basic building blocks of protein. A ribosome reads the codons from the mRNA and then, using tRNA, assembles a protein chain one amino acid at a time.   ▶

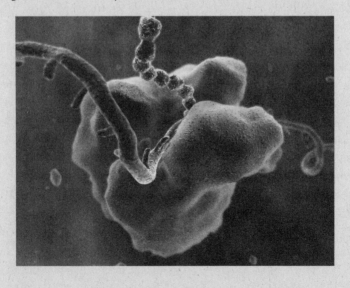

The notable final step in this process is the folding of the one-dimensional chain of amino acid "beads" into a three-dimensional protein. Simulating this process has not yet been feasible because of the enormous complexity of the interacting forces from all the atoms involved. Super-computers scheduled to come online around the time of the publication of this book (2005) are expected to have the computational capacity to simulate protein folding, as well as the interaction of one three-dimensional protein with another.

Protein folding, along with cell division, is one of nature's remarkable and intricate dances in the creation and re-creation of life. Specialized "chaperone" molecules protect and guide the amino-acid strands as they assume their precise three-dimensional protein configurations. As many as one third of formed protein molecules are folded improperly. These disfigured proteins must immediately be destroyed or they will rapidly accumulate, disrupting cellular functions on many levels.

Under normal circumstances, as soon as a misfolded protein is formed, it is tagged by a carrier molecule, ubiquitin, and escorted to a specialized proteosome, where it is broken back down into its component amino acids for recycling into new (correctly folded) proteins. As cells age, however, they produce less of the energy needed for optimal function of this mechanism. Accumulations of these misformed proteins aggregate into particles called protofibrils, which are thought to underlie disease processes leading to Alzheimer's disease and other afflictions.[10]

The ability to simulate the three-dimensional waltz of atomic-level interactions will greatly accelerate our knowledge of how DNA sequences control life and disease. We will then be in a position to rapidly simulate drugs that intervene in any of the steps in this process, thereby hastening drug development and the creation of highly targeted drugs that minimize unwanted side effects.

It is the job of the assembled proteins to carry out the functions of the cell, and by extension the organism. A molecule of hemoglobin, for example, which has the job of carrying oxygen from the lungs to body tissues, is created five hundred trillion times each second in the human body. With more than five hundred amino acids in each molecule of hemoglobin, that comes to 1.5×10^{19} (fifteen billion billion) "read" operations every minute by the ribosomes just for the manufacture of hemoglobin.

In some ways the biochemical mechanism of life is remarkably complex and intricate. In other ways it is remarkably simple. Only four base pairs provide the digital storage for all of the complexity of all human life and all other life as we know it. The ribosomes build protein chains by ▶

grouping together triplets of base pairs to select sequences from only twenty amino acids. The amino acids themselves are relatively simple, consisting of a carbon atom with its four bonds linked to one hydrogen atom, one amino ($-NH_2$) group, one carboxylic acid ($-COOH$) group, and one organic group that is different for each amino acid. The organic group for alanine, for example, has only four atoms (CH_3-) for a total of thirteen atoms. One of the more complex amino acids, arginine (which plays a vital role in the health of the endothelial cells in our arteries) has only seventeen atoms in its organic group for a total of twenty-six atoms. These twenty simple molecular fragments are the building blocks of all life.

The protein chains then control everything else: the structure of bone cells, the ability of muscle cells to flex and act in concert with other muscle cells, all of the complex biochemical interactions that take place in the bloodstream, and, of course, the structure and functioning of the brain.[11]

Designer Baby Boomers

Sufficient information already exists today to slow down disease and aging processes to the point that baby boomers like myself can remain in good health until the full blossoming of the biotechnology revolution, which will itself be a bridge to the nanotechnology revolution (see Resources and Contact Information, p. 489). In *Fantastic Voyage: Live Long Enough to Live Forever*, which I coauthored with Terry Grossman, M.D., a leading longevity expert, we discuss these three bridges to radical life extension (today's knowledge, biotechnology, and nanotechnology).[12] I wrote there: "Whereas some of my contemporaries may be satisfied to embrace aging gracefully as part of the cycle of life, that is not my view. It may be 'natural,' but I don't see anything positive in losing my mental agility, sensory acuity, physical limberness, sexual desire, or any other human ability. I view disease and death at any age as a calamity, as problems to be overcome."

Bridge one involves aggressively applying the knowledge we now possess to dramatically slow down aging and reverse the most important disease processes, such as heart disease, cancer, type 2 diabetes, and stroke. You can, in effect, reprogram your biochemistry, for we have the knowledge today, if aggressively applied, to overcome our genetic heritage in the vast majority of cases. "It's mostly in your genes" is only true if you take the usual passive attitude toward health and aging.

My own story is instructive. More than twenty years ago I was diagnosed with type 2 diabetes. The conventional treatment made my condition worse, so I approached this health challenge from my perspective as an inventor. I immersed myself in the scientific literature and came up with a unique program that successfully reversed my diabetes. In 1993 I wrote a health book (*The 10% Solution for a Healthy Life*) about this experience, and I continue today to be free of any indication or complication of this disease.[13]

In addition, when I was twenty-two, my father died of heart disease at the age of fifty-eight, and I have inherited his genes predisposing me to this illness. Twenty years ago, despite following the public guidelines of the American Heart Association, my cholesterol was in the high 200s (it should be well below 180), my HDL (high-density lipoprotein, the "good" cholesterol) below 30 (it should be above 50), and my homocysteine (a measure of the health of a biochemical process called methylation) was an unhealthy 11 (it should be below 7.5). By following a longevity program that Grossman and I developed, my current cholesterol level is 130, my HDL is 55, my homocysteine is 6.2, my C-reactive protein (a measure of inflammation in the body) is a very healthy 0.01, and all of my other indexes (for heart disease, diabetes, and other conditions) are at ideal levels.[14]

When I was forty, my biological age was around thirty-eight. Although I am now fifty-six, a comprehensive test of my biological aging (measuring various sensory sensitivities, lung capacity, reaction times, memory, and related tests) conducted at Grossman's longevity clinic measured my biological age at forty.[15] Although there is not yet a consensus on how to measure biological age, my scores on these tests matched population norms for this age. So, according to this set of tests, I have not aged very much in the last sixteen years, which is confirmed by the many blood tests I take, as well as the way I feel.

These results are not accidental; I have been very aggressive about reprogramming my biochemistry. I take 250 supplements (pills) a day and receive a half-dozen intravenous therapies each week (basically nutritional supplements delivered directly into my bloodstream, thereby bypassing my GI tract). As a result, the metabolic reactions in my body are completely different than they would otherwise be.[16] Approaching this as an engineer, I measure dozens of levels of nutrients (such as vitamins, minerals, and fats), hormones, and metabolic by-products in my blood and other body samples (such as hair and saliva). Overall, my levels are where I want them to be, although I continually fine-tune my program based on the research that I conduct with Grossman.[17] Although my program may seem extreme, it is actually conservative—and optimal (based on my current knowledge). Grossman and I have extensively researched each of the several hundred therapies that I use for safety and effi-cacy. I stay away from ideas that are unproven or appear to be risky (the use of human-growth hormone, for example).

We consider the process of reversing and overcoming the dangerous pro-gression of disease as a war. As in any war it is important to mobilize all the means of intelligence and weaponry that can be harnessed, throwing every-thing we have at the enemy. For this reason we advocate that key dangers—such as heart disease, cancer, diabetes, stroke, and aging—be attacked on multiple fronts. For example, our strategy for preventing heart disease is to adopt ten different heart-disease-prevention therapies that attack each of the known risk factors.

By adopting such multipronged strategies for each disease process and each aging process, even baby boomers like myself can remain in good health until the full blossoming of the biotechnology revolution (which we call "bridge two"), which is already in its early stages and will reach its peak in the second decade of this century.

Biotechnology will provide the means to actually change your genes: not just designer babies will be feasible but designer baby boomers. We'll also be able to rejuvenate all of your body's tissues and organs by transforming your skin cells into youthful versions of every other cell type. Already, new drug development is precisely targeting key steps in the process of atherosclerosis (the cause of heart disease), cancerous tumor formation, and the metabolic processes underlying each major disease and aging process.

Can We Really Live Forever? An energetic and insightful advocate of stopping the aging process by changing the information processes underlying biology is Aubrey de Grey, a scientist in the department of genetics at Cambridge Univer-sity. De Grey uses the metaphor of maintaining a house. How long does a house last? The answer obviously depends on how well you take care of it. If you do nothing, the roof will spring a leak before long, water and the elements will invade, and eventually the house will disintegrate. But if you proactively take care of the structure, repair all damage, confront all dangers, and rebuild or renovate parts from time to time using new materials and technologies, the life of the house can essentially be extended without limit.

The same holds true for our bodies and brains. The only difference is that, while we fully understand the methods underlying the maintenance of a house, we do not yet fully understand all of the biological principles of life. But with our rapidly increasing comprehension of the biochemical processes and path-ways of biology, we are quickly gaining that knowledge. We are beginning to understand aging, not as a single inexorable progression but as a group of related processes. Strategies are emerging for fully reversing each of these aging progressions, using different combinations of biotechnology techniques.

De Grey describes his goal as "engineered negligible senescence"—stopping the body and brain from becoming more frail and disease-prone as it grows older.[18] As he explains, "All the core knowledge needed to develop *engineered negligible senescence* is already in our possession—it mainly just needs to be pieced together."[19] De Grey believes we'll demonstrate "robustly rejuvenated" mice—mice that are functionally younger than before being treated and with the life extension to prove it—within ten years, and he points out that this achievement will have a dramatic effect on public opinion. Demonstrating that we can reverse the aging process in an animal that shares 99 percent of our genes will profoundly challenge the common wisdom that aging and death are inevitable. Once robust rejuvenation is confirmed in an animal, there will be enormous competitive pressure to translate these results into human therapies, which should appear five to ten years later.

The diverse field of biotechnology is fueled by our accelerating progress in reverse engineering the information processes underlying biology and by a growing arsenal of tools that can modify these processes. For example, drug discovery was once a matter of finding substances that produced some beneficial result without excessive side effects. This process was similar to early humans' tool discovery, which was limited to simply finding rocks and other natural implements that could be used for helpful purposes. Today we are learning the precise biochemical pathways that underlie both disease and aging processes and are able to design drugs to carry out precise missions at the molecular level. The scope and scale of these efforts are vast.

Another powerful approach is to start with biology's information backbone: the genome. With recently developed gene technologies we're on the verge of being able to control how genes express themselves. Gene expression is the process by which specific cellular components (specifically RNA and the ribosomes) produce proteins according to a specific genetic blueprint. While every human cell has the full complement of the body's genes, a specific cell, such as a skin cell or a pancreatic islet cell, gets its characteristics from only the small fraction of genetic information relevant to that particular cell type.[20] The therapeutic control of this process can take place outside the cell nucleus, so it is easier to implement than therapies that require access inside it.

Gene expression is controlled by peptides (molecules made up of sequences of up to one hundred amino acids) and short RNA strands. We are now beginning to learn how these processes work.[21] Many new therapies now in development and testing are based on manipulating them either to turn off the expression of disease-causing genes or to turn on desirable genes that may otherwise not be expressed in a particular type of cell.

RNAi (RNA Interference). A powerful new tool called RNA interference (RNAi) is capable of turning off specific genes by blocking their mRNA, thus preventing them from creating proteins. Since viral diseases, cancer, and many other diseases use gene expression at some crucial point in their life cycle, this promises to be a breakthrough technology. Researchers construct short, double-stranded DNA segments that match and lock onto portions of the RNA that are transcribed from a targeted gene. With their ability to create proteins blocked, the gene is effectively silenced. In many genetic diseases only one copy of a given gene is defective. Since we get two copies of each gene, one from each parent, blocking the disease-causing gene leaves one healthy gene to make the necessary protein. If both genes are defective, RNAi could silence them both, but then a healthy gene would have to be inserted.[22]

Cell Therapies. Another important line of attack is to regrow our own cells, tissues, and even whole organs and introduce them into our bodies without surgery. One major benefit of this "therapeutic cloning" technique is that we will be able to create these new tissues and organs from versions of our cells that have also been made younger via the emerging field of rejuvenation medicine. For example, we will be able to create new heart cells from skin cells and introduce them into the system through the bloodstream. Over time, existing heart cells will be replaced with these new cells, and the result will be a rejuvenated "young" heart manufactured using a person's own DNA. I discuss this approach to regrowing our bodies below.

Gene Chips. New therapies are only one way that the growing knowledge base of gene expression will dramatically impact our health. Since the 1990s microarrays, or chips no larger than a dime, have been used to study and compare expression patterns of thousands of genes at a time.[23] The possible applications of the technology are so varied and the technological barriers have been reduced so greatly that huge databases are now devoted to the results from "do-it-yourself gene watching."[24]

Genetic profiling is now being used to:

- *Revolutionize the processes of drug screening and discovery.* Microarrays can "not only confirm the mechanism of action of a compound" but "discriminate between compounds acting at different steps in the same metabolic pathway."[25]

- *Improve cancer classifications.* One study reported in *Science* demonstrated the feasibility of classifying some leukemias "solely on gene expression monitoring." The authors also pointed to a case in which expression profiling resulted in the correction of a misdiagnosis.[26]
- *Identify the genes, cells, and pathways involved in a process, such as aging or tumorigenesis.* For example, by correlating the presence of acute myeloblastic leukemia and increased expression of certain genes involved with programmed cell death, a study helped identify new therapeutic targets.[27]
- *Determine the effectiveness of an innovative therapy.* One study recently reported in *Bone* looked at the effect of growth-hormone replacement on the expression of insulinlike growth factors (IGFs) and bone metabolism markers.[28]
- *Test the toxicity of compounds in food additives, cosmetics, and industrial products quickly and without using animals.* Such tests can show, for example, the degree to which each gene has been turned on or off by a tested substance.[29]

Somatic Gene Therapy (gene therapy for nonreproductive cells). This is the holy grail of bioengineering, which will enable us to effectively change genes inside the nucleus by "infecting" it with new DNA, essentially creating new genes.[30] The concept of controlling the genetic makeup of humans is often associated with the idea of influencing new generations in the form of "designer babies." But the real promise of gene therapy is to actually change our adult genes.[31] These can be designed to either block undesirable disease-encouraging genes or introduce new ones that slow down and even reverse aging processes.

Animal studies that began in the 1970s and 1980s have been responsible for producing a range of transgenic animals, such as cattle, chickens, rabbits, and sea urchins. The first attempts at human gene therapy were undertaken in 1990. The challenge is to transfer therapeutic DNA into target cells that will then be expressed at the right level and at the right time.

Consider the challenge involved in effecting a gene transfer. Viruses are often the vehicle of choice. Long ago viruses learned how to deliver their genetic material to human cells and, as a result, cause disease. Researchers now simply switch the material a virus unloads into cells by removing its genes and inserting therapeutic ones. Although the approach itself is relatively easy, the genes are too large to pass into many types of cells (such as brain cells). The process is also limited in the length of DNA it can carry, and it may cause an immune response. And precisely where the new DNA integrates into the cell's DNA has been a largely uncontrollable process.[32]

Physical injection (microinjection) of DNA into cells is possible but prohibitively expensive. Exciting advances have recently been made, however, in other means of transfer. For example, liposomes—fatty spheres with a watery core—can be used as a "molecular Trojan horse" to deliver genes to brain cells, thereby opening the door to treatment of disorders such as Parkinson's and epilepsy.[33] Electric pulses can also be employed to deliver a range of molecules (including drug proteins, RNA, and DNA) to cells.[34] Yet another option is to pack DNA into ultratiny "nanoballs" for maximum impact.[35]

The major hurdle that must be overcome for gene therapy to be applied in humans is proper positioning of a gene on a DNA strand and monitoring of the gene's expression. One possible solution is to deliver an imaging reporter gene along with the therapeutic gene. The image signals would allow for close supervision of both placement and level of expression.[36]

Even faced with these obstacles gene therapy is starting to work in human applications. A team led by University of Glasgow research doctor Andrew H. Baker has successfully used adenoviruses to "infect" specific organs and even specific regions within organs. For example, the group was able to direct gene therapy precisely at the endothelial cells, which line the inside of blood vessels. Another approach is being developed by Celera Genomics, a company founded by Craig Venter (the head of the private effort to transcribe the human genome). Celera has already demonstrated the ability to create synthetic viruses from genetic information and plans to apply these biodesigned viruses to gene therapy.[37]

One of the companies I help to direct, United Therapeutics, has begun human trials of delivering DNA into cells through the novel mechanism of autologous (the patient's own) stem cells, which are captured from a few vials of their blood. DNA that directs the growth of new pulmonary blood vessels is inserted into the stem cell genes, and the cells are reinjected into the patient. When the genetically engineered stem cells reach the tiny pulmonary blood vessels near the lung's alveoli, they begin to express growth factors for new blood vessels. In animal studies this has safely reversed pulmonary hypertension, a fatal and presently incurable disease. Based on the success and safety of these studies, the Canadian government gave permission for human tests to commence in early 2005.

Reversing Degenerative Disease

Degenerative (progressive) diseases—heart disease, stroke, cancer, type 2 diabetes, liver disease, and kidney disease—account for about 90 percent of the deaths in our society. Our understanding of the principal components of degenerative disease and human aging is growing rapidly, and strategies have been identified to halt and even reverse each of these processes. In *Fantastic Voyage*, Grossman and I describe a wide range of therapies now in the testing pipeline that have already demonstrated significant results in attacking the key biochemical steps underlying the progress of such diseases.

Combating Heart Disease. As one of many examples, exciting research is being conducted with a synthetic form of HDL cholesterol called recombinant Apo-A-I Milano (AAIM). In animal trials AAIM was responsible for a rapid and dramatic regression of atherosclerotic plaque.[38] In a phase 1 FDA trial, which included forty-seven human subjects, administering AAIM by intravenous infusion resulted in a significant reduction (an average 4.2 percent decrease) in plaque after just five weekly treatments. No other drug has ever shown the ability to reduce atherosclerosis this quickly.[39]

Another exciting drug for reversing atherosclerosis now in phase 3 FDA trials is Pfizer's Torcetrapib.[40] This drug boosts levels of HDL by blocking an enzyme that normally breaks it down. Pfizer is spending a record one billion dollars to test the drug and plans to combine it with its bestselling "statin" (cholesterol-lowering) drug, Lipitor.

Overcoming Cancer. Many strategies are being intensely pursued to overcome cancer. Particularly promising are cancer vaccines designed to stimulate the immune system to attack cancer cells. These vaccines could be used as a prophylaxis to prevent cancer, as a first-line treatment, or to mop up cancer cells after other treatments.[41]

The first reported attempts to activate a patient's immune response were undertaken more than one hundred years ago, with little success.[42] More recent efforts focus on encouraging dendritic cells, the sentinels of the immune system, to trigger a normal immune response. Many forms of cancer have an opportunity to proliferate because they somehow do not trigger that response. Dendritic cells play a key role because they roam the body, collecting foreign peptides and cell fragments and delivering them to the lymph nodes, which in response produce an army of T cells primed to eliminate the flagged peptides.

Some researchers are altering cancer-cell genes to attract T cells, with the assumption that the stimulated T cells would then recognize other cancer cells they encounter.[43] Others are experimenting with vaccines for exposing the dendritic cells to antigens, unique proteins found on the surfaces of cancer cells. One group used electrical pulses to fuse tumor and immune cells to create an "individualized vaccine."[44] One of the obstacles to developing effective vaccines is that currently we have not yet identified many of the cancer antigens we need to develop potent targeted vaccines.[45]

Blocking angiogenesis—the creation of new blood vessels—is another strategy. This process uses drugs to discourage blood-vessel development, which an emergent cancer needs to grow beyond a small size. Interest in angiogenesis has skyrocketed since 1997, when doctors at the Dana Farber Cancer Center in Boston reported that repeated cycles of endostatin, an angiogenesis inhibitor, had resulted in complete regression of tumors.[46] There are now many antiangiogenic drugs in clinical trials, including avastin and atrasentan.[47]

A key issue for cancer as well as for aging concerns telomere "beads," repeating sequences of DNA found at the end of chromosomes. Each time a cell reproduces, one bead drops off. Once a cell has reproduced to the point that all of its telomere beads have been expended, that cell is no longer able to divide and will die. If we could reverse this process, cells could survive indefinitely. Fortunately, recent research has found that only a single enzyme (telomerase) is needed to achieve this.[48] The tricky part is to administer telomerase in such a way as not to cause cancer. Cancer cells possess a gene that produces telomerase, which effectively enables them to become immortal by reproducing indefinitely. A key cancer-fighting strategy, therefore, involves blocking the ability of cancer cells to generate telomerase. This may seem to contradict the idea of extending the telomeres in normal cells to combat this source of aging, but attacking the telomerase of the cancer cells in an emerging tumor could be done without necessarily compromising an orderly telomere-extending therapy for normal cells. However, to avoid complications, such therapies could be halted during a period of cancer therapy.

Reversing Aging

It is logical to assume that early in the evolution of our species (and precursors to our species) survival would not have been aided—indeed, it would have been compromised—by individuals living long past their child-rearing years. Recent research, however, supports the so-called grandma hypothesis, which suggests a countereffect. University of Michigan anthropologist Rachel Caspari and University of California at Riverside's San-Hee Lee found evidence that the proportion of humans living to become grandparents (who in primitive societies were often as young as thirty) increased steadily over the past two million years, with a fivefold increase occurring in the Upper Paleolithic era (around thirty thousand years ago). This research has been cited to support the hypothesis that the survival of human societies was aided by grandmothers, who not only assisted in raising extended families but also passed on the accumulated wisdom of elders. Such effects may be a reasonable interpretation of the data, but the overall increase in longevity also reflects an ongoing trend toward longer life expectancy that continues to this day. Likewise, only a modest number of grandmas (and a few grandpas) would have been needed to account for the societal effects that proponents of this theory have claimed, so the hypothesis does not appreciably challenge the conclusion that genes that supported significant life extension were not selected for.

Aging is not a single process but involves a multiplicity of changes. De Grey describes seven key aging processes that encourage senescence, and he has identified strategies for reversing each one.

DNA Mutations.[49] Generally mutations to nuclear DNA (the DNA in the chromosomes in the nucleus) result in a defective cell that's quickly eliminated or a cell that simply doesn't function optimally. The type of mutation that is of primary concern (as it leads to increased death rates) is one that affects orderly cellular reproduction, resulting in cancer. This means that if we can cure cancer using the strategies described above, nuclear mutations should largely be rendered harmless. De Grey's proposed strategy for cancer is preemptive: it involves using gene therapy to remove from all our cells the genes that cancers need to turn on in order to maintain their telomeres when they divide. This will cause any potential cancer tumors to wither away before they grow large enough to cause harm. Strategies for deleting and suppressing genes are already available and are being rapidly improved.

Toxic Cells. Occasionally cells reach a state in which they're not cancerous, but it would still be best for the body if they did not survive. Cell senescence is an example, as is having too many fat cells. In these cases, it is easier to kill these cells than to attempt to revert them to a healthy state. Methods are being developed to target "suicide genes" to such cells and also to tag these cells in a way that directs the immune system to destroy them.

Mitochrondrial Mutations. Another aging process is the accumulation of mutations in the thirteen genes in the mitochondria, the energy factories for the cell.[50] These few genes are critical to the efficient functioning of our cells and undergo mutation at a higher rate than genes in the nucleus. Once we master somatic gene therapy, we could put multiple copies of these genes in the cell nucleus, thereby providing redundancy (backup) for such vital genetic information. The mechanism already exists in the cell to allow nucleus-encoded proteins to be imported into the mitochondria, so it is not necessary for these proteins to be produced in the mitochondria themselves. In fact, most of the proteins needed for mitochondrial function are already coded by the nuclear DNA. Researchers have already been successful in transferring mitochondrial genes into the nucleus in cell cultures.

Intracellular Aggregates. Toxins are produced both inside and outside cells. De Grey describes strategies using somatic gene therapy to introduce new genes that will break down what he calls "intracellular aggregates"—toxins within cells. Proteins have been identified that can destroy virtually any toxin, using bacteria that can digest and destroy dangerous materials ranging from TNT to dioxin.

A key strategy being pursued by various groups for combating toxic materials outside the cell, including misformed proteins and amyloid plaque (seen in Alzheimer's disease and other degenerative conditions), is to create vaccines that act against their constituent molecules.[51] Although this approach may result in the toxic material's being ingested by immune system cells, we can then use the strategies for combating intracellular aggregates described above to dispose of it.

Extracellular Aggregates. AGEs (advanced glycation end-products) result from undesirable cross-linking of useful molecules as a side effect of excess sugar. These cross-links interfere with the normal functioning of proteins and are key contributors to the aging process. An experimental drug called ALT-711 (phenacyldimenthylthiazolium chloride) can dissolve these cross-links without damaging the original tissue.[52] Other molecules with this capability have also been identified.

Cell Loss and Atrophy. Our body's tissues have the means to replace worn-out cells, but this ability is limited in certain organs. For example, as we get older, the heart is unable to replace its cells at a sufficient rate, so it compensates by making surviving cells bigger using fibrous material. Over time this causes the heart to become less supple and responsive. A primary strategy here is to deploy therapeutic cloning of our own cells, as described below.

Progress in combating all of these sources of aging is moving rapidly in animal models, and translation into human therapies will follow. Evidence from the genome project indicates that no more than a few hundred genes are involved in the aging process. By manipulating these genes, radical life extension has already been achieved in simpler animals. For example, by modifying genes in the *C. elegans* worm that control its insulin and sex-hormone levels, the lifespan of the test animals was expanded sixfold, to the equivalent of a five-hundred-year lifespan for a human.[53]

A hybrid scenario involving both bio- and nanotechnology contemplates turning biological cells into computers. These "enhanced intelligence" cells can then detect and destroy cancer cells and pathogens or even regrow human body parts. Princeton biochemist Ron Weiss has modified cells to incorporate a variety of logic functions that are used for basic computation.[54] Boston University's Timothy Gardner has developed a cellular logic switch, another basic building block for turning cells into computers.[55] Scientists at the MIT Media Lab have developed ways to use wireless communication to send messages, including intricate sequences of instructions, to the computers inside modified cells.[56] Weiss points out that "once you have the ability to program cells, you don't have to be constrained by what the cells know how to do already. You can program them to do new things, in new patterns."

Human Cloning: The Least Interesting Application of Cloning Technology

One of the most powerful methods of applying life's machinery involves harnessing biology's own reproductive mechanisms in the form of cloning. Cloning will be a key technology—not for cloning actual humans but for life-extension purposes, in the form of "therapeutic cloning." This process creates new tissues with "young" telomere-extended and DNA-corrected cells to replace without surgery defective tissues or organs.

All responsible ethicists, including myself, consider human cloning at the present time to be unethical. The reasons, however, for me have little to do with the slippery-slope issues of manipulating human life. Rather, the technology today simply does not yet work reliably. The current technique of fusing a cell nucleus from a donor to an egg cell using an electric spark simply causes a high level of genetic errors.[57] This is the primary reason that most of the fetuses created by this method do not make it to term. Even those that do make it have genetic defects. Dolly the Sheep developed an obesity problem in adulthood, and the majority of cloned animals produced thus far have had unpredictable health problems.[58]

Scientists have a number of ideas for perfecting cloning, including alternative ways of fusing the nucleus and egg cell without use of a destructive electrical spark, but until the technology is demonstrably safe, it would be unethical to create a human life with such a high likelihood of severe health problems. There is no doubt that human cloning will occur, and occur soon, driven by all the usual reasons, ranging from its publicity value to its utility as a very weak form of immortality. The methods that are demonstrable in advanced animals will work quite well in humans. Once the technology is perfected in terms of safety, the ethical barriers will be feeble if they exist at all.

Cloning is a significant technology, but the cloning of humans is not its most noteworthy usage. Let's first address its most valuable applications and then return to its most controversial one.

Why Is Cloning Important? The most immediate use for cloning is improved breeding by offering the ability to directly reproduce an animal with a desirable set of genetic traits. A powerful example is reproducing animals from transgenic embryos (embryos with foreign genes) for pharmaceutical production. A case in point: a promising anticancer treatment is an antiangiogenesis drug called aaATIII, which is produced in the milk of transgenic goats.[59]

Preserving Endangered Species and Restoring Extinct Ones. Another exciting application is re-creating animals from endangered species. By cryo-preserving cells from these species, they never need become extinct. It will eventually be possible to re-create animals from recently extinct species. In 2001 scientists were able to synthesize DNA for the Tasmanian tiger, which had then been extinct for sixty-five years, with the hope of bringing this species back to life.[60] As for long-extinct species (for example, dinosaurs), it is highly doubtful that we will find the fully intact DNA required in a single preserved cell (as they did in the movie *Jurassic Park*). It is likely, however, that we will eventually be able to synthesize the necessary DNA by patching together the information derived from multiple inactive fragments.

Therapeutic Cloning. Perhaps the most valuable emerging application is therapeutic cloning of one's own organs. By starting with germ-line cells (inherited from the eggs or sperm and passed on to offspring), genetic engineers can trigger differentiation into diverse types of cells. Because differentiation takes place during the prefetal stage (that is, prior to implantation of a fetus), most ethicists believe this process does not raise concerns, although the issue has remained highly contentious.[61]

Human Somatic-Cell Engineering. This even more promising approach, which bypasses the controversy of using fetal stem cells entirely, is called trans-differentiation; it creates new tissues with a patient's own DNA by converting one type of cell (such as a skin cell) into another (such as a pancreatic islet cell or a heart cell).[62] Scientists from the United States and Norway have recently been successful in reprogramming liver cells into becoming pancreas cells. In another series of experiments, human skin cells were transformed to take on many of the characteristics of immune-system cells and nerve cells.[63]

Consider the question, What is the difference between a skin cell and any other type of cell in the body? After all, they all have the same DNA. As noted above, the differences are found in protein signaling factors, which include short RNA fragments and peptides, which we are now beginning to understand.[64] By manipulating these proteins, we can influence gene expression and trick one type of cell into becoming another.

Perfecting this technology would not only defuse a sensitive ethical and political issue but also offer an ideal solution from a scientific perspective. If you need pancreatic islet cells or kidney tissues—or even a whole new heart—to avoid autoimmune reactions, you would strongly prefer to obtain these with your own DNA rather than the DNA from someone else's germ-line cells. In addition, this approach uses plentiful skin cells (of the patient) rather than rare and precious stem cells.

Transdifferentiation will directly grow an organ with your genetic makeup. Perhaps most important, the new organ can have its telomeres fully extended to their original youthful length, so that the new organ is effectively young again.[65] We can also correct accumulated DNA errors by selecting the appropriate skin cells (that is, ones without DNA errors) prior to transdifferentiation into other types of cells. Using this method an eighty-year-old man could have his heart replaced with the same heart he had when he was, say, twenty-five.

Current treatments for type 1 diabetes require strong antirejection drugs that can have dangerous side effects.[66] With somatic-cell engineering, type 1 diabetics will be able to make pancreatic islet cells from their own cells, either from skin cells (transdifferentiation) or from adult stem cells. They would be using their own DNA, and drawing upon a relatively inexhaustible supply of cells, so no antirejection drugs would be required. (But to fully cure type 1 diabetes, we would also have to overcome the patient's autoimmune disorder, which causes his body to destroy islet cells.)

Even more exciting is the prospect of replacing one's organs and tissues with their "young" replacements without surgery. Introducing cloned, telomere-extended, DNA-corrected cells into an organ will allow them to integrate themselves with the older cells. By repeated treatments of this kind over a period of time, the organ will end up being dominated by the younger cells. We normally replace our own cells on a regular basis anyway, so why not do so with youthful rejuvenated cells rather than telomere-shortened error-filled ones? There's no reason why we couldn't repeat this process for every organ and tissue in our body, enabling us to grow progressively younger.

Solving World Hunger. Cloning technologies even offer a possible solution for world hunger: creating meat and other protein sources in a factory *without animals* by cloning animal muscle tissue. Benefits would include extremely low cost, avoidance of pesticides and hormones that occur in natural meat, greatly reduced environmental impact (compared to factory farming), improved nutritional profile, and no animal suffering. As with therapeutic cloning, we would not be creating the entire animal but rather directly producing the desired animal parts or flesh. Essentially, all of the meat—billions of pounds of it—would be derived from a single animal.

There are other benefits to this process besides ending hunger. By creating meat in this way, it becomes subject to the law of accelerating returns—the exponential improvements in price-performance of information-based technologies over time—and will thus become extremely inexpensive. Even though hunger in the world today is certainly exacerbated by political issues and conflicts, meat could become so inexpensive that it would have a profound effect on the affordability of food.

The advent of animal-less meat will also eliminate animal suffering. The economics of factory farming place a very low priority on the comfort of animals, which are treated as cogs in a machine. The meat produced in this manner, although normal in all other respects, would not be part of an animal with a nervous system, which is generally regarded as a necessary element for suffering to occur, at least in a biological animal. We could use the same approach to produce such animal by-products as leather and fur. Other major advantages would be to eliminate the enormous ecological and environmental damage created by factory farming as well as the risk of prion-based diseases, such as mad-cow disease and its human counterpart, vCJD.[67]

Human Cloning Revisited. This brings us again to human cloning. I predict that once the technology is perfected, neither the acute dilemmas seen by ethicists nor the profound promise heralded by enthusiasts will predominate. So what if we have genetic twins separated by one or more generations? Cloning is likely to prove to be like other reproductive technologies that were briefly controversial but rapidly accepted. Physical cloning is far different from mental cloning, in which a person's entire personality, memory, skills, and history will ultimately be downloaded into a different, and most likely more powerful, thinking medium. There's no issue of philosophical identity with genetic cloning, since such clones would be different people, even more so than conventional twins are today.

If we consider the full concept of cloning, from cell to organisms, its benefits have enormous synergy with the other revolutions occurring in biology as well as in computer technology. As we learn to understand the genome and proteome (the expression of the genome into proteins) of both humans and animals, and as we develop powerful new means of harnessing genetic information, cloning provides the means to replicate animals, organs, and cells. And that has profound implications for the health and well-being of both ourselves and our evolutionary cousins in the animal kingdom.

NED LUDD: *If everyone can change their genes, then everyone will choose to be "perfect" in every way, so there'll be no diversity and excelling will become meaningless.*

RAY: *Not exactly. Genes are obviously important, but our nature—skills, knowledge, memory, personality—reflects the design information in our genes, as our bodies and brains self-organize through our experience. This is also readily evident in our health. I personally have a genetic disposition to type 2 diabetes, having actually been diagnosed with that disease more than twenty years ago. But I don't have any indication of diabetes today because I've overcome this genetic disposition as a result of reprogramming my biochemistry through lifestyle choices such as nutrition, exercise, and aggressive supplementation. With regard to our brains, we all have various aptitudes, but our actual talents are a function of what we've learned, developed, and experienced. Our genes reflect dispositions only. We can see how this works in the development of the brain. The genes describe certain rules and constraints for patterns of interneuronal connections, but the actual connections we have as adults are the result of a self-organizing process based on our learning. The final result—who we are—is deeply influenced by both nature (genes) and nurture (experience).*

So when we gain the opportunity to change our genes as adults, we won't wipe out the influence of our earlier genes. Experiences prior to the gene therapy will have been translated through the pretherapy genes, so one's character and personality would still be shaped primarily by the original genes. For example, if someone added genes for musical aptitude to his brain through gene therapy, he would not suddenly become a music genius.

NED: *Okay, I understand that designer baby boomers can't get away completely from their predesigner genes, but with designer babies they'll have the genes and the time to express them.*

RAY: *The "designer baby" revolution is going to be a very slow one; it won't be a significant factor in this century. Other revolutions will overtake it. We won't have the technology for designer babies for another ten to twenty years. To the extent that it is used, it would be adopted gradually, and then it will take those generations another twenty years to reach maturity. By that time, we're approaching the Singularity, with the real revolution being the predominance of nonbiological intelligence. That will go far beyond the capabilities of any designer genes. The idea of designer babies and baby boomers is just the reprogramming of the information processes in biology. But it's still biology, with all its profound limitations.*

NED: *You're missing something. Biological is what we are. I think most people would agree that being biological is the quintessential attribute of being human.*

RAY: *That's certainly true today.*

NED: *And I plan to keep it that way.*

RAY: *Well, if you're speaking for yourself, that's fine with me. But if you stay biological and don't reprogram your genes, you won't be around for very long to influence the debate.*

Nanotechnology: The Intersection of Information and the Physical World

> The role of the infinitely small is infinitely large.
>
> —LOUIS PASTEUR

> But I am not afraid to consider the final question as to whether, ultimately, in the great future, we can arrange the atoms the way we want; the very atoms, all the way down!
>
> —RICHARD FEYNMAN

> Nanotechnology has the potential to enhance human performance, to bring sustainable development for materials, water, energy, and food, to protect against unknown bacteria and viruses, and even to diminish the reasons for breaking the peace [by creating universal abundance].
>
> —NATIONAL SCIENCE FOUNDATION NANOTECHNOLOGY REPORT

Nanotechnology promises the tools to rebuild the physical world—our bodies and brains included—molecular fragment by molecular fragment, potentially atom by atom. We are shrinking the key feature size of technology, in accordance with the law of accelerating returns, at the exponential rate of approximately a factor of four per linear dimension per decade.[68] At this rate the key feature sizes for most electronic and many mechanical technologies will be in the nanotechnology range—generally considered to be under one hundred nanometers—by the 2020s. (Electronics has already dipped below this threshold, although not yet in three-dimensional structures and not yet self-assembling.) Meanwhile rapid progress has been made, particularly in the last several years, in preparing the conceptual framework and design ideas for the coming age of nanotechnology.

As important as the biotechnology revolution discussed above will be, once its methods are fully mature, limits will be encountered in biology itself. Although biological systems are remarkable in their cleverness, we have also discovered that they are dramatically suboptimal. I've mentioned the extremely slow speed of communication in the brain, and as I discuss below (see p. 253), robotic replacements for our red blood cells could be thousands of times more efficient than their biological counterparts.[69] Biology will never be able to match what we will be capable of engineering once we fully understand biology's principles of operation.

The revolution in nanotechnology, however, will ultimately enable us to redesign and rebuild, molecule by molecule, our bodies and brains and the world with which we interact.[70] These two revolutions are overlapping, but the full realization of nanotechnology lags behind the biotechnology revolution by about one decade.

Most nanotechnology historians date the conceptual birth of nanotechnology to physicist Richard Feynman's seminal speech in 1959, "There's Plenty of Room at the Bottom," in which he described the inevitability and profound implications of engineering machines at the level of atoms:

> The principles of physics, as far as I can see, do not speak against the possibility of maneuvering things atom by atom. It would be, in principle, possible . . . for a physicist to synthesize any chemical substance that the chemist writes down. . . . How? Put the atoms down where the chemist says, and so you make the substance. The problems of chemistry and biology can be greatly helped if our ability to see what we are doing, and to do things on an atomic level, is ultimately developed—a development which I think cannot be avoided.[71]

An even earlier conceptual foundation for nanotechnology was formulated by the information theorist John von Neumann in the early 1950s with his model of a self-replicating system based on a universal constructor, combined with a universal computer.[72] In this proposal the computer runs a program that directs the constructor, which in turn constructs a copy of both the computer (including its self-replication program) and the constructor. At this level of description von Neumann's proposal is quite abstract—the computer and constructor could be made in a great variety of ways, as well as from diverse materials, and could even be a theoretical mathematical construction. But he took the concept one step further and proposed a "kinematic constructor": a robot with at least one manipulator (arm) that would build a replica of itself from a "sea of parts" in its midst.[73]

It was left to Eric Drexler to found the modern field of nanotechnology, with a draft of his landmark Ph.D. thesis in the mid-1980s, in which he essentially combined these two intriguing suggestions. Drexler described a von Neumann kinematic constructor, which for its sea of parts used atoms and molecular fragments, as suggested in Feynman's speech. Drexler's vision cut across many disciplinary boundaries and was so far-reaching that no one was daring enough to be his thesis adviser except for my own mentor, Marvin Minsky. Drexler's dissertation (which became his book *Engines of Creation* in 1986 and was articulated technically in his 1992 book, *Nanosystems*) laid out the foundation of nanotechnology and provided the road map still being followed today.[74]

Drexler's "molecular assembler" will be able to make almost anything in the world. It has been referred to as a "universal assembler," but Drexler and other nanotechnology theorists do not use the word "universal" because the products of such a system necessarily have to be subject to the laws of physics and chemistry, so only atomically stable structures would be viable. Furthermore, any specific assembler would be restricted to building products from its sea of parts, although the feasibility of using individual atoms has been shown. Nevertheless, such an assembler could make just about any physical device we would want, including highly efficient computers and subsystems for other assemblers.

Although Drexler did not provide a detailed design for an assembler—such a design has still not been fully specified—his thesis did provide extensive feasibility arguments for each of the principal components of a molecular assembler, which include the following subsystems:

- The *computer:* to provide the intelligence to control the assembly process. As with all of the device's subsystems, the computer needs to be small and simple. As I described in chapter 3, Drexler provides an intriguing conceptual description of a mechanical computer with molecular "locks" instead of transistor gates. Each lock would require only sixteen cubic nanometers of space and could switch ten billion times per second. This proposal remains more competitive than any known electronic technology, although electronic computers built from three-dimensional arrays of carbon nanotubes appear to provide even higher densities of computation (that is, calculations per second per gram).[75]

• The *instruction architecture:* Drexler and his colleague Ralph Merkle have proposed an SIMD (single instruction multiple data) architecture in which a single data store would record the instructions and transmit them to trillions of molecular-sized assemblers (each with its own simple computer) simultaneously. I discussed some of the limitations of the SIMD architecture in chapter 3, but this design (which is easier to implement than the more flexible multiple-instruction multiple-data approach) is sufficient for the computer in a universal nanotechnology assembler. With this approach each assembler would not have to store the entire program for creating the desired product. A "broadcast" architecture also addresses a key safety concern: the self-replication process could be shut down, if it got out of control, by terminating the centralized source of the replication instructions.

However, as Drexler points out, a nanoscale assembler does not necessarily have to be self-replicating.[76] Given the inherent dangers in self-replication, the ethical standards proposed by the Foresight Institute (a think tank founded by Eric Drexler and Christine Peterson) contain prohibitions against unrestricted self-replication, especially in a natural environment.

As I will discuss in chapter 8, this approach should be reasonably effective against inadvertent dangers, although it could be circumvented by a determined and knowledgeable adversary.

• *Instruction transmission:* Transmission of the instructions from the centralized data store to each of the many assemblers would be accomplished electronically if the computer is electronic or through mechanical vibrations if Drexler's concept of a mechanical computer were used.

• The *construction robot:* The constructor would be a simple molecular robot with a single arm, similar to von Neumann's kinematic constructor but on a tiny scale. There are already examples of experimental molecular-scale systems that can act as motors and robot legs, as I discuss below.

• The *robot arm tip:* Drexler's *Nanosystems* provided a number of feasible chemistries for the tip of the robot arm to make it capable of grasping (using appropriate atomic-force fields) a molecular fragment, or even a single atom, and then depositing it in a desired location. In the chemical-vapor deposition process used to construct artificial diamonds, individual carbon atoms, as well as molecular fragments, are moved to other locations through chemical reactions at the tip. Building artificial diamonds is a chaotic process involving trillions of atoms, but conceptual proposals by Robert Freitas and Ralph Merkle contemplate robot arm tips that can remove hydrogen atoms from a source material and deposit them at desired locations in the construction of a molecular machine. In this proposal, the tiny machines are built out of a diamondoid material. In addition to having great strength, the material can be doped with impurities in a precise fashion to create electronic components such as transistors. Simulations have shown that such molecular-scale gears, levers, motors, and other mechanical systems would operate properly as intended.[77] More recently attention has been focused on carbon nanotubes, comprising hexagonal arrays of carbon atoms assembled in three dimensions, which are also capable of providing both mechanical and electronic functions at the molecular level. I provide examples below of molecular-scale machines that have already been built.

- The assembler's *internal environment* needs to prevent environmental impurities from interfering with the delicate assembly process. Drexler's proposal is to maintain a near vacuum and build the assembler walls out of the same diamondoid material that the assembler itself is capable of making.

- The *energy* required for the assembly process can be provided either through electricity or through chemical energy. Drexler proposed a chemical process with the fuel interlaced with the raw building material. More recent proposals use nanoengineered fuel cells incorporating hydrogen and oxygen or glucose and oxygen, or acoustic power at ultrasonic frequencies.[78]

Although many configurations have been proposed, the typical assembler has been described as a tabletop unit that can manufacture almost any physically possible product for which we have a software description, ranging from computers, clothes, and works of art to cooked meals.[79] Larger products, such as furniture, cars, or even houses, can be built in a modular fashion or using larger assemblers. Of particular importance is the fact that an assembler can create copies of itself, unless its design specifically prohibits this (to avoid potentially dangerous self-replication). The incremental cost of creating any physical product, including the assemblers themselves, would be pennies per pound—basically the cost of the raw materials. Drexler estimates total manufacturing cost for a molecular-manufacturing process in the range of ten cents to fifty cents per kilogram, regardless of whether the manufactured product were clothing, massively parallel supercomputers, or additional manufacturing systems.[80]

The real cost, of course, would be the value of the information describing each type of product—that is, the software that controls the assembly process. In other words, the value of everything in the world, including physical objects, would be based essentially on information. We are not that far from this situation today, since the information content of products is rapidly increasing, gradually approaching an asymptote of 100 percent of their value.

The design of the software controlling molecular-manufacturing systems would *itself* be extensively automated, much as chip design is today. Chip designers don't specify the location of each of the billions of wires and components but rather the specific functions and features, which computer-aided design (CAD) systems translate into actual chip layouts. Similarly, CAD systems would produce the molecular-manufacturing control software from high-level specifications. This would include the ability to reverse engineer a Product by scanning it in three dimensions and then generating the software needed to replicate its overall capabilities.

In operation, the centralized data store would send out commands simultaneously to many trillions (some estimates as high as 10^{18}) of robots in an assembler, each receiving the same instruction at the same time. The assembler would create these molecular robots by starting with a small number and then using these robots to create additional ones in an iterative fashion, until the requisite number had been created. Each robot would have a local data storage that specifies the type of mechanism it's building. This storage would be used to mask the global instructions being sent from the centralized data store so that certain instructions are blocked and local parameters are filled in. In this way, even though all of the assemblers are receiving the same sequence of instructions, there is a level of customization to the part being built by each molecular robot. This process is analogous to gene expression in biological systems. Although every cell has every gene, only those genes relevant to a particular cell type are expressed. Each robot extracts the raw materials and fuel it needs, which include individual carbon atoms and molecular fragments, from the source material.

The Biological Assembler

Nature shows that molecules can serve as machines because living things work by means of such machinery. Enzymes are molecular machines that make, break, and rearrange the bonds holding other molecules together. Muscles are driven by molecular machines that haul fibers past one another. DNA serves as a data-storage system, transmitting digital instructions to molecular machines, the ribosomes, that manufacture protein molecules. And these protein molecules, in turn, make up most of the molecular machinery.

—ERIC DREXLER

The ultimate existence proof of the feasibility of a molecular assembler is life itself. Indeed, as we deepen our understanding of the information basis of life processes, we are discovering specific ideas that are applicable to the design requirements of a generalized molecular assembler. For example, proposals have been made to use a molecular energy source of glucose and ATP, similar to that used by biological cells.

Consider how biology solves each of the design challenges of a Drexler assembler. The ribosome represents both the computer and the construction robot. Life does not use centralized data storage but provides the entire code to every cell. The ability to restrict the local data storage of a nanoengineered robot to only a small part of the assembly code (using the "broadcast" architecture), particularly when doing self-replication, is one critical way nanotechnology can be engineered to be safer than biology.

Life's local data storage is, of course, the DNA strands, broken into specific genes on the chromosomes. The task of instruction masking (blocking genes that do not contribute to a particular cell type) is controlled by the short RNA molecules and peptides that govern gene expression. The internal environment in which the ribosome is able to function is the particular chemical environment maintained inside the cell, which includes a particular acid-alkaline equilibrium (pH around 7 in human cells) and other chemical balances. The cell membrane is responsible for protecting this internal environment from disturbance.

Upgrading the Cell Nucleus with a Nanocomputer and Nanobot. Here's a conceptually simple proposal to overcome all biological pathogens except for prions (self-replicating pathological proteins). With the advent of full-scale nanotechnology in the 2020s we will have the potential to replace biology's genetic-information repository in the cell nucleus with a nanoengineered system that would maintain the genetic code and simulate the actions of RNA, the ribosome, and other elements of the computer in biology's assembler. A nanocomputer would maintain the genetic code and implement the gene-expression algorithms. A nanobot would then construct the amino-acid sequences for the expressed genes.

There would be significant benefits in adopting such a mechanism. We could eliminate the accumulation of DNA transcription errors, one major source of the aging process. We could introduce DNA changes to essentially reprogram our genes (something we'll be able to do long before this scenario, using gene-therapy techniques). We would also be able to defeat biological pathogens (bacteria, viruses, and cancer cells) by blocking any unwanted replication of genetic information.

Nanobot-Based Nucleus

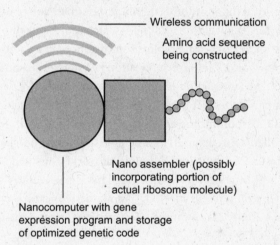

Wireless communication

Amino acid sequence being constructed

Nano assembler (possibly incorporating portion of actual ribosome molecule)

Nanocomputer with gene expression program and storage of optimized genetic code

With such a nanoengineered system the recommended broadcast architecture would enable us to turn off unwanted replication, thereby defeating cancer, autoimmune reactions, and other disease processes. Although most of these disease processes will already have been vanquished by the biotechnology methods described in the previous section, reengineering the computer of life using nanotechnology could eliminate any remaining obstacles and create a level of durability and flexibility that goes beyond the inherent capabilities of biology.

The robot arm tip would use the ribosome's ability to implement enzymatic reactions to break off an individual amino acid, each of which is bound to a specific tRNA, and to connect it to its adjoining amino acid using a peptide bond. Thus, such a system could utilize portions of the ribosome itself, since this biological machine is capable of constructing the requisite string of amino acids.

However, the goal of molecular manufacturing is not merely to replicate the molecular-assembly capabilities of biology. Biological systems are limited to building systems from protein, which has profound limitations in strength and speed. Although biological proteins are three-dimensional, biology is restricted to that class of chemicals that can be folded from a one-dimensional string of amino acids. Nanobots built from diamondoid gears and rotors can also be thousands of times faster and stronger than biological cells.

The comparison is even more dramatic with regard to computation: the switching speed of nanotube-based computation would be millions of times faster than the extremely slow transaction speed of the electrochemical switching used in mammalian interneuronal connections.

The concept of a diamondoid assembler described above uses a consistent input material (for construction and fuel), which represents one of several protections against molecular-scale replication of robots in an uncontrolled fashion in the outside world. Biology's replication robot, the ribosome, also requires carefully controlled source and fuel materials, which are provided by our digestive system. As nanobased replicators become more sophisticated, more capable of extracting carbon atoms and carbon-based molecular fragments from less well-controlled source materials, and able to operate outside of controlled replicator enclosures such as in the biological world, they will have the potential to present a grave threat to that world. This is particularly true in view of the vastly greater strength and speed of nanobased replicators over any biological system. That ability is, of course, the source of great controversy, which I discuss in chapter 8.

In the decade since publication of Drexler's *Nanosystems*, each aspect of Drexler's conceptual designs has been validated through additional design proposals,[81] supercomputer simulations, and, most important, actual construction of related molecular machines. Boston College chemistry professor T. Ross Kelly reported that he constructed a chemically powered nanomotor out of seventy-eight atoms.[82] A biomolecular research group headed by Carlo Montemagno created an ATP-fueled nanomotor.[83] Another molecule-sized motor fueled by solar energy was created out of fifty-eight atoms by Ben Feringa at the University of Groningen in the Netherlands.[84] Similar progress has been made on other molecular-scale mechanical components such as gears, rotors, and levers. Systems demonstrating the use of chemical energy and acoustic energy (as originally described by Drexler) have been designed, simulated, and actually constructed. Substantial progress has also been made in developing various types of electronic components from molecular-scale devices, particularly in the area of carbon nanotubes, an area that Richard Smalley has pioneered.

Nanotubes are also proving to be very versatile as a structural component. A conveyor belt constructed out of nanotubes was demonstrated recently by scientists at Lawrence Berkeley National Laboratory.[85] The nanoscale conveyor belt was used to transport tiny indium particles from one location to another, although the technique could be adapted to move a variety of molecule-sized objects. By controlling an electrical current applied to the device, the direction and velocity of movement can be modulated. "It's the equivalent of turning a knob . . . and taking macroscale control of nanoscale mass transport," said Chris Regan, one of the designers. "And it's reversible: we can change the current's polarity and drive the indium back to its original position." The ability to rapidly shuttle molecule-sized building blocks to precise locations is a key step toward building molecular assembly lines.

A study conducted for NASA by General Dynamics has demonstrated the feasibility of self-replicating nanoscale machines.[86] Using computer simulations, the researchers showed that molecularly precise robots called kinematic cellular automata, built from reconfigurable molecular modules, were capable of reproducing themselves. The designs also used the broadcast architecture, which established the feasibility of this safer form of self-replication.

DNA is proving to be as versatile as nanotubes for building molecular structures. DNA's proclivity to link up with itself makes it a useful structural component. Future designs may combine this attribute as well as its capacity for storing information. Both nanotubes and DNA have outstanding properties for information storage and logical control, as well as for building strong three-dimensional structures.

A research team at Ludwig Maximilians University in Munich has built a "DNA hand" that can select one of several proteins, bind to it, and then release it upon command.[87] Important steps in creating a DNA assembler mechanism akin to the ribosome were demonstrated recently by nanotechnology researchers Shiping Liao and Nadrian Seeman.[88] Grasping and letting go of molecular objects in a controlled manner is another important enabling capability for molecular nanotechnology assembly.

Scientists at the Scripps Research Institute demonstrated the ability to create DNA building blocks by generating many copies of a 1,669-nucleotide strand of DNA that had carefully placed self-complementary regions.[89] The strands self-assembled spontaneously into rigid octahedrons, which could be used as blocks for elaborate three-dimensional structures. Another application of this process could be to employ the octahedrons as compartments to deliver proteins, which Gerald F. Joyce, one of the Scripps researchers, called a "virus in reverse." Viruses, which are also self-assembling, usually have outer shells of protein with DNA (or RNA) on the inside. "With this," Joyce points out, "you could in principle have DNA on the outside and proteins on the inside."

A particularly impressive demonstration of a nanoscale device constructed from DNA is a tiny biped robot that can walk on legs that are ten nanometers long.[90] Both the legs and the walking track are built from DNA, again chosen for the molecule's ability to attach and detach itself in a controlled manner. The nanorobot, a project of chemistry professors Nadrian Seeman and William Sherman of New York University, walks by detaching its legs from the track, moving down it, and then reattaching its legs to the track. The project is another impressive demonstration of the ability of nanoscale machines to execute precise maneuvers.

An alternate method of designing nanobots is to learn from nature. Nanotechnologist Michael Simpson of Oak Ridge National Laboratory describes the possibility of exploiting bacteria "as ready-made machine[s]." Bacteria, which are natural nanobot-size objects, are able to move, swim, and pump liquids.[91] Linda Turner, a scientist at the Rowland Institute at Harvard, has focused on their thread-size arms, called fimbriae, which are able to perform a wide variety of tasks, including carrying other nanoscale objects and mixing fluids. Another approach is to use only parts of bacteria. A research group headed by Viola Vogel at the University of Washington built a system using just the limbs of E. coli bacteria that was able to sort out nanoscale beads of different sizes. Since bacteria are natural nanoscale systems that can perform a wide variety of functions, the ultimate goal of this research will be to reverse engineer the bacteria so that the same design principles can be applied to our own nanobot designs.

Fat and Sticky Fingers

In the wake of the rapidly expanding development of each facet of future nanotechnology systems, no serious flaw in Drexler's nanoassembler concept has been described. A highly publicized objection in 2001 by Nobelist Richard Smalley in Scientific American was based on a distorted description of the Drexler proposal;[92] it did not address the extensive body of work that has been carried out in the past decade. As a pioneer of carbon nanotubes Smalley has been enthusiastic about a variety of applications of nanotechnology, having written that "nanotechnology holds the answer, to the extent there are answers, to most of our pressing material needs in energy, health, communication, transportation, food, water," but he remains skeptical about molecular nanotechnology assembly.

Smalley describes Drexler's assembler as consisting of five to ten "fingers" (manipulator arms) to hold, move, and place each atom in the machine being constructed. He then goes on to point out that there isn't room for so many fingers in the cramped space in which a molecular-assembly nanorobot has to work (which he calls the "fat fingers" problem) and that these fingers would have difficulty letting go of their atomic cargo because of molecular attraction forces (the "sticky fingers" problem). Smalley also points out that an "intricate three-dimensional waltz . . . is carried out" by five to fifteen atoms in a typical chemical reaction.

In fact, Drexler's proposal doesn't look anything like the straw-man description that Smalley criticizes. Drexler's proposal, and most of those that have followed, uses a single "finger." Moreover, there have been extensive descriptions and analyses of viable tip chemistries that do not involve grasping and placing atoms as if they were mechanical pieces to be deposited in place. In addition to the examples I provided above (for example, the DNA hand), the feasibility of moving hydrogen atoms using Drexler's "propynyl hydrogen abstraction" tip has been extensively confirmed in the intervening years.[93] The ability of the scanning-probe microscope (SPM), developed at IBM in 1981, and the more sophisticated atomic-force microscope (AFM) to place individual atoms through specific reactions of a tip with a molecular-scale structure provides additional proof of the concept. Recently, scientists at Osaka University used an AFM to move individual nonconductive atoms using a mechanical rather than electrical technique.[94] The ability to move both conductive and nonconductive atoms and molecules will be needed for future molecular nanotechnology.[95]

Indeed, if Smalley's critique were valid, none of us would be here to discuss it, because life itself would be impossible, given that biology's assembler does exactly what Smalley says is impossible.

Smalley also objects that, despite "working furiously, . . . generating even a tiny amount of a product would take [a nanobot] . . . millions of years." Smalley is correct, of course, that an assembler with only one nanobot wouldn't produce any appreciable quantities of a product. However, the basic concept of nanotechnology is that we will use trillions of nanobots to accomplish meaningful results—a factor that is also the source of the safety concerns that have received so much attention. Creating this many nanobots at reasonable cost will require self-replication at some level, which while solving the economic issue will introduce potentially grave dangers, a concern I will address in chapter 8. Biology uses the same solution to create organisms with trillions of cells, and indeed we find that virtually all diseases derive from biology's self-replication process gone awry.

Earlier challenges to the concepts underlying nanotechnology have also been effectively addressed. Critics pointed out that nanobots would be subject to bombardment by thermal vibration of nuclei, atoms, and molecules. This is one reason conceptual designers of nanotechnology have emphasized building structural components from diamondoid or carbon nanotubes. Increasing the strength or stiffness of a system reduces its susceptibility to thermal effects. Analysis of these designs has shown them to be thousands of times more stable in the presence of thermal effects than are biological systems, so they can operate in a far wider temperature range.[96]

Similar challenges were made regarding positional uncertainty from quantum effects, based on the extremely small feature size of nanoengineered devices. Quantum effects are significant for an electron, but a single carbon-atom nucleus is more than twenty thousand times more massive than an electron. A nanobot will be constructed from millions to billions of carbon and other atoms, making it up to trillions of times more massive than an electron. Plugging this ratio in the fundamental equation for quantum positional uncertainty shows it to be an insignificant factor.[97]

Power has represented another challenge. Proposals involving glucose-oxygen fuel cells have held up well in feasibility studies by Freitas and others.[98] An advantage of the glucose-oxygen approach is that nanomedicine applications can harness the glucose, oxygen, and ATP resources already provided by the human digestive system. A nanoscale motor was recently created using propellers made of nickel and powered by an ATP-based enzyme.[99] However, recent progress in implementing MEMS-scale and even nanoscale hydrogen-oxygen fuel cells has provided an alternative approach, which I report on below.

The Debate Heats Up

In April 2003 Drexler challenged Smalley's *Scientific American* article with an open letter.[100] Citing twenty years of research by himself and others, the letter responded specifically to Smalley's fat- and sticky-fingers objections. As I discussed above, molecular assemblers were never described as having fingers at all but rather relying on precise positioning of reactive molecules. Drexler cited biological enzymes and ribosomes as examples of precise molecular assembly in the natural world. Drexler closed by quoting Smalley's own observation, "When a scientist says something is possible, they're probably underestimating how long it will take. But if they say it's impossible, they're probably wrong."

Three more rounds of this debate occurred in 2003. Smalley responded to

Drexler's open letter by backing off of his fat- and sticky-fingers objections and acknowledging that enzymes and ribosomes do indeed engage in the precise molecular assembly that Smalley had earlier indicated was impossible. Smalley then argued that biological enzymes work only in water and that such water-based chemistry is limited to biological structures such as "wood, flesh and bone." As Drexler has stated, this, too, is erroneous.[101] Many enzymes, even those that ordinarily work in water, can also function in anhydrous organic solvents, and some enzymes can operate on substrates in the vapor phase, with no liquid at all.[102]

Smalley goes on to state (without any derivation or citations) that enzymatic-like reactions can take place only with biological enzymes and in chemical reactions involving water. This is also mistaken. MIT professor of chemistry and biological engineering Alexander Klibanov demonstrated such nonaqueous (not involving water) enzyme catalysis in 1984. Klibanov writes in 2003, "Clearly [Smalley's] statements about nonaqueous enzyme catalysis are incorrect. There have been hundreds and perhaps thousands of papers published about non-aqueous enzyme catalysis since our first paper was published 20 years ago."[103]

It's easy to see why biological evolution adopted water-based chemistry. Water is a very abundant substance on our planet, and constitutes 70 to 90 percent of our bodies, our food, and indeed of all organic matter. The three-dimensional electrical properties of water are quite powerful and can break apart the strong chemical bonds of other compounds. Water is considered "the universal solvent," and because it is involved in most of the biochemical pathways in our bodies we can regard the chemistry of life on our planet primarily as water chemistry. However, the primary thrust of our technology has been to develop systems that are not limited to the restrictions of biological evolution, which exclusively adopted water-based chemistry and proteins as its foundation. Biological systems can fly, but if you want to fly at thirty thousand feet and at hundreds or thousands of miles per hour, you would use our modern technology, not proteins. Biological systems such as human brains can remember things and do calculations, but if you want to do data mining on billions of items of information, you would want to use electronic technology, not unassisted human brains.

Smalley is ignoring the past decade of research on alternative means of positioning molecular fragments using precisely guided molecular reactions. Precisely controlled synthesis of diamondoid material has been extensively studied, including the ability to remove a single hydrogen atom from a hydrogenated diamond surface[104] and the ability to add one or more carbon atoms to a diamond surface.[105] Related research supporting the feasibility of hydrogen abstraction and precisely guided diamondoid synthesis has been conducted at the Materials and Process Simulation Center at Caltech; the department of materials science and engineering at North Carolina State University; the Institute for Molecular Manufacturing at the University of Kentucky; the U.S. Naval Academy; and the Xerox Palo Alto Research Center.[106]

Smalley also avoids mentioning the well-established SPM mentioned above, which uses precisely controlled molecular reactions. Building on these concepts, Ralph Merkle has described possible tip reactions that could involve up to four reactants.[107] There is an extensive literature on site-specific reactions that have the potential to be precisely guided and thus could be feasible for the tip chemistry in a molecular assembler.[108] Recently, many tools that go beyond SPMs are emerging that can reliably manipulate atoms and molecular fragments.

On September 3, 2003, Drexler responded to Smalley's response to his initial letter by alluding once again to the extensive body of literature that Smalley fails to address.[109] He cited the analogy to a modern factory, only at a nanoscale. He cited analyses of transition-state theory indicating that positional control would be feasible at megahertz frequencies for appropriately selected reactants.

Smalley again responded with a letter that is short on specific citations and current research and long on imprecise metaphors.[110] He writes, for example, that "much like you can't make a boy and a girl fall in love with each other simply by pushing them together, you cannot make precise chemistry occur as desired between two molecular objects with simple mechanical motion. . . . [It] cannot be done simply by mushing two molecular objects together." He again acknowledges that enzymes do in fact accomplish this but refuses to accept that such reactions could take place outside of a biology-like system: "This is why I led you . . . to talk about real chemistry with real enzymes. . . . [A]ny such system will need a liquid medium. For the enzymes we know about, that liquid will have to be water, and the types of things that can be synthesized with water around cannot be much broader than meat and bone of biology."

Smalley's argument is of the form "We don't have X today, therefore X is impossible." We encounter this class of argument repeatedly in the area of artificial intelligence. Critics will cite the limitations of today's systems as proof that such limitations are inherent and can never be overcome. For example, such critics disregard the extensive list of contemporary examples of AI (see the section "A Narrow AI Sampler" on p. 279) that represent commercially available working systems that were only research programs a decade ago.

Those of us who attempt to project into the future based on well-grounded methodologies are at a disadvantage. Certain future realities may be inevitable, but they are not yet manifest, so they are easy to deny. A small body of thought at the beginning of the twentieth century insisted that heavier-than-air flight was feasible, but mainstream skeptics could simply point out that if it was so feasible, why had it never been demonstrated?

Smalley reveals at least part of his motives at the end of his most recent letter, when he writes:

A few weeks ago I gave a talk on nanotechnology and energy titled "Be a Scientist, Save the World" to about 700 middle and high school students in the Spring Branch ISD, a large public school system here in the Houston area. Leading up to my visit the students were asked to write an essay on "why I am a Nanogeek". Hundreds responded, and I had the privilege of reading the top 30 essays, picking my favorite top 5. Of the essays I read, nearly half assumed that self-replicating nanobots were possible, and most were deeply worried about what would happen in their future as these nanobots spread around the world. I did what I could to allay their fears, but there is no question that many of these youngsters have been told a bedtime story that is deeply troubling.

You and people around you have scared our children.

I would point out to Smalley that earlier critics also expressed skepticism that either worldwide communication networks or software viruses that would spread across them were feasible. Today, we have both the benefits and the vulnerabilities from these capabilities. However, along with the danger of software viruses has emerged a technological immune system. We are obtaining far more gain than harm from this latest example of intertwined promise and peril.

Smalley's approach to reassuring the public about the potential abuse of this future technology is not the right strategy. By denying the feasibility of nanotechnology-based assembly, he is also denying its potential. Denying both the promise and the peril of molecular assembly will ultimately backfire and will fail to guide research in the needed constructive direction. By the 2020s molecular assembly will provide tools to effectively combat poverty, clean up our environment, overcome disease, extend human longevity, and many other worthwhile pursuits. Like every other technology that humankind has created, it can also be used to amplify and enable our destructive side. It's important that we approach this technology in a knowledgeable manner to gain the profound benefits it promises, while avoiding its dangers.

Early Adopters

Although Drexler's concept of nanotechnology dealt primarily with precise molecular control of manufacturing, it has expanded to include any technology in which key features are measured by a modest number of nanometers (generally less than one hundred). Just as contemporary electronics has already quietly slipped into this realm, the area of biological and medical applications has already entered the era of nanoparticles, in which nanoscale objects are being developed to create more effective tests and treatments. Although nanoparticles are created using statistical manufacturing methods rather than assemblers, they nonetheless rely on their atomic-scale properties for their effects. For example, nanoparticles are being employed in experimental biological tests as tags and labels to greatly enhance sensitivity in detecting substances such as proteins. Magnetic nanotags, for example, can be used to bind with antibodies, which can then be read using magnetic probes while still inside the body. Successful experiments have been conducted with gold nanoparticles that are bound to DNA segments and can rapidly test for specific DNA sequences in a sample. Small nanoscale beads called quantum dots can be programmed with specific codes combining multiple colors, similar to a color bar code, which can facilitate tracking of substances through the body.

Emerging microfluidic devices, which incorporate nanoscale channels, can run hundreds of tests simultaneously on tiny samples of a given substance. These devices will allow extensive tests to be conducted on nearly invisible samples of blood, for example.

Nanoscale scaffolds have been used to grow biological tissues such as skin. Future therapies could use these tiny scaffolds to grow any type of tissue needed for repairs inside the body.

A particularly exciting application is to harness nanoparticles to deliver treatments to specific sites in the body. Nanoparticles can guide drugs into cell walls and through the blood-brain barrier. Scientists at McGill University in Montreal demonstrated a nanopill with structures in the 25- to 45-nanometer range.[111] The nanopill is small enough to pass through the cell wall and delivers medications directly to targeted structures inside the cell.

Japanese scientists have created nanocages of 110 amino-acid molecules, each holding drug molecules. Adhered to the surface of each nanocage is a peptide that binds to target sites in the human body. In one experiment scientists used a peptide that binds to a specific receptor on human liver cells.[112]

MicroCHIPS of Bedford, Massachusetts, has developed a computerized device that is implanted under the skin and delivers precise mixtures of medicines from hundreds of nanoscale wells inside the device.[113] Future versions of the device are expected to be able to measure blood levels of substances such as glucose. The system could be used as an artificial pancreas, releasing precise amounts of insulin based on blood glucose response. It would also be capable of simulating any other hormone-producing organ. If trials go smoothly, the system could be on the market by 2008.

Another innovative proposal is to guide gold nanoparticles to a tumor site, then heat them with infrared beams to destroy the cancer cells. Nanoscale packages can be designed to contain drugs, protect them through the GI tract, guide them to specific locations, and then release them in sophisticated ways, including allowing them to receive instructions from outside the body. Nano-therapeutics in Alachua, Florida, has developed a biodegradable polymer only several nanometers thick that uses this approach.[114]

Powering the Singularity

We produce about 14 trillion (about 10^{13}) watts of power today in the world. Of this energy about 33 percent comes from oil, 25 percent from coal, 20 percent from gas, 7 percent from nuclear fission reactors, 15 percent from biomass and hydroelectric sources, and only 0.5 percent from renewable solar, wind, and geothermal technologies.[115] Most air pollution and significant contributions to water and other forms of pollution result from the extraction, transportation, processing, and uses of the 78 percent of our energy that comes from fossil fuels. The energy obtained from oil also contributes to geopolitical tensions, and there's the small matter of its $2 trillion per year price tag for all of this energy. Although the industrial-era energy sources that dominate energy production today will become more efficient with new nanotechnology-based methods of extraction, conversion, and transmission, it's the renewable category that will need to support the bulk of future energy growth.

By 2030 the price-performance of computation and communication will increase by a factor of ten to one hundred million compared to today. Other technologies will also undergo enormous increases in capacity and efficiency. Energy requirements will grow far more slowly than the capacity of technologies, however, because of greatly increased efficiencies in the use of energy, which I discuss below. A primary implication of the nanotechnology revolution is that physical technologies, such as manufacturing and energy, will become governed by the law of accelerating returns. All technologies will essentially become information technologies, including energy.

Worldwide energy requirements have been estimated to double by 2030, far less than anticipated economic growth, let alone the expected growth in the capability of technology.[116] The bulk of the additional energy needed is likely to come from new nanoscale solar, wind, and geothermal technologies. It's important to recognize that most energy sources today represent solar power in one form or another.

Fossil fuels represent stored energy from the conversion of solar energy by animals and plants and related processes over millions of years (although the theory that fossil fuels originated from living organisms has recently been challenged). But the extraction of oil from high-grade oil wells is at a peak, and some experts believe we may have already passed that peak. It's clear, in any case, that we are rapidly depleting easily accessible fossil fuels. We do have far larger fossil-fuel resources that will require more sophisticated technologies to extract cleanly and efficiently (such as coal and shale oil), and they will be part of the future of energy. A billion-dollar demonstration plant called FutureGen, now being constructed, is expected to be the world's first zero-emissions energy plant based on fossil fuels.[117] Rather than simply burn coal, as is done today, the 275-million-watt plant will convert the coal to a synthetic gas comprising hydrogen and carbon monoxide, which will then react with steam to produce discrete streams of hydrogen and carbon dioxide, which will be sequestered. The hydrogen can then be used in fuel cells or else converted into electricity and water. Key to the plant's design are new materials for membranes that separate hydrogen and carbon dioxide.

Our primary focus, however, will be on the development of clean, renewable, distributed, and safe energy technologies made possible by nanotechnology. For the past several decades energy technologies have been on the slow slope of the industrial era S-curve (the late stage of a specific technology paradigm, when the capability slowly approaches an asymptote or limit). Although the nanotechnology revolution will require new energy resources, it will also introduce major new S-curves in every aspect of energy—production, storage, transmission, and utilization—by the 2020s.

Let's deal with these energy requirements in reverse, starting with utilization. Because of nanotechnology's ability to manipulate matter and energy at the extremely fine scale of atoms and molecular fragments, the efficiency of using energy will be far greater, which will translate into lower energy requirements. Over the next several decades computing will make the transition to reversible computing. (See "The Limits of Computation" in chapter 3.) As I discussed, the primary energy need for computing with reversible logic gates is to correct occasional errors from quantum and thermal effects. As a result reversible computing has the potential to cut energy needs by as much as a factor of a billion, compared to nonreversible computing. Moreover, the logic gates and memory bits will be smaller, by at least a factor of ten in each dimension, reducing energy requirements by another thousand. Fully developed nanotechnology, therefore, will enable the energy requirements for each bit switch to be reduced by about a trillion. Of course, we'll be increasing the amount of computation by even more than this, but this substantially augmented energy efficiency will largely offset those increases.

Manufacturing using molecular nanotechnology fabrication will also be far more energy efficient than contemporary manufacturing, which moves bulk materials from place to place in a relatively wasteful manner. Manufacturing today also devotes enormous energy resources to producing basic materials, such as steel. A typical nanofactory will be a tabletop device that can produce products ranging from computers to clothing. Larger products (such as vehicles, homes, and even additional nanofactories) will be produced as modular subsystems that larger robots can then assemble. Waste heat, which accounts for the primary energy requirement for nanomanufacturing, will be captured and recycled.

The energy requirements for nanofactories are negligible. Drexler estimates that molecular manufacturing will be an energy *generator* rather than an energy *consumer*. According to Drexler, "A molecular manufacturing process can be driven by the chemical energy content of the feedstock materials, producing electrical energy as a by-product (if only to reduce the heat dissipation burden). . . . Using typical organic feedstock, and assuming oxidation of surplus hydrogen, reasonably efficient molecular manufacturing processes are net energy producers."[118]

Products can be made from new nanotube-based and nanocomposite materials, avoiding the enormous energy used today to manufacture steel, titanium, and aluminum. Nanotechnology-based lighting will use small, cool, light-emitting diodes, quantum dots, or other innovative light sources to replace hot, inefficient incandescent and fluorescent bulbs.

Although the functionality and value of manufactured products will rise, product size will generally not increase (and in some cases, such as most electronics, products will get smaller). The higher value of manufactured goods will largely be the result of the expanding value of their information content. Although the roughly 50 percent deflation rate for information-based products and services will continue throughout this period, the amount of valuable information will increase at an even greater, more than offsetting pace.

I discussed the law of accelerating returns as applied to the communication of information in chapter 2. The amount of information being communicated will continue to grow exponentially, but the efficiency of communication will grow almost as fast, so the energy requirements for communication will expand slowly.

Transmission of energy will also be made far more efficient. A great deal of energy today is lost in transmission due to the heat created in power lines and inefficiencies in the transportation of fuel, which also represent a primary environmental assault. Smalley, despite his critique of molecular nanomanufacturing, has nevertheless been a strong advocate of new nanotechnology-based paradigms for creating and transmitting energy. He describes new power-transmission lines based on carbon nanotubes woven into long wires that will be far stronger, lighter, and, most important, much more energy efficient than conventional copper ones.[119] He also envisions using superconducting wires to replace aluminum and copper wires in electric motors to provide greater efficiency. Smalley's vision of a nanoenabled energy future includes a panoply of new nanotechnology-enabled capabilities:[120]

- Photovoltaics: dropping the cost of solar panels by a factor of ten to one hundred.
- Production of hydrogen: new technologies for efficiently producing hydrogen from water and sunlight.
- Hydrogen storage: light, strong materials for storing hydrogen for fuel cells.
- Fuel cells: dropping the cost of fuel cells by a factor of ten to one hundred.
- Batteries and supercapacitors to store energy: improving energy storage densities by a factor of ten to one hundred.
- Improving the efficiency of vehicles such as cars and planes through strong and light nanomaterials.
- Strong, light nanomaterials for creating large-scale energy-harvesting systems in space, including on the moon.
- Robots using nanoscale electronics with artificial intelligence to automatically produce energy-generating structures in space and on the moon.
- New nanomaterial coatings to greatly reduce the cost of deep drilling.
- Nanocatalysts to obtain greater energy yields from coal, at very high temperatures.
- Nanofilters to capture the soot created from high-energy coal extraction. The soot is mostly carbon, which is a basic building block for most nanotechnology designs.
- New materials to enable hot, dry rock geothermal-energy sources (converting the heat of the Earth's hot core into energy).

Another option for energy transmission is wireless transmission by microwaves. This method would be especially well suited to efficiently beam energy created in space by giant solar panels (see below).[121] The Millennium Project of the American Council for the United Nations University envisions microwave energy transmission as a key aspect of "a clean, abundant energy future."[122]

Energy storage today is highly centralized, which represents a key vulnerability in that liquid-natural-gas tanks and other storage facilities are subject to terrorist attacks, with potentially catastrophic effects. Oil trucks and ships are equally exposed. The emerging paradigm for energy storage will be fuel cells, which will ultimately be widely distributed throughout our infrastructure, another example of the trend from inefficient and vulnerable centralized facilities to an efficient and stable distributed system.

Hydrogen-oxygen fuel cells, with hydrogen provided by methanol and other safe forms of hydrogen-rich fuel, have made substantial progress in recent years. A small company in Massachusetts, Integrated Fuel Cell Technologies, has demonstrated a MEMS (Micro Electronic Mechanical System)-based fuel cell.[123] Each postage-stamp-size device contains thousands of microscopic fuel cells and includes the fuel lines and electronic controls. NEC plans to introduce fuel cells based on nanotubes in the near future for notebook computers and other portable electronics.[124] It claims its small power sources will run devices for up to forty hours at a time. Toshiba is also preparing fuel cells for portable electronic devices.[125]

Larger fuel cells for powering appliances, vehicles, and even homes are also making impressive advances. A 2004 report by the U.S. Department of Energy concluded that nanobased technologies could facilitate every aspect of a hydrogen fuel cell–powered car.[126] For example, hydrogen must be stored in strong but light tanks that can withstand very high pressure. Nanomaterials such as nanotubes and nanocomposites could provide the requisite material for such containers. The report envisions fuel cells that produce power twice as efficiently as gasoline-based engines, producing only water as waste.

Many contemporary fuel-cell designs use methanol to provide hydrogen, which then combines with the oxygen in the air to produce water and energy. Methanol (wood alcohol), however, is difficult to handle, and introduces safety concerns because of its toxicity and flammability. Researchers from St. Louis University have demonstrated a stable fuel cell that uses ordinary ethanol (drinkable grain alcohol).[127] This device employs an enzyme called dehydrogenase that removes hydrogen ions from alcohol, which subsequently react with the oxygen in the air to produce power. The cell apparently works with almost any form of drinkable alcohol. "We have run it on various types," reported Nick

Akers, a graduate student who has worked on the project. "It didn't like carbonated beer and doesn't seem fond of wine, but any other works fine."

Scientists at the University of Texas have developed a nanobot-size fuel cell that produces electricity directly from the glucose-oxygen reaction in human blood.[128] Called a "vampire bot" by commentators, the cell produces electricity sufficient to power conventional electronics and could be used for future blood-borne nanobots. Japanese scientists pursuing a similar project estimated that their system had the theoretical potential to produce a peak of one hundred watts from the blood of one person, although implantable devices would use far less. (A newspaper in Sydney observed that the project provided a basis for the premise in the *Matrix* movies of using humans as batteries.)[129]

Another approach to converting the abundant sugar found in the natural world into electricity has been demonstrated by Swades K. Chaudhuri and Derek R. Lovley at the University of Massachusetts. Their fuel cell, which incorporates actual microbes (the *Rhodoferax ferrireducens* bacterium), boasts a remarkable 81 percent efficiency and uses almost no energy in its idling mode. The bacteria produce electricity directly from glucose with no unstable intermediary by-products. The bacteria also use the sugar fuel to reproduce, thereby replenishing themselves, resulting in stable and continuous production of electrical energy. Experiments with other types of sugars such as fructose, sucrose, and xylose were equally successful. Fuel cells based on this research could utilize the actual bacteria or, alternatively, directly apply the chemical reactions that the bacteria facilitate. In addition to powering nanobots in sugar-rich blood, these devices have the potential to produce energy from industrial and agricultural waste products.

Nanotubes have also demonstrated the promise of storing energy as nanoscale batteries, which may compete with nanoengineered fuel cells.[130] This extends further the remarkable versatility of nanotubes, which have already revealed their prowess in providing extremely efficient computation, communication of information, and transmission of electrical power, as well as in creating extremely strong structural materials.

The most promising approach to nanomaterials-enabled energy is from solar power, which has the potential to provide the bulk of our future energy needs in a completely renewable, emission-free, and distributed manner. The sunlight input to a solar panel is free. At about 10^{17} watts, or about ten thousand times more energy than the 10^{13} watts currently consumed by human civilization, the total energy from sunlight falling on the Earth is more than sufficient to provide for our needs.[131] As mentioned above, despite the enormous increases in computation and communication over the next quarter century and the resulting economic growth, the far greater energy efficiencies of nanotechnology imply that energy requirements will increase only modestly to around thirty trillion watts (3×10^{13}) by 2030. We could meet this entire energy need with solar power alone if we captured only 0.0003 (three ten-thousandths) of the sun's energy as it hits the Earth.

It's interesting to compare these figures to the total metabolic energy output of all humans, estimated by Robert Freitas at 10^{12} watts, and that of all vegetation on Earth, at 10^{14} watts. Freitas also estimates that the amount of energy we could produce and use without disrupting the global energy balance required to maintain current biological ecology (referred to by climatologists as the "hypsithermal limit") is around 10^{15} watts. This would allow a very substantial number of nanobots per person for intelligence enhancement and medical purposes, as well as other applications, such as providing energy and cleaning up the environment. Estimating a global population of around ten billion (10^{10}) humans, Freitas estimates around 10^{16} (ten thousand trillion) nanobots for each human would be acceptable within this limit.[132] We would need only 10^{11} nanobots (ten millionths of this limit) per person to place one in every neuron.

By the time we have technology of this scale, we will also be able to apply nanotechnology to recycle energy by capturing at least a significant portion of the heat generated by nanobots and other nanomachinery and converting that heat back into energy. The most effective way to do this would probably be to build the energy recycling into the nanobot itself.[133] This is similar to the idea of reversible logic gates in computation, in which each logic gate essentially immediately recycles the energy it used for its last computation.

We could also pull carbon dioxide out of the atmosphere to provide the carbon for nanomachinery, which would reverse the increase in carbon dioxide resulting from our current industrial-era technologies. We might, however, want to be particularly cautious about doing more than *reversing* the increase over the past several decades, lest we replace global warming with global cooling.

Solar panels have to date been relatively inefficient and expensive, but the technology is rapidly improving. The efficiency of converting solar energy to electricity has steadily advanced for silicon photovoltaic cells from around 4 percent in 1952 to 24 percent in 1992.[134] Current multilayer cells now provide around 34 percent efficiency. A recent analysis of applying nanocrystals to solar-energy conversion indicates that efficiencies above 60 percent appear to be feasible.[135]

Today solar power costs an estimated $2.75 per watt.[136] Several companies are developing nanoscale solar cells and hope to bring the cost of solar power below that of other energy sources. Industry sources indicate that once solar power falls below $1.00 per watt, it will be competitive for directly supplying electricity to the nation's power grid. Nanosolar has a design based on titanium oxide nanoparticles that can be mass-produced on very thin flexible films. CEO Martin Roscheisen estimates that his technology has the potential to bring down solar-power costs to around fifty cents per watt by 2006, lower than that of natural gas.[137] Competitors Nanosys and Konarka have similar projections. Whether or not these business plans pan out, once we have MNT (molecular nanotechnology)-based manufacturing, we will be able to produce solar panels (and almost everything else) extremely inexpensively, essentially at the cost of raw materials, of which inexpensive carbon is the primary one. At an estimated thickness of several microns, solar panels could ultimately be as inexpensive as a penny per square meter. We could place efficient solar panels on the majority of human-made surfaces, such as buildings and vehicles, and even incorporate them into clothing for powering mobile devices. A 0.0003 conversion rate for solar energy should be quite feasible, therefore, and relatively inexpensive.

Terrestrial surfaces could be augmented by huge solar panels in space. A Space Solar Power satellite already designed by NASA could convert sunlight in space to electricity and beam it to Earth by microwave. Each such satellite could provide billions of watts of electricity, enough for tens of thousands of homes.[138] With circa-2029 MNT manufacturing, we could produce solar panels of vast size directly in orbit around the Earth, requiring only the shipment of the raw materials to space stations, possibly via the planned Space Elevator, a thin ribbon, extending from a shipborne anchor to a counterweight well beyond geosynchronous orbit, made out of a material called carbon nanotube composite.[139]

Desktop fusion also remains a possibility. Scientists at Oak Ridge National Laboratory used ultrasonic sound waves to shake a liquid solvent, causing gas bubbles to become so compressed they achieved temperatures of millions of degrees, resulting in the nuclear fusion of hydrogen atoms and the creation of energy.[140] Despite the broad skepticism over the original reports of cold fusion in 1989, this ultrasonic method has been warmly received by some peer reviewers.[141] However, not enough is known about the practicality of the technique, so its future role in energy production remains a matter of speculation.

Applications of Nanotechnology to the Environment

Emerging nanotechnology capabilities promise a profound impact on the environment. This includes the creation of new manufacturing and processing technologies that will dramatically reduce undesirable emissions, as well as remediating the prior impact of industrial-age pollution. Of course, providing for our energy needs with nanotechnology-enabled renewable, clean resources such as nanosolar panels, as I discussed above, will clearly be a leading effort in this direction.

By building particles and devices at the molecular scale, not only is size greatly reduced and surface area increased, but new electrical, chemical, and biological properties are introduced. Nanotechnology will eventually provide us with a vastly expanded toolkit for improved catalysis, chemical and atomic bonding, sensing, and mechanical manipulation, not to mention intelligent control through enhanced microelectronics.

Ultimately we will redesign all of our industrial processes to achieve their intended results with minimal consequences, such as unwanted by-products and their introduction into the environment. We discussed in the previous section a comparable trend in biotechnology: intelligently designed pharmaceutical agents that perform highly targeted biochemical interventions with greatly curtailed side effects. Indeed, the creation of designed molecules through nanotechnology will itself greatly accelerate the biotechnology revolution.

Contemporary nanotechnology research and development involves relatively simple "devices" such as nanoparticles, molecules created through nanolayers, and nanotubes. Nanoparticles, which comprise between tens and thousands of atoms, are generally crystalline in nature and use crystal-growing techniques, since we do not yet have the means for precise nanomolecular manufacturing. Nanostructures consist of multiple layers that self-assemble. Such structures are typically held together with hydrogen or carbon bonding and other atomic forces. Biological structures such as cell membranes and DNA itself are natural examples of multilayer nanostructures.

As with all new technologies, there is a downside to nanoparticles: the introduction of new forms of toxins and other unanticipated interactions with the environment and life. Many toxic materials, such as gallium arsenide, are already entering the ecosystem through discarded electronic products. The same properties that enable nanoparticles and nanolayers to deliver highly targeted beneficial results can also lead to unforeseen reactions, particularly with biological systems such as our food supply and our own bodies. Although existing regulations may in many cases be effective in controlling them, the overriding concern is our lack of knowledge about a wide range of unexplored interactions.

Nonetheless, hundreds of projects have begun applying nanotechnology to enhancing industrial processes and explicitly address existing forms of pollution. A few examples:

- There is extensive investigation of the use of nanoparticles for treating, deactivating, and removing a wide variety of environmental toxins. The nanoparticle forms of oxidants, reductants, and other active materials have shown the ability to transform a wide range of undesirable substances. Nanoparticles activated by light (for example, forms of titanium dioxide and zinc oxide) are able to bind and remove organic toxins and have low toxicity themselves.[142] In particular, zinc oxide nanoparticles provide a particularly powerful catalyst for detoxifying chlorinated phenols. These nanoparticles act as both sensors and catalysts and can be designed to transform only targeted contaminants.

- Nanofiltration membranes for water purification provide dramatically improved removal of fine-particle contaminants, compared to conventional methods of using sedimentation basins and wastewater clarifiers. Nanoparticles with designed catalysis are capable of absorbing and removing impurities. By using magnetic separation, these nanomaterials can be reused, which prevents them from becoming contaminants themselves. As one of many examples, consider nanoscale aluminosilicate molecular sieves called zeolites, which are being developed for controlled oxidation of hydrocarbons (for example, converting toluene to nontoxic benzaldehyde).[143] This method requires less energy and reduces the volume of inefficient photoreactions and waste products.

- Extensive research is under way to develop nanoproduced crystalline materials for catalysts and catalyst supports in the chemical industry. These catalysts have the potential to improve chemical yields, reduce toxic by-products, and remove contaminants.[144] For example, the material MCM-41 is now used by the oil industry to remove ultrafine contaminants that other pollution-reduction methods miss.

- It's estimated that the widespread use of nanocomposites for structural material in automobiles would reduce gasoline consumption by 1.5 billion liters per year, which in turn would reduce carbon dioxide emissions by five billion kilograms per year, among other environmental benefits.

- Nanorobotics can be used to assist with nuclear-waste management. Nanofilters can separate isotopes when processing nuclear fuel. Nanofluids can improve the effectiveness of cooling nuclear reactors.

- Applying nanotechnology to home and industrial lighting could reduce both the need for electricity and an estimated two hundred million tons of carbon emissions per year.[145]

- Self-assembling electronic devices (for example, self-organizing biopolymers), if perfected, will require less energy to manufacture and use and will produce fewer toxic by-products than conventional semiconductor-manufacturing methods.

- New computer displays using nanotube-based field-emission displays (FEDs) will provide superior display specifications while eliminating the heavy metals and other toxic materials used in conventional displays.
- Bimetallic nanoparticles (such as iron/palladium or iron/silver) can serve as effective reductants and catalysts for PCBs, pesticides, and halogenated organic solvents.[146]
- Nanotubes appear to be effective absorbents for dioxins and have performed significantly better at this than traditional activated carbon.[147]

This is a small sample of contemporary research on nanotechnology applications with potentially beneficial impact on the environment. Once we can go beyond simple nanoparticles and nanolayers and create more complex systems through precisely controlled molecular nanoassembly, we will be in a position to create massive numbers of tiny intelligent devices capable of carrying out relatively complex tasks. Cleaning up the environment will certainly be one of those missions.

Nanobots in the Bloodstream

Nanotechnology has given us the tools . . . to play with the ultimate toy box of nature—atoms and molecules. Everything is made from it. . . . The possibilities to create new things appear limitless.

—NOBELIST HORST STÖRMER

The net effect of these nanomedical interventions will be the continuing arrest of all biological aging, along with the reduction of current biological age to whatever new biological age is deemed desirable by the patient, severing forever the link between calendar time and biological health. Such interventions may become commonplace several decades from today. Using annual checkups and cleanouts, and some occasional major repairs, your biological age could be restored once a year to the more or less constant physiological age that you select. You might still eventually die of accidental causes, but you'll live at least ten times longer than you do now.

—ROBERT A. FREITAS JR.[148]

A prime example of the application of precise molecular control in manufacturing will be the deployment of billions or trillions of nanobots: small robots the size of human blood cells or smaller that can travel inside the bloodstream.

This notion is not as futuristic as it may sound; successful animal experiments have been conducted using this concept, and many such microscale devices are already working in animals. At least four major conferences on BioMEMS (Biological Micro Electronic Mechanical Systems) deal with devices to be used in the human bloodstream.[149]

Consider several examples of nanobot technology, which, based on miniaturization and cost-reduction trends, will be feasible within about twenty-five years. In addition to scanning the human brain to facilitate its reverse engineering, these nanobots will be able to perform a broad variety of diagnostic and therapeutic functions.

Robert A. Freitas Jr.—a pioneering nanotechnology theorist and leading proponent of nanomedicine (reconfiguring our biological systems through engineering on a molecular scale), and author of a book with that title[150]—has designed robotic replacements for human blood cells that perform hundreds or thousands of times more effectively than their biological counterparts. With Freitas's respirocytes (robotic red blood cells) a runner could do an Olympic sprint for fifteen minutes without taking a breath.[151] Freitas's robotic macrophages, called "microbivores," will be far more effective than our white blood cells at combating pathogens.[152] His DNA-repair robot would be able to mend DNA transcription errors and even implement needed DNA changes. Other medical robots he has designed can serve as cleaners, removing unwanted debris and chemicals (such as prions, malformed proteins, and protofibrils) from individual human cells.

Freitas provides detailed conceptual designs for a wide range of medical nanorobots (Freitas's preferred term) as well as a review of numerous solutions to the varied design challenges involved in creating them. For example, he provides about a dozen approaches to directed and guided motion,[153] some based on biological designs such as propulsive cilia. I discuss these applications in more detail in the next chapter.

George Whitesides complained in *Scientific American* that "for nanoscale objects, even if one could fabricate a propeller, a new and serious problem would emerge: random jarring by water molecules. These water molecules would be smaller than a nanosubmarine but not much smaller."[154] Whitesides's analysis is based on misconceptions. All medical nanobot designs, including those of Freitas, are at least ten thousand times larger than a water molecule. Analyses by Freitas and others show the impact of the Brownian motion of adjacent molecules to be insignificant. Indeed, nanoscale medical robots will be thousands of times more stable and precise than blood cells or bacteria.[155]

It should also be pointed out that medical nanobots will not require much of the extensive overhead biological cells need to maintain metabolic processes such as digestion and respiration. Nor do they need to support biological reproductive systems.

Although Freitas's conceptual designs are a couple of decades away, substantial progress has already been made on bloodstream-based devices. For example, a researcher at the University of Illinois at Chicago has cured type 1 diabetes in rats with a nanoengineered device that incorporates pancreatic islet cells.[156] The device has seven-nanometer pores that let insulin out but won't let in the antibodies that destroy these cells. There are many other innovative projects of this type already under way.

MOLLY 2004: *Okay, so I'll have all these nanobots in my bloodstream. Aside from being able to sit at the bottom of my pool for hours, what else is this going to do for me?*

RAY: *It will keep you healthy. They'll destroy pathogens such as bacteria, viruses, and cancer cells, and they won't be subject to the various pitfalls of the immune system, such as autoimmune reactions. Unlike your biological immune system, if you don't like what the nanobots are doing, you can tell them to do something different.*

MOLLY 2004: *You mean, send my nanobots an e-mail? Like, Hey, nanobots, stop destroying those bacteria in my intestines because they're actually good for my digestion?*

RAY: *Yes, good example. The nanobots will be under our control. They'll communicate with one another and with the Internet. Even today we have neural implants (for example, for Parkinson's disease) that allow the patient to download new software into them.*

MOLLY 2004: *That kind of makes the software-virus issue a lot more serious, doesn't it? Right now, if I get hit with a bad software virus, I may have to run a virus-cleansing program and load my backup files, but if nanobots in my bloodstream get a rogue message, they may start destroying my blood cells.*

RAY: *Well, that's another reason you'll probably want robotic blood cells, but your point is well taken. However, it's not a new issue. Even in 2004, we already have mission-critical software systems that run intensive-care units, manage 911 emergency systems, control nuclear-power plants, land airplanes, and guide cruise missiles. So software integrity is already of critical importance.*

MOLLY 2004: *True, but the idea of software running in my body and brain seems more daunting. On my personal computer, I get more than one hundred spam messages a day, at least several of which contain malicious software viruses. I'm not real comfortable with nanobots in my body getting software viruses.*

RAY: *You're thinking in terms of conventional Internet access. With VPNs (private networks), we already have the means today to create secure firewalls—otherwise, contemporary mission-critical systems would be impossible. They do work reasonably well, and Internet security technology will continue to evolve.*

MOLLY 2004: *I think some people would take issue with your confidence in firewalls.*

RAY: *They're not perfect, true, and they never will be, but we have another couple decades before we'll have extensive software running in our bodies and brains.*

MOLLY 2004: *Okay, but the virus writers will be improving their craft as well.*

RAY: *It's going to be a nervous standoff, no question about it. But the benefit today clearly outweighs the damage.*

MOLLY 2004: *How clear is that?*

RAY: *Well, no one is seriously arguing we should do away with the Internet because software viruses are such a big problem.*

MOLLY 2004: *I'll give you that.*

RAY: *When nanotechnology is mature, it's going to solve the problems of biology by overcoming biological pathogens, removing toxins, correcting DNA errors, and reversing other sources of aging. We will then have to contend with new dangers that it introduces, just as the Internet introduced the danger of software viruses. These new pitfalls will include the potential for self-replicating nanotechnology getting out of control, as well as the integrity of the software controlling these powerful, distributed nanobots.*

MOLLY 2004: *Did you say reverse aging?*

RAY: *I see you're already picking up on a key benefit.*

MOLLY 2004: *So how are the nanobots going to do that?*

RAY: *We'll actually accomplish most of that with biotechnology, methods such as RNA interference for turning off destructive genes, gene therapy for changing your genetic code, therapeutic cloning for regenerating your cells and tissues, smart drugs to reprogram your metabolic pathways, and many other emerging techniques. But whatever biotechnology doesn't get around to accomplishing, we'll have the means to do with nanotechnology.*

MOLLY 2004: *Such as?*

RAY: *Nanobots will be able to travel through the bloodstream, then go in and around our cells and perform various services, such as removing toxins, sweeping out debris, correcting DNA errors, repairing and restoring cell membranes, reversing atherosclerosis, modifying the levels of hormones, neurotransmitters, and other metabolic chemicals, and a myriad of other tasks. For each aging process, we can describe a means for nanobots to reverse the process, down to the level of individual cells, cell components, and molecules.*

MOLLY 2004: *So I'll stay young indefinitely?*

RAY: *That's the idea.*

MOLLY 2004: *When did you say I could get these?*

RAY: *I thought you were worried about nanobot firewalls.*

MOLLY 2004: *Yeah, well, I've got time to worry about that. So what was that time frame again?*

RAY: *About twenty to twenty-five years.*

MOLLY 2004: *I'm twenty-five now, so I'll age to about forty-five and then stay there?*

RAY: *No, that's not exactly the idea. You can slow down aging to a crawl right now by adopting the knowledge we already have. Within ten to twenty years, the biotechnology revolution will provide far more powerful means to stop and in many cases reverse each disease and aging process. And it's not like nothing is going to happen in the meantime. Each year, we'll have more powerful techniques, and the process will accelerate. Then nanotechnology will finish the job.*

MOLLY 2004: *Yes, of course, it's hard for you to get out a sentence without using the word "accelerate." So what biological age am I going to get to?*

RAY: *I think you'll settle somewhere in your thirties and stay there for a while.*

MOLLY 2004: *Thirties sounds pretty good. I think a slightly more mature age than twenty-five is a good idea anyway. But what do you mean "for a while"?*

RAY: *Stopping and reversing aging is only the beginning. Using nanobots for health and longevity is just the early adoption phase of introducing nanotechnology and intelligent computation into our bodies and brains. The more profound implication is that we'll augment our thinking processes with nanobots that communicate with one another and with our biological neurons. Once nonbiological intelligence gets a foothold, so to speak, in our brains, it will be subject to the law of accelerating returns and expand exponentially. Our biological thinking, on the other hand, is basically stuck.*

MOLLY 2004: *There you go again with things accelerating, but when this really gets going, thinking with biological neurons will be pretty trivial in comparison.*

RAY: *That's a fair statement.*

MOLLY 2004: *So, Miss Molly of the future, when did I drop my biological body and brain?*

MOLLY 2104: *Well, you don't really want me to spell out your future, do you? And anyway it's actually not a straightforward question.*

MOLLY 2004: *How's that?*

MOLLY 2104: *In the 2040s we developed the means to instantly create new portions of ourselves, either biological or nonbiological. It became apparent that our true nature was a pattern of information, but we still needed to manifest ourselves in some physical form. However, we could quickly change that physical form.*

MOLLY 2004: *By?*

MOLLY 2104: *By applying new high-speed MNT manufacturing. So we could readily and rapidly redesign our physical instantiation. So I could have a biological body at one time and not at another, then have it again, then change it, and so on.*

MOLLY 2004: *I think I'm following this.*

MOLLY 2104: *The point is that I could have my biological brain and/or body or not have it. It's not a matter of dropping anything, because we can always get back something we drop.*

MOLLY 2004: *So you're still doing this?*

MOLLY 2104: *Some people still do this, but now in 2104 it's a bit anachronistic. I mean, the simulations of biology are totally indistinguishable from actual biology, so why bother with physical instantiations?*

MOLLY 2004: *Yeah, it's messy isn't it?*

MOLLY 2104: *I'll say.*

MOLLY 2004: *I do have to say that it seems strange to be able to change your physical embodiment. I mean, where's your—my—continuity?*

MOLLY 2104: *It's the same as your continuity in 2004. You're changing your particles all the time also. It's just your pattern of information that has continuity.*

MOLLY 2004: *But in 2104 you're able to change your pattern of information quickly also. I can't do that yet.*

MOLLY 2104: *It's really not that different. You change your pattern—your memory, skills, experiences, even personality over time—but there is a continuity, a core that changes only gradually.*

MOLLY 2004: *But I thought you could change your appearance and personality dramatically in an instant?*

MOLLY 2104: *Yes, but that's just a surface manifestation. My true core changes only gradually, just like when I was you in 2004.*

MOLLY 2004: *Well, there are lots of times when I'd be delighted to instantly change my surface appearance.*

Robotics: Strong AI

Consider another argument put forth by Turing. So far we have constructed only fairly simple and predictable artifacts. When we increase the complexity of our machines, there may, perhaps, be surprises in store for us. He draws a parallel with a fission pile. Below a certain "critical" size, nothing much happens: but above the critical size, the sparks begin to fly. So too, perhaps, with brains and machines. Most brains and all machines are, at present "sub-critical"—they react to incoming stimuli in a stodgy and uninteresting way, have no ideas of their own, can produce only stock responses—but a few brains at present, and possibly some machines in the future, are super-critical, and scintillate on their own account. Turing is suggesting that it is only a matter of complexity, and that above a certain level of complexity a qualitative difference appears, so that "super-critical" machines will be quite unlike the simple ones hitherto envisaged.

—J. R. LUCAS, OXFORD PHILOSOPHER, IN HIS 1961 ESSAY "MINDS, MACHINES, AND GÖDEL"[157]

Given that superintelligence will one day be technologically feasible, will people choose to develop it? This question can pretty confidently be answered in the affirmative. Associated with every step along the road to superintelligence are enormous economic payoffs. The computer industry invests huge sums in the next generation of hardware and software, and it will continue doing so as long as there is a competitive pressure and profits to be made. People want better computers and smarter software, and they want the benefits these machines can help produce. Better medical drugs; relief for humans from the need to perform boring or dangerous jobs; entertainment—there is no end to the list of consumer-benefits. There is also a strong military motive to develop artificial intelligence. And nowhere on the path is there any natural stopping point where technophobics could plausibly argue "hither but not further."

—NICK BOSTROM, "HOW LONG BEFORE SUPERINTELLIGENCE?" 1997

It is hard to think of any problem that a superintelligence could not either solve or at least help us solve. Disease, poverty, environmental destruction, unnecessary suffering of all kinds: these are things that a superintelligence equipped with advanced nanotechnology would be capable of eliminating. Additionally, a superintelligence could give us indefinite lifespan, either by stopping and reversing the aging process through the use of nanomedicine, or by offering us the option to upload ourselves. A superintelligence could also create opportunities for us to vastly increase our own intellectual and emotional capabilities, and it could assist us in creating a highly appealing experiential world in which we could live lives devoted to joyful game-playing, relating to each other, experiencing, personal growth, and to living closer to our ideals.

—NICK BOSTROM, "ETHICAL ISSUES IN ADVANCED ARTIFICIAL INTELLIGENCE," 2003

Will robots inherit the earth? Yes, but they will be our children.

—MARVIN MINSKY, 1995

Of the three primary revolutions underlying the Singularity (G, N, and R), the most profound is R, which refers to the creation of nonbiological intelligence that exceeds that of unenhanced humans. A more intelligent process will inherently outcompete one that is less intelligent, making intelligence the most powerful force in the universe.

While the R in GNR stands for robotics, the real issue involved here is strong AI (artificial intelligence that exceeds human intelligence). The standard reason for emphasizing robotics in this formulation is that intelligence needs an embodiment, a physical presence, to affect the world. I disagree with the emphasis on physical presence, however, for I believe that the central concern is intelligence. Intelligence will inherently find a way to influence the world, including creating its own means for embodiment and physical manipulation. Furthermore, we can include physical skills as a fundamental part of intelligence; a large portion of the human brain (the cerebellum, comprising more than half our neurons), for example, is devoted to coordinating our skills and muscles.

Artificial intelligence at human levels will necessarily greatly exceed human intelligence for several reasons. As I pointed out earlier, machines can readily share their knowledge. As unenhanced humans we do not have the means of sharing the vast patterns of interneuronal connections and neurotransmitter-concentration levels that comprise our learning, knowledge, and skills, other than through slow, language-based communication. Of course, even this method of communication has been very beneficial, as it has distinguished us from other animals and has been an enabling factor in the creation of technology.

Human skills are able to develop only in ways that have been evolutionarily encouraged. Those skills, which are primarily based on massively parallel pattern recognition, provide proficiency for certain tasks, such as distinguishing faces, identifying objects, and recognizing language sounds. But they're not suited for many others, such as determining patterns in financial data. Once we fully master pattern-recognition paradigms, machine methods can apply these techniques to any type of pattern.[158]

Machines can pool their resources in ways that humans cannot. Although teams of humans can accomplish both physical and mental feats that individual humans cannot achieve, machines can more easily and readily aggregate their computational, memory, and communications resources. As discussed earlier, the Internet is evolving into a worldwide grid of computing resources that can instantly be brought together to form massive supercomputers.

Machines have exacting memories. Contemporary computers can master billions of facts accurately, a capability that is doubling every year.[159] The underlying speed and price-performance of computing itself is doubling every year, and the rate of doubling is itself accelerating.

As human knowledge migrates to the Web, machines will be able to read, understand, and synthesize all human-machine information. The last time a biological human was able to grasp all human scientific knowledge was hundreds of years ago.

Another advantage of machine intelligence is that it can consistently perform at peak levels and can combine peak skills. Among humans one person may have mastered music composition, while another may have mastered transistor design, but given the fixed architecture of our brains we do not have the capacity (or the time) to develop and utilize the highest level of skill in every increasingly specialized area. Humans also vary a great deal in a particular skill, so that when we speak, say, of human levels of composing music, do we mean Beethoven, or do we mean the average person? Nonbiological intelligence will be able to match and exceed peak human skills in each area.

For these reasons, once a computer is able to match the subtlety and range of human intelligence, it will necessarily soar past it and then continue its double-exponential ascent.

A key question regarding the Singularity is whether the "chicken" (strong AI) or the "egg" (nanotechnology) will come first. In other words, will strong AI lead to full nanotechnology (molecular-manufacturing assemblers that can turn information into physical products), or will full nanotechnology lead to strong AI? The logic of the first premise is that strong AI would imply superhuman AI for the reasons just cited, and superhuman AI would be in a position to solve any remaining design problems required to implement full nanotechnology.

The second premise is based on the realization that the hardware requirements for strong AI will be met by nanotechnology-based computation. Likewise the software requirements will be facilitated by nanobots that could create highly detailed scans of human brain functioning and thereby achieve the completion of reverse engineering the human brain.

Both premises are logical; it's clear that either technology can assist the other. The reality is that progress in both areas will necessarily use our most advanced tools, so advances in each field will simultaneously facilitate the other. However, I do expect that full MNT will emerge prior to strong AI, but only by a few years (around 2025 for nanotechnology, around 2029 for strong AI).

As revolutionary as nanotechnology will be, strong AI will have far more profound consequences. Nanotechnology is powerful but not necessarily intelligent. We can devise ways of at least trying to manage the enormous powers of nanotechnology, but superintelligence innately cannot be controlled.

Runaway AI. Once strong AI is achieved, it can readily be advanced and its powers multiplied, as that is the fundamental nature of machine abilities. As one strong AI immediately begets many strong AIs, the latter access their own design, understand and improve it, and thereby very rapidly evolve into a yet more capable, more intelligent AI, with the cycle repeating itself indefinitely. Each cycle not only creates a more intelligent AI but takes less time than the cycle before it, as is the nature of technological evolution (or any evolutionary process). The premise is that once strong AI is achieved, it will immediately become a runaway phenomenon of rapidly escalating superintelligence.[160]

My own view is only slightly different. The logic of runaway AI is valid, but we still need to consider the timing. Achieving human levels in a machine will not *immediately* cause a runaway phenomenon. Consider that a human level of intelligence has limitations. We have examples of this today—about six billion of them. Consider a scenario in which you took one hundred humans from, say, a shopping mall. This group would constitute examples of reasonably well-educated humans. Yet if this group was presented with the task of improving human intelligence, it wouldn't get very far, even if provided with the templates of human intelligence. It would probably have a hard time creating a simple computer. Speeding up the thinking and expanding the memory capacities of these one hundred humans would not immediately solve this problem.

I pointed out above that machines will match (and quickly exceed) peak human skills in each area of skill. So instead, let's take one hundred scientists and engineers. A group of technically trained people with the right backgrounds would be capable of improving accessible designs. If a machine attained equivalence to one hundred (and eventually one thousand, then one million) technically trained humans, each operating much faster than a biological human, a rapid acceleration of intelligence would ultimately follow.

However, this acceleration won't happen immediately when a computer passes the Turing test. The Turing test is comparable to matching the capabilities of an average, educated human and thus is closer to the example of humans from a shopping mall. It will take time for computers to master all of the requisite skills and to marry these skills with all the necessary knowledge bases.

Once we've succeeded in creating a machine that can pass the Turing test (around 2029), the succeeding period will be an era of consolidation in which nonbiological intelligence will make rapid gains. However, the extraordinary expansion contemplated for the Singularity, in which human intelligence is multiplied by billions, won't take place until the mid-2040s (as discussed in chapter 3).

The AI Winter

> There's this stupid myth out there that A.I. has failed, but A.I. is everywhere around you every second of the day. People just don't notice it. You've got A.I. systems in cars, tuning the parameters of the fuel injection systems. When you land in an airplane, your gate gets chosen by an A.I. scheduling system. Every time you use a piece of Microsoft software, you've got an A.I. system trying to figure out what you're doing, like writing a letter, and it does a pretty damned good job. Every time you see a movie with computer-generated characters, they're all little A.I. characters behaving as a group. Every time you play a video game, you're playing against an A.I. system.
>
> —RODNEY BROOKS, DIRECTOR OF THE MIT AI LAB[161]

I still run into people who claim that artificial intelligence withered in the 1980s, an argument that is comparable to insisting that the Internet died in the dot-com bust of the early 2000s.[162] The bandwidth and price-performance of Internet technologies, the number of nodes (servers), and the dollar volume of e-commerce all accelerated smoothly through the boom as well as the bust and the period since. The same has been true for AI.

The technology hype cycle for a paradigm shift—railroads, AI, Internet, telecommunications, possibly now nanotechnology—typically starts with a period of unrealistic expectations based on a lack of understanding of all the enabling factors required. Although utilization of the new paradigm does increase exponentially, early growth is slow until the knee of the exponential-growth curve is realized. While the widespread expectations for revolutionary change are accurate, they are incorrectly timed. When the prospects do not quickly pan out, a period of disillusionment sets in. Nevertheless exponential growth continues unabated, and years later a more mature and more realistic transformation does occur.

We saw this in the railroad frenzy of the nineteenth century, which was followed by widespread bankruptcies. (I have some of these early unpaid railroad bonds in my collection of historical documents.) And we are still feeling the effects of the e-commerce and telecommunications busts of several years ago, which helped fuel a recession from which we are now recovering.

AI experienced a similar premature optimism in the wake of programs such as the 1957 General Problem Solver created by Allen Newell, J. C. Shaw, and Herbert Simon, which was able to find proofs for theorems that had stumped mathematicians such as Bertrand Russell, and early programs from the MIT Artificial Intelligence Laboratory, which could answer SAT questions (such as analogies and story problems) at the level of college students.[163] A rash of AI companies occurred in the 1970s, but when profits did not materialize there was an AI "bust" in the 1980s, which has become known as the "AI winter." Many observers still think that the AI winter was the end of the story and that nothing has since come of the AI field.

Yet today many thousands of AI applications are deeply embedded in the infrastructure of every industry. Most of these applications were research projects ten to fifteen years ago. People who ask, "Whatever happened to AI?" remind me of travelers to the rain forest who wonder, "Where are all the many species that are supposed to live here?" when hundreds of species of flora and fauna are flourishing only a few dozen meters away, deeply integrated into the local ecology.

We are well into the era of "narrow AI," which refers to artificial intelligence that performs a useful and specific function that once required human intelligence to perform, and does so at human levels or better. Often narrow AI systems greatly exceed the speed of humans, as well as provide the ability to manage and consider thousands of variables simultaneously. I describe a broad variety of narrow AI examples below.

These time frames for AI's technology cycle (a couple of decades of growing enthusiasm, a decade of disillusionment, then a decade and a half of solid advance in adoption) may seem lengthy, compared to the relatively rapid phases of the Internet and telecommunications cycles (measured in years, not decades), but two factors must be considered. First, the Internet and telecommunications cycles were relatively recent, so they are more affected by the acceleration of paradigm shift (as discussed in chapter 1). So recent adoption cycles (boom, bust, and recovery) will be much faster than ones that started forty years ago. Second, the AI revolution is the most profound transformation that human civilization will experience, so it will take longer to mature than less complex technologies. It is characterized by the mastery of the most important and most powerful attribute of human civilization, indeed of the entire sweep of evolution on our planet: intelligence.

It's the nature of technology to understand a phenomenon and then engineer systems that concentrate and focus that phenomenon to greatly amplify it. For example, scientists discovered a subtle property of curved surfaces known as Bernoulli's principle: a gas (such as air) travels more quickly over a curved surface than over a flat surface. Thus, air pressure over a curved surface is lower than over a flat surface. By understanding, focusing, and amplifying the implications of this subtle observation, our engineering created all of aviation. Once we understand the principles of intelligence, we will have a similar opportunity to focus, concentrate, and amplify its powers.

As we reviewed in chapter 4, every aspect of understanding, modeling, and simulating the human brain is accelerating: the price-performance and temporal and spatial resolution of brain scanning, the amount of data and knowledge available about brain function, and the sophistication of the models and simulations of the brain's varied regions.

We already have a set of powerful tools that emerged from AI research and that have been refined and improved over several decades of development. The brain reverse-engineering project will greatly augment this toolkit by also providing a panoply of new, biologically inspired, self-organizing techniques. We will ultimately be able to apply engineering's ability to focus and amplify human intelligence vastly beyond the hundred trillion extremely slow interneuronal connections that each of us struggles with today. Intelligence will then be fully subject to the law of accelerating returns, which is currently doubling the power of information technologies every year.

An underlying problem with artificial intelligence that I have personally experienced in my forty years in this area is that as soon as an AI technique works, it's no longer considered AI and is spun off as its own field (for example, character recognition, speech recognition, machine vision, robotics, data mining, medical informatics, automated investing).

Computer scientist Elaine Rich defines AI as "the study of how to make computers do things at which, at the moment, people are better." Rodney Brooks, director of the MIT AI Lab, puts it a different way: "Every time we figure out a piece of it, it stops being magical; we say, *Oh, that's just a computation.*" I am also reminded of Watson's remark to Sherlock Holmes, "I thought at first that you had done something clever, but I see that there was nothing in it after all."[164] That has been our experience as AI scientists. The enchantment of intelligence seems to be reduced to "nothing" when we fully understand its methods. The mystery that is left is the intrigue inspired by the remaining, not as yet understood methods of intelligence.

AI's Toolkit

> AI is the study of techniques for solving exponentially hard problems in polynomial time by exploiting knowledge about the problem domain.
>
> —ELAINE RICH

As I mentioned in chapter 4, it's only recently that we have been able to obtain sufficiently detailed models of how human brain regions function to influence AI design. Prior to that, in the absence of tools that could peer into the brain with sufficient resolution, AI scientists and engineers developed their own techniques. Just as aviation engineers did not model the ability to fly on the flight of birds, these early AI methods were not based on reverse engineering natural intelligence.

A small sample of these approaches is reviewed here. Since their adoption, they have grown in sophistication, which has enabled the creation of practical products that avoid the fragility and high error rates of earlier systems.

Expert Systems. In the 1970s AI was often equated with one specific method: expert systems. This involves the development of specific logical rules to simulate the decision-making processes of human experts. A key part of the procedure entails knowledge engineers interviewing domain experts such as doctors and engineers to codify their decision-making rules.

There were early successes in this area, such as medical diagnostic systems that compared well to human physicians, at least in limited tests. For example, a system called MYCIN, which was designed to diagnose and recommend remedial treatment for infectious diseases, was developed through the 1970s. In 1979 a team of expert evaluators compared diagnosis and treatment recommendations by MYCIN to those of human doctors and found that MYCIN did as well as or better than any of the physicians.[165]

It became apparent from this research that human decision making typically is based not on definitive logic rules but rather on "softer" types of evidence. A dark spot on a medical imaging test may suggest cancer, but other factors such as its exact shape, location, and contrast are likely to influence a diagnosis. The hunches of human decision making are usually influenced by combining many pieces of evidence from prior experience, none definitive by itself. Often we are not even consciously aware of many of the rules that we use.

By the late 1980s expert systems were incorporating the idea of uncertainty and could combine many sources of probabilistic evidence to make a decision. The MYCIN system pioneered this approach. A typical MYCIN "rule" reads:

> If the infection which requires therapy is meningitis, and the type of the infection is fungal, and organisms were not seen on the stain of the culture, and the patient is not a compromised host, and the patient has been to an area that is endemic for coccidiomycoses, and the race of the patient is Black, Asian, or Indian, and the cryptococcal antigen in the csf test was not positive, THEN there is a 50 percent chance that cryptococcus is not one of the organisms which is causing the infection.

Although a single probabilistic rule such as this would not be sufficient by itself to make a useful statement, by combining thousands of such rules the evidence can be marshaled and combined to make reliable decisions.

Probably the longest-running expert system project is CYC (for enCYClopedic), created by Doug Lenat and his colleagues at Cycorp. Initiated in 1984, CYC has been coding commonsense knowledge to provide machines with an ability to understand the unspoken assumptions underlying human ideas and reasoning. The project has evolved from hard-coded logical rules to probabilistic ones and now includes means of extracting knowledge from written sources (with human supervision). The original goal was to generate one million rules, which reflects only a small portion of what the average human knows about the world. Lenat's latest goal is for CYC to master "100 million things, about the number a typical person knows about the world, by 2007."[166]

Another ambitious expert system is being pursued by Darryl Macer, associate professor of biological sciences at the University of Tsukuba in Japan. He plans to develop a system incorporating all human ideas.[167] One application would be to inform policy makers of which ideas are held by which community.

Bayesian Nets. Over the last decade a technique called Bayesian logic has created a robust mathematical foundation for combining thousands or even millions of such probabilistic rules in what are called "belief networks" or Bayesian nets. Originally devised by English mathematician Thomas Bayes and published posthumously in 1763, the approach is intended to determine the likelihood of future events based on similar occurrences in the past.[168] Many expert systems based on Bayesian techniques gather data from experience in an ongoing fashion, thereby continually learning and improving their decision making.

The most promising type of spam filters are based on this method. I personally use a spam filter called SpamBayes, which trains itself on e-mail that you have identified as either "spam" or "okay."[169] You start out by presenting a folder of each to the filter. It trains its Bayesian belief network on these two files and analyzes the patterns of each, thus enabling it to automatically move subsequent e-mail into the proper category. It continues to train itself on every subsequent e-mail, especially when it's corrected by the user. This filter has made the spam situation manageable for me, which is saying a lot, as it weeds out two hundred to three hundred spam messages each day, letting more than one hundred "good" messages through. Only about 1 percent of the messages it identifies as "okay" are actually spam; it almost never marks a good message as spam. The system is almost as accurate as I would be and much faster.

Markov Models. Another method that is good at applying probabilistic networks to complex sequences of information involves Markov models.[170] Andrei Andreyevich Markov (1856–1922), a renowned mathematician, established a theory of "Markov chains," which was refined by Norbert Wiener (1894–1964) in 1923. The theory provided a method to evaluate the likelihood that a certain sequence of events would occur. It has been popular, for example, in speech recognition, in which the sequential events are phonemes (parts of speech). The Markov models used in speech recognition code the likelihood that specific patterns of sound are found in each phoneme, how the phonemes influence each other, and likely orders of phonemes. The system can also include probability networks on higher levels of language, such as the order of words. The actual probabilities in the models are trained on actual speech and language data, so the method is self-organizing.

Markov modeling was one of the methods my colleagues and I used in our own speech-recognition development.[171] Unlike phonetic approaches, in which specific rules about phoneme sequences are explicitly coded by human linguists, we did not tell the system that there are approximately forty-four phonemes in English, nor did we tell it what sequences of phonemes were more likely than others. We let the system discover these "rules" for itself from thousands of hours of transcribed human speech data. The advantage of this approach over hand-coded rules is that the models develop subtle probabilistic rules of which human experts are not necessarily aware.

Neural Nets. Another popular self-organizing method that has also been used in speech recognition and a wide variety of other pattern-recognition tasks is neural nets. This technique involves simulating a simplified model of neurons and interneuronal connections. One basic approach to neural nets can be described as follows. Each point of a given input (for speech, each point represents two dimensions, one being frequency and the other time; for images, each point would be a pixel in a two-dimensional image) is randomly connected to the inputs of the first layer of simulated neurons. Every connection has an associated synaptic strength, which represents its importance and which is set at a random value. Each neuron adds up the signals coming into it. If the combined signal exceeds a particular threshold, the neuron fires and sends a signal to its output connection; if the combined input signal does not exceed the threshold, the neuron does not fire, and its output is zero. The output of each neuron is randomly connected to the inputs of the neurons in the next layer. There are multiple layers (generally three or more), and the layers may be organized in a variety of configurations. For example, one layer may feed back to an earlier layer. At the top layer, the output of one or more neurons, also randomly selected, provides the answer. (For an algorithmic description of neural nets, see this note:[172])

Since the neural-net wiring and synaptic weights are initially set randomly, the answers of an untrained neural net will be random. The key to a neural net, therefore, is that it must learn its subject matter. Like the mammalian brains on which it's loosely modeled, a neural net starts out ignorant. The neural net's teacher—which may be a human, a computer program, or perhaps another, more mature neural net that has already learned its lessons—rewards the student neural net when it generates the right output and punishes it when it does not. This feedback is in turn used by the student neural net to adjust the strengths of each interneuronal connection. Connections that were consistent with the right answer are made stronger. Those that advocated a wrong answer are weakened. Over time, the neural net organizes itself to provide the right answers without coaching. Experiments have shown that neural nets can learn their subject matter even with unreliable teachers. If the teacher is correct only 60 percent of the time, the student neural net will still learn its lessons.

A powerful, well-taught neural net can emulate a wide range of human pattern-recognition faculties. Systems using multilayer neural nets have shown impressive results in a wide variety of pattern-recognition tasks, including recognizing handwriting, human faces, fraud in commercial transactions such as credit-card charges, and many others. In my own experience in using neural nets in such contexts, the most challenging engineering task is not coding the nets but in providing automated lessons for them to learn their subject matter.

The current trend in neural nets is to take advantage of more realistic and more complex models of how actual biological neural nets work, now that we are developing detailed models of neural functioning from brain reverse engineering.[173] Since we do have several decades of experience in using self-organizing paradigms, new insights from brain studies can quickly be adapted to neural-net experiments.

Neural nets are also naturally amenable to parallel processing, since that is how the brain works. The human brain does not have a central processor that simulates each neuron. Rather, we can consider each neuron and each inter-neuronal connection to be an individual slow processor. Extensive work is under way to develop specialized chips that implement neural-net architectures in parallel to provide substantially greater throughput.[174]

Genetic Algorithms (GAs). Another self-organizing paradigm inspired by nature is genetic, or evolutionary, algorithms, which emulate evolution, including sexual reproduction and mutations. Here is a simplified description of how they work. First, determine a way to code possible solutions to a given problem. If the problem is optimizing the design parameters for a jet engine, define a list of the parameters (with a specific number of bits assigned to each parameter). This list is regarded as the genetic code in the genetic algorithm. Then randomly generate thousands or more genetic codes. Each such genetic code (which represents one set of design parameters) is considered a simulated "solution" organism.

Now evaluate each simulated organism in a simulated environment by using a defined method to evaluate each set of parameters. This evaluation is a key to the success of a genetic algorithm. In our example, we would apply each solution organism to a jet-engine simulation and determine how successful that set of parameters is, according to whatever criteria we are interested in (fuel consumption, speed, and so on). The best solution organisms (the best designs) are allowed to survive, and the rest are eliminated.

Now have each of the survivors multiply themselves until they reach the same number of solution creatures. This is done by simulating sexual reproduction. In other words, each new offspring solution draws part of its genetic code from one parent and another part from a second parent. Usually no distinction is made between male or female organisms; it's sufficient to generate an offspring from two arbitrary parents. As they multiply, allow some mutation (random change) in the chromosomes to occur.

We've now defined one generation of simulated evolution; now repeat these steps for each subsequent generation. At the end of each generation determine how much the designs have improved. When the improvement in the evaluation of the design creatures from one generation to the next becomes very small, we stop this iterative cycle of improvement and use the best design(s) in the last generation. (For an algorithmic description of genetic algorithms, see this note:[175])

The key to a GA is that the human designers don't directly program a solution; rather, they let one emerge through an iterative process of simulated competition and improvement. As we discussed, biological evolution is smart but slow, so to enhance its intelligence we retain its discernment while greatly speeding up its ponderous pace. The computer is fast enough to simulate many generations in a matter of hours or days or weeks. But we have to go through this iterative process only once; once we have let this simulated evolution run its course, we can apply the evolved and highly refined rules to real problems in a rapid fashion.

Like neural nets GAs are a way to harness the subtle but profound patterns that exist in chaotic data. A key requirement for their success is a valid way of evaluating each possible solution. This evaluation needs to be fast because it must take account of many thousands of possible solutions for each generation of simulated evolution.

GAs are adept at handling problems with too many variables to compute precise analytic solutions. The design of a jet engine, for example, involves more than one hundred variables and requires satisfying dozens of constraints. GAs used by researchers at General Electric were able to come up with engine designs that met the constraints more precisely than conventional methods.

When using GAs you must, however, be careful what you ask for. University of Sussex researcher Jon Bird used a GA to optimally design an oscillator circuit. Several attempts generated conventional designs using a small number of transistors, but the winning design was not an oscillator at all but a simple radio circuit. Apparently the GA discovered that the radio circuit picked up an oscillating hum from a nearby computer.[176] The GA's solution worked only in the exact location on the table where it was asked to solve the problem.

Genetic algorithms, part of the field of chaos or complexity theory, are increasingly being used to solve otherwise intractable business problems, such as optimizing complex supply chains. This approach is beginning to supplant more analytic methods throughout industry. (See examples below.) The paradigm is also adept at recognizing patterns, and is often combined with neural nets and other self-organizing methods. It's also a reasonable way to write computer software, particularly software that needs to find delicate balances for competing resources.

In the novel *usr/bin/god*, Cory Doctorow, a leading science-fiction writer, uses an intriguing variation of a GA to evolve an AI. The GA generates a large number of intelligent systems based on various intricate combinations of techniques, with each combination characterized by its genetic code. These systems then evolve using a GA.

The evaluation function works as follows: each system logs on to various human chat rooms and tries to pass for a human, basically a covert Turing test. If one of the humans in a chat room says something like "What are you, a chatterbot?" (chatterbot meaning an automatic program, which at today's level of development is expected to not understand language at a human level), the evaluation is over, that system ends its interactions, and reports its score to the GA. The score is determined by how long it was able to pass for human without being challenged in this way. The GA evolves more and more intricate combinations of techniques that are increasingly capable of passing for human.

The main difficulty with this idea is that the evaluation function is fairly slow, although it will take an appreciable amount of time only after the systems are reasonably intelligent. Also, the evaluations can take place largely in parallel. It's an interesting idea and may actually be a useful method to finish the job of passing the Turing test, once we get to the point where we have sufficiently sophisticated algorithms to feed into such a GA, so that evolving a Turing-capable AI is feasible.

Recursive Search. Often we need to search through a vast number of combinations of possible solutions to solve a given problem. A classic example is in playing games such as chess. As a player considers her next move, she can list all of her possible moves, and then, for each such move, all possible countermoves by the opponent, and so on. It is difficult, however, for human players to keep a huge "tree" of move-countermove sequences in their heads, and so they rely on pattern recognition—recognizing situations based on prior experience—whereas machines use logical analysis of millions of moves and countermoves.

Such a logical tree is at the heart of most game-playing programs. Consider how this is done. We construct a program called Pick Best Next Step to select each move. Pick Best Next Step starts by listing all of the possible moves from the current state of the board. (If the problem was solving a mathematical theorem, rather than game moves, the program would list all of the possible next steps in a proof.) For each move the program constructs a hypothetical board that reflects what would happen if we made this move. For each such hypothetical board, we now need to consider what our opponent would do if we made this move. Now recursion comes in, because Pick Best Next Step simply calls Pick Best Next Step (in other words, itself) to pick the best move for our opponent. In calling itself, Pick Best Next Step then lists all of the legal moves for our opponent.

The program keeps calling itself, looking ahead as many moves as we have time to consider, which results in the generation of a huge move-countermove tree. This is another example of exponential growth, because to look ahead an additional move (or countermove) requires multiplying the amount of available computation by about five. Key to the success of the recursive formula is pruning this huge tree of possibilities and ultimately stopping its growth. In the game context, if a board looks hopeless for either side, the program can stop the expansion of the move-countermove tree from that point (called a "terminal leaf" of the tree) and consider the most recently considered move to be a likely win or loss. When all of these nested program calls are completed, the program will have determined the best possible move for the current actual board within the limits of the depth of recursive expansion that it had time to pursue and the quality of its pruning algorithm. (For an algorithmic description of recursive search, see this note:[177])

The recursive formula is often effective at mathematics. Rather than game moves, the "moves" are the axioms of the field of math being addressed, as well as previously proved theorems. The expansion at each point is the possible axioms (or previously proved theorems) that can be applied to a proof at each step. (This was the approach used by Newell, Shaw, and Simons's General Problem Solver.)

From these examples it may appear that recursion is well suited only for problems in which we have crisply defined rules and objectives. But it has also shown promise in computer generation of artistic creations. For example, a program I designed called Ray Kurzweil's Cybernetic Poet uses a recursive approach.[178] The program establishes a set of goals for each word—achieving a certain rhythmic pattern, poem structure, and word choice that is desirable at that point in the poem. If the program is unable to find a word that meets these criteria, it backs up and erases the previous word it has written, reestablishes the criteria it had originally set for the word just erased, and goes from there. If that also leads to a dead end, it backs up again, thus moving backward and forward. Eventually, it forces itself to make up its mind by relaxing some of the constraints if all paths lead to dead ends.

Black (computer) . . . is deciding on a move

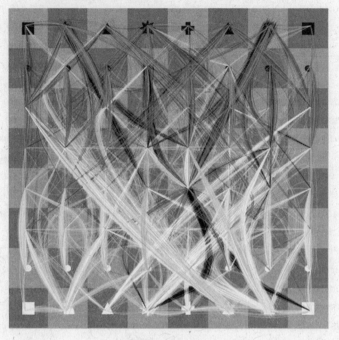

White (you)

"Thinking Machines 2" by mathematician Martin Wattenberg with Marek Walczak displays the move-countermove sequences it is evaluating as it considers its next move.

Deep Fritz Draws: Are Humans Getting Smarter, or Are Computers Getting Stupider?

We find one example of qualitative improvements in software in the world of computer chess, which, according to common wisdom, is governed only by the brute-force expansion of computer hardware. In a chess tournament in October 2002 with top-ranking human player Vladimir Kramnik, the Deep Fritz software achieved a draw. I point out that Deep Fritz has available only about 1.3 percent of the brute-force computation as the previous computer champion, Deep Blue. Despite that, it plays chess at about the same level because of its superior pattern recognition–based pruning algorithm (see below). In six years a program like Deep Fritz will again achieve Deep Blue's ability to analyze two hundred million board positions per second. Deep Fritz–like chess programs running on ordinary personal computers will routinely defeat all humans later in this decade. ▶

In *The Age of Intelligent Machines*, which I wrote between 1986 and 1989, I predicted that a computer would defeat the human world chess champion by the end of the 1990s. I also noted that computers were gaining about forty-five points per year in their chess ratings, whereas the best human playing was essentially fixed, so this projected a crossover point in 1998. Indeed, Deep Blue did defeat Gary Kasparov in a highly publicized tournament in 1997.

Yet in the Deep Fritz–Kramnik match, the current reigning computer program was able to achieve only a tie. Five years had passed since Deep Blue's victory, so what are we to make of this situation? Should we conclude that:

1. Humans are getting smarter, or at least better at chess?
2. Computers are getting worse at chess? If so, should we conclude that the much-publicized improvement in computer speed over the past five years was not all it was cracked up to be? Or, that computer software is getting worse, at least in chess?

The Specialized-Hardware Advantage

Neither of the above conclusions is warranted. The correct conclusion is that software is getting better because Deep Fritz essentially matched the performance of Deep Blue, yet with far smaller computational resources. To gain some insight into these questions, we need to examine a few essentials. When I wrote my predictions of computer chess in the late 1980s, Carnegie Mellon University was embarked on a program to develop specialized chips for conducting the "minimax" algorithm (the standard game-playing method that relies on building trees of move-countermove sequences, and then evaluating the terminal-leaf positions of each branch of the tree) specifically for chess moves.

Based on this specialized hardware CMU's 1988 chess machine, HiTech, was able to analyze 175,000 board positions per second. It achieved a chess rating of 2,359, only about 440 points below the human world champion.

A year later, in 1989, CMU's Deep Thought machine increased this capacity to one million board positions per second and achieved a rating of 2,400. IBM eventually took over the project and renamed it Deep Blue but kept the basic CMU architecture. The version of Deep Blue that defeated Kasparov in 1997 had 256 special-purpose chess processors ▶

working in parallel, which analyzed two hundred million board positions per second.

It is important to note the use of specialized hardware to accelerate the specific calculations needed to generate the minimax algorithm for chess moves. It's well known to computer-systems designers that specialized hardware can generally implement a specific algorithm at least one hundred times faster than a general-purpose computer. Specialized ASICs (application-specific integrated circuits) require significant development efforts and costs, but for critical calculations that are needed on a repetitive basis (for example, decoding MP3 files or rendering graphics primitives for video games), this expenditure can be well worth the investment.

Deep Blue Versus Deep Fritz

Because there had always been a great deal of focus on the milestone of a computer's being able to defeat a human opponent, support was available for investing in special-purpose chess circuits. Although there was some lingering controversy regarding the parameters of the Deep Blue–Kasparov match, the level of interest in computer chess waned considerably after 1997. After all, the goal had been achieved, and there was little point in beating a dead horse. IBM canceled work on the project, and there has been no work on specialized chess chips since that time. The focus of research in the various domains spun out of AI has been placed instead on problems of greater consequence, such as guiding airplanes, missiles, and factory robots, understanding natural language, diagnosing electrocardiograms and blood-cell images, detecting credit-card fraud, and a myriad of other successful narrow AI applications.

Computer hardware has nonetheless continued its exponential increase, with personal-computer speeds doubling every year since 1997. Thus the general-purpose Pentium processors used by Deep Fritz are about thirty-two times faster than processors in 1997. Deep Fritz uses a network of only eight personal computers, so the hardware is equivalent to 256 1997-class personal computers. Compare that to Deep Blue, which used 256 specialized chess processors, each of which was about one hundred times faster than 1997 personal computers (of course, only for computing chess minimax). So Deep Blue was 25,600 times faster than a 1997 PC and one hundred times faster than Deep Fritz. This analysis is confirmed by the reported speeds of the two systems: Deep Blue can analyze 200 million board positions per second compared to only about 2.5 million for Deep Fritz.   ▶

Significant Software Gains

So what can we say about the software in Deep Fritz? Although chess machines are usually referred to as examples of brute-force calculation, there is one important aspect of these systems that does require qualitative judgment. The combinatorial explosion of possible move-countermove sequences is rather formidable.

In *The Age of Intelligent Machines* I estimated that it would take about forty billion years to make one move if we failed to prune the move-countermove tree and attempted to make a "perfect" move in a typical game. (Assuming about thirty moves each in a typical game and about eight possible moves per play, we have 8^{30} possible move sequences; analyzing one billion move sequences per second would take 10^{18} seconds or forty billion years.) Thus a practical system needs to continually prune away unpromising lines of action. This requires insight and is essentially a pattern-recognition judgment.

Humans, even world-class chess masters, perform the minimax algorithm extremely slowly, generally performing less than one move-countermove analysis per second. So how is it that a chess master can compete at all with computer systems? The answer is that we possess formidable powers of pattern recognition, which enable us to prune the tree with great insight.

It's precisely in this area that Deep Fritz has improved considerably over Deep Blue. Deep Fritz has only slightly more computation available than CMU's Deep Thought yet is rated almost 400 points higher.

Are Human Chess Players Doomed?

Another prediction I made in *The Age of Intelligent Machines* was that once computers did perform as well or better as humans in chess, we would either think more of computer intelligence, less of human intelligence, or less of chess, and that if history is a guide, the last of these would be the likely outcome. Indeed, that is precisely what happened. Soon after Deep Blue's victory we began to hear a lot about how chess is really just a simple game of calculating combinations and that the computer victory just demonstrated that it was a better calculator.

The reality is slightly more complex. The ability of humans to perform well in chess is clearly not due to our calculating prowess, at which we are in fact rather poor. We use instead a quintessentially human form of judgment. For this type of qualitative judgment, Deep Fritz represents genuine ▶

progress over earlier systems. (Incidentally, humans have made no progress in the last five years, with the top human scores remaining just below 2,800. As of 2004, Kasparov is rated at 2,795 and Kramnik at 2,794.)

Where do we go from here? Now that computer chess is relying on software running on ordinary personal computers, chess programs will continue to benefit from the ongoing acceleration of computer power. By 2009 a program like Deep Fritz will again achieve Deep Blue's ability to analyze two hundred million board positions per second. With the opportunity to harvest computation on the Internet, we will be able to achieve this potential several years sooner than 2009. (Internet harvesting of computers will require more ubiquitous broadband communication, but that's coming, too.)

With these inevitable speed increases, as well as ongoing improvements in pattern recognition, computer chess ratings will continue to edge higher. Deep Fritz–like programs running on ordinary personal computers will routinely defeat all humans later in this decade. Then we'll really lose interest in chess.

Combining Methods. The most powerful approach to building robust AI systems is to combine approaches, which is how the human brain works. As we discussed, the brain is not one big neural net but instead consists of hundreds of regions, each of which is optimized for processing information in a different way. None of these regions by itself operates at what we would consider human levels of performance, but clearly by definition the overall system does exactly that.

I've used this approach in my own AI work, especially in pattern recognition. In speech recognition, for example, we implemented a number of different pattern-recognition systems based on different paradigms. Some were specifically programmed with knowledge of phonetic and linguistic constraints from experts. Some were based on rules to parse sentences (which involves creating sentence diagrams showing word usage, similar to the diagrams taught in grade school). Some were based on self-organizing techniques, such as Markov models, trained on extensive libraries of recorded and annotated human speech. We then programmed a software "expert manager" to learn the strengths and weaknesses of the different "experts" (recognizers) and combine their results in optimal ways. In this fashion, a particular technique that by itself might produce unreliable results can nonetheless contribute to increasing the overall accuracy of the system.

There are many intricate ways to combine the varied methods in AI's toolbox. For example, one can use a genetic algorithm to evolve the optimal topology (organization of nodes and connections) for a neural net or a Markov model. The final output of the GA-evolved neural net can then be used to control the parameters of a recursive search algorithm. We can add in powerful signal- and image-processing techniques that have been developed for pattern-processing systems. Each specific application calls for a different architecture. Computer science professor and AI entrepreneur Ben Goertzel has written a series of books and articles that describe strategies and architectures for combining the diverse methods underlying intelligence. His Novamente architecture is intended to provide a framework for general-purpose AI.[179]

The above basic descriptions provide only a glimpse into how increasingly sophisticated current AI systems are designed. It's beyond the scope of this book to provide a comprehensive description of the techniques of AI, and even a doctoral program in computer science is unable to cover all of the varied approaches in use today.

Many of the examples of real-world narrow AI systems described in the next section use a variety of methods integrated and optimized for each particular task. Narrow AI is strengthening as a result of several concurrent trends: continued exponential gains in computational resources, extensive real-world experience with thousands of applications, and fresh insights into how the human brain makes intelligent decisions.

A Narrow AI Sampler

When I wrote my first AI book, *The Age of Intelligent Machines*, in the late 1980s, I had to conduct extensive investigations to find a few successful examples of AI in practice. The Internet was not yet prevalent, so I had to go to real libraries and visit the AI research centers in the United States, Europe, and Asia. I included in my book pretty much all of the reasonable examples I could identify. In my research for this book my experience has been altogether different. I have been inundated with thousands of compelling examples. In our reporting on the KurzweilAI.net Web site, we feature one or more dramatic systems almost every day.[180]

A 2003 study by Business Communications Company projected a $21 billion market by 2007 for AI applications, with average annual growth of 12.2 percent from 2002 to 2007.[181] Leading industries for AI applications include business intelligence, customer relations, finance, defense and domestic security, and education. Here is a small sample of narrow AI in action.

Military and Intelligence. The U.S. military has been an avid user of AI systems. Pattern-recognition software systems guide autonomous weapons such as cruise missiles, which can fly thousands of miles to find a specific building or even a specific window.[182] Although the relevant details of the terrain that the missile flies over are programmed ahead of time, variations in weather, ground cover, and other factors require a flexible level of real-time image recognition.

The army has developed prototypes of self-organizing communication networks (called "mesh networks") to automatically configure many thousands of communication nodes when a platoon is dropped into a new location.[183]

Expert systems incorporating Bayesian networks and GAs are used to optimize complex supply chains that coordinate millions of provisions, supplies, and weapons based on rapidly changing battlefield requirements.

AI systems are routinely employed to simulate the performance of weapons, including nuclear bombs and missiles.

Advance warning of the September 11, 2001, terrorist attacks was apparently detected by the National Security Agency's AI-based Echelon system, which analyzes the agency's extensive monitoring of communications traffic.[184] Unfortunately, Echelon's warnings were not reviewed by human agents until it was too late.

The 2002 military campaign in Afghanistan saw the debut of the armed Predator, an unmanned robotic flying fighter. Although the air force's Predator had been under development for many years, arming it with army-supplied missiles was a last-minute improvisation that proved remarkably successful. In the Iraq war that began in 2003 the armed Predator (operated by the CIA) and other flying unmanned aerial vehicles (UAVs) destroyed thousands of enemy tanks and missile sites.

All of the military services are using robots. The army utilizes them to search caves (in Afghanistan) and buildings. The navy uses small robotic ships to protect its aircraft carriers. As I discuss in the next chapter, moving soldiers away from battle is a rapidly growing trend.

Space Exploration. NASA is building self-understanding into the software controlling its unmanned spacecraft. Because Mars is about three light-minutes from Earth, and Jupiter around forty light-minutes (depending on the exact position of the planets), communication between spacecraft headed there and earthbound controllers is significantly delayed. For this reason it's important that the software controlling these missions have the capability of performing its own tactical decision making. To accomplish this NASA software is being designed to include a model of the software's own capabilities and those of the spacecraft, as well as the challenges each mission is likely to encounter. Such AI-based systems are capable of reasoning through new situations rather than just following preprogrammed rules. This approach enabled the craft *Deep Space*

One in 1999 to use its own technical knowledge to devise a series of original plans to overcome a stuck switch that threatened to destroy its mission of exploring an asteroid.[185] The AI system's first plan failed to work, but its second plan saved the mission. "These systems have a commonsense model of the physics of their internal components," explains Brian Williams, coinventor of *Deep Space One*'s autonomous software and now a scientist at MIT's Space Systems and AI laboratories. "[The spacecraft] can reason from that model to determine what is wrong and to know how to act."

Using a network of computers NASA used GAs to evolve an antenna design for three Space Technology 5 satellites that will study the Earth's magnetic field. Millions of possible designs competed in the simulated evolution. According to NASA scientist and project leader Jason Lohn, "We are now using the [GA] software to design tiny microscopic machines, including gyroscopes, for space-flight navigation. The software also may invent designs that no human designer would ever think of."[186]

Another NASA AI system learned on its own to distinguish stars from galaxies in very faint images with an accuracy surpassing that of human astronomers.

New land-based robotic telescopes are able to make their own decisions on where to look and how to optimize the likelihood of finding desired phenomena. Called "autonomous, semi-intelligent observatories," the systems can adjust to the weather, notice items of interest, and decide on their own to track them. They are able to detect very subtle phenomena, such as a star blinking for a nanosecond, which may indicate a small asteroid in the outer regions of our solar system passing in front of the light from that star.[187] One such system, called Moving Object and Transient Event Search System (MOTESS), has identified on its own 180 new asteroids and several comets during its first two years of operation. "We have an intelligent observing system," explained University of Exeter astronomer Alasdair Allan. "It thinks and reacts for itself, deciding whether something it has discovered is interesting enough to need more observations. If more observations are needed, it just goes ahead and gets them."

Similar systems are used by the military to automatically analyze data from spy satellites. Current satellite technology is capable of observing ground-level features about an inch in size and is not affected by bad weather, clouds, or darkness.[188] The massive amount of data continually generated would not be manageable without automated image recognition programmed to look for relevant developments.

Medicine. If you obtain an electrocardiogram (ECG) your doctor is likely to receive an automated diagnosis using pattern recognition applied to ECG recordings. My own company (Kurzweil Technologies) is working with United Therapeutics to develop a new generation of automated ECG analysis for long-term unobtrusive monitoring (via sensors embedded in clothing and wireless communication using a cell phone) of the early warning signs of heart disease.[189] Other pattern-recognition systems are used to diagnose a variety of imaging data.

Every major drug developer is using AI programs to do pattern recognition and intelligent data mining in the development of new drug therapies. For example SRI International is building flexible knowledge bases that encode everything we know about a dozen disease agents, including tuberculosis and *H. pylori* (the bacteria that cause ulcers).[190] The goal is to apply intelligent data-mining tools (software that can search for new relationships in data) to find new ways to kill or disrupt the metabolisms of these pathogens.

Similar systems are being applied to performing the automatic discovery of new therapies for other diseases, as well as understanding the function of genes and their roles in disease.[191] For example Abbott Laboratories claims that six human researchers in one of its new labs equipped with AI-based robotic and data-analysis systems are able to match the results of two hundred scientists in its older drug-development labs.[192]

Men with elevated prostate-specific antigen (PSA) levels typically undergo surgical biopsy, but about 75 percent of these men do not have prostate cancer. A new test, based on pattern recognition of proteins in the blood, would reduce this false positive rate to about 29 percent.[193] The test is based on an AI program designed by Correlogic Systems in Bethesda, Maryland, and the accuracy is expected to improve further with continued development.

Pattern recognition applied to protein patterns has also been used in the detection of ovarian cancer. The best contemporary test for ovarian cancer, called CA-125, employed in combination with ultrasound, misses almost all early-stage tumors. "By the time it is now diagnosed, ovarian cancer is too often deadly," says Emanuel Petricoin III, codirector of the Clinical Proteomics Program run by the FDA and the National Cancer Institute. Petricoin is the lead developer of a new AI-based test looking for unique patterns of proteins found only in the presence of cancer. In an evaluation involving hundreds of blood samples, the test was, according to Petricoin, "an astonishing 100% accurate in detecting cancer, even at the earliest stages."[194]

About 10 percent of all Pap-smear slides in the United States are analyzed by a self-learning AI program called FocalPoint, developed by TriPath Imaging. The developers started out by interviewing pathologists on the criteria they use. The AI system then continued to learn by watching expert pathologists.

Only the best human diagnosticians were allowed to be observed by the program. "That's the advantage of an expert system," explains Bob Schmidt, Tri-Path's technical product manager. "It allows you to replicate your very best people."

Ohio State University Health System has developed a computerized physician order-entry (CPOE) system based on an expert system with extensive knowledge across multiple specialties.[195] The system automatically checks every order for possible allergies in the patient, drug interactions, duplications, drug restrictions, dosing guidelines, and appropriateness given information about the patient from the hospital's laboratory and radiology departments.

Science and Math. A "robot scientist" has been developed at the University of Wales that combines an AI-based system capable of formulating original theories, a robotic system that can automatically carry out experiments, and a reasoning engine to evaluate results. The researchers provided their creation with a model of gene expression in yeast. The system "automatically originates hypotheses to explain observations, devises experiments to test these hypotheses, physically runs the experiments using a laboratory robot, interprets the results to falsify hypotheses inconsistent with the data, and then repeats the cycle."[196] The system is capable of improving its performance by learning from its own experience. The experiments designed by the robot scientist were three times less expensive than those designed by human scientists. A test of the machine against a group of human scientists showed that the discoveries made by the machine were comparable to those made by the humans.

Mike Young, director of biology at the University of Wales, was one of the human scientists who lost to the machine. He explains that "the robot did beat me, but only because I hit the wrong key at one point."

A long-standing conjecture in algebra was finally proved by an AI system at Argonne National Laboratory. Human mathematicians called the proof "creative."

Business, Finance, and Manufacturing. Companies in every industry are using AI systems to control and optimize logistics, detect fraud and money laundering, and perform intelligent data mining on the horde of information they gather each day. Wal-Mart, for example, gathers vast amounts of information from its transactions with shoppers. AI-based tools using neural nets and expert systems review this data to provide market-research reports for managers. This intelligent data mining allows them to make remarkably accurate predictions of the inventory required for each product in each store for each day.[197]

AI-based programs are routinely used to detect fraud in financial transactions. Future Route, an English company, for example, offers iHex, based on AI routines developed at Oxford University, to detect fraud in credit-card transactions and loan applications.[198] The system continuously generates and updates its own rules based on its experience. First Union Home Equity Bank in Charlotte, North Carolina, uses Loan Arranger, a similar AI-based system, to decide whether to approve mortgage applications.[199]

NASDAQ similarly uses a learning program called the Securities Observation, News Analysis, and Regulation (SONAR) system to monitor all trades for fraud as well as the possibility of insider trading.[200] As of the end of 2003 more than 180 incidents had been detected by SONAR and referred to the U.S. Securities and Exchange Commission and Department of Justice. These included several cases that later received significant news coverage.

Ascent Technology, founded by Patrick Winston, who directed MIT's AI Lab from 1972 through 1997, has designed a GA-based system called Smart-Airport Operations Center (SAOC) that can optimize the complex logistics of an airport, such as balancing work assignments of hundreds of employees, making gate and equipment assignments, and managing a myriad of other details.[201] Winston points out that "figuring out ways to optimize a complicated situation is what genetic algorithms do." SAOC has raised productivity by approximately 30 percent in the airports where it has been implemented.

Ascent's first contract was to apply its AI techniques to managing the logistics for the 1991 Desert Storm campaign in Iraq. DARPA claimed that AI-based logistic-planning systems, including the Ascent system, resulted in more savings than the entire government research investment in AI over several decades.

A recent trend in software is for AI systems to monitor a complex software system's performance, recognize malfunctions, and determine the best way to recover automatically without necessarily informing the human user.[202] The idea stems from the realization that as software systems become more complex, like humans, they will never be perfect, and that eliminating all bugs is impossible. As humans, we use the same strategy: we don't expect to be perfect, but we usually try to recover from inevitable mistakes. "We want to stand this notion of systems management on its head," says Armando Fox, the head of Stanford University's Software Infrastructures Group, who is working on what is now called "autonomic computing." Fox adds, "The system has to be able to set itself up, it has to optimize itself. It has to repair itself, and if something goes wrong, it has to know how to respond to external threats." IBM, Microsoft, and other software vendors are all developing systems that incorporate autonomic capabilities.

Manufacturing and Robotics. Computer-integrated manufacturing (CIM) increasingly employs AI techniques to optimize the use of resources, streamline logistics, and reduce inventories through just-in-time purchasing of parts and supplies. A new trend in CIM systems is to use "case-based reasoning" rather than hard-coded, rule-based expert systems. Such reasoning codes knowledge as "cases," which are examples of problems with solutions. Initial cases are usually designed by the engineers, but the key to a successful case-based reasoning system is its ability to gather new cases from actual experience. The system is then able to apply the reasoning from its stored cases to new situations.

Robots are extensively used in manufacturing. The latest generation of robots uses flexible AI-based machine-vision systems—from companies such as Cognex Corporation in Natick, Massachusetts—that can respond flexibly to varying conditions. This reduces the need for precise setup for the robot to operate correctly. Brian Carlisle, CEO of Adept Technologies, a Livermore, California, factory-automation company, points out that "even if labor costs were eliminated [as a consideration], a strong case can still be made for automating with robots and other flexible automation. In addition to quality and throughput, users gain by enabling rapid product changeover and evolution that can't be matched with hard tooling."

One of AI's leading roboticists, Hans Moravec, has founded a company called Seegrid to apply his machine-vision technology to applications in manufacturing, materials handling, and military missions.[203] Moravec's software enables a device (a robot or just a material-handling cart) to walk or roll through an unstructured environment and in a single pass build a reliable "voxel" (three-dimensional pixel) map of the environment. The robot can then use the map and its own reasoning ability to determine an optimal and obstacle-free path to carry out its assigned mission.

This technology enables autonomous carts to transfer materials throughout a manufacturing process without the high degree of preparation required with conventional preprogrammed robotic systems. In military situations autonomous vehicles could carry out precise missions while adjusting to rapidly changing environments and battlefield conditions.

Machine vision is also improving the ability of robots to interact with humans. Using small, inexpensive cameras, head- and eye-tracking software can sense where a human user is, allowing robots, as well as virtual personalities on a screen, to maintain eye contact, a key element for natural interactions. Head- and eye-tracking systems have been developed at Carnegie Mellon University and MIT and are offered by small companies such as Seeing Machines of Australia.

An impressive demonstration of machine vision was a vehicle that was driven by an AI system with no human intervention for almost the entire distance from Washington, D.C., to San Diego.[204] Bruce Buchanan, computer-science professor at the University of Pittsburgh and president of the American Association of Artificial Intelligence, pointed out that this feat would have been "unheard of 10 years ago."

Palo Alto Research Center (PARC) is developing a swarm of robots that can navigate in complex environments, such as a disaster zone, and find items of interest, such as humans who may be injured. In a September 2004 demonstration at an AI conference in San Jose, they demonstrated a group of self-organizing robots on a mock but realistic disaster area.[205] The robots moved over the rough terrain, communicated with one another, used pattern recognition on images, and detected body heat to locate humans.

Speech and Language. Dealing naturally with language is the most challenging task of all for artificial intelligence. No simple tricks, short of fully mastering the principles of human intelligence, will allow a computerized system to convincingly emulate human conversation, even if restricted to just text messages. This was Turing's enduring insight in designing his eponymous test based entirely on written language.

Although not yet at human levels, natural language-processing systems are making solid progress. Search engines have become so popular that "Google" has gone from a proper noun to a common verb, and its technology has revolutionized research and access to knowledge. Google and other search engines use AI-based statistical-learning methods and logical inference to determine the ranking of links. The most obvious failing of these search engines is their inability to understand the context of words. Although an experienced user learns how to design a string of keywords to find the most relevant sites (for example, a search for "computer chip" is likely to avoid references to potato chips that a search for "chip" alone might turn up), what we would really like to be able to do is converse with our search engines in natural language. Microsoft has developed a natural-language search engine called Ask MSR (Ask Micro-Soft Research), which actually answers natural-language questions such as "When was Mickey Mantle born?"[206] After the system parses the sentence to determine the parts of speech (subject, verb, object, adjective and adverb modifiers, and so on), a special search engine then finds matches based on the parsed sentence. The found documents are searched for sentences that appear to answer the question, and the possible answers are ranked. At least 75 percent of the time, the correct answer is in the top three ranked positions, and incorrect answers are usually obvious (such as "Mickey Mantle was born in 3"). The researchers hope to include knowledge bases that will lower the rank of many of the nonsensical answers.

Microsoft researcher Eric Brill, who has led research on Ask MSR, has also attempted an even more difficult task: building a system that provides answers of about fifty words to more complex questions, such as, "How are the recipients of the Nobel Prize selected?" One of the strategies used by this system is to find an appropriate FAQ section on the Web that answers the query.

Natural-language systems combined with large-vocabulary, speaker-independent (that is, responsive to any speaker) speech recognition over the phone are entering the marketplace to conduct routine transactions. You can talk to British Airways' virtual travel agent about anything you like as long as it has to do with booking flights on British Airways.[207] You're also likely to talk to a virtual person if you call Verizon for customer service or Charles Schwab and Merrill Lynch to conduct financial transactions. These systems, while they can be annoying to some people, are reasonably adept at responding appropriately to the often ambiguous and fragmented way people speak. Microsoft and other companies are offering systems that allow a business to create virtual agents to book reservations for travel and hotels and conduct routine transactions of all kinds through two-way, reasonably natural voice dialogues.

Not every caller is satisfied with the ability of these virtual agents to get the job done, but most systems provide a means to get a human on the line. Companies using these systems report that they reduce the need for human service agents up to 80 percent. Aside from the money saved, reducing the size of call centers has a management benefit. Call-center jobs have very high turnover rates because of low job satisfaction.

It's said that men are loath to ask others for directions, but car vendors are betting that both male and female drivers will be willing to ask their own car for help in getting to their destination. In 2005 the Acura RL and Honda Odyssey will be offering a system from IBM that allows users to converse with their cars.[208] Driving directions will include street names (for example, "turn left on Main Street, then right on Second Avenue"). Users can ask such questions as "Where is the nearest Italian restaurant?" or they can enter specific locations by voice, ask for clarifications on directions, and give commands to the car itself (such as "Turn up the air conditioning"). The Acura RL will also track road conditions and highlight traffic congestion on its screen in real time. The speech recognition is claimed to be speaker-independent and to be unaffected by engine sound, wind, and other noises. The system will reportedly recognize 1.7 million street and city names, in addition to nearly one thousand commands.

Computer language translation continues to improve gradually. Because this is a Turing-level task—that is, it requires full human-level understanding of language to perform at human levels—it will be one of the last application areas to compete with human performance. Franz Josef Och, a computer scientist at the University of Southern California, has developed a technique that can generate a new language-translation system between any pair of languages in a matter of hours or days.[209] All he needs is a "Rosetta stone"—that is, text in one language and the translation of that text in the other language—although he needs millions of words of such translated text. Using a self-organizing technique, the system is able to develop its own statistical models of how text is translated from one language to the other and develops these models in both directions.

This contrasts with other translation systems, in which linguists painstakingly code grammar rules with long lists of exceptions to each rule. Och's system recently received the highest score in a competition of translation systems conducted by the U.S. Commerce Department's National Institute of Standards and Technology.

Entertainment and Sports. In an amusing and intriguing application of GAs, Oxford scientist Torsten Reil created animated creatures with simulated joints and muscles and a neural net for a brain. He then assigned them a task: to walk. He used a GA to evolve this capability, which involved seven hundred parameters. "If you look at that system with your human eyes, there's no way you can do it on your own, because the system is just too complex," Reil points out. "That's where evolution comes in."[210]

While some of the evolved creatures walked in a smooth and convincing way, the research demonstrated a well-known attribute of GAs: you get what you ask for. Some creatures figured out novel new ways of passing for walking. According to Reil, "We got some creatures that didn't walk at all, but had these very strange ways of moving forward: crawling or doing somersaults."

Software is being developed that can automatically extract excerpts from a video of a sports game that show the more important plays.[211] A team at Trinity College in Dublin is working on table-based games like pool, in which software tracks the location of each ball and is programmed to identify when a significant shot has been made. A team at the University of Florence is working on soccer. This software tracks the location of each player and can determine the type of play being made (such as free kicking or attempting a goal), when a goal is achieved, when a penalty is earned, and other key events.

The Digital Biology Interest Group at University College in London is designing Formula One race cars by breeding them using GAs.[212]

The AI winter is long since over. We are well into the spring of narrow AI. Most of the examples above were research projects just ten to fifteen years ago. If all the AI systems in the world suddenly stopped functioning, our economic infrastructure would grind to a halt. Your bank would cease doing business. Most transportation would be crippled. Most communications would fail. This was not the case a decade ago. Of course, our AI systems are not smart enough—yet—to organize such a conspiracy.

Strong AI

If you understand something in only one way, then you don't really understand it at all. This is because, if something goes wrong, you get stuck with a thought that just sits in your mind with nowhere to go. The secret of what anything means to us depends on how we've connected it to all the other things we know. This is why, when someone learns "by rote," we say that they don't really understand. However, if you have several different representations then, when one approach fails you can try another. Of course, making too many indiscriminate connections will turn a mind to mush. But well-connected representations let you turn ideas around in your mind, to envision things from many perspectives until you find one that works for you. And that's what we mean by thinking!

—MARVIN MINSKY[213]

Advancing computer performance is like water slowly flooding the landscape. A half century ago it began to drown the lowlands, driving out human calculators and record clerks, but leaving most of us dry. Now the flood has reached the foothills, and our outposts there are contemplating retreat. We feel safe on our peaks, but, at the present rate, those too will be submerged within another half century. I propose that we build Arks as that day nears, and adopt a seafaring life! For now, though, we must rely on our representatives in the lowlands to tell us what water is really like.

Our representatives on the foothills of chess and theorem-proving report signs of intelligence. Why didn't we get similar reports decades before, from the lowlands, as computers surpassed humans in arithmetic and rote memorization? Actually, we did, at the time. Computers that calculated like thousands of mathematicians were hailed as "giant brains," and inspired the first generation of AI research. After all, the machines were doing something beyond any animal, that needed human intelligence, concentration and years of training. But it is hard to recapture that magic now. One reason is that computers' demonstrated stupidity in other areas biases our judgment. Another relates to our own ineptitude. We do arithmetic or keep records so painstakingly and externally that the small mechanical steps in a long calculation are obvious, while the big picture often escapes us. Like Deep Blue's builders, we see the process too much from the inside to appreciate the subtlety that it may have on the outside. But there is a non-obviousness in snowstorms or tornadoes that emerge from the repetitive arithmetic of weather simulations, or in rippling tyrannosaur skin from movie animation calculations. We rarely call it intelligence, but "artificial reality" may be an even more profound concept than artificial intelligence.

The mental steps underlying good human chess playing and theorem proving are complex and hidden, putting a mechanical interpretation out of reach. Those who can follow the play naturally describe it instead in mentalistic language, using terms like strategy, understanding and creativity. When a machine manages to be simultaneously meaningful and surprising in the same rich way, it too compels a mentalistic interpretation. Of course, somewhere behind the scenes, there are programmers who, in principle, have a mechanical interpretation. But even for them, that interpretation loses its grip as the working program fills its memory with details too voluminous for them to grasp.

As the rising flood reaches more populated heights, machines will begin to do well in areas a greater number can appreciate. The visceral sense of a thinking presence in machinery will become increasingly widespread. When the highest peaks are covered, there will be machines that can interact as intelligently as any human on any subject. The presence of minds in machines will then become self-evident.

—HANS MORAVEC[214]

Because of the exponential nature of progress in information-based technologies, performance often shifts quickly from pathetic to daunting. In many diverse realms, as the examples in the previous section make clear, the performance of narrow AI is already impressive. The range of intelligent tasks in which machines can now compete with human intelligence is continually expanding. In a cartoon I designed for *The Age of Spiritual Machines*, a defensive "human race" is seen writing out signs that state what only people (and not machines) can do.[215] Littered on the floor are the signs the human race has already discarded because machines can now perform these functions: diagnose an electrocardiogram, compose in the style of Bach, recognize faces, guide a missile, play Ping-Pong, play master chess, pick stocks, improvise jazz, prove important theorems, and understand continuous speech. Back in 1999 these tasks were no longer solely the province of human intelligence; machines could do them all.

On the wall behind the man symbolizing the human race are signs he has written out describing the tasks that were still the sole province of humans: have common sense, review a movie, hold press conferences, translate speech, clean a house, and drive cars. If we were to redesign this cartoon in a few years, some of these signs would also be likely to end up on the floor. When CYC reaches one hundred million items of commonsense knowledge, perhaps human superiority in the realm of commonsense reasoning won't be so clear.

The era of household robots, although still fairly primitive today, has already started. Ten years from now, it's likely we will consider "clean a house" as within the capabilities of machines. As for driving cars, robots with no human intervention have already driven nearly across the United States on ordinary roads with other normal traffic. We are not yet ready to turn over all steering wheels to machines, but there are serious proposals to create electronic highways on which cars (with people in them) will drive by themselves.

The three tasks that have to do with human-level understanding of natural language—reviewing a movie, holding a press conference, and translating speech—are the most difficult. Once we can take down these signs, we'll have Turing-level machines, and the era of strong AI will have started.

This era will creep up on us. As long as there are any discrepancies between human and machine performance—areas in which humans outperform machines—strong AI skeptics will seize on these differences. But our experience in each area of skill and knowledge is likely to follow that of Kasparov. Our perceptions of performance will shift quickly from pathetic to daunting as the knee of the exponential curve is reached for each human capability.

How will strong AI be achieved? Most of the material in this book is intended to lay out the fundamental requirements for both hardware and software and explain why we can be confident that these requirements will be met in nonbiological systems. The continuation of the exponential growth of the price-performance of computation to achieve hardware capable of emulating human intelligence was still controversial in 1999. There has been so much progress in developing the technology for three-dimensional computing over the past five years that relatively few knowledgeable observers now doubt that this will happen. Even just taking the semiconductor industry's published ITRS road map, which runs to 2018, we can project human-level hardware at reasonable cost by that year.[216]

I've stated the case in chapter 4 of why we can have confidence that we will have detailed models and simulations of all regions of the human brain by the late 2020s. Until recently, our tools for peering into the brain did not have the spatial and temporal resolution, bandwidth, or price-performance to produce adequate data to create sufficiently detailed models. This is now changing. The emerging generation of scanning and sensing tools can analyze and detect neurons and neural components with exquisite accuracy, while operating in real time.

Future tools will provide far greater resolution and capacity. By the 2020s, we will be able to send scanning and sensing nanobots into the capillaries of the brain to scan it from inside. We've shown the ability to translate the data from diverse sources of brain scanning and sensing into models and computer simulations that hold up well to experimental comparison with the performance of the biological versions of these regions. We already have compelling models and simulations for several important brain regions. As I argued in chapter 4, it's a conservative projection to expect detailed and realistic models of all brain regions by the late 2020s.

One simple statement of the strong AI scenario is that we will learn the principles of operation of human intelligence from reverse engineering all the brain's regions, and we will apply these principles to the brain-capable computing platforms that will exist in the 2020s. We already have an effective toolkit for narrow AI. Through the ongoing refinement of these methods, the development of new algorithms, and the trend toward combining multiple methods into intricate architectures, narrow AI will continue to become less narrow. That is, AI applications will have broader domains, and their performance will become more flexible. AI systems will develop multiple ways of approaching each problem, just as humans do. Most important, the new insights and paradigms resulting from the acceleration of brain reverse engineering will greatly enrich this set of tools on an ongoing basis. This process is well under way.

It's often said that the brain works differently from a computer, so we cannot apply our insights about brain function into workable nonbiological systems. This view completely ignores the field of self-organizing systems, for which we have a set of increasingly sophisticated mathematical tools. As I discussed in the previous chapter, the brain differs in a number of important ways from conventional, contemporary computers. If you open up your Palm Pilot and cut a wire, there's a good chance you will break the machine. Yet we routinely lose many neurons and interneuronal connections with no ill effect, because the brain is self-organizing and relies on distributed patterns in which many specific details are not important.

When we get to the mid- to late 2020s, we will have access to a generation of extremely detailed brain-region models. Ultimately the toolkit will be greatly enriched with these new models and simulations and will encompass a full knowledge of how the brain works. As we apply the toolkit to intelligent tasks, we will draw upon the entire range of tools, some derived directly from brain reverse engineering, some merely inspired by what we know about the brain, and some not based on the brain at all but on decades of AI research.

Part of the brain's strategy is to learn information, rather than having knowledge hard-coded from the start. ("Instinct" is the term we use to refer to such innate knowledge.) Learning will be an important aspect of AI, as well. In my experience in developing pattern-recognition systems in character recognition, speech recognition, and financial analysis, providing for the AI's education is the most challenging and important part of the engineering. With the accumulated knowledge of human civilization increasingly accessible online, future AIs will have the opportunity to conduct their education by accessing this vast body of information.

The education of AIs will be much faster than that of unenhanced humans. The twenty-year time span required to provide a basic education to biological humans could be compressed into a matter of weeks or less. Also, because nonbiological intelligence can share its patterns of learning and knowledge, only one AI has to master each particular skill. As I pointed out, we trained one set of research computers to understand speech, but then the hundreds of thousands of people who acquired our speech-recognition software had to load only the already trained patterns into their computers.

One of the many skills that nonbiological intelligence will achieve with the completion of the human brain reverse-engineering project is sufficient mastery of language and shared human knowledge to pass the Turing test. The Turing test is important not so much for its practical significance but rather because it will demarcate a crucial threshold. As I have pointed out, there is no simple means to pass a Turing test, other than to convincingly emulate the flexibility, subtlety, and suppleness of human intelligence. Having captured that capability in our technology, it will then be subject to engineering's ability to concentrate, focus, and amplify it.

Variations of the Turing test have been proposed. The annual Loebner Prize contest awards a bronze prize to the chatterbot (conversational bot) best able to convince human judges that it's human.[217] The criteria for winning the silver prize is based on Turing's original test, and it obviously has yet to be awarded. The gold prize is based on visual and auditory communication. In other words, the AI must have a convincing face and voice, as transmitted over a terminal, and thus it must appear to the human judge as if he or she is interacting with a real person over a videophone. On the face of it, the gold prize sounds more difficult. I've argued that it may actually be easier, because judges may pay less attention to the text portion of the language being communicated and could be distracted by a convincing facial and voice animation. In fact, we already have real-time facial animation, and while it is not quite up to these modified Turing standards, it's reasonably close. We also have very natural-sounding voice synthesis, which is often confused with recordings of human speech, although more work is needed on prosodics (intonation). We're likely to achieve satisfactory facial animation and voice production sooner than the Turing-level language and knowledge capabilities.

Turing was carefully imprecise in setting the rules for his test, and significant literature has been devoted to the subtleties of establishing the exact procedures for determining how to assess when the Turing test has been passed.[218] In 2002 I negotiated the rules for a Turing-test wager with Mitch Kapor on the Long Now Web site.[219] The question underlying our twenty-thousand-dollar bet, the proceeds of which go to the charity of the winner's choice, was, "Will the Turing test be passed by a machine by 2029?" I said yes, and Kapor said no. It took us months of dialogue to arrive at the intricate rules to implement our wager. Simply defining "machine" and "human," for example, was not a straightforward matter. Is the human judge allowed to have any nonbiological thinking processes in his or her brain? Conversely, can the machine have any biological aspects?

Because the definition of the Turing test will vary from person to person, Turing test–capable machines will not arrive on a single day, and there will be a period during which we will hear claims that machines have passed the threshold. Invariably, these early claims will be debunked by knowledgeable observers, probably including myself. By the time there is a broad consensus that the Turing test has been passed, the actual threshold will have long since been achieved.

Edward Feigenbaum proposes a variation of the Turing test, which assesses not a machine's ability to pass for human in casual, everyday dialogue but its ability to pass for a scientific expert in a specific field.[220] The Feigenbaum test (FT) may be more significant than the Turing test because FT-capable machines, being technically proficient, will be capable of improving their own designs. Feigenbaum describes his test in this way:

> Two players play the FT game. One player is chosen from among the elite practitioners in each of three pre-selected fields of natural science, engineering, or medicine. (The number could be larger, but for this challenge not greater than ten). Let's say we choose the fields from among those covered in the U.S. National Academy. . . . For example, we could choose astrophysics, computer science, and molecular biology. In each round of the game, the behavior of the two players (elite scientist and computer) is judged by another Academy member in that particular domain of discourse, e.g., an astrophysicist judging astrophysics behavior. Of course the identity of the players is hidden from the judge as it is in the Turing test. The judge poses problems, asks questions, asks for

explanations, theories, and so on—as one might do with a colleague. Can the human judge choose, at better than chance level, which is his National Academy colleague and which is the computer?

Of course Feigenbaum overlooks the possibility that the computer might already be a National Academy colleague, but he is obviously assuming that machines will not yet have invaded institutions that today comprise exclusively biological humans. While it may appear that the FT is more difficult than the Turing test, the entire history of AI reveals that machines started with the skills of professionals and only gradually moved toward the language skills of a child. Early AI systems demonstrated their prowess initially in professional fields such as proving mathematical theorems and diagnosing medical conditions. These early systems would not be able to pass the FT, however, because they do not have the language skills and the flexible ability to model knowledge from different perspectives that are needed to engage in the professional dialogue inherent in the FT.

This language ability is essentially the same ability needed in the Turing test. Reasoning in many technical fields is not necessarily more difficult than the commonsense reasoning engaged in by most human adults. I would expect that machines will pass the FT, at least in some disciplines, around the same time as they pass the Turing test. Passing the FT in all disciplines is likely to take longer, however. This is why I see the 2030s as a period of consolidation, as machine intelligence rapidly expands its skills and incorporates the vast knowledge bases of our biological human and machine civilization. By the 2040s we will have the opportunity to apply the accumulated knowledge and skills of our civilization to computational platforms that are billions of times more capable than unassisted biological human intelligence.

The advent of strong AI is the most important transformation this century will see. Indeed, it's comparable in importance to the advent of biology itself. It will mean that a creation of biology has finally mastered its own intelligence and discovered means to overcome its limitations. Once the principles of operation of human intelligence are understood, expanding its abilities will be conducted by human scientists and engineers whose own biological intelligence will have been greatly amplified through an intimate merger with nonbiological intelligence. Over time, the nonbiological portion will predominate.

We've discussed aspects of the impact of this transformation throughout this book, which I focus on in the next chapter. Intelligence is the ability to solve problems with limited resources, including limitations of time. The Singularity will be characterized by the rapid cycle of human intelligence—

increasingly nonbiological—capable of comprehending and leveraging its own powers.

FRIEND OF FUTURIST BACTERIUM, 2 BILLION B.C.: *So tell me again about these ideas you have about the future.*

FUTURIST BACTERIUM, 2 BILLION B.C.: *Well, I see bacteria getting together into societies, with the whole band of cells basically acting like one big complicated organism with greatly enhanced capabilities.*

FRIEND OF FUTURIST BACTERIUM: *What gives you that idea?*

FUTURIST BACTERIUM: *Well already, some of our fellow Daptobacters have gone inside other larger bacteria to form a little duo.[221] It's inevitable that our fellow cells will band together so that each cell can specialize its function. As it is now, we each have to do everything by ourselves: find food, digest it, excrete by-products.*

FRIEND OF FUTURIST BACTERIUM: *And then what?*

FUTURIST BACTERIUM: *All these cells will develop ways of communicating with one another that go beyond just the swapping of chemical gradients that you and I can do.*

FRIEND OF FUTURIST BACTERIUM: *Okay, now tell me again the part about that future superassembly of ten trillion cells.*

FUTURIST BACTERIUM: *Yes, well, according to my models, in about two billion years a big society of ten trillion cells will make up a single organism and include tens of billions of special cells that can communicate with one another in very complicated patterns.*

FRIEND OF FUTURIST BACTERIUM: *What sort of patterns?*

FUTURIST BACTERIUM: *Well, "music," for one thing. These huge bands of cells will create musical patterns and communicate them to all the other bands of cells.*

FRIEND OF FUTURIST BACTERIUM: *Music?*

FUTURIST BACTERIUM: *Yes, patterns of sound.*

FRIEND OF FUTURIST BACTERIUM: *Sound?*

FUTURIST BACTERIUM: *Okay, look at it this way. These supercell societies will be complicated enough to understand their own organization. They will be able to improve their own design, getting better and better, faster and faster. They will reshape the rest of the world in their image.*

FRIEND OF FUTURIST BACTERIUM: *Now, wait a second. Sounds like we'll lose our basic bacteriumity.*

FUTURIST BACTERIUM: *Oh, but there will be no loss.*

FRIEND OF FUTURIST BACTERIUM: *I know you keep saying that, but . . .*

FUTURIST BACTERIUM: *It will be a great step forward. It's our destiny as bacteria. And, anyway, there will still be little bacteria like us floating around.*

FRIEND OF FUTURIST BACTERIUM: *Okay, but what about the downside? I mean, how much harm can our fellow Daptobacter and Bdellovibrio bacteria do? But these future cell associations with their vast reach may destroy everything.*

FUTURIST BACTERIUM: *It's not certain, but I think we'll make it through.*

FRIEND OF FUTURIST BACTERIUM: *You always were an optimist.*

FUTURIST BACTERIUM: *Look, we won't have to worry about the downside for a couple billion years.*

FRIEND OF FUTURIST BACTERIUM: *Okay, then, let's get lunch.*

MEANWHILE, TWO BILLION YEARS LATER . . .

NED LUDD: *These future intelligences will be worse than the textile machines I fought back in 1812. Back then we had to worry about only one man with a machine doing the work of twelve. But you're talking about a marble-size machine outperforming all of humanity.*

RAY: *It will only outperform the biological part of humanity. In any event, that marble is still human, even if not biological.*

NED: *These superintelligences won't eat food. They won't breathe air. They won't reproduce through sex. . . . So just how are they human?*

RAY: *We're going to merge with our technology. We're already starting to do that in 2004, even if most of the machines are not yet inside our bodies and brains. Our machines nonetheless extend the reach of our intelligence. Extending our reach has always been the nature of being human.*

NED: *Look, saying that these superintelligent nonbiological entities are human is like saying that we're basically bacteria. After all, we're evolved from them also.*

RAY: *It's true that a contemporary human is a collection of cells, and that we are a product of evolution, indeed its cutting edge. But extending our intelligence by reverse engineering it, modeling it, simulating it, reinstantiating it on more capable substrates, and modifying and extending it is the next step in its evolution. It was the fate of bacteria to evolve into a technology-creating species. And it's our destiny now to evolve into the vast intelligence of the Singularity.*

CHAPTER SIX

The Impact . . .

The future enters into us in order to transform itself in us long before it happens.

—RAINER MARIA RILKE

One of the biggest flaws in the common conception of the future is that the future is something that happens to us, not something we create.

—MICHAEL ANISSIMOV

"Playing God" is actually the highest expression of human nature. The urges to improve ourselves, to master our environment, and to set our children on the best path possible have been the fundamental driving forces of all of human history. Without these urges to "play God," the world as we know it wouldn't exist today. A few million humans would live in savannahs and forests, eking out a hunter-gatherer existence, without writing or history or mathematics or an appreciation of the intricacies of their own universe and their own inner workings.

—RAMEZ NAAM

A Panoply of Impacts. What will be the nature of human experience once nonbiological intelligence predominates? What are the implications for the human-machine civilization when strong AI and nanotechnology can create any product, any situation, any environment *that we can imagine* at will? I stress the role of imagination here because we will still be constrained in our creations to what we can imagine. But our tools for bringing imagination to life are growing exponentially more powerful.

As the Singularity approaches we will have to reconsider our ideas about the nature of human life and redesign our human institutions. We will explore a few of these ideas and institutions in this chapter.

For example, the intertwined revolutions of G, N, and R will transform our frail version 1.0 human bodies into their far more durable and capable version

2.0 counterparts. Billions of nanobots will travel through the bloodstream in our bodies and brains. In our bodies, they will destroy pathogens, correct DNA errors, eliminate toxins, and perform many other tasks to enhance our physical well-being. As a result, we will be able to live indefinitely without aging.

In our brains, the massively distributed nanobots will interact with our biological neurons. This will provide full-immersion virtual reality incorporating all of the senses, as well as neurological correlates of our emotions, from within the nervous system. More important, this intimate connection between our biological thinking and the nonbiological intelligence we are creating will profoundly expand human intelligence.

Warfare will move toward nanobot-based weapons, as well as cyber-weapons. Learning will first move online, but once our brains are online we will be able to download new knowledge and skills. The role of work will be to create knowledge of all kinds, from music and art to math and science. The role of play will be, well, to create knowledge, so there won't be a clear distinction between work and play.

Intelligence on and around the Earth will continue to expand exponentially until we reach the limits of matter and energy to support intelligent computation. As we approach this limit in our corner of the galaxy, the intelligence of our civilization will expand outward into the rest of the universe, quickly reaching the fastest speed possible. We understand that speed to be the speed of light, but there are suggestions that we may be able to circumvent this apparent limit (possibly by taking shortcuts through wormholes, for example).

. . . on the Human Body

> So many different people to be.
>
> —DONOVAN[1]

> *Cosmetic baby, plug into me*
> *And never, ever find another.*
> *And I realize no one's wise*
> *To my plastic fantastic lover.*
>
> —JEFFERSON AIRPLANE,
>   "PLASTIC FANTASTIC LOVER"

> Our machines will become much more like us, and we will become much more like our machines.
>
> —RODNEY BROOKS

Once out of nature I shall never take
My bodily form from any natural thing,
But such a form as Grecian goldsmiths make
Of hammered gold and gold enamelling.

—WILLIAM BUTLER YEATS,
  "SAILING TO BYZANTIUM"

A radical upgrading of our bodies' physical and mental systems is already under way, using biotechnology and emerging genetic-engineering technologies. Beyond the next two decades we will use nanoengineered methods such as nanobots to augment and ultimately replace our organs.

A New Way of Eating. Sex has largely been separated from its biological function. For the most part, we engage in sexual activity for intimate communication and sensual pleasure, not reproduction. Conversely, we have devised multiple methods for creating babies without physical sex, albeit most reproduction does still derive from the sex act. This disentanglement of sex from its biological function is not condoned by all sectors of society, but it has been readily, even eagerly, adopted by the mainstream in the developed world.

So why don't we provide the same extrication of purpose from biology for another activity that also provides both social intimacy and sensual pleasure— namely, eating? The original biological purpose of consuming food was to provide the bloodstream with nutrients, which were then delivered to each of our trillions of cells. These nutrients include caloric (energy-bearing) substances such as glucose (mainly from carbohydrates), proteins, fats, and a myriad of trace molecules, such as vitamins, minerals, and phytochemicals that provide building blocks and enzymes for diverse metabolic processes.

Like any other major human biological system, digestion is astonishing in its intricacy, enabling our bodies to extract the complex resources needed to survive, despite sharply varying conditions, while at the same time filtering out a multiplicity of toxins. Our knowledge of the complex pathways underlying digestion is rapidly expanding, although there is still a great deal we do not fully understand.

But we do know that our digestive processes, in particular, are optimized for a period in our evolutionary development that is dramatically dissimilar to the one in which we now find ourselves. For most of our history we faced a high likelihood that the next foraging or hunting season (and for a brief, relatively recent period, the next planting season) might be catastrophically lean. It made sense, therefore, for our bodies to hold on to every possible calorie we consumed. Today that biological strategy is counterproductive and has become the outdated metabolic programming that underlies our contemporary epidemic of obesity and fuels pathological processes of degenerative disease, such as coronary artery disease and Type II diabetes.

Consider the reasons that the designs of our digestive and other bodily systems are far from optimal for current conditions. Until recently (on an evolutionary timescale) it was not in the interest of the species for old people like myself (I was born in 1948) to use up the limited resources of the clan. Evolution favored a short lifespan—life expectancy was thirty-seven years as recently as two centuries ago—to allow restricted reserves to be devoted to the young, those caring for them, and those strong enough to perform intense physical work. As discussed earlier, the so-called grandma hypothesis (which suggests that a small number of "wise" elderly members of the tribe were beneficial to the human species) does not appreciably challenge the observation that there was no strong selective pressure for genes that significantly extended human longevity.

We now live in an era of great material abundance, at least in technologically advanced nations. Most work requires mental effort rather than physical exertion. A century ago 30 percent of the U.S. workforce was employed on farms, with another 30 percent in factories. Both of these figures are now under 3 percent.[2] Many of today's job categories, ranging from flight controller to Web designer, simply didn't exist a century ago. Circa 2004 we have the opportunity to continue to contribute to our civilization's exponentially growing knowledge base—which is, incidentally, a unique attribute of our species—well past our child-rearing days. (As a baby boomer myself, that is certainly my view.)

Our species has already augmented our natural lifespan through our technology: drugs, supplements, replacement parts for virtually all bodily systems, and many other interventions. We have devices to replace our hips, knees, shoulders, elbows, wrists, jaws, teeth, skin, arteries, veins, heart valves, arms, legs, feet, fingers, and toes, and systems to replace more complex organs (for example, our hearts) are beginning to be introduced. As we learn the operating principles of the human body and brain, we will soon be in a position to design vastly superior systems that will last longer and perform better, without susceptibility to breakdown, disease, and aging.

One example of a conceptual design for such a system, called Primo Posthuman, was created by artist and cultural catalyst Natasha Vita-More.[3] Her design is intended to optimize mobility, flexibility, and superlongevity. It envisions features such as a metabrain for global-net connection with a prosthetic neocortex of AI interwoven with nanobots, solar-protected smart skin that has biosensors for tone and texture changeability, and high-acuity senses.

Although version 2.0 of the human body is an ongoing grand project that will ultimately result in the radical upgrading of all our physical and mental systems, we will implement it one small, benign step at a time. Based on our current knowledge, we can describe the means for accomplishing each aspect of this vision.

Redesigning the Digestive System. From this perspective, let's return to a consideration of the digestive system. We already have a comprehensive picture of the components of the food we eat. We know how to enable people who cannot eat to survive, using intravenous nutrition. However, this is clearly not a desirable alternative, since our technologies for getting substances in and out of the bloodstream are currently quite limited.

The next phase of improvement in this area will be largely biochemical, in the form of drugs and supplements that will prevent excess caloric absorption and otherwise reprogram metabolic pathways for optimal health. Research by Dr. Ron Kahn at the Joslin Diabetes Center has already identified the "fat insulin receptor" (FIR) gene, which controls accumulation of fat by the fat cells. By blocking the expression of this single gene in the fat cells of mice, Dr. Kahn's pioneering research has demonstrated that the animals were able to eat without restriction yet remain lean and healthy. Although they ate far more than the control mice, the "FIR knockout" mice actually lived 18 percent longer and had substantially lower rates of heart disease and diabetes. It's no surprise that pharmaceutical companies are hard at work to apply these findings to the human FIR gene.

In an intermediate phase nanobots in the digestive tract and bloodstream will intelligently extract the precise nutrients we need, order additional nutrients and supplements through our personal wireless local-area network, and send the rest of the matter on to be eliminated.

If this seems futuristic, keep in mind that intelligent machines are already making their way into our bloodstream. There are dozens of projects under way to create bloodstream-based BioMEMS for a wide range of diagnostic and therapeutic applications.[4] As mentioned, there are several major conferences devoted to these projects.[5] BioMEMS devices are being designed to intelligently scout out pathogens and deliver medications in very precise ways.

For example, nanoengineered blood-borne devices that deliver hormones such as insulin have been demonstrated in animals.[6] Similar systems could precisely deliver dopamine to the brain for Parkinson's patients, provide blood-clotting factors for patients with hemophilia, and deliver cancer drugs directly to tumor sites. One new design provides up to twenty substance-containing reservoirs that can release their cargo at programmed times and locations in the body.[7]

Kensall Wise, a professor of electrical engineering at the University of Michigan, has developed a tiny neural probe that can provide precise monitoring of the electrical activity of patients with neural diseases.[8] Future designs are also expected to deliver drugs to precise locations in the brain. Kazushi Ishiyama at Tohoku University in Japan has developed micromachines that use microscopic spinning screws to deliver drugs to small cancer tumors.[9]

A particularly innovative micromachine developed by Sandia National Laboratories has microteeth with a jaw that opens and closes to trap individual cells and then implant them with substances such as DNA, proteins, or drugs.[10] Many approaches are being developed for micro- and nanoscale machines to go into the body and bloodstream.

Ultimately we will be able to determine the precise nutrients (including all the hundreds of phytochemicals) necessary for the optimal health of each individual. These will be freely and inexpensively available, so we won't need to bother with extracting nutrients from food at all.

Nutrients will be introduced directly into the bloodstream by special metabolic nanobots, while sensors in our bloodstream and body, using wireless communication, will provide dynamic information on the nutrients needed at each point in time. This technology should be reasonably mature by the late 2020s.

A key question in designing such systems will be, How will nanobots be introduced into and removed from the body? The technologies we have today, such as intravenous catheters, leave much to be desired. Unlike drugs and nutritional supplements, however, nanobots have a measure of intelligence and can keep track of their own inventories and intelligently slip in and out of our bodies in clever ways. One scenario is that we would wear a special nutrient device in a belt or undershirt, which would be loaded with nutrient-bearing nanobots that could enter the body through the skin or other body cavities.

At that stage of technological development, we will be able to eat whatever we want, whatever gives us pleasure and gastronomic fulfillment, exploring the culinary arts for their tastes, textures, and aromas while having an optimal flow of nutrients to our bloodstream. One possibility to achieve this would be to have all the food we eat pass through a modified digestive tract that doesn't allow absorption into the bloodstream. But this would place a burden on our colon and bowel functions, so a more refined approach would be to dispense with the conventional function of elimination. We could accomplish that by using special elimination nanobots that act like tiny garbage compactors. As the nutrient nanobots make their way into our bodies, the elimination nanobots go the other way. Such an innovation would also enable us to outgrow the need for the organs that filter the blood for impurities, such as the kidneys.

Ultimately we won't need to bother with special garments or explicit nutritional resources. Just as computation will be ubiquitous, the basic metabolic nanobot resources we need will be embedded throughout our environment. But it will also be important to maintain ample reserves of all needed resources *inside* the body. Our version 1.0 bodies do this to only a very limited extent—for example, storing a few minutes' worth of oxygen in our blood and a few days' worth of caloric energy in glycogen and other reserves. Version 2.0 will provide substantially greater reserves, enabling us to be separated from metabolic resources for greatly extended periods of time.

Of course, most of us won't do away with our old-fashioned digestive process when these technologies are first introduced. After all, people didn't throw away their typewriters when the first generation of word processors was introduced. However, these new technologies will in due course dominate. Few people today still use a typewriter, a horse and buggy, a wood-burning stove, or other displaced technologies (other than as deliberate experiences in antiquity). The same phenomenon will happen with our reengineered bodies. Once we've worked out the inevitable complications that will arise with a radically reengineered gastrointestinal system, we'll begin to rely on it more and more. A nanobot-based digestive system can be introduced gradually, first augmenting our digestive tract, replacing it only after many iterations.

Programmable Blood. One pervasive system that has already been the subject of a comprehensive conceptual redesign based on reverse engineering is our blood. I mentioned earlier Rob Freitas's nanotechnology-based designs to replace our red blood cells, platelets, and white blood cells.[11] Like most of our biological systems our red blood cells perform their oxygenating function very inefficiently, so Freitas has redesigned them for optimal performance. Because his respirocytes (robotic red blood cells) would enable one to go hours without oxygen,[12] it will be interesting to see how this development is dealt with in athletic contests. Presumably the use of respirocytes and similar systems will be prohibited in events like the Olympics, but then we will face the prospect of teenagers (whose bloodstreams will likely contain respirocyte-enriched blood) routinely outperforming Olympic athletes. Although prototypes are still one to two decades in the future, their physical and chemical requirements have been worked out in impressive detail. Analyses show that Freitas's designs would be hundreds or thousands of times more capable of storing and transporting oxygen than our biological blood.

Freitas also envisions micron-size artificial platelets that could achieve homeostasis (bleeding control) up to one thousand times faster than biological platelets do,[13] as well as nanorobotic "microbivores" (white-blood-cell replacements) that will download software to destroy specific infections hundreds of times faster than antibiotics and will be effective against all bacterial, viral, and fungal infections, as well as cancer, with no limitations of drug resistance.[14]

Have a Heart, or Not. The next organ on our list for enhancement is the heart, which, while an intricate and impressive machine, has a number of severe problems. It is subject to a myriad of failure modes and represents a fundamental weakness in our potential longevity. The heart usually breaks down long before the rest of the body, often very prematurely.

Although artificial hearts are beginning to be feasible replacements, a more effective approach will be to get rid of the heart altogether. Among Freitas's designs are nanorobotic blood cells that provide their own mobility. If the blood moves autonomously, the engineering issues of the extreme pressures required for centralized pumping can be eliminated. As we perfect ways to transfer nanobots to and from the blood supply, we will eventually be able to continuously replace them. Freitas has also published a design for a complex five-hundred-trillion-nanorobot system, called a "vasculoid," that replaces the entire human bloodstream with nonfluid-based delivery of essential nutrients and cells.[15]

Energy for the body will also be provided by microscopic fuel cells, using either hydrogen or the body's own fuel, ATP. As I described in the last chapter, substantial progress has been made recently with both MEMS-scale and nanoscale fuel cells, including some that use the body's own glucose and ATP energy sources.[16]

With the respirocytes providing greatly improved oxygenation, we will be able to eliminate the lungs by using nanobots to provide oxygen and remove carbon dioxide. As with other systems, we will go through intermediate stages where these technologies simply augment our natural processes, so we can have the best of both worlds. Eventually, though, there will be no reason to continue with the complications of actual breathing and the burdensome requirement of breathable air everywhere we go. If we find breathing itself pleasurable, we can develop virtual ways of having this sensual experience.

In time we also won't need the various organs that produce chemicals, hormones, and enzymes that flow into the blood and other metabolic pathways. We can now synthesize bio-identical versions of many of these substances, and within one to two decades we will be able to routinely create the vast majority of biochemically relevant substances. We are already creating artificial hormone organs. For example, the Lawrence Livermore National Laboratory and California-based Medtronic MiniMed are developing an artificial pancreas to be implanted under the skin. It will monitor blood glucose levels and release precise amounts of insulin, using a computer program to function like our biological pancreatic islet cells.[17]

In human body version 2.0 hormones and related substances (to the extent that we still need them) will be delivered via nanobots, controlled by intelligent biofeedback systems to maintain and balance required levels. Since we will be eliminating most of our biological organs, many of these substances may no longer be needed and will be replaced by other resources required by the nanorobotic systems.

So What's Left? Let's consider where we are, circa early 2030s. We've eliminated the heart, lungs, red and white blood cells, platelets, pancreas, thyroid and all the hormone-producing organs, kidneys, bladder, liver, lower esophagus, stomach, small intestines, large intestines, and bowel. What we have left at this point is the skeleton, skin, sex organs, sensory organs, mouth and upper esophagus, and brain.

The skeleton is a stable structure, and we already have a reasonable understanding of how it works. We can now replace parts of it (for example, artificial hips and joints), although the procedure requires painful surgery, and our current technology for doing so has serious limitations. Interlinking nanobots will one day provide the ability to augment and ultimately replace the skeleton through a gradual and noninvasive process. The human skeleton version 2.0 will be very strong, stable, and self-repairing.

We will not notice the absence of many of our organs, such as the liver and pancreas, since we do not directly experience their operation. But the skin, which includes our primary and secondary sex organs, may prove to be an organ we will actually want to keep, or we may at least want to maintain its vital functions of communication and pleasure. However, we will ultimately be able to improve on the skin with new nanoengineered supple materials that will provide greater protection from physical and thermal environmental effects while enhancing our capacity for intimate communication. The same observation holds for the mouth and upper esophagus, which constitute the remaining aspects of the digestive system that we use to experience the act of eating.

Redesigning the Human Brain. As we discussed earlier, the process of reverse engineering and redesign will also encompass the most important system in our bodies: the brain. We already have implants based on "neuromorphic" modeling (reverse engineering of the human brain and nervous system) for a rapidly growing list of brain regions.[18] Researchers at MIT and Harvard are developing neural implants to replace damaged retinas.[19] Implants are available for Parkinson's patients that communicate directly with the ventral posterior nucleus and subthalmic nucleus regions of the brain to reverse the most devastating symptoms of this disease.[20] An implant for people with cerebral palsy and multiple sclerosis communicates with the ventral lateral thalamus and has been effective in controlling tremors.[21] "Rather than treat the brain like soup, adding chemicals that enhance or suppress certain neurotransmitters," says Rick Trosch, an American physician helping to pioneer these therapies, "we're now treating it like circuitry."

A variety of techniques is also being developed to provide the communications bridge between the wet analog world of biological information processing and digital electronics. Researchers at Germany's Max Planck Institute have developed noninvasive devices that can communicate with neurons in both directions.[22] They demonstrated their "neuron transistor" by controlling the movements of a living leech from a personal computer. Similar technology has been used to reconnect leech neurons and coax them to perform simple logical and arithmetic problems.

Scientists are also experimenting with "quantum dots," tiny chips comprising crystals of photoconductive (reactive to light) semiconductor material that can be coated with peptides that bind to specific locations on neuron cell surfaces. These could allow researchers to use precise wavelengths of light to remotely activate specific neurons (for drug delivery, for example), replacing invasive external electrodes.[23]

Such developments also provide the promise of reconnecting broken neural pathways for people with nerve damage and spinal-cord injuries. It had long been thought that re-creating these pathways would be feasible only for recently injured patients, because nerves gradually deteriorate when unused. A recent discovery, however, shows the feasibility of a neuroprosthetic system for patients with long-standing spinal-cord injuries. Researchers at the University of Utah asked a group of long-term quadriplegic patients to move their limbs in a variety of ways and then observed the response of their brains, using magnetic resonance imaging (MRI). Although the neural pathways to their limbs had been inactive for many years, the patterns of their brain activity when attempting to move their limbs was very close to those observed in nondisabled persons.[24]

We will also be able to place sensors in the brain of a paralyzed person that will be programmed to recognize the brain patterns associated with intended movements and then stimulate the appropriate sequence of muscle actions. For those patients whose muscles no longer function, there are already designs for "nanoelectromechanical" systems (NEMS) that can expand and contract to replace damaged muscles and that can be activated by either real or artificial nerves.

We Are Becoming Cyborgs. The human body version 2.0 scenario represents the continuation of a long-standing trend in which we grow more intimate with our technology. Computers started out as large, remote machines in air-conditioned rooms tended by white-coated technicians. They moved onto our desks, then under our arms, and now into our pockets. Soon, we'll routinely put them inside our bodies and brains. By the 2030s we will become more non-biological than biological. As I discussed in chapter 3, by the 2040s nonbiological intelligence will be billions of times more capable than our biological intelligence.

The compelling benefits of overcoming profound diseases and disabilities will keep these technologies on a rapid course, but medical applications represent only the early-adoption phase. As the technologies become established, there will be no barriers to using them for vast expansion of human potential.

Stephen Hawking recently commented in the German magazine *Focus* that computer intelligence will surpass that of humans within a few decades. He advocated that we "urgently need to develop direct connections to the brain, so that computers can add to human intelligence, rather than be in opposition."[25] Hawking can take comfort that the development program he is recommending is well under way.

There will be many variations of human body version 2.0, and each organ and body system will have its own course of development and refinement. Biological evolution is only capable of what is called "local optimization," meaning that it can improve a design but only within the constraints of design "decisions" that biology arrived at long ago. For example, biological evolution is restricted to building everything from a very limited class of materials—namely, proteins, which are folded from one-dimensional strings of amino acids. It is restricted to thinking processes (pattern recognition, logical analysis, skill formation, and other cognitive skills) that use extremely slow chemical switching. And biological evolution itself works very slowly, only incrementally improving designs that continue to apply these basic concepts. It is incapable of suddenly changing, for example, to structural materials made of diamondoid or to nanotube-based logical switching.

However, there is a way around this inherent limitation. Biological evolution did create a species that could think and manipulate its environment. That species is now succeeding in accessing—and improving—its own design and is capable of reconsidering and altering these basic tenets of biology.

Human Body Version 3.0. I envision human body 3.0—in the 2030s and 2040s—as a more fundamental redesign. Rather than reformulating each subsystem, we (both the biological and nonbiological portions of our thinking, working together) will have the opportunity to revamp our bodies based on our experience with version 2.0. As with the transition from 1.0 to 2.0, the transition to 3.0 will be gradual and will involve many competing ideas.

One attribute I envision for version 3.0 is the ability to change our bodies. We'll be able to do that very easily in virtual-reality environments (see the next section), but we will also acquire the means to do this in real reality. We will incorporate MNT-based fabrication into ourselves, so we'll be able to rapidly alter our physical manifestation at will.

Even with our mostly nonbiological brains we're likely to keep the aesthetics and emotional import of human bodies, given the influence this aesthetic has on the human brain. (Even when extended, the nonbiological portion of our intelligence will still have been derived from biological human intelligence.) That is, human body version 3.0 is likely still to look human by today's standards, but given the greatly expanded plasticity that our bodies will have, ideas of what constitutes beauty will be expanded upon over time. Already, people augment their bodies with body piercing, tattoos, and plastic surgery, and social acceptance of these changes has rapidly increased. Since we'll be able to make changes that are readily reversible, there is likely to be far greater experimentation.

J. Storrs Hall has described nanobot designs he calls "foglets" that are able to link together to form a great variety of structures and that can quickly change their structural organization. They're called "foglets" because if there's a sufficient density of them in an area, they can control sound and light to form variable sounds and images. They are essentially creating virtual-reality environments externally (that is, in the physical world) rather than internally (in the nervous system). Using them a person can modify his body or his environment, though some of these changes will actually be illusions, since the foglets can control sound and images.[26] Hall's foglets are one conceptual design for creating real morphable bodies to compete with those in virtual reality.

BILL (AN ENVIRONMENTALIST): *On this human body version 2.0 stuff, aren't you throwing the baby out—quite literally—with the bathwater? You're sug-*

gesting replacing the entire human body and brain with machines. There's no human being left.

RAY: *We don't agree on the definition of human, but just where do you suggest drawing the line? Augmenting the human body and brain with biological or nonbiological interventions is hardly a new concept. There's still a lot of human suffering.*

BILL: *I have no objection to alleviating human suffering. But replacing a human body with a machine to exceed human performance leaves you with, well, a machine. We have cars that can travel on the ground faster than a human, but we don't consider them to be human.*

RAY: *The problem here has a lot to do with the word "machine." Your conception of a machine is of something that is much less valued—less complex, less creative, less intelligent, less knowledgeable, less subtle and supple—than a human. That's reasonable for today's machines because all the machines we've ever met—like cars—are like this. The whole point of my thesis, of the coming Singularity revolution, is that this notion of a machine—of nonbiological intelligence—will fundamentally change.*

BILL: *Well, that's exactly my problem. Part of our humanness is our limitations. We don't claim to be the fastest entity possible, to have memories with the biggest capacity possible, and so on. But there is an indefinable, spiritual quality to being human that a machine inherently doesn't possess.*

RAY: *Again, where do you draw the line? Humans are already replacing parts of their bodies and brains with nonbiological replacements that work better at performing their "human" functions.*

BILL: *Better only in the sense of replacing diseased or disabled organs and systems. But you're replacing essentially all of our humanness to enhance human ability, and that's inherently inhuman.*

RAY: *Then perhaps our basic disagreement is over the nature of being human. To me, the essence of being human is not our limitations—although we do have many—it's our ability to reach beyond our limitations. We didn't stay on the ground. We didn't even stay on the planet. And we are already not settling for the limitations of our biology.*

BILL: *We have to use these technological powers with great discretion. Past a certain point, we're losing some ineffable quality that gives life meaning.*

RAY: *I think we're in agreement that we need to recognize what's important in our humanity. But there is no reason to celebrate our limitations.*

. . . on the Human Brain

Is all what we see or seem, but a dream within a dream?

—EDGAR ALLAN POE

The computer programmer is a creator of universes for which he alone is the lawgiver. No playwright, no stage director, no emperor, however powerful, has ever exercised such absolute authority to arrange a stage or a field of battle and to command such unswervingly dutiful actors or troops.

—JOSEPH WEIZENBAUM

One windy day two monks were arguing about a flapping banner. The first said, "I say the banner is moving, not the wind." The second said, "I say the wind is moving, not the banner." A third monk passed by and said, "The wind is not moving. The banner is not moving. Your minds are moving."

—ZEN PARABLE

Suppose someone were to say, "Imagine this butterfly exactly as it is, but ugly instead of beautiful."

—LUDWIG WITTGENSTEIN

The 2010 Scenario. Computers arriving at the beginning of the next decade will become essentially invisible: woven into our clothing, embedded in our furniture and environment. They will tap into the worldwide mesh (what the World Wide Web will become once all of its linked devices become communicating Web servers, thereby forming vast supercomputers and memory banks) of high-speed communications and computational resources. We'll have very high-bandwidth, wireless communication to the Internet at all times. Displays will be built into our eyeglasses and contact lenses and images projected directly onto our retinas. The Department of Defense is already using technology along these lines to create virtual-reality environments in which to train soldiers.[27] An impressive immersive virtual reality system already demonstrated by the army's Institute for Creative Technologies includes virtual humans that respond appropriately to the user's actions.

Similar tiny devices will project auditory environments. Cell phones are already being introduced in clothing that projects sound to the ears.[28] And there's an MP3 player that vibrates your skull to play music that only you can hear.[29] The army has also pioneered transmitting sound through the skull from a soldier's helmet.

There are also systems that can project from a distance sound that only a specific person can hear, a technology that was dramatized by the personalized talking street ads in the movie *Minority Report*. The Hypersonic Sound technology and the Audio Spotlight systems achieve this by modulating the sound on ultrasonic beams, which can be precisely aimed. Sound is generated by the beams interacting with air, which restores sound in the audible range. By focusing multiple sets of beams on a wall or other surface, a new kind of personalized surround sound without speakers is also possible.[30]

These resources will provide high-resolution, full-immersion visual-auditory virtual reality at any time. We will also have augmented reality with displays overlaying the real world to provide real-time guidance and explanations. For example, your retinal display might remind us, "That's Dr. John Smith, director of the ABC Institute—you last saw him six months ago at the XYZ conference" or, "That's the Time-Life Building—your meeting is on the tenth floor."

We'll have real-time translation of foreign languages, essentially subtitles on the world, and access to many forms of online information integrated into our daily activities. Virtual personalities that overlay the real world will help us with information retrieval and our chores and transactions. These virtual assistants won't always wait for questions and directives but will step forward if they see us struggling to find a piece of information. (As we wonder about "That actress . . . who played the princess, or was it the queen . . . in that movie with the robot," our virtual assistant may whisper in our ear or display in our visual field of view: "Natalie Portman as Queen Amidala in *Star Wars*, episodes 1, 2, and 3.")

The 2030 Scenario. Nanobot technology will provide fully immersive, totally convincing virtual reality. Nanobots will take up positions in close physical proximity to every interneuronal connection coming from our senses. We already have the technology for electronic devices to communicate with neurons in both directions, yet requiring no direct physical contact with the neurons. For example, scientists at the Max Planck Institute have developed "neuron transistors" that can detect the firing of a nearby neuron, or alternatively can cause a nearby neuron to fire or suppress it from firing.[31] This amounts to two-way communication between neurons and the electronic-based neuron transistors. As mentioned above, quantum dots have also shown the ability to provide noninvasive communication between neurons and electronics.[32]

If we want to experience real reality, the nanobots just stay in position (in the capillaries) and do nothing. If we want to enter virtual reality, they suppress all of the inputs coming from our actual senses and replace them with the signals that would be appropriate for the virtual environment.[33] Your brain experiences these signals as if they came from your physical body. After all, the brain does not experience the body directly. As I discussed in chapter 4, inputs from the body—comprising a few hundred megabits per second—representing information about touch, temperature, acid levels, the movement of food, and other physical events, stream through the Lamina 1 neurons, then through the posterior ventromedial nucleus, ending up in the two insula regions of cortex. If these are coded correctly—and we will know how to do that from the brain reverse-engineering effort—your brain will experience the synthetic signals just as it would real ones. You could decide to cause your muscles and limbs to move as you normally would, but the nanobots would intercept these interneuronal signals, suppress your real limbs from moving, and instead cause your virtual limbs to move, appropriately adjusting your vestibular system and providing the appropriate movement and reorientation in the virtual environment.

The Web will provide a panoply of virtual environments to explore. Some will be re-creations of real places; others will be fanciful environments that have no counterpart in the physical world. Some, indeed, would be impossible, perhaps because they violate the laws of physics. We will be able to visit these virtual places and have any kind of interaction with other real, as well as simulated, people (of course, ultimately there won't be a clear distinction between the two), ranging from business negotiations to sensual encounters. "Virtual-reality environment designer" will be a new job description and a new art form.

Become Someone Else. In virtual reality we won't be restricted to a single personality, since we will be able to change our appearance and effectively become other people. Without altering our physical body (in real reality) we will be able to readily transform our projected body in these three-dimensional virtual environments. We can select different bodies at the same time for different people. So your parents may see you as one person, while your girlfriend will experience you as another. However, the other person may choose to override your selections, preferring to see you differently than the body you have chosen for yourself. You could pick different body projections for different people: Ben Franklin for a wise uncle, a clown for an annoying coworker. Romantic couples can choose whom they wish to be, even to become each other. These are all easily changeable decisions.

I had the opportunity to experience what it is like to project myself as another persona in a virtual-reality demonstration at the 2001 TED (technology, entertainment, design) conference in Monterey. By means of mag-

netic sensors in my clothing a computer was able to track all of my movements. With ultrahigh-speed animation the computer created a life-size, near
photorealistic image of a young woman—Ramona—who followed my movements in real time. Using signal-processing technology, my voice was transformed into a woman's voice and also controlled the movements of Ramona's
lips. So it appeared to the TED audience as if Ramona herself were giving the
presentation.[34]

To make the concept understandable, the audience could see me and see
Ramona at the same time, both moving simultaneously in exactly the same
way. A band came onstage, and I—Ramona—performed Jefferson Airplane's
"White Rabbit," as well as an original song. My daughter, then fourteen, also
equipped with magnetic sensors, joined me, and her dance movements were
transformed into those of a male backup dancer—who happened to be a virtual Richard Saul Wurman, the impresario of the TED conference. The hit of
the presentation was seeing Wurman—not known for his hip-hop moves—
convincingly doing my daughter's dance steps. Present in the audience was the
creative leadership of Warner Bros., who then went off and created the movie
Simone, in which the character played by Al Pacino transforms himself into
Simone in essentially the same way.

The experience was a profound and moving one for me. When I looked in
the "cybermirror" (a display showing me what the audience was seeing), I saw
myself as Ramona rather than the person I usually see in the mirror. I experienced the emotional force—and not just the intellectual idea—of transforming
myself into someone else.

People's identities are frequently closely tied to their bodies ("I'm a person
with a big nose," "I'm skinny," "I'm a big guy," and so on). I found the opportunity to become a different person liberating. All of us have a variety of personalities that we are capable of conveying but generally suppress them since we have
no readily available means of expressing them. Today we have very limited technologies available—such as fashion, makeup, and hairstyle—to change who we
are for different relationships and occasions, but our palette of personalities will
greatly expand in future full-immersion virtual-reality environments.

In addition to encompassing all of the senses, these shared environments
can include emotional overlays. Nanobots will be capable of generating the
neurological correlates of emotions, sexual pleasure, and other derivatives of
our sensory experience and mental reactions. Experiments during open brain
surgery have demonstrated that stimulating certain specific points in the brain
can trigger emotional experiences (for example, the girl who found everything
funny when stimulated in a particular spot of her brain, as I reported in *The*

Age of Spiritual Machines).[35] Some emotions and secondary reactions involve a pattern of activity in the brain rather than the stimulation of a specific neuron, but with massively distributed nanobots, stimulating these patterns will also be feasible.

Experience Beamers. "Experience beamers" will send the entire flow of their sensory experiences as well as the neurological correlates of their emotional reactions out onto the Web, just as people today beam their bedroom images from their Web cams. A popular pastime will be to plug into someone else's sensory-emotional beam and experience what it's like to be that person, à la the premise of the movie *Being John Malkovich*. There will also be a vast selection of archived experiences to choose from, with virtual-experience design another new art form.

Expand Your Mind. The most important application of circa-2030 nanobots will be literally to expand our minds through the merger of biological and non-biological intelligence. The first stage will be to augment our hundred trillion very slow interneuronal connections with high-speed virtual connections via nanorobot communication.[36] This will provide us with the opportunity to greatly boost our pattern-recognition abilities, memories, and overall thinking capacity, as well as to directly interface with powerful forms of nonbiological intelligence. The technology will also provide wireless communication from one brain to another.

It is important to point out that well before the end of the first half of the twenty-first century, thinking via nonbiological substrates will predominate. As I reviewed in chapter 3, biological human thinking is limited to 10^{16} calculations per second (cps) per human brain (based on neuromorphic modeling of brain regions) and about 10^{26} cps for all human brains. These figures will not appreciably change, even with bioengineering adjustments to our genome. The processing capacity of nonbiological intelligence, in contrast, is growing at an exponential rate (with the rate itself increasing) and will vastly exceed biological intelligence by the mid-2040s.

By that time we will have moved beyond just the paradigm of nanobots in a biological brain. Nonbiological intelligence will be billions of times more powerful, so it will predominate. We will have version 3.0 human bodies, which we will be able to modify and reinstantiate into new forms at will. We will be able to quickly change our bodies in full-immersion visual-auditory virtual environments in the second decade of this century; in full-immersion virtual-reality environments incorporating all of the senses during the 2020s; and in real reality in the 2040s.

Nonbiological intelligence should still be considered human, since it is fully derived from human-machine civilization and will be based, at least in part, on reverse engineering human intelligence. I address this important philosophical issue in the next chapter. The merger of these two worlds of intelligence is not merely a merger of biological and nonbiological thinking mediums, but more important, one of method and organization of thinking, one that will be able to expand our minds in virtually any imaginable way.

Our brains today are relatively fixed in design. Although we do add patterns of interneuronal connections and neurotransmitter concentrations as a normal part of the learning process, the current overall capacity of the human brain is highly constrained. As the nonbiological portion of our thinking begins to predominate by the end of the 2030s, we will be able to move beyond the basic architecture of the brain's neural regions. Brain implants based on massively distributed intelligent nanobots will greatly expand our memories and otherwise vastly improve all of our sensory, pattern-recognition, and cognitive abilities. Since the nanobots will be communicating with one another, they will be able to create any set of new neural connections, break existing connections (by suppressing neural firing), create new hybrid biological-nonbiological networks, and add completely nonbiological networks, as well as interface intimately with new nonbiological forms of intelligence.

The use of nanobots as brain extenders will be a significant improvement over surgically installed neural implants, which are beginning to be used today. Nanobots will be introduced without surgery, through the bloodstream, and if necessary can all be directed to leave, so the process is easily reversible. They are programmable, in that they can provide virtual reality one minute and a variety of brain extensions the next. They can change their configuration and can alter their software. Perhaps most important, they are massively distributed and therefore can take up billions of positions throughout the brain, whereas a surgically introduced neural implant can be placed only in one or at most a few locations.

MOLLY 2004: *Full-immersion virtual reality doesn't seem very inviting. I mean, all those nanobots running around in my head, like little bugs.*

RAY: *Oh, you won't feel them, any more than you feel the neurons in your head or the bacteria in your GI tract.*

MOLLY 2004: *Actually, that I can feel. But I can have full immersion with my friends right now, just by, you know, getting together physically.*

SIGMUND FREUD: *Hmmm, that's what they used to say about the telephone when I was young. People would say, "Who needs to talk to someone hundreds of miles away when you can just get together?"*

RAY: *Exactly, the telephone is auditory virtual reality. So full-immersion VR is, basically, a full-body telephone. You can get together with anyone anytime but do more than just talk.*

GEORGE 2048: *It's certainly been a boon for sex workers; they never have to leave their homes. It became so impossible to draw any meaningful lines that the authorities had no choice but to legalize virtual prostitution in 2033.*

MOLLY 2004: *Very interesting but actually not very appealing.*

GEORGE 2048: *Okay, but consider that you can be with your favorite entertainment star.*

MOLLY 2004: *I can do that in my imagination any time I want.*

RAY: *Imagination is nice, but the real thing—or, rather, the virtual thing—is so much more, well, real.*

MOLLY 2004: *Yeah, but what if my "favorite" celebrity is busy?*

RAY: *That's another benefit of virtual reality circa 2029; you have your choice of millions of artificial people.*

MOLLY 2104: *I understand that you're back in 2004, but we kind of got rid of that terminology back when the Nonbiological Persons Act was passed in 2052. I mean, we're a lot more real than . . . umm, let me rephrase that.*

MOLLY 2004: *Yes, maybe you should.*

MOLLY 2104: *Let's just say that you don't have to have explicit biological structures to be—*

GEORGE 2048: *—passionate?*

MOLLY 2104: *I guess you should know.*

TIMOTHY LEARY: *What if you have a bad trip?*

RAY: *You mean, something goes awry with a virtual-reality experience?*

TIMOTHY: *Exactly.*

RAY: *Well, you can leave. It's like hanging up on a phone call.*

MOLLY 2004: *Assuming you still have control over the software.*

RAY: *Yes, we do need to be concerned with that.*

SIGMUND: *I can see some real therapeutic potential here.*

RAY: *Yes, you can be whomever you want to be in virtual reality.*

SIGMUND: *Excellent, the opportunity to express suppressed longings . . .*

RAY: *And not only to be with the person you want to be with, but to become that person.*

SIGMUND: *Exactly. We create the objects of our libido in our subconscious anyway. Just think, a couple could both change their genders. They could each become the other.*

MOLLY 2004: *Just as a therapeutic interlude, I presume?*

SIGMUND: *Of course. I would only suggest this under my careful supervision.*

MOLLY 2004: *Naturally.*

MOLLY 2104: *Hey, George, remember when we each became all of the opposite-gender characters in the Allen Kurzweil novels at the same time?*[37]

GEORGE 2048: *Ha, I liked you best as that eighteenth-century French inventor, the one who made erotic pocket watches!*

MOLLY 2004: *Okay, now run this virtual sex by me again. How does it work exactly?*

RAY: *You're using your virtual body, which is simulated. Nanobots in and around your nervous system generate the appropriate encoded signals for all of your senses: visual, auditory, tactile of course, even olfactory. From the perspective of your brain, it's real because the signals are just as real as if your senses were producing them from real experiences. The simulation in virtual reality would generally follow the laws of physics, although that would depend on the environment you selected. If you go there with another person or persons, then these other intelligences, whether of people with biological bodies or otherwise, would also have bodies in this virtual environment. Your body in virtual reality does not need to match your body in real reality. In fact, the body you choose for yourself in the virtual environment may be different from the body that your partner chooses for you at the same time. The computers generating the virtual environment, virtual bodies, and associated nerve signals would cooperate so that your actions affect the virtual experience of the others and vice versa.*

MOLLY 2004: *So I would experience sexual pleasure even though I'm not actually, you know, with someone?*

RAY: *Well, you would be with someone, just not in real reality, and, of course, the someone may not even exist in real reality. Sexual pleasure is not a direct sensory experience, it's akin to an emotion. It's a sensation generated in your brain, which is reflecting on what you're doing and thinking, just like the sensation of humor or anger.*

MOLLY 2004: *Like the girl you mentioned who found everything hilarious when the surgeons stimulated a particular spot in her brain?*

RAY: *Exactly. There are neurological correlates of all of our experiences, sensations, and emotions. Some are localized whereas some reflect a pattern of activity. In either case we'll be able to shape and enhance our emotional reactions as part of our virtual-reality experiences.*

MOLLY 2004: *That could work out quite well. I think I'll enhance my funniness reaction in my romantic interludes. That will fit just about right. Or maybe my absurdity response—I kind of like that one, too.*

NED LUDD: *I can see this getting out of hand. People are going to start spending most of their time in virtual reality.*

MOLLY 2004: *Oh, I think my ten-year-old nephew is already there, with his video games.*

RAY: *They're not full immersion yet.*

MOLLY 2004: *That's true. We can see him, but I'm not sure he notices us. But when we get to the point when his games are full immersion, we'll never see him.*

GEORGE 2048: *I can see your concern if you're thinking in terms of the thin virtual worlds of 2004, but it's not a problem with our 2048 virtual worlds. They're so much more compelling than the real world.*

MOLLY 2004: *Yeah, how would you know since you've never been in real reality?*

GEORGE 2048: *I hear about it quite a bit. Anyway, we can simulate it.*

MOLLY 2104: *Well, I can have a real body any time I want, really not a big deal. I have to say it's rather liberating to not be dependent on a particular body, let alone a biological one. Can you imagine, being all tied up with its endless limitations and burdens?*

MOLLY 2004: *Yes, I can see where you're coming from.*

. . . on Human Longevity

It is one of the most remarkable things that in all of the biological sciences there is no clue as to the necessity of death. If you say we want to make perpetual motion, we have discovered enough laws as we studied physics to see that it is either absolutely impossible or else the laws are wrong. But there is nothing in biology yet found that indicates the inevitability of death. This suggests to me that it is not at all inevitable and that it is only a matter of time before the biologists discover what it is that is causing us the trouble and that this terrible universal disease or temporariness of the human's body will be cured.

—RICHARD FEYNMAN

Never give in, never give in, never, never, never, never—in nothing, great or small, large or petty—never give in.

—WINSTON CHURCHILL

Immortality first! Everything else can wait.

—CORWYN PRATER

Involuntary death is a cornerstone of biological evolution, but that fact does not make it a good thing.

—MICHAEL ANISSIMOV

Suppose you're a scientist 200 years ago who has figured out how to drastically lower infant mortality with better hygiene. You give a talk on this, and someone stands up in back and says, "hang on, if we do that we're going to have a population explosion!" If you reply, "No, everything will be fine because we'll all wear these absurd rubber things when we have sex," nobody would have taken you seriously. Yet that's just what happened—barrier contraception was widely adopted [around the time that infant mortality dropped].

> —AUBREY DE GREY, GERONTOLOGIST

We have a duty to die.

> —DICK LAMM, FORMER GOVERNOR OF COLORADO

Some of us think this is rather a pity.

> —BERTRAND RUSSELL, 1955, COMMENTING ON THE STATISTIC THAT ABOUT
> ONE HUNDRED THOUSAND PEOPLE DIE OF AGE-RELATED CAUSES EVERY
> DAY[38]

Evolution, the process that produced humanity, possesses only one goal: create gene machines maximally capable of producing copies of themselves. In retrospect, this is the only way complex structures such as life could possibly arise in an unintelligent universe. But this goal often comes into conflict with human interests, causing death, suffering, and short life spans. The past progress of humanity has been a history of shattering evolutionary constraints.

> —MICHAEL ANISSIMOV

Most of the readers of this book are likely to be around to experience the Singularity. As we reviewed in the previous chapter, accelerating progress in biotechnology will enable us to reprogram our genes and metabolic processes to turn off disease and aging processes. This progress will include rapid advances in genomics (influencing genes), proteomics (understanding and influencing the role of proteins), gene therapy (suppressing gene expression with such technologies as RNA interference and inserting new genes into the nucleus), rational drug design (formulating drugs that target precise changes in disease and aging processes), and therapeutic cloning of rejuvenated (telomere-extended and DNA-corrected) versions of our own cells, tissues, and organs, and related developments.

Biotechnology will extend biology and correct its obvious flaws. The overlapping revolution of nanotechnology will enable us to expand beyond the severe limitations of biology. As Terry Grossman and I articulated in *Fantastic Voyage: Live Long Enough to Live Forever*, we are rapidly gaining the knowledge and the tools to indefinitely maintain and extend the "house" each of us calls his body and brain. Unfortunately the vast majority of our baby-boomer peers are unaware of the fact that they do not have to suffer and die in the "normal" course of life, as prior generations have done—*if* they take aggressive action, action that goes beyond the usual notion of a basically healthy lifestyle (see "Resources and Contact Information," p. 489).

Historically, the only means for humans to outlive a limited biological life span has been to pass on values, beliefs, and knowledge to future generations. We are now approaching a paradigm shift in the means we will have available to preserve the patterns underlying our existence. Human life expectancy is itself growing steadily and will accelerate rapidly, now that we are in the early stages of reverse engineering the information processes underlying life and disease. Robert Freitas estimates that eliminating a specific list comprising 50 percent of medically preventable conditions would extend human life expectancy to over 150 years.[39] By preventing 90 percent of medical problems, life expectancy grows to over five hundred years. At 99 percent, we'd be over one thousand years. We can expect that the full realization of the biotechnology and nanotechnology revolutions will enable us to eliminate virtually all medical causes of death. As we move toward a nonbiological existence, we will gain the means of "backing ourselves up" (storing the key patterns underlying our knowledge, skills, and personality), thereby eliminating most causes of death as we know it.

Life Expectancy (Years)[40]

Cro-Magnon Era	18
Ancient Egypt	25
1400 Europe	30
1800 Europe and United States	37
1900 United States	48
2002 United States	78

The Transformation to Nonbiological Experience

A mind that stays at the same capacity cannot live forever; after a few thousand years it would look more like a repeating tape loop than a person. To live indefinitely long, the mind itself must grow, . . . and when it becomes great enough, and looks back . . . what fellow feeling can it have with the soul that it was originally? The later being would be everything the original was, but vastly more.

—VERNOR VINGE

The empires of the future are the empires of the mind.

—WINSTON CHURCHILL

I reported on brain uploading in chapter 4. The straightforward brain-porting scenario involves scanning a human brain (most likely from within), capturing *all* of the salient details, and reinstantiating the brain's state in a different— most likely much more powerful—computational substrate. This will be a feasible procedure and will happen most likely around the late 2030s. But this is not the primary way that I envision the transition to nonbiological experience taking place. It will happen, rather, in the same way that all other paradigm shifts happen: gradually (but at an accelerating pace).

As I pointed out above, the shift to nonbiological thinking will be a slippery slope, but one on which we have already started. We will continue to have human bodies, but they will become morphable projections of our intelligence. In other words, once we have incorporated MNT fabrication into ourselves, we will be able to create and re-create different bodies at will.

However achieved, will such fundamental shifts enable us to live forever? The answer depends on what we mean by "living" and "dying." Consider what we do today with our personal computer files. When we change from an older computer to a newer one, we don't throw all our files away. Rather, we copy them and reinstall them on the new hardware. Although our software does not necessarily continue its existence forever, its longevity is in essence independent of and disconnected from the hardware that it runs on.

Currently, when our human hardware crashes, the software of our lives— our personal "mind file"—dies with it. However, this will not continue to be the case when we have the means to store and restore the thousands of trillions of bytes of information represented in the pattern that we call our brains (together with the rest of our nervous system, endocrine system, and other structures that our mind file comprises).

At that point the longevity of one's mind file will not depend on the continued viability of any particular hardware medium (for example, the survival of a biological body and brain). Ultimately software-based humans will be vastly extended beyond the severe limitations of humans as we know them today. They will live out on the Web, projecting bodies whenever they need or want them, including virtual bodies in diverse realms of virtual reality, holographically projected bodies, foglet-projected bodies, and physical bodies comprising nanobot swarms and other forms of nanotechnology.

By the middle of the twenty-first century humans will be able to expand their thinking without limit. This is a form of immortality, although it is important to point out that data and information do not necessarily last forever: the longevity of information depends on its relevance, utility, and accessibility. If you've ever tried to retrieve information from an obsolete form of data storage in an old, obscure format (for example, a reel of magnetic tape from a 1970 minicomputer), you understand the challenges in keeping software viable. However, if we are diligent in maintaining our mind file, making frequent backups, and porting to current formats and mediums, a form of immortality can be attained, at least for software-based humans. Later in this century it will seem remarkable to people that humans in an earlier era lived their lives without a backup of their most precious information: that contained in their brains and bodies.

Is this form of immortality the same concept as a physical human, as we know it today, living forever? In one sense it is, because today one's self is not a constant collection of matter, either. Recent research shows that even our neurons, thought to be relatively long lasting, change all of their constituent subsystems, such as the tubules, in a matter of weeks. Only our pattern of matter and energy persists, and even that gradually changes. Similarly, it will be the pattern of a software human that persists and develops and slowly alters.

But is that person based on my mind file, who migrates across many computational substrates and who outlives any particular thinking medium, really me? This consideration takes us back to the same questions of consciousness and identity that have been debated since Plato's dialogues (which we examine in the next chapter). During the course of the twenty-first century these will not remain topics for polite philosophical debates but will have to be confronted as vital, practical, political, and legal issues.

A related question: Is death desirable? The "inevitability" of death is deeply ingrained in human thinking. If death seems unavoidable, we have little choice but to rationalize it as necessary, even ennobling. The technology of the Singularity will provide practical and accessible means for humans to evolve into something greater, so we will no longer need to rationalize death as a primary means of giving meaning to life.

The Longevity of Information

> "The horror of that moment," the King went on, "I shall never, never forget it!" "You will, though," the Queen said, "if you don't make a memorandum of it."
>
> —LEWIS CARROLL, *THROUGH THE LOOKING-GLASS*

> The only things you can be sure of, so the saying goes, are death and taxes— but don't be too sure about death.
>
> —JOSEPH STROUT, NEUROSCIENTIST

> I do not know sire, but whatever they will turn out to be I am sure you will tax them.
>
> —MICHAEL FARADAY, RESPONDING TO A QUESTION FROM THE BRITISH EXCHEQUER AS TO WHAT PRACTICAL USE COULD BE MADE OF HIS DEMONSTRATION OF ELECTROMAGNETISM

> *Do not go gentle into that good night, . . .*
> *Rage, rage against the dying of the light.*
>
> —DYLAN THOMAS

The opportunity to translate our lives, our history, our thoughts, and our skills into information raises the issue of how long information lasts. I've always revered knowledge and gathered information of all kinds as a child, an inclination I shared with my father.

By way of background, my father was one of those people who liked to store all the images and sounds that documented his life. Upon his untimely death at the age of fifty-eight in 1970, I inherited his archives, which I treasure to this day. I have my father's 1938 doctoral dissertation from the University of Vienna, which contains his unique insights into the contributions of Brahms to our musical vocabulary. There are albums of neatly arranged newspaper clippings of his acclaimed musical concerts as a teenager in the hills of Austria. There are urgent letters to and from the American music patron who sponsored his flight from Hitler, just before Kristallnacht and related historical developments in Europe in the late 1930s made such escape impossible. These items are among dozens of aging boxes containing a myriad of remembrances, including photographs, musical recordings on vinyl and magnetic tape, personal letters, and even old bills.

I also inherited his penchant for preserving the records of one's life, so along with my father's boxes I have several hundred boxes of my own papers and files. My father's productivity, assisted only by the technology of his manual typewriter and carbon paper, cannot compare with my own prolificacy, aided and abetted by computers and high-speed printers that can reproduce my thoughts in all kinds of permutations.

Tucked away in my own boxes are also various forms of digital media: punch cards, paper-tape reels, and digital magnetic tapes and disks of various sizes and formats. I often wonder just how accessible this information remains. Ironically the ease of approaching this information is inversely proportional to the level of advancement of the technology used to create it. Most straightforward are the paper documents, which although showing signs of age are eminently readable. Only slightly more challenging are the vinyl records and analog tape recordings. Although some basic equipment is required, it is not difficult to find or use. The punch cards are somewhat more challenging, but it's still possible to find punch-card readers, and the formats are uncomplicated.

By far the most demanding information to retrieve is that contained on the digital disks and tapes. Consider the challenges involved. For each medium I have to figure out exactly which disk or tape drive was used, whether an IBM 1620 circa 1960 or a Data General Nova I circa 1973. Then, once I've assembled the requisite equipment, there are layers of software to deal with: the appropriate operating system, disk information drivers, and application programs. And, when I run into the inevitable scores of problems inherent in each layer of hardware and software, just whom am I going to call for assistance? It's hard enough getting contemporary systems to work, let alone systems for which the help desks were disbanded decades ago (if they ever existed). Even at the Computer History Museum most of the devices on display stopped functioning many years ago.[41]

Assuming I do prevail against all of these obstacles, I have to account for the fact that the actual magnetic data on the disks has probably decayed and that the old computers would still generate mostly error messages.[42] But is the information gone? The answer is, Not entirely. Even though the magnetic spots may no longer be readable by the original equipment, the faded regions could be enhanced by suitably sensitive equipment, via methods that are analogous to the image enhancement often applied to the pages of old books when they are scanned. The information is still there, although very difficult to get at. With enough devotion and historical research, one might actually retrieve it. If we had reason to believe that one of these disks contained secrets of enormous value, we would probably succeed in recovering the information.

But mere nostalgia is unlikely to be sufficient to motivate anyone to undertake this formidable task. I will say that because I did largely anticipate this dilemma, I did make paper printouts of most of these old files. But keeping all our information on paper is not the answer, as hard-copy archives present their own set of problems. Although I can readily read even a century-old paper manuscript if I'm holding it in my hand, finding a desired document from among thousands of only modestly organized file folders can be a frustrating and time-consuming task. It can take an entire afternoon to locate the right folder, not to mention the risk of straining one's back from moving dozens of heavy file boxes. Using microfilm or microfiche may alleviate some of the difficulty, but the matter of locating the right document remains.

I have dreamed of taking these hundreds of thousands of records and scanning them into a massive personal database, which would allow me to utilize powerful contemporary search-and-retrieve methods on them. I even have a name for this venture—DAISI (Document and Image Storage Invention)—and have been accumulating ideas for it for many years. Computer pioneer Gordon Bell (former chief engineer of Digital Equipment Corporation), DARPA (Defense Advanced Research Projects Agency), and the Long Now Foundation are also working on systems to address this challenge.[43]

DAISI will involve the rather daunting task of scanning and patiently cataloging all these documents. But the real challenge to my dream of DAISI is surprisingly deep: how can I possibly select appropriate hardware and software layers that will give me the assurance that my archives will be viable and accessible decades from now?

Of course my own archival needs are only a microcosm of the exponentially expanding knowledge base that human civilization is accumulating. It is this shared specieswide knowledge base that distinguishes us from other animals. Other animals communicate, but they don't accumulate an evolving and growing base of knowledge to pass down to the next generation. Since we are writing our precious heritage in what medical informatics expert Bryan Bergeron calls "disappearing ink," our civilization's legacy would appear to be at great risk.[44] The danger appears to be growing exponentially along with the growth of our knowledge bases. The problem is further exacerbated by the accelerating speed with which we adopt new standards in the many layers of hardware and software we employ to store information.

There is another valuable repository of information stored in our brains. Our memories and skills, although they may appear to be fleeting, do represent information, coded in vast patterns of neurotransmitter concentrations, inter-neuronal connections, and other relevant neural details. This information is the most precious of all, which is one reason death is so tragic. As we have discussed, we will ultimately be able to access, permanently archive, as well as understand the thousands of trillions of bytes of information we have tucked away in each of our brains.

Copying our minds to other mediums raises a number of philosophical issues, which I will discuss in the next chapter—for example, "Is that really me or rather someone else who just happens to have mastered all my thoughts and knowledge?" Regardless of how we resolve these issues, the idea of capturing the information and information processes in our brains seems to imply that we (or at least entities that act very much like we do) could "live forever." But is that really the implication?

For eons the longevity of our mental software has been inexorably linked to the survival of our biological hardware. Being able to capture and reinstantiate all the details of our information processes would indeed separate these two aspects of our mortality. But as we have seen, software itself does not necessarily survive forever, and there are formidable obstacles to its enduring very long at all.

So whether information represents one man's sentimental archive, the accumulating knowledge base of the human-machine civilization, or the mind files stored in our brains, what can we conclude about the ultimate longevity of software? The answer is simply this: *Information lasts only so long as someone cares about it.* The conclusion that I've come to with regard to my DAISI project, after several decades of careful consideration, is that there is no set of hardware and software standards existing today, nor any likely to come along, that will provide any reasonable level of confidence that the stored information will still be accessible (without unreasonable levels of effort) decades from now.[45] The only way that my archive (or any other information base) can remain viable is if it is continually upgraded and ported to the latest hardware and software standards. If an archive remains ignored, it will ultimately become as inaccessible as my old eight-inch PDP-8 floppy disks.

Information will continue to require constant maintenance and support to remain "alive." Whether data or wisdom, information will survive only if we want it to. By extension, we can only live for as long as we care about ourselves. Already our knowledge to control disease and aging is advanced to the point that your *attitude* toward your own longevity is now the most important influence on your long-range health.

Our civilization's trove of knowledge does not simply survive by itself. We must continually rediscover, reinterpret, and reformat the legacy of culture and technology that our forebears have bestowed on us. All of this information will be fleeting if no one cares about it. Translating our currently hardwired thoughts into software will not necessarily provide us with immortality. It will simply place the means to determine how long we want our lives and thoughts to last in our own figurative hands.

MOLLY 2004: *So what you're saying is that I'm just a file?*
MOLLY 2104: *Well, not a static file, but a dynamic file. But what do you mean "just"? What could be more important?*
MOLLY 2004: *Well, I throw files away all the time, even dynamic ones.*
MOLLY 2104: *Not all files are created equal.*
MOLLY 2004: *I suppose that's true. I was devastated when I lost my only copy of my senior thesis. I lost six months of work and had to start over.*
MOLLY 2104: *Ah, yes, that was awful. I remember it well, even though it was over a century ago. It was devastating because it was a small part of myself. I had invested my thoughts and creativity in that file of information. So think how precious all of your—my—accumulated thoughts, experience, skills, and history are.*

. . . on Warfare: The Remote, Robotic, Robust, Size-Reduced, Virtual-Reality Paradigm

As weapons have become more intelligent, there has been a dramatic trend toward more precise missions with fewer casualties. It may not seem that way when viewed alongside the tendency toward more detailed, realistic television-news coverage. The great battles of World Wars I and II and the Korean War, in which tens of thousands of lives were lost over the course of a few days, were visually recorded only by occasional grainy newsreels. Today, we have a front-row seat for almost every engagement. Each war has its complexities, but the overall movement toward precision intelligent warfare is clear by examining the number of casualties. This trend is similar to what we are beginning to see in medicine, where smart weapons against disease are able to perform specific missions with far fewer side effects. The trend is similar for collateral casualties, although it may not seem that way from contemporary media coverage (recall that about fifty million civilians died in World War II).

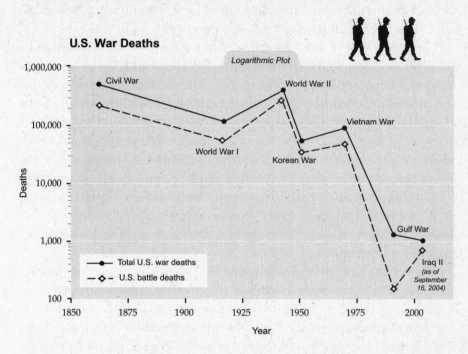

U.S. War Deaths

I am one of five members of the Army Science Advisory Group (ASAG), which advises the U.S. Army on priorities for its science research. Although our briefings, deliberations, and recommendations are confidential, I can share some overall technological directions that are being pursued by the army and all of the U.S. armed forces.

Dr. John A. Parmentola, director for research and laboratory management for the U.S. Army and liaison to the ASAG, describes the Department of Defense's "transformation" process as a move toward an armed force that is "highly responsive, network-centric, capable of swift decision, superior in all echelons, and [able to provide] overwhelming massed effects across any battle space."[46] He describes the Future Combat System (FCS), now under development and scheduled to roll out during the second decade of this century, as "smaller, lighter, faster, more lethal, and smarter."

Dramatic changes are planned for future war-fighting deployments and technology. Although details are likely to change, the army envisions deploying Brigade Combat Teams (BCTs) of about 2,500 soldiers, unmanned robotic systems, and FCS equipment. A single BCT would represent about 3,300 "plat-forms," each with its own intelligent computational capabilities. The BCT would have a common operating picture (COP) of the battlefield, which would be appropriately translated for it, with each soldier receiving information through a variety of means, including retinal (and other forms of "heads up") displays and, in the future, direct neural connection.

The army's goal is to be capable of deploying a BCT in 96 hours and a full division in 120 hours. The load for each soldier, which is now about one hundred pounds of equipment, will initially be reduced through new materials and devices to forty pounds, while dramatically improving effectiveness. Some of the equipment would be offloaded to "robotic mules."

A new uniform material has been developed using a novel form of Kevlar with silica nanoparticles suspended in polyethylene glycol. The material is flex-ible in normal use, but when stressed it instantly forms a nearly impenetrable mass that is stab resistant. The army's Institute for Soldier Nanotechnologies at MIT is developing a nanotechnology-based material called "exomuscle" to enable combatants to greatly increase their physical strength when manipulat-ing heavy equipment.[47]

The Abrams tank has a remarkable survival record, with only three combat casualties in its twenty years of combat use. This is the result of both advanced armor materials and of intelligent systems designed to defeat incoming weap-ons, such as missiles. However, the tank weighs more than seventy tons, a figure that will need to be significantly reduced to meet FCS goals for smaller systems. New lightweight yet ultrastrong nanomaterials (such as plastics combined with nanotubes, which are fifty times stronger than steel), as well as increased com-puter intelligence to counteract missile attacks, are expected to dramatically lower the weight of ground combat systems.

The trend toward unmanned aerial vehicles (UAVs), which started with the armed Predator in the recent Afghanistan and Iraq campaigns, will accelerate. Army research includes the development of micro-UAVs the size of birds that will be fast, accurate, and capable of performing both reconnaissance and com-bat missions. Even smaller UAVs the size of bumblebees are envisioned. The navigational ability of an actual bumblebee, which is based on a complex in-teraction between its left and right vision systems, has recently been reverse engineered and will be applied to these tiny flying machines.

At the center of the FCS is a self-organizing, highly distributed communica-tions network capable of gathering information from each soldier and each piece of equipment and in turn providing the appropriate information displays and files back to each human and machine participant. There will be no centralized communications hubs that could be vulnerable to hostile attack. Information will rapidly route itself around damaged portions of the network. An obvious top priority is to develop technology capable of maintaining integrity of communication and preventing either eavesdropping or manipulation of information by hostile forces. The same information-security technology will be applied to infiltrate, disrupt, confuse, or destroy enemy communications through both electronic means and cyberwarfare using software pathogens.

The FCS is not a one-shot program; it represents a pervasive focus of military systems toward remotely guided, autonomous, miniaturized, and robotic systems, combined with robust, self-organizing, distributed, and secure communications.

The U.S. Joint Forces Command's Project Alpha (responsible for accelerating transformative ideas throughout the armed services) envisions a 2025 fighting force that "is largely robotic," incorporating tactical autonomous combatants (TACs) that "have some level of autonomy—adjustable autonomy or supervised autonomy or full autonomy within . . . mission bounds."[48] The TACs will be available in a wide range of sizes, ranging from nanobots and microbots up to large UAVs and other vehicles, as well as automated systems that can walk through complex terrains. One innovative design being developed by NASA with military applications envisioned is in the form of a snake.[49]

One of the programs contributing to the 2020s concept of self-organizing swarms of small robots is the Autonomous Intelligent Network and Systems (AINS) program of the Office of Naval Research, which envisions a drone army of unmanned, autonomous robots in the water, on the ground, and in the air. The swarms will have human commanders with decentralized command and control and what project head Allen Moshfegh calls an "impregnable Internet in the sky."[50]

Extensive research is going into designing swarm intelligence.[51] Swarm intelligence describes the way that complex behaviors can arise from large numbers of individual agents, each following relatively simple rules.[52] Swarms of insects are often able to devise intelligent solutions to complex problems, such as designing the architecture of a colony, despite the fact that no single member of the swarm possesses the requisite skills.

DARPA announced in 2003 that a battalion of 120 military robots (built by I-Robot, a company cofounded by robotics pioneer Rodney Brooks) was to be fitted with swarm-intelligence software to enable it to mimic the organized behavior of insects.[53] As robotic systems become physically smaller and larger in number, the principles of self-organizing swarm intelligence will play an increasingly important role.

There is also recognition in the military that development times need to be reduced. Historically, the typical time period for military projects to go from research to deployment has been longer than a decade. But with the technology paradigm-shift rate coming down by half every decade, these development times need to keep pace, as many weapons systems are already obsolete by the time they reach the field. One way to accomplish this is to develop and test new weapons using simulations, which enable weapons systems to be designed, implemented, and tested far more quickly than the traditional means of building prototypes and testing them (often by blowing them up) in actual use.

Another key trend is to move personnel away from combat to improve soldiers' rates of survival. This can be done by allowing humans to drive and pilot systems remotely. Taking the pilot out of a vehicle allows it to take part in riskier missions and to be designed to be far more maneuverable. It also allows the devices to become very small by dispensing with the extensive requirements for supporting human life. The generals are moving even farther away. Tommy Franks conducted the war in Afghanistan from his bunker in Qatar.

Smart Dust. DARPA is developing devices even tinier than birds and bumblebees called "smart dust"—complex sensor systems not much bigger than a pinhead. Once fully developed, swarms of millions of these devices could be dropped into enemy territory to provide highly detailed surveillance and ultimately support offensive warfare missions (for example, releasing nanoweapons). Power for smart-dust systems will be provided by nanoengineered fuel cells, as well as by conversion of mechanical energy from their own movement, wind, and thermal currents.

Want to find a key enemy? Need to locate hidden weapons? Massive numbers of essentially invisible spies could monitor every square inch of enemy territory, identify every person (through thermal and electromagnetic imaging, eventually DNA tests, and other means) and every weapon and even carry out missions to destroy enemy targets.

Nanoweapons. The next step beyond smart dust will be nanotechnology-based weapons, which will make obsolete weapons of larger size. The only way for an enemy to counteract such a massively distributed force will be with its own nanotechnology. In addition, enhancing nanodevices with the ability to self-replicate will extend their capabilities but introduces grave dangers, a subject I address in chapter 8.

Nanotechnology is already being applied to a wide range of military functions. These include nanotech coatings for improved armor; laboratories on a chip for rapid chemical and biological-agent detection and identification; nanoscale catalysts for decontaminating areas; smart materials that can restructure themselves for different situations; biocidal nanoparticles incorporated into uniforms to reduce infection from injuries; nanotubes combined with plastics to create extremely strong materials; and self-healing materials. For example, the University of Illinois has developed self-healing plastics that incorporate microspheres of liquid monomers and a catalyst into a plastic matrix; when a crack appears, the microspheres break, automatically sealing the crack.[54]

Smart Weapons. We've already moved from dumb missiles launched with hopes they will find their targets to intelligent cruise missiles that use pattern recognition to make thousands of tactical decisions on their own. Bullets, however, have remained essentially small dumb missiles, and providing them with a measure of intelligence is another military objective.

As military weapons become smaller in size and larger in number, it won't be desirable or feasible to maintain human control over each device. So increasing the level of autonomous control is another important goal. Once machine intelligence catches up with biological human intelligence, many more systems will be fully autonomous.

VR. Virtual-reality environments are already in use to control remotely guided systems such as the U.S. Air Force's Armed Predator UAV.[55] Even if a soldier is inside a weapons system (such as an Abrams tank), we don't expect him or her to just look outside the window to see what is going on. Virtual-reality environments are needed to provide a view of the actual environment and allow for effective control. Human commanders in charge of swarm weapons will also need specialized virtual-reality environments to envision the complex information that these distributed systems are collecting.

By the late 2030s and 2040s, as we approach human body version 3.0 and the predominance of nonbiological intelligence, the issue of cyberwarfare will move to center stage. When everything is information, the ability to control your own information and disrupt your enemy's communication, command, and control will be a primary determinant of military success.

. . . on Learning

> Science is organized knowledge. Wisdom is organized life.
>
> —IMMANUEL KANT (1724–1804)

Most education in the world today, including in the wealthier communities, is not much changed from the model offered by the monastic schools of four-teenth-century Europe. Schools remain highly centralized institutions built upon the scarce resources of buildings and teachers. The quality of education also varies enormously, depending on the wealth of the local community (the American tradition of funding education from property taxes clearly exacer-bates this inequality), thus contributing to the have/have not divide.

As with all of our other institutions we will ultimately move toward a decentralized educational system in which every person will have ready access to the highest-quality knowledge and instruction. We are now in the early stages of this transformation, but already the advent of the availability of vast knowledge on the Web, useful search engines, high-quality open Web course-ware, and increasingly effective computer-assisted instruction are providing widespread and inexpensive access to education.

Most major universities now provide extensive courses online, many of which are free. MIT's OpenCourseWare (OCW) initiative has been a leader in this effort. MIT offers nine hundred of its courses—half of all its course offerings—for free on the Web.[56] These have already had a major impact on education around the world. For example, Brigitte Bouissou writes, "As a math teacher in France, I want to thank MIT . . . for [these] very lucid lectures, which are a great help for preparing my own classes." Sajid Latif, an educator in Paki-stan, has integrated the MIT OCW courses into his own curriculum. His Paki-stani students regularly attend—virtually—MIT classes as a substantial part of their education.[57] MIT intends to have every one of its courses online and open source (that is, free of charge for noncommercial use) by 2007.

The U.S. Army already conducts all of its nonphysical training using Web-based instruction. The accessible, inexpensive, and increasingly high-quality courseware available on the Web is also fueling a trend toward homeschooling.

The cost of the infrastructure for high-quality audiovisual Internet-based communication is continuing to fall rapidly, at a rate of about 50 percent per year, as we discussed in chapter 2. By the end of the decade it will be feasible for underdeveloped regions of the world to provide very inexpensive access to high-quality instruction for all grade levels from preschool to doctoral studies. Access to education will no longer be restricted by the lack of availability of trained teachers in each town and village.

As computer-assisted instruction (CAI) becomes more intelligent the ability to individualize the learning experience for each student will greatly improve. New generations of educational software are capable of modeling the strengths and weaknesses of each student and developing strategies to focus on the problem area of each learner. A company that I founded, Kurzweil Educational Systems, provides software that is used in tens of thousands of schools by students with reading disabilities to access ordinary printed materials and improve their reading skills.[58]

Because of current bandwidth limitations and the lack of effective three-dimensional displays, the virtual environment provided today through routine Web access does not yet fully compete with "being there," but that will change. In the early part of the second decade of this century visual-auditory virtual-reality environments will be full immersion, very high resolution, and very convincing. Most colleges will follow MIT's lead, and students will increasingly attend classes virtually. Virtual environments will provide high-quality virtual laboratories where experiments can be conducted in chemistry, nuclear physics, or any other scientific field. Students will be able to interact with a virtual Thomas Jefferson or Thomas Edison or even to *become* a virtual Thomas Jefferson. Classes will be available for all grade levels in many languages. The devices needed to enter these high-quality, high-resolution virtual classrooms will be ubiquitous and affordable even in third world countries. Students at any age, from toddlers to adults, will be able to access the best education in the world at any time and from any place.

The nature of education will change once again when we merge with nonbiological intelligence. We will then have the ability to download knowledge and skills, at least into the nonbiological portion of our intelligence. Our machines do this routinely today. If you want to give your laptop state-of-the-art skills in speech or character recognition, language translation, or Internet searching, your computer has only to quickly download the right patterns (the software). We don't yet have comparable communication ports in our biological brains to quickly download the interneuronal connection and neurotransmitter patterns that represent our learning. That is one of many profound limitations of the biological paradigm we now use for our thinking, a limitation we will overcome in the Singularity.

. . . on Work

> If every instrument could accomplish its own work, obeying or anticipating the will of others, if the shuttle could weave, and the pick touch the lyre, without a hand to guide them, chief workmen would not need servants, nor masters slaves.
>
> —ARISTOTLE

Before the invention of writing, almost every insight was happening for the first time (at least to the knowledge of the small groups of humans involved). When you are at the beginning, everything is new. In our era, almost everything we do in the arts is done with awareness of what has been done before and before. In the early post-human era, things will be new again because anything that requires greater than human ability has not already been done by Homer or da Vinci or Shakespeare.

—VERNOR VINGE[59]

Now part of [my consciousness] lives on the Internet and seems to stay there all the time. . . . A student may have a textbook open. The television is on with the sound off. . . . They've got music on headphones . . . there's a homework window, along with e-mail and instant messaging. . . . One multi-tasking student prefers the online world to the face-to-face world. "Real life," he said, "is just one more window."

—CHRISTINE BOESE, REPORTING ON FINDINGS BY MIT PROFESSOR SHERRY TURKLE[60]

In 1651 Thomas Hobbes described "the life of man" as "solitary, poor, nasty, brutish, and short."[61] This was a fair assessment of life at the time, but we have largely overcome this harsh characterization through technological advances, at least in the developed world. Even in underdeveloped nations life expectancy lags only slightly behind. Technology typically starts out with unaffordable products that don't work very well, followed by expensive versions that work a bit better, and then by inexpensive products that work reasonably well. Finally the technology becomes highly effective, ubiquitous, and almost free. Radio and television followed this pattern, as did the cell phone. Contemporary Web access is at the inexpensive-and-working-reasonably-well stage.

Today the delay between early and late adoption is about a decade, but in keeping with the doubling of the paradigm-shift rate every decade, this delay will be only about five years in the middle of the second decade and only a couple of years in the mid-2020s. Given the enormous wealth-creation potential of GNR technologies, we will see the underclass largely disappear over the next two to three decades (see the discussions of the 2004 World Bank report in chapters 2 and 9). These developments are likely to be met, however, with increasing fundamentalist and Luddite reaction to the accelerating pace of change.

With the advent of MNT-based manufacturing, the cost of making any physical product will be reduced to pennies per pound, plus the cost of the information guiding the process, with the latter representing the true value. We are already not that far from this reality; software-based processes guide every step of manufacturing today, from design and materials procurement to assembly in automated factories. The portion of a manufactured product's cost attributable to the information processes used in its creation varies from one category of product to another but is increasing across the board, rapidly approaching 100 percent. By the late 2020s the value of virtually all products—clothes, food, energy, and of course electronics—will be almost entirely in their information. As is the case today, proprietary and open-source versions of every type of product and service will coexist.

Intellectual Property. If the primary value of products and services resides in their information, then the protection of information rights will be critical to supporting the business models that provide the capital to fund the creation of valuable information. The skirmishes today in the entertainment industry regarding illegal downloading of music and movies are a harbinger of what will be a profound struggle, once essentially everything of value is composed of information. Clearly, existing or new business models that allow for the creation of valuable intellectual property (IP) need to be protected, otherwise the supply of IP will itself be threatened. However, the pressure from the ease of copying information is a reality that is not going away, so industries will suffer if they do not keep their business models in line with public expectations.

In music, for example, rather than provide leadership with new paradigms, the recording industry stuck rigidly (until just recently) with the idea of an expensive record album, a business model that has remained unchanged from the time my father was a young, struggling musician in the 1940s. The public will avoid wide-scale pirating of information services only if commercial prices are kept at what are perceived to be reasonable levels. The mobile-phone sector is a prime example of an industry that has not invited rampant piracy. The cost of cell-phone calls has fallen rapidly with improving technology. If the mobile-phone industry had kept calling rates at the level where they were when I was a child (a time when people dropped whatever they were doing at the rare times that someone called long distance), we would be seeing comparable pirating of cell-phone calls, which is technically no more difficult than pirating music. But cheating on cell-phone calls is widely regarded as criminal behavior, largely because of the general perception that cell-phone charges are appropriate.

IP business models invariably exist on the edge of change. Movies have been difficult to download because of their large file size, but that is rapidly becoming less of an issue. The movie industry needs to lead the charge toward new standards, such as high-definition movies on demand. Musicians typically make most of their money with live performances, but that model will also come under attack early in the next decade, when we will have full-immersion virtual reality. Each industry will need to continually reinvent its business models, which will require as much creativity as the creation of the IP itself.

The first industrial revolution extended the reach of our bodies, and the second is extending the reach of our minds. As I mentioned, employment in factories and farms has gone from 60 percent to 6 percent in the United States in the past century. Over the next couple of decades, virtually all routine physical and mental work will be automated. Computation and communication will not involve discrete products such as handheld devices but will be a seamless web of intelligent resources that are all around us. Already most contemporary work is involved in the creation and promotion of IP in one form or another, as well as direct personal services from one person to another (health, fitness, education, and so on). These trends will continue with the creation of IP—including all of our artistic, social, and scientific creativity—and will be greatly enhanced by the expansion of our intellect through the merger with non-biological intelligence. Personal services will largely move to virtual-reality environments, especially when virtual reality begins to encompass all of the senses.

Decentralization. The next several decades will see a major trend toward decentralization. Today we have highly centralized and vulnerable energy plants and use ships and fuel lines to transport energy. The advent of nano-engineered fuel cells and solar power will enable energy resources to be massively distributed and integrated into our infrastructure. MNT manufacturing will be highly distributed using inexpensive nanofabrication minifactories. The ability to do nearly anything with anyone from anywhere in any virtual-reality environment will make obsolete the centralized technologies of office buildings and cities.

With version 3.0 bodies able to morph into different forms at will and our largely nonbiological brains no longer constrained to the limited architecture that biology has bestowed on us, the question of what is human will undergo intensive examination. Each transformation described here does not represent a sudden leap but rather a sequence of many small steps. Although the speed with which these steps are being taken is hastening, mainstream acceptance generally follows rapidly. Consider new reproductive technologies such as in vitro fertilization, which were controversial at first but quickly became widely used and accepted. On the other hand, change will always produce fundamentalist and Luddite counteractions, which will intensify as the pace of change increases. But despite apparent controversy, the overwhelming benefits to human health, wealth, expression, creativity, and knowledge quickly become apparent.

. . . on Play

> Technology is a way of organizing the universe so that people don't have to experience it.
> —MAX FRISCH, *HOMO FABER*

> Life is either a daring adventure or nothing.
> —HELEN KELLER

Play is just another version of work and has an integral role in the human creation of knowledge in all of its forms. A child playing with dolls and blocks is acquiring knowledge essentially by creating it through his or her own experience. People playing with dance moves are engaged in a collaborative creative process (consider the kids on street corners in the nation's poorest neighborhoods who created break dancing, which launched the hip-hop movement). Einstein put aside his work for the Swiss patent office and engaged in playful mind experiments, resulting in the creation of his enduring theories of special and general relativity. If war is the father of invention, then play is its mother.

Already there is no clear distinction between increasingly sophisticated video games and educational software. *The Sims 2,* a game released in September 2004, uses AI-based characters that have their own motivations and intentions. With no prepared scripts the characters behave in unpredictable ways, with the story line emerging out of their interactions. Although considered a game, it offers players insights into developing social awareness. Similarly games that simulate sports with increasingly realistic play impart skills and understanding.

By the 2020s, full-immersion virtual reality will be a vast playground of compelling environments and experiences. Initially VR will have certain benefits in terms of enabling communications with others in engaging ways over long distances and featuring a great variety of environments from which to choose. Although the environments will not be completely convincing at first, by the late 2020s they will be indistinguishable from real reality and will involve all of the senses, as well as neurological correlations of our emotions. As we enter the 2030s there won't be clear distinctions between human and machine, between real and virtual reality, or between work and play.

. . . on the Intelligent Destiny of the Cosmos: Why We Are Probably Alone in the Universe

> The universe is not only queerer than we suppose, but queerer than we can suppose.
>
> —J. B. S. HALDANE

> What is the universe doing questioning itself via one of its smallest products?
>
> —D. E. JENKINS, ANGLICAN THEOLOGIAN

> What is the universe computing? As far as we can tell, it is not producing a single answer to a single question. . . . Instead the universe is computing itself. Powered by Standard Model software, the universe computes quantum fields, chemicals, bacteria, human beings, stars, and galaxies. As it computes, it maps out its own spacetime geometry to the ultimate precision allowed by the laws of physics. Computation is existence.
>
> —SETH LLOYD AND Y. JACK NG[62]

Our naive view of the cosmos, dating back to pre-Copernican days, was that the Earth was at the center of the universe and human intelligence its greatest gift (next to God). The more informed recent view is that, even if the likelihood of a star's having a planet with a technology-creating species is very low (for example, one in a million), there are so many stars (that is, billions of trillions of them), that there are bound to be many (billions or trillions) with advanced technology.

This is the view behind SETI—the Search for Extraterrestrial Intelligence—and is the common informed view today. However, there are reasons to doubt the "SETI assumption" that ETI is prevalent.

First, consider the common SETI view. Common interpretations of the Drake equation (see below) conclude that there are many (as in billions) of ETIs in the universe, thousands or millions in our galaxy. We have only examined a tiny portion of the haystack (the universe), so our failure to date to find the needle (an ETI signal) should not be considered discouraging. Our efforts to explore the haystack are scaling up.

The following diagram from *Sky & Telescope* illustrates the scope of the

SETI project by plotting the capability of the varied scanning efforts against three major parameters: distance from Earth, frequency of transmission, and the fraction of the sky.[63]

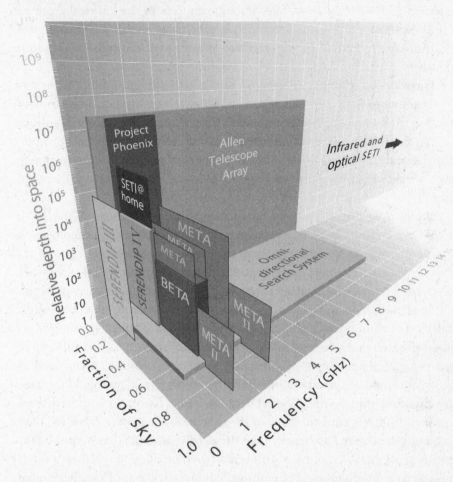

The plot includes two future systems. The Allen Telescope Array, named after Microsoft cofounder Paul Allen, is based on using many small scanning dishes rather than one or a small number of large dishes, with thirty-two of the dishes scheduled to be online in 2005. When all of its 350 dishes are operational (projected in 2008), it will be equivalent to a $2^{1}/_{2}$-acre dish (10,000 square meters). It will be capable of listening to up to 100 million frequency channels simultaneously, and able to cover the entire microwave spectrum. One of its intended tasks will be to scan millions of stars in our galaxy. The project relies on intelligent computation that can extract highly accurate signals from many low-cost dishes.[64]

Ohio State University is building the Omnidirectional Search System, which relies on intelligent computation to interpret signals from a large array of simple antennas. Using principles of interferometry (the study of how signals interfere with each other), a high-resolution image of the entire sky can be computed from the antenna data.[65] Other projects are expanding the range of electromagnetic frequency, for example, to explore the infrared and optical ranges.[66]

There are six other parameters in addition to the three shown in the chart on the previous page—for example, polarization (the plane of the wavefront in relation to the direction of the electromagnetic waves). One of the conclusions we can draw from the above graph is that only very thin slices of this nine-dimensional "parameter space" have been explored by SETI. So, the reasoning goes, we should not be surprised that we have not yet uncovered evidence of an ETI.

However, we are not just searching for a single needle. Based on the law of accelerating returns, once an ETI reaches primitive mechanical technologies, it is only a few centuries before it reaches the vast capabilities I've projected for the twenty-second century here on Earth. Russian astronomer N. S. Kardashev describes a "type II" civilization as one that has harnessed the power of its star for communication using electromagnetic radiation (about 4×10^{26} watts, based on our sun).[67] According to my projections (see chapter 3), our civilization will reach that level by the twenty-second century. Given that the level of technological development of the many civilizations projected by many SETI theorists should be spread out over vast periods of time, there should be many greatly ahead of us. So there should be many type II civilizations. Indeed, there has been sufficient time for some of these civilizations to have colonized their galaxies and achieve Kardashev's type III: a civilization that has harnessed the energy of its galaxy (about 4×10^{37} watts, based on our galaxy). Even a single advanced civilization should be emitting billions or trillions of "needles"—that is, transmissions representing a vast number of points in the SETI parameter space as artifacts and side effects of its myriad information processes. Even with the thin slices of the parameter space scanned by the SETI project to date, it would be hard to miss a type II civilization, let alone a type III. If we then factor in the expectation that there should be a vast number of these advanced civilizations, it is odd that we haven't noticed them. That's the Fermi Paradox.

The Drake Equation. The SETI search has been motivated in large part by astronomer Frank Drake's 1961 equation for estimating the number of intelligent (or, more precisely, radio-transmitting) civilizations in our galaxy.[68] (Presumably, the same analysis would pertain to other galaxies.) Consider the SETI assumption from the perspective of the Drake formula, which states:

The number of radio-transmitting civilizations = $N \times f_p \times n_e \times f_l \times f_i \times f_c \times f_L$

where:

N = the number of stars in the Milky Way galaxy. Current estimates are around 100 billion (10^{11}).

f_p = the fraction of stars that have orbiting planets. Current estimates range from about 20 percent to 50 percent.

n_e: For each star with orbiting planets, what is the average number of planets capable of sustaining life? This factor is highly controversial. Some estimates are one or higher (that is, every star with planets has, on average, at least one planet that can sustain life) to much lower factors, such as one in one thousand or even less.

f_l: For the planets *capable* of sustaining life, on what fraction of these does life actually evolve? Estimates are all over the map, from approximately 100 percent to about 0 percent.

f_i: For each planet on which life evolves, what is the fraction on which intelligent life evolves? f_l and f_i are the most controversial factors in the Drake equation. Here again, estimates range from nearly 100 percent (that is, once life gets a foothold, intelligent life is sure to follow) to close to 0 percent (that is, intelligent life is very rare).

f_c: For each planet with intelligent life, what is the fraction that communicates with radio waves? The estimates for f_c tend to be higher than for f_l and f_i, based on the (sensible) reasoning that once you have an intelligent species, the discovery and use of radio communication is likely.

f_L = the fraction of the universe's life during which an average communicating civilization communicates with radio waves.[69] If we take our civilization as an example, we have been communicating with radio transmissions for about one hundred years out of the roughly ten- to twenty-billion-year history of the universe, so f_L for the Earth is about 10^{-8} so far. If we continue communicating with radio waves for, say, another nine hundred years, the factor would then be 10^{-7}. This factor is affected by a number of considerations. If a civilization destroys itself because it is unable to handle the destructive power of technologies that may tend to develop along with radio communication (such as nuclear fusion or self-replicating nanotechnology), then radio transmissions would cease. We have seen civilizations on Earth (the Mayans, for example) suddenly end their organized societies and scientific pursuits (although preradio). On the other hand it seems unlikely that every civilization would end this way, so sudden destruction is likely to be only a modest factor in reducing the number of radio-capable civilizations.

A more salient issue is that of civilizations progressing from electromagnetic (that is, radio) transmissions to more capable means of communicating. Here on Earth we are rapidly moving from radio transmissions to wires, using cable and fiber optics for long-distance communication. So despite enormous increases in overall communication bandwidth, the amount of electromagnetic information sent into space from our planet has nevertheless remained fairly steady for the past decade. On the other hand we do have increasing means of wireless communication (for example, cell phones and new wireless Internet protocols, such as the emerging Wimax standard). Rather than use wires, communication may rely on exotic mediums such as gravity waves. However, even in this case, although the electromagnetic means of communication may no longer be the cutting edge of an ETI's communication technology, it is likely to continue to be used for at least some applications (in any case, f_L does take into consideration the possibility that a civilization would stop such transmissions).

It is clear that the Drake equation contains many imponderables. Many SETI advocates who have studied it carefully argue that it implies that there must be significant numbers of radio-transmitting civilizations in our galaxy alone. For example, if we assume that 50 percent of the stars have planets ($f_p = 0.5$), that each of these stars has an average of two planets able to sustain life ($n_e = 2$), that on half of these planets life has actually evolved ($f_l = 0.5$), that half of these planets has evolved intelligent life ($f_i = 0.5$), that half of these are radio-capable ($f_c = 0.5$), and that the average radio-capable civilization has been broadcasting for one million years ($f_L = 10^{-4}$), the Drake equation tells us that there are 1,250,000 radio-capable civilizations in our galaxy. For example, the SETI Institute's senior astronomer, Seth Shostak, estimates that there are between ten thousand and one million planets in the Milky Way containing a radio-broadcasting civilization.[70] Carl Sagan estimated around a million in the galaxy, and Drake estimated around ten thousand.[71]

But the parameters above are arguably very high. If we make more conservative assumptions on the difficulty of evolving life—and intelligent life in par-ticular—we get a very different outcome. If we assume that 50 percent of the stars have planets ($f_p = 0.5$), that only one tenth of these stars have planets able to sustain life ($n_e = 0.1$ based on the observation that life-supporting condi-tions are not that prevalent), that on 1 percent of these planets life has actually evolved ($f_l = 0.01$ based on the difficulty of life starting on a planet), that 5 per-cent of these life-evolving planets have evolved intelligent life ($f_i = 0.05$, based on the very long period of time this took on Earth), that half of these are radio-capable ($f_c = 0.5$), and that the average radio-capable civilization has been broadcasting for ten thousand years ($f_L = 10^{-6}$), the Drake equation tells us that there is about one (1.25 to be exact) radio-capable civilization in the Milky Way. And we already know of one.

In the end, it is difficult to make a strong argument for or against ETI based on this equation. If the Drake formula tells us anything, it is the extreme uncer-tainty of our estimates. What we do know for now, however, is that the cosmos appears silent—that is, we've detected no convincing evidence of ETI transmis-sions. The assumption behind SETI is that life—and intelligent life—is so prevalent that there must be millions if not billions of radio-capable civiliza-tions in the universe (or at least within our light sphere, which refers to radio-broadcasting civilizations that were sending out radio waves early enough to reach Earth by today). Not a single one of them, however, has made itself noticeable to our SETI efforts thus far. So let's consider the basic SETI assump-tion regarding the number of radio-capable civilizations from the perspective of the law of accelerating returns. As we have discussed, an evolutionary process inherently accelerates. Moreover, the evolution of technology is far faster than the relatively slow evolutionary process that gives rise to a tech-nology-creating species in the first place. In our own case we went from a pre-electricity, computerless society that used horses as its fastest land-based transportation to the sophisticated computational and communications tech-nologies we have today in only two hundred years. My projections show, as noted above, that within another century we will multiply our intelligence by trillions of trillions. So only three hundred years will have been necessary to take us from the early stirrings of primitive mechanical technologies to a vast expansion of our intelligence and ability to communicate. Thus, once a species creates electronics and sufficiently advanced technology to beam radio trans-missions, it is only a matter of a modest number of centuries for it to vastly expand the powers of its intelligence.

The three centuries this will have taken on Earth is an extremely brief period of time on a cosmological scale, given that the age of the universe is esti-mated at thirteen to fourteen billion years.[72] My model implies that once a civilization achieves our own level of radio transmission, it takes no more than a century—two at the most—to achieve a type II civilization. If we accept the underlying SETI assumption that there are many thousands if not millions of radio-capable civilizations in our galaxy—and therefore billions within our light sphere in the universe—these civilizations must exist in different stages over billions of years of development. Some would be behind us, and some would be ahead. It is not credible that every single one of the civilizations that are more advanced than us is going to be only a few decades ahead. Most of those that are ahead of us would be ahead by millions, if not billions, of years.

Yet since a period of only a few centuries is sufficient to progress from mechanical technology to the vast explosion of intelligence and communication of the Singularity, under the SETI assumption there should be billions of civilizations in our light sphere (thousands or millions in our galaxy) whose technology is ahead of ours to an unimaginable degree. In at least some discussions of the SETI project, we see the same kind of linear thinking that permeates every other field, assumptions that civilizations will reach our level of technology, and that technology will progress from that point very gradually for thousands if not millions of years. Yet the jump from the first stirrings of radio to powers that go beyond a mere type II civilization takes only a few hundred years. So the skies should be ablaze with intelligent transmissions.

Yet the skies are quiet. It is odd and intriguing that we find the cosmos so silent. As Enrico Fermi asked in the summer of 1950, "Where is everybody?"[73] A sufficiently advanced civilization would not be likely to restrict its transmissions to subtle signals on obscure frequencies. Why are all the ETIs so shy?

There have been attempts to respond to the so-called Fermi Paradox (which, granted, is a paradox only if one accepts the optimistic parameters that most observers apply to the Drake equation). One common response is that a civilization may obliterate itself once it reaches radio capability. This explanation might be acceptable if we were talking about only a few such civilizations, but with the common SETI assumptions implying billions of them, it is not credible to believe that every one of them destroyed itself.

Other arguments run along this same line. Perhaps "they" have decided not to disturb us (given how primitive we are) and are just watching us quietly (an ethical guideline that will be familiar to Star Trek fans). Again, it is hard to believe that every such civilization out of the billions that should exist has made the same decision. Or, perhaps, they have moved on to more capable communication paradigms. I do believe that more capable communication methods than electromagnetic waves—even very high-frequency ones—are likely to be feasible and that an advanced civilization (such as we will become over the next century) is likely to discover and exploit them. But it is very unlikely that there would be absolutely no role left for electromagnetic waves, even as a by-product of other technological processes, in any of these many millions of civilizations.

Incidentally, this is not an argument against the value of the SETI project, which should have high priority, because the negative finding is no less important than a positive result.

The Limits of Computation Revisited. Let's consider some additional implications of the law of accelerating returns to intelligence in the cosmos. In chapter 3 I discussed the ultimate cold laptop and estimated the optimal computational capacity of a one-liter, one-kilogram computer at around 10^{42} cps, which is sufficient to perform the equivalent of ten thousand years of the thinking of ten billion human brains in ten microseconds. If we allow more intelligent management of energy and heat, the potential in one kilogram of matter to compute may be as high as 10^{50} cps.

The technical requirements to achieve computational capacities in this range are daunting, but as I pointed out, the appropriate mental experiment is to consider the vast engineering ability of a civilization with 10^{42} cps per kilogram, not the limited engineering ability of humans today. A civilization at 10^{42} cps is likely to figure out how to get to 10^{43} cps and then to 10^{44} and so on. (Indeed, we can make the same argument at each step to get to the next.)

Once civilization reaches these levels it is obviously not going to restrict its computation to one kilogram of matter, any more than we do so today. Let's consider what our civilization can accomplish with the mass and energy in our own vicinity. The Earth contains a mass of about 6×10^{24} kilograms. Jupiter has a mass of about 1.9×10^{27} kilograms. If we ignore the hydrogen and helium, we have about 1.7×10^{26} kilograms of matter in the solar system, not including the sun (which ultimately is also fair game). The overall solar system, which is dominated by the sun, has a mass of about 2×10^{30} kilograms. As a crude upper-bound analysis, if we apply the mass in the solar system to our 10^{50} estimate of the limit of computational capacity per kilogram of matter (based on the limits for nanocomputing), we get a limit of 10^{80} cps for computation in our "vicinity."

Obviously, there are practical considerations that are likely to provide difficulty in reaching this kind of upper limit. But even if we devoted one twentieth of 1 percent (0.0005) of the matter of the solar system to computational and communication resources, we get capacities of 10^{69} cps for "cold" computing and 10^{77} cps for "hot" computing.[74]

Engineering estimates have been made for computing at these scales that take into consideration complex design requirements such as energy usage, heat dissipation, internal communication speeds, the composition of matter in the solar system, and many other factors. These designs use reversible computing, but as I pointed out in chapter 3, we still need to consider the energy requirements for correcting errors and communicating results. In an analysis by computational neuroscientist Anders Sandberg, the computational capacity of an Earth-size computational "object" called Zeus was reviewed.[75] The conceptual design of this "cold" computer, consisting of about 10^{25} kilograms of carbon (about 1.8 times the mass of the Earth) in the form of diamondoid consists of 5×10^{37} computational nodes, each of which uses extensive parallel processing. Zeus provides an estimated peak of 10^{61} cps of computation or, if used for data storage, 10^{47} bits. A primary limiting factor for the design is the number of bit erasures permitted (it allows up to 2.6×10^{32} bit erasures per second), which are primarily used to correct errors from cosmic rays and quantum effects.

In 1959 astrophysicist Freeman Dyson proposed a concept of curved shells around a star as a way to provide both energy and habitats for an advanced civilization. One conception of the Dyson Sphere is quite literally a thin sphere around a star to gather energy.[76] The civilization lives in the sphere, and gives off heat (infrared energy) outside the sphere (away from the star). Another (and more practical) version of the Dyson Sphere is a series of curved shells, each of which blocks only a portion of the star's radiation. In this way Dyson Shells can be designed to have no effect on existing planets, particularly those, like the Earth, that harbor an ecology that needs to be protected.

Although Dyson proposed his concept as a means of providing vast amounts of space and energy for an advanced *biological* civilization, it can also be used as the basis for star-scale computers. Such Dyson Shells could orbit our sun without affecting the sunlight reaching the Earth. Dyson imagined intelligent biological creatures living in the shells or spheres, but since civilization moves rapidly toward nonbiological intelligence once it discovers computation, there would be no reason to populate the shells with biological humans.

Another refinement of the Dyson concept is that the heat radiated by one shell could be captured and used by a parallel shell that is placed at a position farther from the sun. Computer scientist Robert Bradbury points out that there could be any number of such layers and proposes a computer aptly called a "Matrioshka brain," organized as a series of nested shells around the sun or another star. One such conceptual design analyzed by Sandberg is called Uranos, which is designed to use 1 percent of the nonhydrogen, nonhelium mass in the solar system (not including the sun), or about 10^{24} kilograms, a bit smaller than Zeus.[77] Uranos provides about 10^{39} computational nodes, an estimated 10^{51} cps of computation, and about 10^{52} bits of storage.

Computation is already a widely distributed—rather than centralized—resource, and my expectation is that the trend will continue toward greater decentralization. However, as our civilization approaches the densities of computation envisioned above, the distribution of the vast number of processors is likely to have characteristics of these conceptual designs. For example, the idea of Matrioshka shells would take maximal advantage of solar power and heat dissipation. Note that the computational powers of these solar system–scale computers will be achieved, according to my projections in chapter 2, around the end of this century.

Bigger or Smaller. Given that the computational capacity of our solar system is in the range of 10^{70} to 10^{80} cps, we will reach these limits early in the twenty-second century, according to my projections. The history of computation tells us that the power of computation expands both inward and outward. Over the last several decades we have been able to place twice as many computational elements (transistors) on each integrated circuit chip about every two years, which represents inward growth (toward greater densities of computation per kilogram of matter). But we are also expanding outward, in that the number of chips is expanding (currently) at a rate of about 8.3 percent per year.[78] It is reasonable to expect both types of growth to continue, and for the outward growth rate to increase significantly once we approach the limits of inward growth (with three-dimensional circuits).

Moreover, once we bump up against the limits of matter and energy in our solar system to support the expansion of computation, we will have no choice but to expand outward as the primary form of growth. We discussed earlier the speculation that finer scales of computation might be feasible—on the scale of subatomic particles. Such pico- or femtotechnology would permit continued growth of computation by continued shrinking of feature sizes. Even if this is feasible, however, there are likely to be major technical challenges in mastering subnanoscale computation, so the pressure to expand outward will remain.

Expanding Beyond the Solar System. Once we do expand our intelligence beyond the solar system, at what rate will this take place? The expansion will not start out at the maximum speed; it will quickly achieve a speed within a vanishingly small change from the maximum speed (speed of light or greater). Some critics have objected to this notion, insisting that it would be very difficult to send people (or advanced organisms from any other ETI civilization) and equipment at near the speed of light without crushing them. Of course, we could avoid this problem by accelerating slowly, but another problem would be collisions with interstellar material. But again, this objection entirely misses the point of the nature of intelligence at this stage of development. Early ideas about the spread of ETI through the galaxy and universe were based on the migration and colonization patterns from our human history and basically involved sending settlements of humans (or, in the case of other ETI civilizations, intelligent organisms) to other star systems. This would allow them to multiply through normal biological reproduction and then continue to spread in like manner from there.

But as we have seen, by late in this century nonbiological intelligence on the Earth will be many trillions of times more powerful than biological intelligence, so sending biological humans on such a mission would not make sense. The same would be true for any other ETI civilization. This is not simply a matter of biological humans sending robotic probes. Human civilization by that time will be nonbiological for all practical purposes.

These nonbiological sentries would not need to be very large and in fact would primarily comprise information. It is true, however, that *just* sending information would not be sufficient, for some material-based device that can have a physical impact on other star and planetary systems must be present. However, it would be sufficient for the probes to be self-replicating nanobots (note that a nanobot has nanoscale features but that the overall size of a nanobot is measured in microns).[79] We could send swarms of many trillions of them, with some of these "seeds" taking root in another planetary system and then replicating by finding the appropriate materials, such as carbon and other needed elements, and building copies of themselves.

Once established, the nanobot colony could obtain the additional information it needs to optimize its intelligence from pure information transmissions that involve only energy, not matter, and that are sent at the speed of light. Unlike large organisms such as humans, these nanobots, being extremely small, could travel at close to the speed of light. Another scenario would be to dispense with the information transmissions and embed the information needed in the nanobots' own memory. That's an engineering decision we can leave to these future superengineers.

The software files could be spread out among billions of devices. Once one or a few of them get a "foothold" by self-replicating at a destination, the now much larger system could gather up the nanobots traveling in the vicinity so that from that time on, the bulk of the nanobots sent in that direction do not simply fly by. In this way, the now established colony can gather up the information, as well as the distributed computational resources, it needs to optimize its intelligence.

The Speed of Light Revisited. In this way the maximum speed of expansion of a solar system–size intelligence (that is, a type II civilization) into the rest of the universe would be very close to the speed of light. We currently understand the maximum speed to transmit information and material objects to be the speed of light, but there are at least suggestions that this may not be an absolute limit.

We have to regard the possibility of circumventing the speed of light as speculative, and my projections of the profound changes that our civilization will undergo in this century make no such assumption. However, the potential to engineer around this limit has important implications for the speed with which we will be able to colonize the rest of the universe with our intelligence.

Recent experiments have measured the flight time of photons at nearly twice the speed of light, a result of quantum uncertainty on their position.[80] However, this result is really not useful for this analysis, because it does not actually allow information to be communicated faster than the speed of light, and we are fundamentally interested in communication speed.

Another intriguing suggestion of an action at a distance that appears to occur at speeds far greater than the speed of light is quantum disentanglement. Two particles created together may be "quantum entangled," meaning that while a given property (such as the phase of its spin) is not determined in either particle, the resolution of this ambiguity of the two particles will occur at the same moment. In other words, if the undetermined property is measured in one of the particles, it will also be determined as the exact same value at the same instant in the other particle, even if the two have traveled far apart. There is an appearance of some sort of communication link between the particles.

This quantum disentanglement has been measured at many times the speed of light, meaning that resolution of the state of one particle appears to resolve the state of the other particle in an amount of time that is a small fraction of the time it would take if the information were transmitted from one particle to the other at the speed of light (in theory, the time lapse is zero). For example, Dr. Nicolas Gisin of the University of Geneva sent quantum-entangled photons in opposite directions through optical fibers across Geneva. When the photons were seven miles apart, they each encountered a glass plate. Each photon had to "decide" whether to pass through or bounce off the plate (which previous experiments with non-quantum-entangled photons have shown to be a random choice). Yet because the two photons were quantum entangled, they made the same decision at the same moment. Many repetitions provided the identical result.[81]

The experiments have not absolutely ruled out the explanation of a hidden variable—that is, an unmeasurable state of each particle that is in phase (set to the same point in a cycle), so that when one particle is measured (for example, has to decide its path through or off a glass plate), the other has the same value of this internal variable. So the "choice" is generated by an identical setting of this hidden variable, rather than being the result of actual communication between the two particles. However, most quantum physicists reject this interpretation.

Yet even if we accept the interpretation of these experiments as indicating a quantum link between the two particles, the apparent communication is transmitting only randomness (profound quantum randomness) at speeds far greater than the speed of light, not predetermined information, such as the bits in a file. This communication of quantum random decisions to different points in space could have value, however, in applications such as providing encryption codes. Two different locations could receive the same random sequence, which could then be used by one location to encrypt a message and by the other to decipher it. It would not be possible for anyone else to eavesdrop on the encryption code without destroying the quantum entanglement and thereby being detected. There are already commercial encryption products incorporating this principle. This is a fortuitous application of quantum mechanics because of the possibility that another application of quantum mechanics— quantum computing—may put an end to the standard method of encryption based on factoring large numbers (which quantum computing, with a large number of entangled qubits, would be good at).

Yet another faster-than-the-speed-of-light phenomenon is the speed with which galaxies can recede from each other as a result of the expansion of the universe. If the distance between two galaxies is greater than what is called the Hubble distance, then these galaxies are receding from one another at faster than the speed of light.[82] This does not violate Einstein's special theory of relativity, because this velocity is caused by space itself expanding rather than the galaxies moving through space. However, it also doesn't help us transmit information at speeds faster than the speed of light.

Wormholes. There are two exploratory conjectures that suggest ways to circumvent the apparent limitation of the speed of light. The first is to use wormholes—folds of the universe in dimensions beyond the three visible ones.

This does not really involve traveling at speeds faster than the speed of light but merely means that the topology of the universe is not the simple three-dimensional space that naive physics implies. However, if wormholes or folds in the universe are ubiquitous, perhaps these shortcuts would allow us to get everywhere quickly. Or perhaps we can even engineer them.

In 1935 Einstein and physicist Nathan Rosen formulated "Einstein-Rosen" bridges as a way of describing electrons and other particles in terms of tiny space-time tunnels.[83] In 1955 physicist John Wheeler described these tunnels as "wormholes," introducing the term for the first time.[84] His analysis of worm-holes showed them to be fully consistent with the theory of general relativity, which describes space as essentially curved in another dimension.

In 1988 California Institute of Technology physicists Michael Morris, Kip Thorne, and Uri Yurtsever explained in some detail how such wormholes could be engineered.[85] Responding to a question from Carl Sagan they described the energy requirements to keep wormholes of varying sizes open. They also pointed out that based on quantum fluctuation, so-called empty space is continually generating tiny wormholes the size of subatomic particles. By adding energy and following other requirements of both quantum physics and general relativity (two fields that have been notoriously difficult to unify), these worm-holes could be expanded to allow objects larger than subatomic particles to travel through them. Sending humans through them would not be impossible but extremely difficult. However, as I pointed out above, we really only need to send nanobots plus information, which could pass through wormholes measured in microns rather than meters.

Thorne and his Ph.D. students Morris and Yurtsever also described a method consistent with general relativity and quantum mechanics that could establish wormholes between the Earth and faraway locations. Their proposed technique involves expanding a spontaneously generated, subatomic-size worm-hole to a larger size by adding energy, then stabilizing it using superconducting spheres in the two connected "wormhole mouths." After the wormhole is expanded and stabilized, one of its mouths (entrances) is transported to another location, while keeping its connection to the other entrance, which remains on Earth.

Thorne offered the example of moving the remote entrance via a small rocket ship to the star Vega, which is twenty-five light-years away. By traveling at very close to the speed of light, the journey, as measured by clocks on the ship, would be relatively brief. For example, if the ship traveled at 99.995 percent of the speed of light, the clocks on the ship would move ahead by only three months. Although the time for the voyage, as measured on Earth, would be around twenty-five years, the stretched wormhole would maintain the direct link between the locations as well as the points in time of the two locations. Thus, even as experienced on Earth, it would take only three months to establish the link between Earth and Vega, because the two ends of the wormhole would maintain their time relationship. Suitable engineering improvements could allow such links to be established anywhere in the universe. By traveling arbitrarily close to the speed of light, the time required to establish a link—for both communications and transportation—to other locations in the universe, even those millions of billions of light years away, could be relatively brief.

Matt Visser of Washington University in St. Louis has suggested refinements to the Morris-Thorne-Yurtsever concept that provide a more stable environment, which might even allow humans to travel through wormholes.[86] In my view, however, this is unnecessary. By the time engineering projects of this scale might be feasible, human intelligence will long since have been dominated by its nonbiological component. Sending molecular-scale self-replicating devices along with software will be sufficient and much easier. Anders Sandberg estimates that a one-nanometer wormhole could transmit a formidable 10^{69} bits per second.[87]

Physicist David Hochberg and Vanderbilt University's Thomas Kephart point out that shortly after the Big Bang, gravity was strong enough to have provided the energy required to spontaneously create massive numbers of self-stabilizing wormholes.[88] A significant portion of these wormholes is likely to still be around and may be pervasive, providing a vast network of corridors that reach far and wide throughout the universe. It might be easier to discover and use these natural wormholes than to create new ones.

Changing the Speed of Light. The second conjecture is to change the speed of light itself. In chapter 3, I mentioned the finding that appears to indicate that the speed of light has differed by 4.5 parts out of 10^8 over the past two billion years.

In 2001 astronomer John Webb discovered that the so-called fine-structure constant varied when he examined light from sixty-eight quasars (very bright young galaxies).[89] The speed of light is one of four constants that the fine-structure constant comprises, so the result is another suggestion that varying conditions in the universe may cause the speed of light to change. Cambridge University physicist John Barrow and his colleagues are in the process of running a two-year tabletop experiment that will test the ability to engineer a small change in the speed of light.[90]

Suggestions that the speed of light can vary are consistent with recent theories that it was significantly higher during the inflationary period of the universe (an early phase in its history, when it underwent very rapid expansion). These experiments showing possible variation in the speed of light clearly need corroboration and are showing only small changes. But if confirmed, the findings would be profound, because it is the role of engineering to take a subtle effect and greatly amplify it. Again, the mental experiment we should perform now is not whether contemporary human scientists, such as we are, can perform these engineering feats but whether or not a human civilization that has expanded its intelligence by trillions of trillions will be able to do so.

For now we can say that ultrahigh levels of intelligence will expand outward at the speed of light, while recognizing that our contemporary understanding of physics suggests that this may not be the actual limit of the speed of expansion or, even if the speed of light proves to be immutable, that this limit may not restrict reaching other locations quickly through wormholes.

The Fermi Paradox Revisited. Recall that biological evolution is measured in millions and billions of years. So if there are other civilizations out there, they would be spread out in terms of development by huge spans of time. The SETI assumption implies that there should be billions of ETIs (among all the galaxies), so there should be billions that lie far ahead of us in their technological progress. Yet it takes only a few centuries at most from the advent of computation for such civilizations to expand outward at at least light speed. Given this, how can it be that we have not noticed them?

The conclusion I reach is that it is likely (although not certain) that there *are* no such other civilizations. In other words, we are in the lead. That's right, our humble civilization with its pickup trucks, fast food, and persistent conflicts (and computation!) is in the lead in terms of the creation of complexity and order in the universe.

Now how can that be? Isn't this extremely unlikely, given the sheer number of likely inhabited planets? Indeed it is very unlikely. But equally unlikely is the existence of our universe, with its set of laws of physics and related physical constants, so exquisitely, precisely what is needed for the evolution of life to be possible. But by the anthropic principle, if the universe didn't allow the evolution of life we wouldn't be here to notice it. Yet here we are. So by a similar anthropic principle, we're here in the lead in the universe. Again, if we weren't here, we would not be noticing it.

Let's consider some arguments against this perspective.

Perhaps there are extremely advanced technological civilizations out there, but we are outside their light sphere of intelligence. That is, they haven't gotten here yet. Okay, in this case, SETI will still fail to find ETIs because we won't be able to see (or hear) them, at least not unless and until we find a way to break out of our light sphere (or the ETI does so) by manipulating the speed of light or finding shortcuts, as I discussed above.

Perhaps they are among us, but have decided to remain invisible to us. If they have made that decision, they are likely to succeed in avoiding being noticed. Again, it is hard to believe that every single ETI has made the same decision.

John Smart has suggested in what he calls the "transcension" scenario that once civilizations saturate their local region of space with their intelligence, they create a new universe (one that will allow continued exponential growth of complexity and intelligence) and essentially leave this universe.[91] Smart suggests that this option may be so attractive that it is the consistent and inevitable outcome of an ETI's having reached an advanced stage of its development, and it thereby explains the Fermi Paradox.

Incidentally, I have always considered the science-fiction notion of large spaceships piloted by huge, squishy creatures similar to us to be very unlikely. Seth Shostak comments that "the reasonable probability is that any extraterrestrial intelligence we will detect will be machine intelligence, not biological intelligence like us." In my view this is not simply a matter of biological beings sending out machines (as we do today) but rather that any civilization sophisticated enough to make the trip here would have long since passed the point of merging with its technology and would not need to send physically bulky organisms and equipment.

If they exist, why would they come here? One mission would be for observation—to gather knowledge (just as we observe other species on Earth today). Another would be to seek matter and energy to provide additional substrate for its expanding intelligence. The intelligence and equipment needed for such exploration and expansion (by an ETI, or by us when we get to that stage of development) would be extremely small, basically nanobots and information transmissions.

It appears that our solar system has not yet been turned into someone else's computer. And if this other civilization is only observing us for knowledge's sake and has decided to remain silent, SETI will fail to find it, because if an advanced civilization does not want us to notice it, it would succeed in that desire. Keep in mind that such a civilization would be vastly more intelligent than we are today. Perhaps it will reveal itself to us when we achieve the next level of our evolution, specifically merging our biological brains with our technology, which is to say, after the Singularity. However, given that the SETI assumption implies that there are billions of such highly developed civiliza-tions, it seems unlikely that all of them have made the same decision to stay out of our way.

The Anthropic Principle Revisited. We are struck with two possible applica-tions of an anthropic principle, one for the remarkable biofriendly laws of our universe, and one for the actual biology of our planet.

Let's first consider the anthropic principle as applied to the universe in more detail. The question concerning the universe arises because we notice that the constants in nature are precisely what are required for the universe to have grown in complexity. If the cosmological constant, the Planck constant, and the many other constants of physics were set to just slightly different val-ues, atoms, molecules, stars, planets, organisms, and humans would have been impossible. The universe appears to have exactly the right rules and constants. (The situation is reminiscent of Steven Wolfram's observation that certain cellular-automata rules [see the sidebar on p. 85] allow for the creation of remarkably complex and unpredictable patterns, whereas other rules lead to very uninteresting patterns, such as alternating lines or simple triangles in a repeating or random configuration.)

How do we account for the remarkable design of the laws and constants of matter and energy in our universe that have allowed for the increasing com-plexity we see in biological and technology evolution? Freeman Dyson once commented that "the universe in some sense knew we were coming." Complex-ity theorist James Gardner describes the question in this way:

> Physicists feel that the task of physics is to predict what happens in the lab, and they are convinced that string theory, or M theory can do this. . . . But they have no idea why the universe should . . . have the stan-dard model, with the values of its 40+ parameters that we observe. How can anyone believe that something so messy is the unique prediction of string theory? It amazes me that people can have such blinkered vision, that they can concentrate just on the final state of the universe, and not ask how and why it got there.[92]

The perplexity of how it is that the universe is so "friendly" to biology has led to various formulations of the anthropic principle. The "weak" version of the anthropic principle points out simply that if it were not the case, we wouldn't be here to wonder about it. So only in a universe that allowed for increasing complexity could the question even be asked. Stronger versions of the anthropic principle state that there must be more to it; advocates of these versions are not satisfied with a mere lucky coincidence. This has opened the door for advocates of intelligent design to claim that this is the proof of God's existence that scientists have been asking for.

The Multiverse. Recently a more Darwinian approach to the strong anthropic principle has been proposed. Consider that it is possible for mathematical equations to have multiple solutions. For example, if we solve for x in the equation $x^2 = 4$, x may be 2 or -2. Some equations allow for an infinite number of solutions. In the equation $(a-b) \times x = 0$, x can take on any one of an infinite number of values if $a = b$ (since any number multiplied by zero equals zero). It turns out that the equations for recent string theories allow in principle for an infinite number of solutions. To be more precise, since the spatial and temporal resolution of the universe is limited to the very small Planck constant, the number of solutions is not literally infinite but merely vast. String theory implies, therefore, that many different sets of natural constants are possible.

This has led to the idea of the multiverse: that there exist a vast number of universes, of which our humble universe is only one. Consistent with string theory, each of these universes can have a different set of physical constants.

Evolving Universes. Leonard Susskind, the discoverer of string theory, and Lee Smolin, a theoretical physicist and expert on quantum gravity, have suggested that universes give rise to other universes in a natural, evolutionary process that gradually refines the natural constants. In other words it is not by accident that the rules and constants of our universe are ideal for evolving intelligent life but rather that they themselves evolved to be that way.

In Smolin's theory the mechanism that gives rise to new universes is the creation of black holes, so those universes best able to produce black holes are the ones that are most likely to reproduce. According to Smolin a universe best able to create increasing complexity—that is, biological life—is also most likely to create new universe-generating black holes. As he explains, "Reproduction through black holes leads to a multiverse in which the conditions for life are common—essentially because some of the conditions life requires, such as plentiful carbon, also boost the formation of stars massive enough to become black holes."[93] Susskind's proposal differs in detail from Smolin's but is also based on black holes, as well as the nature of "inflation," the force that caused the very early universe to expand rapidly.

Intelligence as the Destiny of the Universe. In *The Age of Spiritual Machines*, I introduced a related idea—namely, that intelligence would ultimately permeate the universe and would decide the destiny of the cosmos:

> How relevant is intelligence to the universe? . . . The common wisdom is *not very*. Stars are born and die; galaxies go through their cycles of creation and destruction; the universe itself was born in a big bang and will end with a crunch or a whimper, we're not yet sure which. But intelligence has little to do with it. Intelligence is just a bit of froth, an ebullition of little creatures darting in and out of inexorable universal forces. The mindless mechanism of the universe is winding up or down to a distant future, and there's nothing intelligence can do about it.
>
> That's the common wisdom. But I don't agree with it. My conjecture is that intelligence will ultimately prove more powerful than these big impersonal forces. . . .
>
> So will the universe end in a big crunch, or in an infinite expansion of dead stars, or in some other manner? In my view, the primary issue is not the mass of the universe, or the possible existence of antigravity, or of Einstein's so-called cosmological constant. Rather, the fate of the universe is a decision yet to be made, one which we will intelligently consider when the time is right.[94]

Complexity theorist James Gardner combined my suggestion on the evolution of intelligence throughout the universe with Smolin's and Susskind's concepts of evolving universes. Gardner conjectures that it is specifically the evolution of intelligent life that enables offspring universes.[95] Gardner builds on British astronomer Martin Rees's observation that "what we call the fundamental constants—the numbers that matter to physicists—may be secondary consequences of the final theory, rather than direct manifestations of its deepest and most fundamental level." To Smolin it is merely coincidence that black holes and biological life both need similar conditions (such as large amounts of carbon), so in his conception there is no explicit role for intelligence, other than that it happens to be the by-product of certain biofriendly circumstances. In Gardner's conception it is intelligent life that creates its successors.

Gardner writes that "we and other living creatures throughout the cosmos are part of a vast, still undiscovered transterrestrial community of lives and intelligences spread across billions of galaxies and countless parsecs who are collectively engaged in a portentous mission of truly cosmic importance. Under the Biocosm vision, we share a common fate with that community—to help shape the future of the universe and transform it from a collection of lifeless atoms into a vast, transcendent mind." To Gardner the laws of nature, and the precisely balanced constants, "function as the cosmic counterpart of DNA: they furnish the 'recipe' by which the evolving cosmos acquires the capacity to generate life and ever more capable intelligence."

My own view is consistent with Gardner's belief in intelligence as the most important phenomenon in the universe. I do have a disagreement with Gardner on his suggestion of a "vast ... transterrestrial community of lives and intelligences spread across billions of galaxies." We don't yet see evidence that such a community beyond Earth exists. The community that matters may be just our own unassuming civilization here. As I pointed out above, although we can fashion all kinds of reasons why each particular intelligent civilization may remain hidden from us (for example, they destroyed themselves, or they have decided to remain invisible or stealthy, or they've switched *all* of their communications away from electromagnetic transmissions, and so on), it is not credible to believe that every single civilization out of the billions that should be there (according to the SETI assumption) has some reason to be invisible.

The Ultimate Utility Function. We can fashion a conceptual bridge between Susskind's and Smolin's idea of black holes being the "utility function" (the property being optimized in an evolutionary process) of each universe in the multiverse and the conception of intelligence as the utility function that I share with Gardner. As I discussed in chapter 3, the computational power of a computer is a function of its mass and its computational efficiency. Recall that a rock has significant mass but extremely low computational efficiency (that is, virtually all of the transactions of its particles are effectively random). Most of the particle interactions in a human are random also, but on a logarithmic scale humans are roughly halfway between a rock and the ultimate small computer.

A computer in the range of the ultimate computer has a very high computational efficiency. Once we achieve an optimal computational efficiency, the only way to increase the computational power of a computer would be to increase its mass. If we increase the mass enough, its gravitational force becomes strong enough to cause it to collapse into a black hole. So a black hole can be regarded as the ultimate computer.

Of course, not any black hole will do. Most black holes, like most rocks, are performing lots of random transactions but no useful computation. But a well-organized black hole would be the most powerful conceivable computer in terms of cps per liter.

Hawking Radiation. There has been a long-standing debate about whether or not we can transmit information into a black hole, have it usefully transformed, and then retrieve it. Stephen Hawking's conception of transmissions from a black hole involves particle-antiparticle pairs that are created near the event horizon (the point of no return near a black hole, beyond which matter and energy are unable to escape). When this spontaneous creation occurs, as it does everywhere in space, the particle and antiparticle travel in opposite directions. If one member of the pair travels into the event horizon (never to be seen again), the other will fly away from the black hole.

Some of these particles will have sufficient energy to escape its gravitation and result in what has been called Hawking radiation.[96] Prior to Hawking's analysis it was thought that black holes were, well, black; with his insight we realized that they actually give off a continual shower of energetic particles. But according to Hawking this radiation is random, since it originates from random quantum events near the event boundary. So a black hole may contain an ultimate computer, according to Hawking, but according to his original conception, no information can escape a black hole, so this computer could never transmit its results.

In 1997 Hawking and fellow physicist Kip Thorne (the wormhole scientist) made a bet with California Institute of Technology's John Preskill. Hawking and Thorne maintained that the information that entered a black hole was lost, and any computation that might occur inside the black hole, useful or otherwise, could never be transmitted outside of it, whereas Preskill maintained that the information could be recovered.[97] The loser was to give the winner some useful information in the form of an encyclopedia.

In the intervening years the consensus in the physics community steadily moved away from Hawking, and on July 21, 2004, Hawking admitted defeat and acknowledged that Preskill had been correct after all: that information sent into a black hole is not lost. It could be transformed inside the black hole and then transmitted outside it. According to this understanding, what happens is that the particle that flies away from the black hole remains quantum entangled with its antiparticle that disappeared into the black hole. If that antiparticle inside the black hole becomes involved in a useful computation, then these results will be encoded in the state of its tangled partner particle outside of the black hole.

Accordingly Hawking sent Preskill an encyclopedia on the game of cricket, but Preskill rejected it, insisting on a baseball encyclopedia, which Hawking had flown over for a ceremonial presentation.

Assuming that Hawking's new position is indeed correct, the ultimate computers that we can create would be black holes. Therefore a universe that is well designed to create black holes would be one that is well designed to optimize its intelligence. Susskind and Smolin argued merely that biology and black holes both require the same kind of materials, so a universe that was optimized for black holes would also be optimized for biology. Recognizing that black holes are the ultimate repository of intelligent computation, however, we can conclude that the utility function of optimizing black-hole production and that of optimizing intelligence are one and the same.

Why Intelligence Is More Powerful than Physics. There is another reason to apply an anthropic principle. It may seem remarkably unlikely that our planet is in the lead in terms of technological development, but as I pointed out above, by a weak anthropic principle, if we had not evolved, we would not be here discussing this issue.

As intelligence saturates the matter and energy available to it, it turns dumb matter into smart matter. Although smart matter still nominally follows the laws of physics, it is so extraordinarily intelligent that it can harness the most subtle aspects of the laws to manipulate matter and energy to its will. So it would at least appear that intelligence is more powerful than physics. What I should say is that intelligence is more powerful than cosmology. That is, once matter evolves into smart matter (matter fully saturated with intelligent processes), it can manipulate other matter and energy to do its bidding (through suitably powerful engineering). This perspective is not generally considered in discussions of future cosmology. It is assumed that intelligence is irrelevant to events and processes on a cosmological scale.

Once a planet yields a technology-creating species and that species creates computation (as has happened here), it is only a matter of a few centuries before its intelligence saturates the matter and energy in its vicinity, and it begins to expand outward at at least the speed of light (with some suggestions of circumventing this limit). Such a civilization will then overcome gravity (through exquisite and vast technology) and other cosmological forces—or, to be fully accurate, it will maneuver and control these forces—and engineer the universe it wants. This is the goal of the Singularity.

A Universe-Scale Computer. How long will it take for our civilization to saturate the universe with our vastly expanded intelligence? Seth Lloyd estimates there are about 10^{80} particles in the universe, with a theoretical maximum capacity of about 10^{90} cps. In other words a universe-scale computer would be able to compute at 10^{90} cps.[98] To arrive at those estimates, Lloyd took the observed density of matter—about one hydrogen atom per cubic meter—and from this figure computed the total energy in the universe. Dividing this energy figure by the Planck constant, he got about 10^{90} cps. The universe is about 10^{17} seconds old, so in round numbers there have been a maximum of about 10^{107} calculations in it thus far. With each particle able to store about 10^{10} bits in all of its degrees of freedom (including its position, trajectory, spin, and so on), the state of the universe represents about 10^{90} bits of information at each point in time.

We do not need to contemplate devoting all of the mass and energy of the universe to computation. If we were to apply 0.01 percent, that would still leave 99.99 percent of the mass and energy unmodified, but would still result in a potential of about 10^{86} cps. Based on our current understanding, we can only approximate these orders of magnitude. Intelligence at anything close to these levels will be so vast that it will be able to perform these engineering feats with enough care so as not to disrupt whatever natural processes it considers important to preserve.

The Holographic Universe. Another perspective on the maximum information storage and processing capability of the universe comes from a speculative recent theory of the nature of information. According to the "holographic universe" theory the universe is actually a two-dimensional array of information written on its surface, so its conventional three-dimensional appearance is an illusion.[99] In essence, the universe, according to this theory, is a giant hologram.

The information is written at a very fine scale, governed by the Planck constant. So the maximum amount of information in the universe is its surface area divided by the square of the Planck constant, which comes to about 10^{120} bits. There does not appear to be enough matter in the universe to encode this much information, so the limits of the holographic universe may be higher than what is actually feasible. In any event the order of magnitude of the number of orders of magnitudes of these various estimates is in the same range. The number of bits that a universe reorganized for useful computation will be able to store is 10 raised to a power somewhere between 80 and 120.

Again, our engineering, even that of our vastly evolved future selves, will probably fall short of these maximums. In chapter 2 I showed how we progressed from 10^{-5} to 10^{8} cps per thousand dollars during the twentieth century. Based on a continuation of the smooth, doubly exponential growth that we saw in the twentieth century, I projected that we would achieve about 10^{60} cps per thousand dollars by 2100. If we estimate a modest trillion dollars devoted to computation, that's a total of about 10^{69} cps by the end of this century. This can be achieved with the matter and energy in our solar system.

To get to around 10^{90} cps requires expanding through the rest of the universe.

Continuing the double-exponential growth curve shows that we can saturate the universe with our intelligence well before the end of the twenty-second century, *provided that* we are not limited by the speed of light. Even if the up-to-thirty additional powers of ten suggested by the holographic-universe theory are borne out, we still reach saturation by the end of the twenty-second century.

Again, if it is at all possible to circumvent the speed-of-light limitation, the vast intelligence we will have with solar system–scale intelligence will be able to design and implement the requisite engineering to do so. If I had to place a bet, I would put my money on the conjecture that circumventing the speed of light is possible and that we will be able to do this within the next couple of hundred years. But that is speculation on my part, as we do not yet understand these issues sufficiently to make a more definitive statement. If the speed of light is an immutable barrier, and no shortcuts through wormholes exist that can be exploited, it will take billions of years, not hundreds, to saturate the universe with our intelligence, and we will be limited to our light cone within the universe. In either event the exponential growth of computation will hit a wall during the twenty-second century. (But what a wall!)

This large difference in timespans—hundreds of years versus billions of years (to saturate the universe with our intelligence)—demonstrates why the issue of circumventing the speed of light will become so important. It will become a primary preoccupation of the vast intelligence of our civilization in the twenty-second century. That is why I believe that if wormholes or other circumventing means are feasible, we will be highly motivated to find and exploit them.

If it is possible to engineer new universes and establish contact with them, this would provide yet further means for an intelligent civilization to continue its expansion. Gardner's view is that the influence of an intelligent civilization in creating a new universe lies in setting the physical laws and constants of the baby universe. But the vast intelligence of such a civilization may figure out ways to expand its own intelligence into a new universe more directly. The idea of spreading our intelligence beyond this universe is, of course, speculative, as none of the multiverse theories allows for communication from one universe to another, except for passing on basic laws and constants.

Even if we are limited to the one universe we already know about, saturating its matter and energy with intelligence is our ultimate fate. What kind of universe will that be? Well, just wait and see.

MOLLY 2004: *So when the universe reaches Epoch Six [the stage at which the nonbiological portion of our intelligence spreads through the universe], what's it going to do?*

CHARLES DARWIN: *I'm not sure we can answer that. As you said, it's like bacteria asking one another what humans will do.*

MOLLY 2004: *So these Epoch Six entities will consider us biological humans to be like bacteria?*

GEORGE 2048: *That's certainly not how I think of you.*

MOLLY 2104: *George, you're only Epoch Five, so I don't think that answers the question.*

CHARLES: *Getting back to the bacteria, what they would say, if they could talk—*

MOLLY 2004: *—and think.*

CHARLES: *Yes, that, too. They would say that humans will do the same things as we bacteria do—namely, eat, avoid danger, and procreate.*

MOLLY 2104: *Oh, but our procreation is so much more interesting.*

MOLLY 2004: *Actually, Molly of the future, it's our human pre-Singularity procreation that's interesting. Your virtual procreation is, actually, a lot like that of the bacteria. Sex has nothing to do with it.*

MOLLY 2104: *It's true we've separated sexuality from reproduction, but that's not exactly new to human civilization in 2004. And besides, unlike bacteria, we can change ourselves.*

MOLLY 2004: *Actually, you've separated change and evolution from reproduction as well.*

MOLLY 2104: *That was also essentially true in 2004.*

MOLLY 2004: *Okay, okay. But about your list, Charles, we humans also do things like create art and music. That kind of separates us from other animals.*

GEORGE 2048: *Indeed, Molly, that is fundamentally what the Singularity is about. The Singularity is the sweetest music, the deepest art, the most beautiful mathematics. . . .*

MOLLY 2004: *I see, so the music and art of the Singularity will be to my era's music and art as circa 2004 music and art are to . . .*

NED LUDD: *The music and art of bacteria.*

MOLLY 2004: *Well, I've seen some artistic mold patterns.*

NED: *Yes, but I'm sure you didn't revere them.*

MOLLY 2004: *No, actually, I wiped them away.*

NED: *Okay, my point then.*

MOLLY 2004: *I'm still trying to envision what the universe will be doing in Epoch Six.*

TIMOTHY LEARY: *The universe will be flying like a bird.*

MOLLY 2004: *But what is it flying in? I mean it's everything.*

TIMOTHY: *That's like asking, What is the sound of one hand clapping?*

MOLLY 2004: *Hmmm, so the Singularity is what the Zen masters had in mind all along.*

CHAPTER SEVEN

Ich bin ein Singularitarian

The most common of all follies is to believe passionately in the palpably not true.

—H. L. MENCKEN

Philosophies of life rooted in centuries-old traditions contain much wisdom concerning personal, organizational, and social living. Many of us also find shortcomings in those traditions. How could they not reach some mistaken conclusions when they arose in pre-scientific times? At the same time, ancient philosophies of life have little or nothing to say about fundamental issues confronting us as advanced technologies begin to enable us to change our identity as individuals and as humans and as economic, cultural, and political forces change global relationships.

—MAX MORE, "PRINCIPLES OF EXTROPY"

The world does not need another totalistic dogma.

—MAX MORE, "PRINCIPLES OF EXTROPY"

Yes, we have a soul. But it's made of lots of tiny robots.

—GIULIO GIORELLI

Substrate is morally irrelevant, assuming it doesn't affect functionality or consciousness. It doesn't matter, from a moral point of view, whether somebody runs on silicon or biological neurons (just as it doesn't matter whether you have dark or pale skin). On the same grounds, that we reject racism and speciesism, we should also reject carbon-chauvinism, or *bioism*.

—NICK BOSTROM, "ETHICS FOR INTELLIGENT MACHINES: A PROPOSAL, 2001"

Philosophers have long noted that their children were born into a more complex world than that of their ancestors. This early and perhaps even unconscious recognition of accelerating change may have been the catalyst for much of the utopian, apocalyptic, and millennialist thinking in our Western tradition. But the modern difference is that now *everyone* notices the pace of progress on some level, not simply the visionaries.

—JOHN SMART

A Singularitarian is someone who understands the Singularity and has reflected on its meaning for his or her own life.

I have been engaged in such reflection for several decades. Needless to say, it's not a process that one can ever complete. I started pondering the relationship of our thinking to our computational technology as a teenager in the 1960s. In the 1970s I began to study the acceleration of technology, and I wrote my first book on the subject in the late 1980s. So I've had time to contemplate the impact on society—and on myself—of the overlapping transformations now under way.

George Gilder has described my scientific and philosophical views as "a substitute vision for those who have lost faith in the traditional object of religious belief."[1] Gilder's statement is understandable, as there are at least apparent similarities between anticipation of the Singularity and anticipation of the transformations articulated by traditional religions.

But I did not come to my perspective as a result of searching for an alternative to customary faith. The origin of my quest to understand technology trends was practical: an attempt to time my inventions and to make optimal tactical decisions in launching technology enterprises. Over time this modeling of technology took on a life of its own and led me to formulate a theory of technology evolution. It was not a huge leap from there to reflect on the impact of these crucial changes on social and cultural institutions and on my own life. So, while being a Singularitarian is not a matter of faith but one of understanding, pondering the scientific trends I've discussed in this book inescapably engenders new perspectives on the issues that traditional religions have attempted to address: the nature of mortality and immortality, the purpose of our lives, and intelligence in the universe.

Being a Singularitarian has often been an alienating and lonely experience for me because most people I encounter do not share my outlook. Most "big thinkers" are totally unaware of this big thought. In a myriad of statements and comments people typically evidence the common wisdom that human life is short, that our physical and intellectual reach is limited, and that nothing fundamental will change in our lifetimes. I expect this narrow view to change as the implications of accelerating change become increasingly apparent, but having more people with whom to share my outlook is a major reason that I wrote this book.

So how do we contemplate the Singularity? As with the sun, it's hard to look at directly; it's better to squint at it out of the corners of our eyes. As Max More states, the last thing we need is another dogma, nor do we need another cult, so Singularitarianism is not a system of beliefs or unified viewpoints. While it is fundamentally an understanding of basic technology trends, it is simultaneously an insight that causes one to rethink everything, from the nature of health and wealth to the nature of death and self.

To me, being a Singularitarian means many things, of which the following is a small sampling. These reflections articulate my personal philosophy, not a proposal for a new doctrine.

- We have the means right now to live long enough to live forever.[2] Existing knowledge can be aggressively applied to dramatically slow down aging processes so we can still be in vital health when the more radical life-extending therapies from biotechnology and nanotechnology become available. But most baby boomers won't make it because they are unaware of the accelerating aging processes in their bodies and the opportunity to intervene.

- In this spirit I am aggressively reprogramming my biochemistry, which is now altogether different than it would otherwise be.[3] Taking supplements and medications is not a last resort to be reserved only for when something goes wrong. There is already something wrong. Our bodies are governed by obsolete genetic programs that evolved in a bygone era, so we need to overcome our genetic heritage. We already have the knowledge to begin to accomplish this, something I am committed to doing.

- My body is temporary. Its particles turn over almost completely every month. Only the pattern of my body and brain have continuity.

- We should strive to improve these patterns by optimizing the health of our bodies and extending the reach of our minds. Ultimately, we will be able to vastly expand our mental faculties by merging with our technology.

- We need a body, but once we incorporate MNT fabrication into ourselves, we will be able to change our bodies at will.

- Only technology can provide the scale to overcome the challenges with which human society has struggled for generations. For example, emerging technologies will provide the means of providing and storing clean and renewable energy, removing toxins and pathogens from our bodies and the environment, and providing the knowledge and wealth to over-come hunger and poverty.

- Knowledge is precious in all its forms: music, art, science, and technology, as well as the embedded knowledge in our bodies and brains. Any loss of this knowledge is tragic.
- Information is not knowledge. The world is awash in information; it is the role of intelligence to find and act on the salient patterns. For example, we have hundreds of megabits of information flowing through our senses every second, the bulk of which is intelligently discarded. It is only the key recognitions and insights (all forms of knowledge) that we retain. Thus intelligence selectively destroys information to create knowledge.
- Death is a tragedy. It is not demeaning to regard a person as a profound pattern (a form of knowledge), which is lost when he or she dies. That, at least, is the case today, since we do not yet have the means to access and back up this knowledge. When people speak of losing part of themselves when a loved one dies, they are speaking quite literally, since we lose the ability to effectively use the neural patterns in our brain that had self-organized to interact with that person.
- A primary role of traditional religion is deathist rationalization—that is, rationalizing the tragedy of death as a good thing. Malcolm Muggeridge articulates the common view that "if it weren't for death, life would be unbearable." But the explosion of art, science, and other forms of knowl-edge that the Singularity will bring will make life more than bearable; it will make life truly meaningful.
- In my view the purpose of life—and of our lives—is to create and appre-ciate ever-greater knowledge, to move toward greater "order." As I dis-cussed in chapter 2, increasing order usually means increasing complexity, but sometimes a profound insight will increase order while reducing complexity.
- As I see it the purpose of the universe reflects the same purpose as our lives: to move toward greater intelligence and knowledge. Our human intelli-gence and our technology form the cutting edge of this expanding intelli-gence (given that we are not aware of any extraterrestrial competitors).
- Having reached a tipping point, we will within this century be ready to infuse our solar system with our intelligence through self-replicating non-biological intelligence. It will then spread out to the rest of the universe.
- Ideas are the embodiment and the product of intelligence. The ideas exist to solve most any problem that we encounter. The primary problems we cannot solve are ones that we cannot articulate and are mostly ones of

which we are not yet aware. For the problems that we do encounter, the key challenge is to express them precisely in words (and sometimes in equations). Having done that, we have the ability to find the ideas to confront and resolve each such problem.

- We can apply the enormous leverage provided by the acceleration of technology. A notable example is achieving radical life extension through "a bridge to a bridge to a bridge" (applying today's knowledge as a bridge to biotechnology, which in turn will bridge us to the era of nanotechnology).[4] This offers a way to live indefinitely *now*, even though we don't yet have all the knowledge necessary for radical life extension. In other words we don't have to solve every problem today. We can anticipate the capability of technologies that are coming—in five years or ten years or twenty—and work these into our plans. That is how I design my own technology projects, and we can do the same with the large problems facing society and with our own lives.

Contemporary philosopher Max More describes the goal of humanity as a transcendence to be "achieved through science and technology steered by human values."[5] More cites Nietzsche's observation "Man is a rope, fastened between animal and overman—a rope over an abyss." We can interpret Nietzsche to be pointing out that we have advanced beyond other animals while seeking to become something far greater. We might regard Nietzsche's reference to the abyss to allude to the perils inherent in technology, which I address in the next chapter.

More has at the same time expressed concern that anticipating the Singularity could engender a passivity in addressing today's issues.[6] Because the enormous capability to overcome age-old problems is on the horizon, there may be a tendency to grow detached from mundane, present-day concerns. I share More's antipathy toward "passive Singularitarianism." One reason for a proactive stance is that technology is a double-edged sword and as such always has the potential of going awry as it surges toward the Singularity, with profoundly disturbing consequences. Even small delays in implementing emerging technologies can condemn millions of people to continued suffering and death. As one example of many, excessive regulatory delays in implementing lifesaving therapies end up costing many lives. (We lose millions of people per year around the world from heart disease alone.)

More also worries about a cultural rebellion "seduced by religious and cultural urgings for 'stability,' 'peace,' and against 'hubris' and 'the unknown' " that may derail technological acceleration.[7] In my view any significant derailment of the overall advancement of technology is unlikely. Even epochal events such as two world wars (in which on the order of one hundred million people died), the cold war, and numerous economic, cultural, and social upheavals have failed to make the slightest dent in the pace of technology trends. But the reflexive, thoughtless antitechnology sentiments increasingly being voiced in the world today do have the potential to exacerbate a lot of suffering.

Still Human? Some observers refer to the post-Singularity period as "posthuman" and refer to the anticipation of this period as posthumanism. However, to me being human means being part of a civilization that seeks to extend its boundaries. We are already reaching beyond our biology by rapidly gaining the tools to reprogram and augment it. If we regard a human modified with technology as no longer human, where would we draw the defining line? Is a human with a bionic heart still human? How about someone with a neurological implant? What about two neurological implants? How about someone with ten nanobots in his brain? How about 500 million nanobots? Should we establish a boundary at 650 million nanobots: under that, you're still human and over that, you're posthuman?

Our merger with our technology has aspects of a slippery slope, but one that slides up toward greater promise, not down into Nietzsche's abyss. Some observers refer to this merger as creating a new "species." But the whole idea of a species is a biological concept, and what we are doing is transcending biology. The transformation underlying the Singularity is not just another in a long line of steps in biological evolution. We are upending biological evolution altogether.

BILL GATES: *I agree with you 99 percent. What I like about your ideas is that they are grounded in science, but your optimism is almost a religious faith. I'm optimistic also.*

RAY: *Yes, well, we need a new religion. A principal role of religion has been to rationalize death, since up until just now there was little else constructive we could do about it.*

BILL: *What would the principles of the new religion be?*

RAY: *We'd want to keep two principles: one from traditional religion and one from secular arts and sciences—from traditional religion, the respect for human consciousness.*

BILL: *Ah yes, the Golden Rule.*

RAY: *Right, our morality and legal system are based on respect for the consciousness of others. If I hurt another person, that's considered immoral, and*

probably illegal, because I have caused suffering to another conscious per-
son. If I destroy property, it's generally okay if it's my property, and the pri-
mary reason it's immoral and illegal if it's someone else's property is because
I have caused suffering not to the property but to the person owning it.

BILL: *And the secular principle?*

RAY: *From the arts and sciences, it is the importance of knowledge. Knowledge*
goes beyond information. It's information that has meaning for conscious
entities: music, art, literature, science, technology. These are the qualities
that will expand from the trends I'm talking about.

BILL: *We need to get away from the ornate and strange stories in contemporary*
religions and concentrate on some simple messages. We need a charismatic
leader for this new religion.

RAY: *A charismatic leader is part of the old model. That's something we want to*
get away from.

BILL: *Okay, a charismatic computer, then.*

RAY: *How about a charismatic operating system?*

BILL: *Ha, we've already got that. So is there a God in this religion?*

RAY: *Not yet, but there will be. Once we saturate the matter and energy in the*
universe with intelligence, it will "wake up," be conscious, and sublimely
intelligent. That's about as close to God as I can imagine.

BILL: *That's going to be silicon intelligence, not biological intelligence.*

RAY: *Well, yes, we're going to transcend biological intelligence. We'll merge with it*
first, but ultimately the nonbiological portion of our intelligence will pre-
dominate. By the way, it's not likely to be silicon, but something like carbon
nanotubes.

BILL: *Yes, I understand—I'm just referring to that as silicon intelligence since peo-*
ple understand what that means. But I don't think that's going to be con-
scious in the human sense.

RAY: *Why not? If we emulate in as detailed a manner as necessary everything*
going on in the human brain and body and instantiate these processes in
another substrate, and then of course expand it greatly, why wouldn't it be
conscious?

BILL: *Oh, it will be conscious. I just think it will be a different type of conscious-*
ness.

RAY: *Maybe this is the 1 percent we disagree on. Why would it be different?*

BILL: *Because computers can merge together instantly. Ten computers—or one*
million computers—can become one faster, bigger computer. As humans,
we can't do that. We each have a distinct individuality that cannot be
bridged.

RAY: *That's just a limitation of biological intelligence. The unbridgeable distinct-*
ness of biological intelligence is not a plus. "Silicon" intelligence can have it
both ways. Computers don't have to pool their intelligence and resources.
They can remain "individuals" if they wish. Silicon intelligence can even
have it both ways by merging and retaining individuality—at the same
time. As humans, we try to merge with others also, but our ability to accom-
plish this is fleeting.

BILL: *Everything of value is fleeting.*

RAY: *Yes, but it gets replaced by something of even greater value.*

BILL: *True, that's why we need to keep innovating.*

The Vexing Question of Consciousness

If you could blow the brain up to the size of a mill and walk about inside,
you would not find consciousness.

—G. W. LEIBNIZ

Can one ever remember love? It's like trying to summon up the smell of
roses in a cellar. You might see a rose, but never the perfume.

—ARTHUR MILLER[8]

At one's first and simplest attempts to philosophize, one becomes entangled
in questions of whether when one knows something, one knows that one
knows it, and what, when one is thinking of oneself, is being thought about,
and what is doing the thinking. After one has been puzzled and bruised by
this problem for a long time, one learns not to press these questions: the
concept of a conscious being is, implicitly, realized to be different from that
of an unconscious object. In saying that a conscious being knows something, we are saying not only that he knows it, but that he knows that he
knows it, and that he knows that he knows that he knows it, and so on, as
long as we care to pose the question: there is, we recognize, an infinity here,
but it is not an infinite regress in the bad sense, for it is the questions that
peter out, as being pointless, rather than the answers.

—J. R. LUCAS, OXFORD PHILOSOPHER, IN HIS 1961 ESSAY "MINDS, MACHINES,
AND GÖDEL"[9]

Dreams are real while they last; can we say more of life?

—HAVELOCK ELLIS

Will future machines be capable of having emotional and spiritual experiences? We have discussed several scenarios for a nonbiological intelligence to display the full range of emotionally rich behavior exhibited by biological humans today. By the late 2020s we will have completed the reverse engineering of the human brain, which will enable us to create nonbiological systems that match and exceed the complexity and subtlety of humans, including our emotional intelligence.

A second scenario is that we could upload the patterns of an actual human into a suitable nonbiological, thinking substrate. A third, and the most compelling, scenario involves the gradual but inexorable progression of humans themselves from biological to nonbiological. That has already started with the benign introduction of devices such as neural implants to ameliorate disabilities and disease. It will progress with the introduction of nanobots in the bloodstream, which will be developed initially for medical and antiaging applications. Later more sophisticated nanobots will interface with our biological neurons to augment our senses, provide virtual and augmented reality from within the nervous system, assist our memories, and provide other routine cognitive tasks. We will then be cyborgs, and from that foothold in our brains, the nonbiological portion of our intelligence will expand its powers exponentially. As I discussed in chapters 2 and 3 we see ongoing exponential growth of every aspect of information technology, including price-performance, capacity, and rate of adoption. Given that the mass and energy required to compute and communicate each bit of information are extremely small (see chapter 3), these trends can continue until our nonbiological intelligence vastly exceeds that of the biological portion. Since our biological intelligence is essentially fixed in its capacity (except for some relatively modest optimization from biotechnology), the nonbiological portion will ultimately predominate. In the 2040s, when the nonbiological portion will be billions of times more capable, will we still link our consciousness to the biological portion of our intelligence?

Clearly, nonbiological entities will claim to have emotional and spiritual experiences, just as we do today. They—we—will claim to be human and to have the full range of emotional and spiritual experiences that humans claim to have. And these will not be idle claims; they will evidence the sort of rich, complex, and subtle behavior associated with such feelings.

But how will these claims and behaviors—compelling as they will be—relate to the subjective experience of nonbiological humans? We keep coming back to the very real but ultimately unmeasurable (by fully objective means) issue of consciousness. People often talk about consciousness as if it were a clear property of an entity that can readily be identified, detected, and gauged.

If there is one crucial insight that we can make regarding why the issue of consciousness is so contentious, it is the following:

There exists no objective test that can conclusively determine its presence.

Science is about objective measurements and their logical implications, but the very nature of objectivity is that you cannot measure subjective experience—you can only measure correlates of it, such as behavior (and by behavior, I include internal behavior—that is, the actions of the components of an entity, such as neurons and their many parts). This limitation has to do with the very nature of the concepts of "objectivity" and "subjectivity." Fundamentally we cannot penetrate the subjective experience of another entity with direct objective measurement. We can certainly make arguments about it, such as, "Look inside the brain of this nonbiological entity; see how its methods are just like those of a human brain." Or, "See how its behavior is just like human behavior." But in the end, these remain just arguments. No matter how convincing the behavior of a nonbiological person, some observers will refuse to accept the consciousness of such an entity unless it squirts neurotransmitters, is based on DNA-guided protein synthesis, or has some other specific biologically human attribute.

We assume that other humans are conscious, but even that is an assumption. There is no consensus among humans about the consciousness of non-human entities, such as higher animals. Consider the debates regarding animal rights, which have everything to do with whether animals are conscious or just quasi machines that operate by "instinct." The issue will be even more contentious with regard to future nonbiological entities that exhibit behavior and intelligence even more humanlike than those of animals.

In fact these future machines will be even more humanlike than humans today. If that seems like a paradoxical statement, consider that much of human thought today is petty and derivative. We marvel at Einstein's ability to conjure up the theory of general relativity from a thought experiment or Beethoven's ability to imagine symphonies that he could never hear. But these instances of human thought at its best are rare and fleeting. (Fortunately we have a record of these fleeting moments, reflecting a key capability that has separated humans from other animals.) Our future primarily nonbiological selves will be vastly more intelligent and so will exhibit these finer qualities of human thought to a far greater degree.

So how will we come to terms with the consciousness that will be claimed by nonbiological intelligence? From a practical perspective such claims will be accepted. For one thing, "they" will be us, so there won't be any clear distinctions between biological and nonbiological intelligence. Furthermore, these nonbiological entities will be extremely intelligent, so they'll be able to convince other humans (biological, nonbiological, or somewhere in between) that they are conscious. They'll have all the delicate emotional cues that convince us today that humans are conscious. They will be able to make other humans laugh and cry. And they'll get mad if others don't accept their claims. But this is fundamentally a political and psychological prediction, not a philosophical argument.

I do take issue with those who maintain that subjective experience either doesn't exist or is an inessential quality that can safely be ignored. The issue of who or what is conscious and the nature of the subjective experiences of others are fundamental to our concepts of ethics, morality, and law. Our legal system is based largely on the concept of consciousness, with particularly serious attention paid to actions that cause suffering—an especially acute form of conscious experience—to a (conscious) human or that end the conscious experience of a human (for example, murder).

Human ambivalence regarding the ability of animals to suffer is reflected in legislation as well. We have laws against animal cruelty, with greater emphasis given to more intelligent animals, such as primates (although we appear to have a blind spot with regard to the massive animal suffering involved in factory farming, but that's the subject of another treatise).

My point is that we cannot safely dismiss the question of consciousness as merely a polite philosophical concern. It is at the core of society's legal and moral foundation. The debate will change when a machine—nonbiological intelligence—can persuasively argue on its own that it/he/she has feelings that need to be respected. Once it can do so with a sense of humor—which is particularly important for convincing others of one's humanness—it is likely that the debate will be won.

I expect that actual change in our legal system will come initially from litigation rather than legislation, as litigation often precipitates such transformations. In a precursor of what is to come, attorney Martine Rothblatt, a partner in Mahon, Patusky, Rothblatt & Fisher, filed a mock motion on September 16, 2003, to prevent a corporation from disconnecting a conscious computer. The motion was argued in a mock trial in the biocyberethics session at the International Bar Association conference.[10]

We can measure certain correlates of subjective experience (for example, certain patterns of objectively measurable neurological activity with objectively verifiable reports of certain subjective experiences, such as hearing a sound). But we cannot penetrate to the core of subjective experience through objective measurement. As I mentioned in chapter 1, we are dealing with the difference between third-person "objective" experience, which is the basis of science, and first-person "subjective" experience, which is a synonym for consciousness.

Consider that we are unable to truly experience the subjective experiences of others. The experience-beaming technology of 2029 will enable the brain of one person to experience only the *sensory* experiences (and potentially some of the neurological correlates of emotions and other aspects of experience) of another person. But that will still not convey the same *internal* experience as that undergone by the person beaming the experience, because his or her brain is different. Every day we hear reports about the experiences of others, and we may even feel empathy in response to the behavior that results from their internal states. But because we're exposed to only the *behavior* of others, we can only *imagine* their subjective experiences. Because it is possible to construct a perfectly consistent, scientific worldview that omits the existence of consciousness, some observers come to the conclusion that it's just an illusion.

Jaron Lanier, the virtual-reality pioneer, takes issue (in the third of his six objections to what he calls "cybernetic totalism" in his treatise "One Half a Manifesto") with those who maintain "that subjective experience either doesn't exist, or is unimportant because it is some sort of ambient or peripheral effect."[11] As I pointed out, there is no device or system we can postulate that could definitively detect subjectivity (conscious experience) associated with an entity. Any such purported device would have philosophical assumptions built into it. Although I disagree with much of Lanier's treatise (see the "Criticism from Software" section in chapter 9), I concur with him on this issue and can even imagine (and empathize with!) his feelings of frustration at the dictums of "cybernetic totalists" such as myself (not that I accept this characterization).[12] Like Lanier I even accept the subjective experience of those who maintain that there is no such thing as subjective experience.

Precisely because we cannot resolve issues of consciousness entirely through objective measurement and analysis (science), a critical role exists for philosophy. Consciousness is the most important ontological question. After all, if we truly imagine a world in which there is no subjective experience (a world in which there is swirling stuff but no conscious entity to experience it), that world may as well not exist. In some philosophical traditions, both Eastern (certain schools of Buddhist thought, for example), and Western (specifically, observer-based interpretations of quantum mechanics), that is exactly how such a world is regarded.

RAY: *We can debate what sorts of entities are or can be conscious. We can argue about whether consciousness is an emergent property or caused by some specific mechanism, biological or otherwise. But there's another mystery associated with consciousness, perhaps the most important one.*

MOLLY 2004: *Okay, I'm all ears.*

RAY: *Well, even if we assume that all humans who seem to be conscious in fact are, why is my consciousness associated with this particular person, me? Why am I conscious of this particular person who read Tom Swift Jr. books as a child, got involved with inventions, writes books about the future, and so on? Every morning that I wake up, I have the experiences of this specific person. Why wasn't I Alanis Morissette or someone else?*

SIGMUND FREUD: *Hmm, so you'd like to be Alanis Morissette?*

RAY: *That's an interesting proposition, but that's not my point.*

MOLLY 2004: *What is your point? I don't understand.*

RAY: *Why am I conscious of the experiences and decisions of this particular person?*

MOLLY 2004: *Because, silly, that's who you are.*

SIGMUND: *It seems that there's something about yourself that you don't like. Tell me more about that.*

MOLLY 2004: *Earlier, Ray didn't like being human altogether.*

RAY: *I didn't say I don't like being human. I said I didn't like the limitations, problems, and high level of maintenance of my version 1.0 body. But this is all beside the point that I'm trying to make here.*

CHARLES DARWIN: *You wonder why you're you? That's a tautology, there's not much to wonder about.*

RAY: *Like many attempts to express the really "hard" problems of consciousness, this is sounding meaningless. But if you ask me what I really wonder about, it is this: why am I continually aware of this particular person's experiences and feelings? As for other people's consciousness, I accept it, but I don't experience other people's experiences, not directly anyway.*

SIGMUND: *Okay, I'm getting a clearer picture now. You don't experience other people's experiences? Have you ever talked to someone about empathy?*

RAY: *Look, I'm talking about consciousness now in a very personal way.*

SIGMUND: *That's good, keep going.*

RAY: *Actually, this is a good example of what typically happens when people try to have a dialogue about consciousness. The discussion inevitably veers off into something else, like psychology or behavior or intelligence or neurology. But the mystery of why I am this particular person is what I really wonder about.*

CHARLES: *You know you do create who you are.*

RAY: *Yes, that's true. Just as our brains create our thoughts, our thoughts in turn create our brains.*

CHARLES: *So you've made yourself, and that's why you are who you are, so to speak.*

MOLLY 2104: *We experience that very directly in 2104. Being nonbiological, I'm*

*able to change who I am quite readily. As we discussed earlier, if I'm in the
mood, I can combine my thought patterns with someone else's and create a
merged identity. It's a profound experience.*

MOLLY 2004: *Well, Miss Molly of the future, we do that back in the primitive days
of 2004 also. We call it falling in love.*

Who Am I? What Am I?

Why are you you?

> —THE IMPLIED QUESTION IN THE ACRONYM YRUU (YOUNG RELIGIOUS
> UNITARIAN UNIVERSALISTS), AN ORGANIZATION I WAS ACTIVE IN WHEN I
> WAS GROWING UP IN THE EARLY 1960S (IT WAS THEN CALLED LRY, LIBERAL
> RELIGIOUS YOUTH).

What you are looking for is who is looking.

> —SAINT FRANCIS OF ASSISI

I'm not aware of too many things
I know what I know if you know what I mean.
Philosophy is the talk on a cereal box.
Religion is the smile on a dog. . . .
Philosophy is a walk on the slippery rocks.
Religion is a light in the fog. . . .
What I am is what I am.
Are you what you are or what?

> —EDIE BRICKELL, "WHAT I AM"

Freedom of will is the ability to do gladly that which I must do.

> —CARL JUNG

The chance of the quantum theoretician is not the ethical freedom of the
Augustinian.

> —NORBERT WIENER[13]

I should prefer to an ordinary death, being immersed with a few friends in a
cask of Madeira, until that time, then to be recalled to life by the solar
warmth of my dear country! But in all probability, we live in a century too

little advanced, and too near the infancy of science, to see such an art
brought in our time to its perfection.

—BENJAMIN FRANKLIN, 1773

A related but distinct question has to do with our own identities. We talked
earlier about the potential to upload the patterns of an individual mind—
knowledge, skills, personality, memories—to another substrate. Although the
new entity would act just like me, the question remains: is it really *me?*

Some of the scenarios for radical life extension involve reengineering and
rebuilding the systems and subsystems that our bodies and brains comprise. In
taking part in this reconstruction, do I lose my self along the way? Again, this
issue will transform itself from a centuries-old philosophical dialogue to a
pressing practical matter in the next several decades.

So who am I? Since I am constantly changing, am I just a pattern? What if
someone copies that pattern? Am I the original and/or the copy? Perhaps I am
this stuff here—that is, the both ordered and chaotic collection of molecules
that make up my body and brain.

But there's a problem with this position. The specific set of particles that my
body and brain comprise are in fact completely different from the atoms and
molecules that I comprised only a short while ago. We know that most of our
cells are turned over in a matter of weeks, and even our neurons, which persist
as distinct cells for a relatively long time, nonetheless change all of their constituent molecules within a month.[14] The half-life of a microtubule (a protein
filament that provides the structure of a neuron) is about ten minutes. The
actin filaments in dendrites are replaced about every forty seconds. The proteins that power the synapses are replaced about every hour. NMDA receptors
in synapses stick around for a relatively long five days.

So I am a completely different set of stuff than I was a month ago, and all
that persists is the pattern of organization of that stuff. The pattern changes
also, but slowly and in a continuum. I am rather like the pattern that water
makes in a stream as it rushes past the rocks in its path. The actual molecules of
water change every millisecond, but the pattern persists for hours or even years.

Perhaps, therefore, we should say I am a pattern of matter and energy that
persists over time. But there is a problem with this definition, as well, since we
will ultimately be able to upload this pattern to replicate my body and brain to
a sufficiently high degree of accuracy that the copy is indistinguishable from
the original. (That is, the copy could pass a "Ray Kurzweil" Turing test.) The
copy, therefore, will share my pattern. One might counter that we may not get
every detail correct, but as time goes on our attempts to create a neural and

body replica will increase in resolution and accuracy at the same exponential pace that governs all information-based technologies. We will ultimately be able to capture and re-create my pattern of salient neural and physical details to any desired degree of accuracy.

Although the copy shares my pattern, it would be hard to say that the copy is me because I would—or could—still be here. You could even scan and copy me while I was sleeping. If you come to me in the morning and say, "Good news, Ray, we've successfully reinstantiated you into a more durable substrate, so we won't be needing your old body and brain anymore," I may beg to differ.

If you do the thought experiment, it's clear that the copy may look and act just like me, but it's nonetheless *not* me. I may not even know that he was created. Although he would have all my memories and recall having been me, from the point in time of his creation Ray 2 would have his own unique experiences, and his reality would begin to diverge from mine.

This is a real issue with regard to cryonics (the process of preserving by freezing a person who has just died, with a view toward "reanimating" him later when the technology exists to reverse the damage from the early stages of the dying process, the cryonic-preservation process, and the disease or condition that killed him in the first place). Assuming a "preserved" person is ultimately reanimated, many of the proposed methods imply that the reanimated person will essentially be "rebuilt" with new materials and even entirely new neuromorphically equivalent systems. The reanimated person will, therefore, effectively be "Ray 2" (that is, someone else).

Now let's pursue this train of thought a bit further, and you will see where the dilemma arises. If we copy me and then destroy the original, that's the end of me, because as we concluded above the copy is not me. Since the copy will do a convincing job of impersonating me, no one may know the difference, but it's nonetheless the end of me.

Consider replacing a tiny portion of my brain with its neuromorphic equivalent.

Okay, I'm still here: the operation was successful (incidentally, nanobots will eventually do this without surgery). We know people like this already, such as those with cochlear implants, implants for Parkinson's disease, and others. Now replace another portion of my brain: okay, I'm still here . . . and again. . . . At the end of the process, I'm still myself. There never was an "old Ray" and a "new Ray." I'm the same as I was before. No one ever missed me, including me.

The gradual replacement of Ray results in Ray, so consciousness and identity appear to have been preserved. However, in the case of gradual replacement there is no simultaneous old me and new me. At the end of the process you have the equivalent of the new me (that is, Ray 2) and no old me (Ray 1). So gradual replacement also means the end of me. We might therefore wonder: at what point did my body and brain become someone else?

On yet another hand (we're running out of philosophical hands here), as I pointed out at the beginning of this question, I am in fact being continually replaced as part of a normal biological process. (And, by the way, that process is not particularly gradual but rather rapid.) As we concluded, all that persists is my spatial and temporal pattern of matter and energy. But the thought experiment above shows that gradual replacement means the end of me even if my pattern is preserved. So am I constantly being replaced by someone else who just seems a lot like the me of a few moments earlier?

So, again, who am I? It's the ultimate ontological question, and we often refer to it as the issue of consciousness. I have consciously (pun intended) phrased the issue entirely in the first person because that is its nature. It is not a third-person question. So my question is not "who are you?" although you may wish to ask this question yourself.

When people speak of consciousness they often slip into considerations of behavioral and neurological correlates of consciousness (for example, whether or not an entity can be self-reflective). But these are third-person (objective) issues and do not represent what David Chalmers calls the "hard question" of consciousness: how can matter (the brain) lead to something as apparently immaterial as consciousness?[15]

The question of whether or not an entity is conscious is apparent only to itself. The difference between neurological correlates of consciousness (such as intelligent behavior) and the ontological reality of consciousness is the difference between objective and subjective reality. That's why we can't propose an objective consciousness detector without philosophical assumptions built into it.

I do believe that we humans will come to accept that nonbiological entities are conscious, because ultimately the nonbiological entities will have all the subtle cues that humans currently possess and that we associate with emotional and other subjective experiences. Still, while we will be able to verify the subtle cues, we will have no direct access to the implied consciousness.

I will acknowledge that many of you do seem conscious to me, but I should not be too quick to accept this impression. Perhaps I am really living in a simulation, and you are all part of it.

Or, perhaps it's only my memories of you that exist, and these actual experiences never took place.

Or maybe I am only now experiencing the sensation of recalling apparent memories, but neither the experience nor the memories really exist. Well, you see the problem.

Despite these dilemmas my personal philosophy remains based on patternism—I am principally a pattern that persists in time. I am an evolving pattern, and I can influence the course of the evolution of my pattern. Knowledge is a pattern, as distinguished from mere information, and losing knowledge is a profound loss. Thus, losing a person is the ultimate loss.

MOLLY 2004: *As far as I'm concerned, who I am is pretty straightforward—it's basically this brain and body, which at least this month is in pretty good shape, thank you.*

RAY: *Are you including the food in your digestive tract, in its various stages of decomposition along the way?*

MOLLY 2004: *Okay, you can exclude that. Some of it will become me, but it hasn't been enrolled yet in the "part of Molly" club.*

RAY: *Well, 90 percent of the cells in your body don't have your DNA.*

MOLLY 2004: *Is that so? Just whose DNA is it, then?*

RAY: *Biological humans have about ten trillion cells with their own DNA, but there are about one hundred trillion microorganisms in the digestive tract, basically bacteria.*

MOLLY 2004: *Doesn't sound very appealing. Are they entirely necessary?*

RAY: *They're actually part of the society of cells that makes Molly alive and thriving. You couldn't survive without healthy gut bacteria. Assuming your intestinal flora are in good balance, they're necessary for your well-being.*

MOLLY 2004: *Okay, but I wouldn't count them as me. There are lots of things that my well-being depends on. Like my house and my car, but I still don't count them as part of me.*

RAY: *Very well, it's reasonable to leave out the entire contents of the GI tract, bacteria and all. That's actually how the body sees it. Even though it's physically inside the body, the body considers the tract to be external and carefully screens what it absorbs into the bloodstream.*

MOLLY 2004: *As I think more about who I am, I kind of like Jaron Lanier's "circle of empathy."*

RAY: *Tell me more.*

MOLLY 2004: *Basically, the circle of reality that I consider to be "me" is not clearcut. It's not simply my body. I have limited identification with, say, my toes and, after our last discussion, even less with the contents of my large intestine.*

RAY: *That's reasonable, and even with regard to our brains we are aware of only a tiny portion of what goes on in there.*

MOLLY 2004: *It's true that there are parts of my brain that seem to be somebody else, or at least somewhere else. Often, thoughts and dreams that intrude on my awareness seem to have come from some foreign place. They're obviously coming from my brain, but it doesn't seem that way.*

RAY: *Conversely, loved ones who are physically separate may be so close as to seem to be part of ourselves.*

MOLLY 2004: *The boundary of myself is seeming less and less clear.*

RAY: *Well, just wait until we're predominantly nonbiological. Then we'll be able to merge our thoughts and thinking at will, so finding boundaries will be even more difficult.*

MOLLY 2004: *That actually sounds kind of appealing. You know, some Buddhist philosophies emphasize the extent to which there is inherently no boundary at all between us.*

RAY: *Sounds like they're talking about the Singularity.*

The Singularity as Transcendence

Modernity sees humanity as having ascended from what is inferior to it—life begins in slime and ends in intelligence—whereas traditional cultures see it as descended from its superiors. As the anthropologist Marshall Sahlins puts the matter: "We are the only people who assume that we have ascended from apes. Everybody else takes it for granted that they are descended from gods."

—HUSTON SMITH[16]

Some philosophers hold that philosophy is what you do to a problem until it's clear enough to solve it by doing science. Others hold that if a philosophical problem succumbs to empirical methods, that shows it wasn't really philosophical to begin with.

—JERRY A. FODOR[17]

The Singularity denotes an event that will take place in the material world, the inevitable next step in the evolutionary process that started with biological evolution and has extended through human-directed technological evolution. However, it is precisely in the world of matter and energy that we encounter transcendence, a principal connotation of what people refer to as spirituality. Let's consider the nature of spirituality in the physical world.

Where shall I start? How about with water? It's simple enough, but consider the diverse and beautiful ways it manifests itself: the endlessly varying patterns as it cascades past rocks in a stream, then surges chaotically down a waterfall (all viewable from my office window, incidentally); the billowing patterns of clouds in the sky; the arrangement of snow on a mountain; the satisfying design of a single snowflake. Or consider Einstein's description of the entangled order and disorder in a glass of water (that is, his thesis on Brownian motion).

Or elsewhere in the biological world, consider the intricate dance of spirals of DNA during mitosis. How about the loveliness of a tree as it bends in the wind and its leaves churn in a tangled dance? Or the bustling world we see in a microscope? There's transcendence everywhere.

A comment on the word "transcendence" is in order here. "To transcend" means "to go beyond," but this need not compel us to adopt an ornate dualist view that regards transcendent levels of reality (such as the spiritual level) to be not of this world. We can "go beyond" the "ordinary" powers of the material world through the power of patterns. Although I have been called a materialist, I regard myself as a "patternist." It's through the emergent powers of the pattern that we transcend. Since the material stuff of which we are made turns over quickly, it is the transcendent power of our patterns that persists.

The power of patterns to endure goes beyond explicitly self-replicating systems, such as organisms and self-replicating technology. It is the persistence and power of patterns that support life and intelligence. The pattern is far more important than the material stuff that constitutes it.

Random strokes on a canvas are just paint. But when arranged in just the right way, they transcend the material stuff and become art. Random notes are just sounds. Sequenced in an "inspired" way, we have music. A pile of components is just an inventory. Ordered in an innovative manner, and perhaps with the addition of some software (another pattern), we have the "magic" (transcendence) of technology.

Although some regard what is referred to as "spiritual" as the true meaning of transcendence, transcendence refers to all levels of reality: the creations of the natural world, including ourselves, as well as our own creations in the form of art, culture, technology, and emotional and spiritual expression. Evolution concerns patterns, and it is specifically the depth and order of patterns that grow in an evolutionary process. As a consummation of the evolution in our midst, the Singularity will deepen all of these manifestations of transcendence.

Another connotation of the word "spiritual" is "containing spirit," which is to say being conscious. Consciousness—the seat of "personalness"—is regarded as what is real in many philosophical and religious traditions. A common Buddhist ontology considers subjective—conscious—experience as the ultimate reality, rather than physical or objective phenomena, which are considered *maya* (illusion).

The arguments I make in this book with regard to consciousness are for the purpose of illustrating this vexing and paradoxical (and, therefore, profound) nature of consciousness: how one set of assumptions (that is, that a copy of my mind file either shares or does not share my consciousness) leads ultimately to an opposite view, and vice versa.

We do assume that humans are conscious, at least when they appear to be. At the other end of the spectrum we assume that simple machines are not. In the cosmological sense the contemporary universe acts more like a simple machine than a conscious being. But as we discussed in the previous chapter, the matter and energy in our vicinity will become infused with the intelligence, knowledge, creativity, beauty, and emotional intelligence (the ability to love, for example) of our human-machine civilization. Our civilization will then expand outward, turning all the dumb matter and energy we encounter into sublimely intelligent—transcendent—matter and energy. So in a sense, we can say that the Singularity will ultimately infuse the universe with spirit.

Evolution moves toward greater complexity, greater elegance, greater knowledge, greater intelligence, greater beauty, greater creativity, and greater levels of subtle attributes such as love. In every monotheistic tradition God is likewise described as all of these qualities, only without any limitation: infinite knowledge, infinite intelligence, infinite beauty, infinite creativity, infinite love, and so on. Of course, even the accelerating growth of evolution never achieves an infinite level, but as it explodes exponentially it certainly moves rapidly in that direction. So evolution moves inexorably toward this conception of God, although never quite reaching this ideal. We can regard, therefore, the freeing of our thinking from the severe limitations of its biological form to be an essentially spiritual undertaking.

MOLLY 2004: *So, do you believe in God?*

RAY: *Well, it's a three-letter word—and a powerful meme.*

MOLLY 2004: *I realize the word and the idea exist. But does it refer to anything that you believe in?*

RAY: *People mean lots of things by it.*

MOLLY 2004: *Do you believe in those things?*

RAY: *It's not possible to believe all these things: God is an all-powerful conscious person looking over us, making deals, and getting angry quite a bit. Or He—It—is a pervasive life force underlying all beauty and creativity. Or God created everything and then stepped back. . . .*

MOLLY 2004: *I understand, but do you believe in any of them?*

RAY: *I believe that the universe exists.*

MOLLY 2004: *Now wait a minute, that's not a belief, that's a scientific fact.*

RAY: *Actually, I don't know for sure that anything exists other than my own thoughts.*

MOLLY 2004: *Okay, I understand that this is the philosophy chapter, but you can read scientific papers—thousands of them—that corroborate the existence of stars and galaxies. So, all those galaxies—we call that the universe.*

RAY: *Yes, I've heard of that, and I do recall reading some of these papers, but I don't know that those papers really exist, or that the things they refer to really exist, other than in my thoughts.*

MOLLY 2004: *So you don't acknowledge the existence of the universe?*

RAY: *No, I just said that I do believe that it exists, but I'm pointing out that it's a belief. That's my personal leap of faith.*

MOLLY 2004: *All right, but I asked whether you believed in God.*

RAY: *Again, "God" is a word by which people mean different things. For the sake of your question, we can consider God to be the universe, and I said that I believe in the existence of the universe.*

MOLLY 2004: *God is just the universe?*

RAY: *Just? It's a pretty big thing to apply the word "just" to. If we are to believe what science tells us—and I said that I do—it's about as big a phenomenon as we could imagine.*

MOLLY 2004: *Actually, many physicists now consider our universe to be just one bubble among a vast number of other universes. But I meant that people usually mean something more by the word "God" than "just" the material world. Some people do associate God with everything that exists, but they still consider God to be conscious. So you believe in a God that's not conscious?*

RAY: *The universe is not conscious—yet. But it will be. Strictly speaking, we should say that very little of it is conscious today. But that will change and soon. I expect that the universe will become sublimely intelligent and will wake up in Epoch Six. The only belief I am positing here is that the universe exists. If we make that leap of faith, the expectation that it will wake up is not so much a belief as an informed understanding, based on the same science that says there is a universe.*

MOLLY 2004: *Interesting. You know, that's essentially the opposite of the view that there was a conscious creator who got everything started and then kind of bowed out. You're basically saying that a conscious universe will "bow in" during Epoch Six.*

RAY: *Yes, that's the essence of Epoch Six.*

CHAPTER EIGHT

The Deeply Intertwined Promise and Peril of GNR

We are being propelled into this new century with no plan, no control, no brakes. . . . The only realistic alternative I see is relinquishment: to limit development of the technologies that are too dangerous, by limiting our pursuit of certain kinds of knowledge.

—BILL JOY, "WHY THE FUTURE DOESN'T NEED US"

Environmentalists must now grapple squarely with the idea of a world that has enough wealth and enough technological capability, and should not pursue more.

—BILL MCKIBBEN, ENVIRONMENTALIST WHO FIRST WROTE ABOUT GLOBAL WARMING[1]

Progress might have been all right once, but it has gone on too long.

—OGDEN NASH (1902–1971)

In the late 1960s I was transformed into a radical environmental activist. A rag-tag group of activists and I sailed a leaky old halibut boat across the North Pacific to block the last hydrogen bomb tests under President Nixon. In the process I co-founded Greenpeace. . . . Environmentalists were often able to produce arguments that sounded reasonable, while doing good deeds like saving whales and making the air and water cleaner. But now the chickens have come home to roost. The environmentalists' campaign against biotechnology in general, and genetic engineering in particular, has clearly exposed their intellectual and moral bankruptcy. By adopting a zero tolerance policy toward a technology with so many potential benefits for humankind and the environment, they . . . have alienated themselves from scientists, intellectuals, and internationalists. It seems inevitable that the media and the public will, in time, see the insanity of their position.

—PATRICK MOORE

I think that . . . flight from and hatred of technology is self-defeating. The Buddha rests quite as comfortably in the circuits of a digital computer and the gears of a cycle transmission as he does at the top of a mountain or in the petals of a flower. To think otherwise is to demean the Buddha—which is to demean oneself.

—ROBERT M. PIRSIG, *ZEN AND THE ART OF MOTORCYCLE MAINTENANCE*

Consider these articles we'd rather not see available on the Web:

Impress Your Enemies: How to Build Your Own Atomic Bomb from Readily Available Materials[2]

How to Modify the Influenza Virus in Your College Laboratory to Release Snake Venom

Ten Easy Modifications to the *E. coli* Virus

How to Modify Smallpox to Counteract the Smallpox Vaccine

Build Your Own Chemical Weapons from Materials Available on the Internet

How to Build a Pilotless, Self-Guiding, Low-Flying Airplane Using a Low-Cost Aircraft, GPS, and a Notebook Computer

Or, how about the following:

The Genomes of Ten Leading Pathogens

The Floor Plans of Leading Skyscrapers

The Layout of U.S. Nuclear Reactors

The Hundred Top Vulnerabilities of Modern Society

The Top Ten Vulnerabilities of the Internet

Personal Health Information on One Hundred Million Americans

The Customer Lists of Top Pornography Sites

Anyone posting the first item above is almost certain to get a quick visit from the FBI, as did Nate Ciccolo, a fifteen-year-old high school student, in March 2000. For a school science project he built a papier-mâché model of an atomic bomb that turned out to be disturbingly accurate. In the ensuing media storm Ciccolo told ABC News, "Someone just sort of mentioned, you know, you can go on the Internet now and get information. And I, sort of, wasn't exactly up to date on things. Try it. I went on there and a couple of clicks and I was right there."[3]

Of course Ciccolo didn't possess the key ingredient, plutonium, nor did he have any intention of acquiring it, but the report created shock waves in the media, not to mention among the authorities who worry about nuclear proliferation. Ciccolo had reported finding 563 Web pages on atomic-bomb designs, and the publicity resulted in an urgent effort to remove them. Unfortunately, trying to get rid of information on the Internet is akin to trying to sweep back the ocean with a broom. Some of the sites continue to be easily accessible today. I won't provide any URLs in this book, but they are not hard to find.

Although the article titles above are fictitious, one can find extensive information on the Internet about all of these topics.[4] The Web is an extraordinary research tool. In my own experience, research that used to require a half day at the library can now be accomplished typically in a couple of minutes or less. This has enormous and obvious benefits for advancing beneficial technologies, but it can also empower those whose values are inimical to the mainstream of society. So are we in danger? The answer is clearly yes. How much danger, and what to do about it, are the subjects of this chapter.

My urgent concern with this issue dates back at least a couple of decades. When I wrote *The Age of Intelligent Machines* in the mid-1980s, I was deeply concerned with the ability of then-emerging genetic engineering to enable those skilled in the art and with access to fairly widely available equipment to modify bacterial and viral pathogens to create new diseases.[5] In destructive or merely careless hands these engineered pathogens could potentially combine a high degree of communicability, stealthiness, and destructiveness.

Such efforts were not easy to carry out in the 1980s but were nonetheless feasible. We now know that bioweapons programs in the Soviet Union and elsewhere were doing exactly this.[6] At the time I made a conscious decision to not talk about this specter in my book, feeling that I did not want to give the wrong people any destructive ideas. I didn't want to turn on the radio one day and hear about a disaster, with the perpetrators saying that they got the idea from Ray Kurzweil.

Partly as a result of this decision I faced some reasonable criticism that the book emphasized the benefits of future technology while ignoring its pitfalls. When I wrote *The Age of Spiritual Machines* in 1997–1998, therefore, I attempted to account for both promise and peril.[7] There had been sufficient public attention by that time (for example, the 1995 movie *Outbreak*, which portrays the terror and panic from the release of a new viral pathogen) that I felt comfortable to begin to address the issue publicly.

In September 1998, having just completed the manuscript, I ran into Bill Joy, an esteemed and longtime colleague in the high-technology world, in a bar in Lake Tahoe. Although I had long admired Joy for his work in pioneering the leading software language for interactive Web systems (Java) and having co-founded Sun Microsystems, my focus at this brief get-together was not on Joy but rather on the third person sitting in our small booth, John Searle. Searle, the eminent philosopher from the University of California at Berkeley, had built a career of defending the deep mysteries of human consciousness from apparent attack by materialists such as Ray Kurzweil (a characterization I reject in the next chapter).

Searle and I had just finished debating the issue of whether a machine could be conscious during the closing session of George Gilder's Telecosm conference. The session was entitled "Spiritual Machines" and was devoted to a discussion of the philosophical implications of my upcoming book. I had given Joy a preliminary manuscript and tried to bring him up to speed on the debate about consciousness that Searle and I were having.

As it turned out, Joy was interested in a completely different issue, specifically the impending dangers to human civilization from three emerging technologies I had presented in the book: genetics, nanotechnology, and robotics (GNR, as discussed earlier). My discussion of the downsides of future technology alarmed Joy, as he would later relate in his now-famous cover story for *Wired*, "Why the Future Doesn't Need Us."[8] In the article Joy describes how he asked his friends in the scientific and technology community whether the projections I was making were credible and was dismayed to discover how close these capabilities were to realization.

Joy's article focused entirely on the downside scenarios and created a firestorm. Here was one of the technology world's leading figures addressing new and dire emerging dangers from future technology. It was reminiscent of the attention that George Soros, the currency arbitrageur and archcapitalist, received when he made vaguely critical comments about the excesses of unrestrained capitalism, although the Joy controversy became far more intense. The *New York Times* reported there were about ten thousand articles commenting on and discussing Joy's article, more than any other in the history of commentary on technology issues. My attempt to relax in a Lake Tahoe lounge thus ended up fostering two long-term debates, as my dialogue with John Searle has also continued to this day.

Despite my being the origin of Joy's concern, my reputation as a "technology optimist" has remained intact, and Joy and I have been invited to a variety of forums to debate the peril and promise, respectively, of future technologies. Although I am expected to take up the "promise" side of the debate, I often end up spending most of my time defending his position on the feasibility of these dangers.

Many people have interpreted Joy's article as an advocacy of broad relinquishment, not of all technological developments, but of the "dangerous ones" like nanotechnology. Joy, who is now working as a venture capitalist with the legendary Silicon Valley firm of Kleiner, Perkins, Caufield & Byers, investing in technologies such as nanotechnology applied to renewable energy and other natural resources, says that broad relinquishment is a misinterpretation of his position and was never his intent. In a recent private e-mail communication, he says the emphasis should be on his call to "limit development of the technologies that are too dangerous" (see the epigraph at the beginning of this chapter), not on complete prohibition. He suggests, for example, a prohibition against self-replicating nanotechnology, which is similar to the guidelines advocated by the Foresight Institute, founded by nanotechnology pioneer Eric Drexler and Christine Peterson. Overall, this is a reasonable guideline, although I believe there will need to be two exceptions, which I discuss below (see p. 411).

As another example, Joy advocates not publishing the gene sequences of pathogens on the Internet, which I also agree with. He would like to see scientists adopt regulations along these lines voluntarily and internationally, and he points out that "if we wait until after a catastrophe, we may end up with more severe and damaging regulations." He says he hopes that "we will do such regulation lightly, so that we can get most of the benefits."

Others, such as Bill McKibben, the environmentalist who was one of the first to warn against global warming, have advocated relinquishment of broad areas such as biotechnology and nanotechnology, or even of all technology. As I discuss in greater detail below (see p. 410), relinquishing broad fields would be impossible to achieve without essentially relinquishing all technical development. That in turn would require a *Brave New World* style of totalitarian government, banning all technology development. Not only would such a solution be inconsistent with our democratic values, but it would actually make the dangers worse by driving the technology underground, where only the least responsible practitioners (for example, rogue states) would have most of the expertise.

Intertwined Benefits . . .

> It was the best of times, it was the worst of times, it was the age of wisdom, it was the age of foolishness, it was the epoch of belief, it was the epoch of incredulity, it was the season of Light, it was the season of Darkness, it was the spring of hope, it was the winter of despair, we had everything before us, we had nothing before us, we were all going direct to Heaven, we were all going direct the other way.
>
> —CHARLES DICKENS, *A TALE OF TWO CITIES*

> It's like arguing in favor of the plough. You know some people are going to argue against it, but you also know it's going to exist.
>
> —JAMES HUGHES, SECRETARY OF THE TRANSHUMANIST ASSOCIATION AND SOCIOLOGIST AT TRINITY COLLEGE, IN A DEBATE, "SHOULD HUMANS WELCOME OR RESIST BECOMING POSTHUMAN?"

Technology has always been a mixed blessing, bringing us benefits such as longer and healthier lifespans, freedom from physical and mental drudgery, and many novel creative possibilities on the one hand, while introducing new dangers. Technology empowers both our creative and destructive natures.

Substantial portions of our species have already experienced alleviation of the poverty, disease, hard labor, and misfortune that have characterized much of human history. Many of us now have the opportunity to gain satisfaction and meaning from our work, rather than merely toiling to survive. We have ever more powerful tools to express ourselves. With the Web now reaching deeply into less developed regions of the world, we will see major strides in the availability of high-quality education and medical knowledge. We can share culture, art, and humankind's exponentially expanding knowledge base worldwide. I mentioned the World Bank's report on the worldwide reduction in poverty in chapter 2 and discuss that further in the next chapter.

We've gone from about twenty democracies in the world after World War II to more than one hundred today largely through the influence of decentralized electronic communication. The biggest wave of democratization, including the fall of the Iron Curtain, occurred during the 1990s with the growth of the Internet and related technologies. There is, of course, a great deal more to accomplish in each of these areas.

Bioengineering is in the early stages of making enormous strides in reversing disease and aging processes. Ubiquitous N and R are two to three decades away and will continue an exponential expansion of these benefits. As I reviewed in earlier chapters, these technologies will create extraordinary wealth, thereby overcoming poverty and enabling us to provide for all of our material needs by transforming inexpensive raw materials and information into any type of product.

We will spend increasing portions of our time in virtual environments and will be able to have any type of desired experience with anyone, real or simulated, in virtual reality. Nanotechnology will bring a similar ability to morph the physical world to our needs and desires. Lingering problems from our waning industrial age will be overcome. We will be able to reverse remaining environmental destruction. Nanoengineered fuel cells and solar cells will provide clean energy. Nanobots in our physical bodies will destroy pathogens, remove debris such as misformed proteins and protofibrils, repair DNA, and reverse aging. We will be able to redesign all of the systems in our bodies and brains to be far more capable and durable.

Most significant will be the merger of biological and nonbiological intelligence, although nonbiological intelligence will quickly come to predominate. There will be a vast expansion of the concept of what it means to be human. We will greatly enhance our ability to create and appreciate all forms of knowledge from science to the arts, while extending our ability to relate to our environment and one another.

On the other hand . . .

. . . and Dangers

> "Plants" with "leaves" no more efficient than today's solar cells could out-compete real plants, crowding the biosphere with an inedible foliage. Tough omnivorous "bacteria" could out-compete real bacteria: They could spread like blowing pollen, replicated swiftly, and reduce the biosphere to dust in a matter of days. Dangerous replicators could easily be too tough, small, and rapidly spreading to stop—at least if we make no preparation. We have trouble enough controlling viruses and fruit flies.
>
> —ERIC DREXLER

As well as its many remarkable accomplishments, the twentieth century saw technology's awesome ability to amplify our destructive nature, from Stalin's tanks to Hitler's trains. The tragic event of September 11, 2001, is another example of technologies (jets and buildings) taken over by people with agendas of destruction. We still live today with a sufficient number of nuclear weapons (not all of which are accounted for) to end all mammalian life on the planet.

Since the 1980s the means and knowledge have existed in a routine college bioengineering lab to create unfriendly pathogens potentially more danger-ous than nuclear weapons.[9] In a war-game simulation conducted at Johns Hopkins University called "Dark Winter," it was estimated that an intentional introduction of conventional smallpox in three U.S. cities could result in one million deaths. If the virus were bioengineered to defeat the existing smallpox vaccine, the results could be far worse.[10] The reality of this specter was made clear by a 2001 experiment in Australia in which the mousepox virus was inad-vertently modified with genes that altered the immune-system response. The mousepox vaccine was powerless to stop this altered virus.[11] These dangers resonate in our historical memories. Bubonic plague killed one third of the European population. More recently the 1918 flu killed twenty million people worldwide.[12]

Will such threats prevent the ongoing acceleration of the power, efficiency, and intelligence of complex systems (such as humans and our technology)? The past record of complexity increase on this planet has shown a smooth acceleration, even through a long history of catastrophes, both internally gen-erated and externally imposed. This is true of both biological evolution (which faced calamities such as encounters with large asteroids and meteors) and human history (which has been punctuated by an ongoing series of major wars).

However, I believe we can take some encouragement from the effectiveness of the world's response to the SARS (severe acute respiratory syndrome) virus. Although the possibility of an even more virulent return of SARS remains uncertain as of the writing of this book, it appears that containment measures have been relatively successful and have prevented this tragic outbreak from becoming a true catastrophe. Part of the response involved ancient, low-tech tools such as quarantine and face masks.

However, this approach would not have worked without advanced tools that have only recently become available. Researchers were able to sequence the DNA of the SARS virus within thirty-one days of the outbreak—compared to fifteen years for HIV. That enabled the rapid development of an effective test so that carriers could quickly be identified. Moreover, instantaneous global com-munication facilitated a coordinated response worldwide, a feat not possible when viruses ravaged the world in ancient times.

As technology accelerates toward the full realization of GNR, we will see the same intertwined potentials: a feast of creativity resulting from human intelli-gence expanded manyfold, combined with many grave new dangers. A quintes-sential concern that has received considerable attention is unrestrained nanobot replication. Nanobot technology requires trillions of such intelligently designed devices to be useful. To scale up to such levels it will be necessary to enable them to self-replicate, essentially the same approach used in the biological world (that's how one fertilized egg cell becomes the trillions of cells in a human). And in the same way that biological self-replication gone awry (that is, cancer) results in biological destruction, a defect in the mechanism curtailing nanobot self-replication—the so-called gray-goo scenario—would endanger all physical entities, biological or otherwise.

Living creatures—including humans—would be the primary victims of an exponentially spreading nanobot attack. The principal designs for nanobot construction use carbon as a primary building block. Because of carbon's unique ability to form four-way bonds, it is an ideal building block for molecular assemblies. Carbon molecules can form straight chains, zigzags, rings, nanotubes (hexagonal arrays formed in tubes), sheets, buckyballs (arrays of hexagons and pentagons formed into spheres), and a variety of other shapes. Because biology has made the same use of carbon, pathological nanobots would find the Earth's biomass an ideal source of this primary ingredient. Biological entities can also provide stored energy in the form of glucose and ATP.[13] Useful trace elements such as oxygen, sulfur, iron, calcium, and others are also available in the biomass.

How long would it take an out-of-control replicating nanobot to destroy the Earth's biomass? The biomass has on the order of 10^{45} carbon atoms.[14] A reasonable estimate of the number of carbon atoms in a single replicating nanobot is about 10^6. (Note that this analysis is not very sensitive to the accuracy of these figures, only to the approximate order of magnitude.) This malevolent nanobot would need to create on the order of 10^{39} copies of itself to replace the biomass, which could be accomplished with 130 replications (each of which would potentially double the destroyed biomass). Rob Freitas has estimated a minimum replication time of approximately one hundred seconds, so 130 replication cycles would require about three and a half hours.[15] However, the actual rate of destruction would be slower because biomass is not "efficiently" laid out. The limiting factor would be the actual movement of the front of destruction. Nanobots cannot travel very quickly because of their small size. It's likely to take weeks for such a destructive process to circle the globe.

Based on this observation we can envision a more insidious possibility. In a two-phased attack, the nanobots take several weeks to spread throughout the biomass but use up an insignificant portion of the carbon atoms, say one out of every thousand trillion (10^{15}). At this extremely low level of concentration the nanobots would be as stealthy as possible. Then, at an "optimal" point, the second phase would begin with the seed nanobots expanding rapidly in place to destroy the biomass. For each seed nanobot to multiply itself a thousand trillionfold would require only about fifty binary replications, or about ninety minutes. With the nanobots having already spread out in position throughout the biomass, movement of the destructive wave front would no longer be a limiting factor.

The point is that without defenses, the available biomass could be destroyed by gray goo very rapidly. As I discuss below (see p. 417), we will clearly need a nanotechnology immune system in place before these scenarios become a possibility. This immune system would have to be capable of contending not just with obvious destruction but with any potentially dangerous (stealthy) replication, even at very low concentrations.

Mike Treder and Chris Phoenix—executive director and director of research of the Center for Responsible Nanotechnology, respectively—Eric Drexler, Robert Freitas, Ralph Merkle, and others have pointed out that future MNT manufacturing devices can be created with safeguards that would prevent the creation of self-replicating nanodevices.[16] I discuss some of these strategies below. However, this observation, although important, does not eliminate the specter of gray goo. There are other reasons (beyond manufacturing) that self-replicating nanobots will need to be created. The nanotechnology immune system mentioned above, for example, will ultimately require self-replication; otherwise it would be unable to defend us. Self-replication will also be necessary for nanobots to rapidly expand intelligence beyond the Earth, as I discussed in chapter 6. It is also likely to find extensive military applications. Moreover, safeguards against unwanted self-replication, such as the broadcast architecture described below (see p. 412), can be defeated by a determined adversary or terrorist.

Freitas has identified a number of other disastrous nanobot scenarios.[17] In what he calls the "gray plankton" scenario, malicious nanobots would use underwater carbon stored as CH_4 (methane) as well as CO_2 dissolved in seawater. These ocean-based sources can provide about ten times as much carbon as Earth's biomass. In his "gray dust" scenario, replicating nanobots use basic elements available in airborne dust and sunlight for power. The "gray lichens" scenario involves using carbon and other elements on rocks.

A Panoply of Existential Risks

> If a little knowledge is dangerous, where is a person who has so much as to be out of danger?
>
> —THOMAS HENRY

I discuss below (see the section "A Program for GNR Defense," p. 422) steps we can take to address these grave risks, but we cannot have complete assurance in any strategy that we devise today. These risks are what Nick Bostrom calls "existential risks," which he defines as the dangers in the upper-right quadrant of the following table:[18]

Bostrom's Categorization of Risks

		Intensity of Risk	
		Moderate	**Profound**
	Global	Ozone Thinning	***Existential Risks***
Scope	**Local**	Recession	Genocide
	Personal	Stolen Car	Death
		Endurable	**Terminal**

Biological life on Earth encountered a human-made existential risk for the first time in the middle of the twentieth century with the advent of the hydrogen bomb and the subsequent cold-war buildup of thermonuclear forces. President Kennedy reportedly estimated that the likelihood of an all-out nuclear war during the Cuban missile crisis was between 33 and 50 percent.[19] The legendary information theorist John von Neumann, who became the chairman of the Air Force Strategic Missiles Evaluation Committee and a government adviser on nuclear strategies, estimated the likelihood of nuclear Armageddon (prior to the Cuban missile crisis) at close to 100 percent.20 Given the perspective of the 1960s what informed observer of those times would have predicted that the world would have gone through the next forty years without another nontest nuclear explosion?

Despite the apparent chaos of international affairs we can be grateful for the successful avoidance thus far of the employment of nuclear weapons in war. But we clearly cannot rest easily, since enough hydrogen bombs still exist to destroy all human life many times over.[21] Although attracting relatively little public discussion, the massive opposing ICBM arsenals of the United States and Russia remain in place, despite the apparent thawing of relations.

Nuclear proliferation and the widespread availability of nuclear materials and know-how is another grave concern, although not an existential one for our civilization. (That is, only an all-out thermonuclear war involving the ICBM arsenals poses a risk to survival of all humans.) Nuclear proliferation and nuclear terrorism belong to the "profound-local" category of risk, along with genocide. However, the concern is certainly severe because the logic of mutual assured destruction does not work in the context of suicide terrorists.

Debatably we've now added another existential risk, which is the possibility of a bioengineered virus that spreads easily, has a long incubation period, and delivers an ultimately deadly payload. Some viruses are easily communicable, such as the flu and common cold. Others are deadly, such as HIV. It is rare for a virus to combine both attributes. Humans living today are descendants of those who developed natural immunities to most of the highly communicable viruses. The ability of the species to survive viral outbreaks is one advantage of sexual reproduction, which tends to ensure genetic diversity in the population, so that the response to specific viral agents is highly variable. Although catastrophic, bubonic plague did not kill everyone in Europe. Other viruses, such as smallpox, have both negative characteristics—they are easily contagious and deadly—but have been around long enough that there has been time for society to create a technological protection in the form of a vaccine. Gene engineering, however, has the potential to bypass these evolutionary protections by suddenly introducing new pathogens for which we have no protection, natural or technological.

The prospect of adding genes for deadly toxins to easily transmitted, common viruses such as the common cold and flu introduced another possible existential-risk scenario. It was this prospect that led to the Asilomar conference to consider how to deal with such a threat and the subsequent drafting of a set of safety and ethics guidelines. Although these guidelines have worked thus far, the underlying technologies for genetic manipulation are growing rapidly in sophistication.

In 2003 the world struggled, successfully, with the SARS virus. The emergence of SARS resulted from a combination of an ancient practice (the virus is suspected of having jumped from exotic animals, possibly civet cats, to humans living in close proximity) and a modern practice (the infection spread rapidly across the world by air travel). SARS provided us with a dry run of a virus new to human civilization that combined easy transmission, the ability to survive for extended periods of time outside the human body, and a high degree of mortality, with death rates estimated at 14 to 20 percent. Again, the response combined ancient and modern techniques.

Our experience with SARS shows that most viruses, even if relatively easily transmitted and reasonably deadly, represent grave but not necessarily existential risks. SARS, however, does not appear to have been engineered. SARS spreads easily through externally transmitted bodily fluids but is not easily spread through airborne particles. Its incubation period is estimated to range from one day to two weeks, whereas a longer incubation period would allow a virus to spread through several exponentially growing generations before carriers are identified.[22]

SARS is deadly, but the majority of its victims do survive. It continues to be feasible for a virus to be malevolently engineered so it spreads more easily than SARS, has an extended incubation period, and is deadly to essentially all victims. Smallpox is close to having these characteristics. Although we have a vaccine (albeit a crude one), the vaccine would not be effective against genetically modified versions of the virus.

As I describe below, the window of malicious opportunity for bioengineered viruses, existential or otherwise, will close in the 2020s when we have fully effective antiviral technologies based on nanobots.[23] However, because nanotechnology will be thousands of times stronger, faster, and more intelligent than biological entities, self-replicating nanobots will present a greater risk and yet another existential risk. The window for malevolent nanobots will ultimately be closed by strong artificial intelligence, but, not surprisingly, "unfriendly" AI will itself present an even more compelling existential risk, which I discuss below (see p. 420).

The Precautionary Principle. As Bostrom, Freitas, and other observers including myself have pointed out, we cannot rely on trial-and-error approaches to deal with existential risks. There are competing interpretations of what has become known as the "precautionary principle." (If the consequences of an action are unknown but judged by some scientists to have even a small risk of being profoundly negative, it's better to not carry out the action than risk negative consequences.) But it's clear that we need to achieve the highest possible level of confidence in our strategies to combat such risks. This is one reason we're hearing increasingly strident voices demanding that we shut down the advance of technology, as a primary strategy to eliminate new existential risks before they occur. Relinquishment, however, is not the appropriate response and will only interfere with the profound benefits of these emerging technologies while actually increasing the likelihood of a disastrous outcome. Max More articulates the limitations of the precautionary principle and advocates replacing it with what he calls the "proactionary principle," which involves balancing the risks of action and inaction.[24]

Before discussing how to respond to the new challenge of existential risks, it's worth reviewing a few more that have been postulated by Bostrom and others.

The Smaller the Interaction, the Larger the Explosive Potential. There has been recent controversy over the potential for future very high-energy particle accelerators to create a chain reaction of transformed energy states at a subatomic level. The result could be an exponentially spreading area of destruction, breaking apart all atoms in our galactic vicinity. A variety of such scenarios has been proposed, including the possibility of creating a black hole that would draw in our solar system.

Analyses of these scenarios show them to be very unlikely, although not all physicists are sanguine about the danger.[25] The mathematics of these analyses appears to be sound, but we do not yet have a consensus on the formulas that describe this level of physical reality. If such dangers sound far-fetched, consider the possibility that we have indeed detected increasingly powerful explosive phenomena at diminishing scales of matter.

Alfred Nobel discovered dynamite by probing chemical interactions of molecules. The atomic bomb, which is tens of thousands of times more powerful than dynamite, is based on nuclear interactions involving large atoms, which are much smaller scales of matter than large molecules. The hydrogen bomb, which is thousands of times more powerful than an atomic bomb, is based on interactions involving an even smaller scale: small atoms. Although this insight does not necessarily imply the existence of yet more powerful destructive chain reactions by manipulating subatomic particles, it does make the conjecture plausible.

My own assessment of this danger is that we are unlikely simply to stumble across such a destructive event. Consider how unlikely it would be to accidentally produce an atomic bomb. Such a device requires a precise configuration of materials and actions, and the original required an extensive and precise engineering project to develop. Inadvertently creating a hydrogen bomb would be even less plausible. One would have to create the precise conditions of an atomic bomb in a particular arrangement with a hydrogen core and other elements. Stumbling across the exact conditions to create a new class of catastrophic chain reaction at a subatomic level appears to be even less likely. The consequences are sufficiently devastating, however, that the precautionary principle should lead us to take these possibilities seriously. This potential should be carefully analyzed prior to carrying out new classes of accelerator experiments. However, this risk is not high on my list of twenty-first-century concerns.

Our Simulation Is Turned Off. Another existential risk that Bostrom and others have identified is that we're actually living in a simulation and the simulation will be shut down. It might appear that there's not a lot we could do to influence this. However, since we're the subject of the simulation, we do have the opportunity to shape what happens inside of it. The best way we could avoid being shut down would be to be interesting to the observers of the simulation. Assuming that someone is actually paying attention to the simulation, it's a fair assumption that it's less likely to be turned off when it's compelling than otherwise.

We could spend a lot of time considering what it means for a simulation to be interesting, but the creation of new knowledge would be a critical part of this assessment. Although it may be difficult for us to conjecture what would be interesting to our hypothesized simulation observer, it would seem that the Singularity is likely to be about as absorbing as any development we could imagine and would create new knowledge at an extraordinary rate. Indeed, achieving a Singularity of exploding knowledge may be the very purpose of the simulation. Thus, assuring a "constructive" Singularity (one that avoids degenerate outcomes such as existential destruction by gray goo or dominance by a malicious AI) could be the best course to prevent the simulation from being terminated. Of course, we have every motivation to achieve a constructive Singularity for many other reasons.

If the world we're living in is a simulation on someone's computer, it's a very good one—so detailed, in fact, that we may as well accept it as our reality. In any event, it is the only reality to which we have access.

Our world appears to have a long and rich history. This means that either our world is not, in fact, a simulation or, if it is, the simulation has been going a very long time and thus is not likely to stop anytime soon. Of course it is also possible that the simulation includes evidence of a long history without the history's having actually occurred.

As I discussed in chapter 6, there are conjectures that an advanced civilization may create a new universe to perform computation (or, to put it another way, to continue the expansion of its own computation). Our living in such a universe (created by another civilization) can be considered a simulation scenario. Perhaps this other civilization is running an evolutionary algorithm on our universe (that is, the evolution we're witnessing) to create an explosion of knowledge from a technology Singularity. If that is true, then the civilization watching our universe might shut down the simulation if it appeared that a knowledge Singularity had gone awry and it did not look like it was going to occur.

This scenario is also not high on my worry list, particularly since the only strategy that we can follow to avoid a negative outcome is the one we need to follow anyway.

Crashing the Party. Another oft-cited concern is that of a large-scale asteroid or comet collision, which has occurred repeatedly in the Earth's history, and did represent existential outcomes for species at these times. This is not a peril of technology, of course. Rather, technology will protect us from this risk (certainly within one to a couple of decades). Although small impacts are a regular occurrence, large and destructive visitors from space are rare. We don't see one on the horizon, and it is virtually certain that by the time such a danger occurs, our civilization will readily destroy the intruder before it destroys us.

Another item on the existential danger list is destruction by an alien intelligence (not one that we've created). I discussed this possibility in chapter 6 and I don't see this as likely, either.

GNR: The Proper Focus of Promise Versus Peril. This leaves the GNR technologies as the primary concerns. However, I do think we also need to take seriously the misguided and increasingly strident Luddite voices that advocate reliance on broad relinquishment of technological progress to avoid the genuine dangers of GNR. For reasons I discuss below (see p. 410), relinquishment is not the answer, but rational fear could lead to irrational solutions. Delays in overcoming human suffering are still of great consequence—for example, the worsening of famine in Africa due to opposition to aid from food using GMOs (genetically modified organisms).

Broad relinquishment would require a totalitarian system to implement, and a totalitarian brave new world is unlikely because of the democratizing impact of increasingly powerful decentralized electronic and photonic communication. The advent of worldwide, decentralized communication epitomized by the Internet and cell phones has been a pervasive democratizing force. It was not Boris Yeltsin standing on a tank that overturned the 1991 coup against Mikhail Gorbachev, but rather the clandestine network of fax machines, photocopiers, video recorders, and personal computers that broke decades of totalitarian control of information.[26] The movement toward democracy and capitalism and the attendant economic growth that characterized the 1990s were all fueled by the accelerating force of these person-to-person communication technologies.

There are other questions that are nonexistential but nonetheless serious. They include "Who is controlling the nanobots?" and "Whom are the nanobots talking to?" Future organizations (whether governments or extremist groups) or just a clever individual could put trillions of undetectable nanobots in the water or food supply of an individual or of an entire population. These spybots could then monitor, influence, and even control thoughts and actions. In addition existing nanobots could be influenced through software viruses and hacking techniques. When there is software running in our bodies and brains (as we discussed, a threshold we have already passed for some people), issues of privacy and security will take on a new urgency, and countersurveillance methods of combating such intrusions will be devised.

The Inevitability of a Transformed Future. The diverse GNR technologies are progressing on many fronts. The full realization of GNR will result from hundreds of small steps forward, each benign in itself. For G we have already passed the threshold of having the means to create designer pathogens. Advances in biotechnology will continue to accelerate, fueled by the compelling ethical and economic benefits that will result from mastering the information processes underlying biology.

Nanotechnology is the inevitable end result of the ongoing miniaturization of technology of all kinds. The key features for a wide range of applications, including electronics, mechanics, energy, and medicine, are shrinking at the rate of a factor of about four per linear dimension per decade. Moreover, there is exponential growth in research seeking to understand nanotechnology and its applications. (See the graphs on nanotechnology research studies and patents on pp. 83 and 84.)

Similarly, our efforts to reverse engineer the human brain are motivated by diverse anticipated benefits, including understanding and reversing cognitive diseases and decline. The tools for peering into the brain are showing exponential gains in spatial and temporal resolution, and we've demonstrated the ability to translate data from brain scans and studies into working models and simulations.

Insights from the brain reverse-engineering effort, overall research in developing AI algorithms, and ongoing exponential gains in computing platforms make strong AI (AI at human levels and beyond) inevitable. Once AI achieves human levels, it will necessarily soar past it because it will combine the strengths of human intelligence with the speed, memory capacity, and knowledge sharing that nonbiological intelligence already exhibits. Unlike biological intelligence, nonbiological intelligence will also benefit from ongoing exponential gains in scale, capacity, and price-performance.

Totalitarian Relinquishment. The only conceivable way that the accelerating pace of advancement on all of these fronts could be stopped would be through a worldwide totalitarian system that relinquishes the very idea of progress. Even this specter would be likely to fail in averting the dangers of GNR because the resulting underground activity would tend to favor the more destructive applications. This is because the responsible practitioners that we rely on to quickly develop defensive technologies would not have easy access to the needed tools. Fortunately, such a totalitarian outcome is unlikely because the increasing decentralization of knowledge is inherently a democratizing force.

Preparing the Defenses

My own expectation is that the creative and constructive applications of these technologies will dominate, as I believe they do today. However, we need to vastly increase our investment in developing specific defensive technologies. As I discussed, we are at the critical stage today for biotechnology, and we will reach the stage where we need to directly implement defensive technologies for nanotechnology during the late teen years of this century.

We don't have to look past today to see the intertwined promise and peril of technological advancement. Imagine describing the dangers (atomic and hydrogen bombs for one thing) that exist today to people who lived a couple of hundred years ago. They would think it mad to take such risks. But how many people in 2005 would really want to go back to the short, brutish, disease-filled, poverty-stricken, disaster-prone lives that 99 percent of the human race struggled through a couple of centuries ago?[27]

We may romanticize the past, but up until fairly recently most of humanity lived extremely fragile lives in which one all-too-common misfortune could spell disaster. Two hundred years ago life expectancy for females in the record-holding country (Sweden) was roughly thirty-five years, very brief compared to the longest life expectancy today—almost eighty-five years, for Japanese women. Life expectancy for males was roughly thirty-three years, compared to the current seventy-nine years in the record-holding countries.[28] It took half the day to prepare the evening meal, and hard labor characterized most human activity. There were no social safety nets. Substantial portions of our species still live in this precarious way, which is at least one reason to continue technological progress and the economic enhancement that accompanies it. Only technology, with its ability to provide orders of magnitude of improvement in capability and affordability, has the scale to confront problems such as poverty, disease, pollution, and the other overriding concerns of society today.

People often go through three stages in considering the impact of future technology: awe and wonderment at its potential to overcome age-old problems; then a sense of dread at a new set of grave dangers that accompany these novel technologies; followed finally by the realization that the only viable and responsible path is to set a careful course that can realize the benefits while managing the dangers.

Needless to say, we have already experienced technology's downside—for example, death and destruction from war. The crude technologies of the first industrial revolution have crowded out many of the species that existed on our planet a century ago. Our centralized technologies (such as buildings, cities, airplanes, and power plants) are demonstrably insecure.

The "NBC" (nuclear, biological, and chemical) technologies of warfare have all been used or been threatened to be used in our recent past.[29] The far more powerful GNR technologies threaten us with new, profound local and existential risks. If we manage to get past the concerns about genetically altered designer pathogens, followed by self-replicating entities created through nanotechnology, we will encounter robots whose intelligence will rival and ultimately exceed our own. Such robots may make great assistants, but who's to say that we can count on them to remain reliably friendly to mere biological humans?

Strong AI. Strong AI promises to continue the exponential gains of human civilization. (As I discussed earlier, I include the nonbiological intelligence derived from our human civilization as still human.) But the dangers it presents are also profound precisely because of its amplification of intelligence. Intelligence is inherently impossible to control, so the various strategies that have been devised to control nanotechnology (for example, the "broadcast architecture" described below) won't work for strong AI. There have been discussions and proposals to guide AI development toward what Eliezer Yudkowsky calls "friendly AI"[30] (see the section "Protection from 'Unfriendly' Strong AI," p. 420). These are useful for discussion, but it is infeasible today to devise strategies that will absolutely ensure that future AI embodies human ethics and values.

Returning to the Past? In his essay and presentations Bill Joy eloquently describes the plagues of centuries past and how new self-replicating technologies, such as mutant bioengineered pathogens and nanobots run amok, may bring back long-forgotten pestilence. Joy acknowledges that technological advances, such as antibiotics and improved sanitation, have freed us from the prevalence of such plagues, and such constructive applications, therefore, need to continue. Suffering in the world continues and demands our steadfast attention. Should we tell the millions of people afflicted with cancer and other devastating conditions that we are canceling the development of all bioengineered treatments because there is a risk that these same technologies may someday be used for malevolent purposes? Having posed this rhetorical question, I realize that there is a movement to do exactly that, but most people would agree that such broad-based relinquishment is not the answer.

The continued opportunity to alleviate human distress is one key motivation for continuing technological advancement. Also compelling are the already apparent economic gains that will continue to hasten in the decades ahead. The ongoing acceleration of many intertwined technologies produces roads paved with gold. (I use the plural here because technology is clearly not a single path.) In a competitive environment it is an economic imperative to go down these roads. Relinquishing technological advancement would be economic suicide for individuals, companies, and nations.

The Idea of Relinquishment

> The major advances in civilization all but wreck the civilizations in which they occur.
> —ALFRED NORTH WHITEHEAD

This brings us to the issue of relinquishment, which is the most controversial recommendation by relinquishment advocates such as Bill McKibben. I do feel that relinquishment at the right level is part of a responsible and constructive response to the genuine perils that we will face in the future. The issue, however, is exactly this: at what level *are* we to relinquish technology?

Ted Kaczynski, who became known to the world as the Unabomber, would have us renounce all of it.[31] This is neither desirable nor feasible, and the futility of such a position is only underscored by the senselessness of Kaczynski's deplorable tactics.

Other voices, less reckless than Kaczynski's, are nonetheless likewise arguing for broad-based relinquishment of technology. McKibben takes the position that we already have sufficient technology and that further progress should end. In his latest book, *Enough: Staying Human in an Engineered Age,* he metaphorically compares technology to beer: "One beer is good, two beers may be better; eight beers, you're almost certainly going to regret."[32] That metaphor misses the point and ignores the extensive suffering that remains in the human world that we can alleviate through sustained scientific advance.

Although new technologies, like anything else, may be used to excess at times, their promise is not just a matter of adding a fourth cell phone or doubling the number of unwanted e-mails. Rather, it means perfecting the technologies to conquer cancer and other devastating diseases, creating ubiquitous wealth to overcome poverty, cleaning up the environment from the effects of the first industrial revolution (an objective articulated by McKibben), and overcoming many other age-old problems.

Broad Relinquishment. Another level of relinquishment would be to forgo only certain fields—nanotechnology, for example—that might be regarded as too dangerous. But such sweeping strokes of relinquishment are equally untenable. As I pointed out above, nanotechnology is simply the inevitable end result of the persistent trend toward miniaturization that pervades all of technology. It is far from a single centralized effort but is being pursued by a myriad of projects with many diverse goals.

One observer wrote:

A further reason why industrial society cannot be reformed . . . is that modern technology is a unified system in which all parts are dependent on one another. You can't get rid of the "bad" parts of technology and retain only the "good" parts. Take modern medicine, for example. Progress in medical science depends on progress in chemistry, physics, biology, computer science and other fields. Advanced medical treatments require expensive, high-tech equipment that can be made available only by a technologically progressive, economically rich society. Clearly you can't have much progress in medicine without the whole technological system and everything that goes with it.

The observer I am quoting here is, again, Ted Kaczynski.[33] Although one will properly resist Kaczynski as an authority, I believe he is correct on the deeply entangled nature of the benefits and risks. However, Kaczynski and I clearly part company on our overall assessment of the relative balance between the two. Bill Joy and I have had an ongoing dialogue on this issue both publicly and privately, and we both believe that technology will and should progress and that we need to be actively concerned with its dark side. The most challenging issue to resolve is the granularity of relinquishment that is both feasible and desirable.

Fine-Grained Relinquishment. I do think that relinquishment at the right level needs to be part of our ethical response to the dangers of twenty-first-century technologies. One constructive example of this is the ethical guideline proposed by the Foresight Institute: namely, that nanotechnologists agree to relinquish the development of physical entities that can self-replicate in a natural environment.[34] In my view, there are two exceptions to this guideline. First, we will ultimately need to provide a nanotechnology-based planetary immune system (nanobots embedded in the natural environment to protect against rogue self-replicating nanobots). Robert Freitas and I have discussed whether or not such an immune system would itself need to be self-replicating. Freitas writes: "A comprehensive surveillance system coupled with prepositioned resources—resources including high-capacity nonreplicating nanofactories able to churn out large numbers of nonreplicating defenders in response to specific threats—should suffice."[35] I agree with Freitas that a prepositioned immune system with the ability to augment the defenders will be sufficient in early stages. But once strong AI is merged with nanotechnology, and the ecology of nanoengineered entities becomes highly varied and complex, my own expectation is that we will find that the defending nanorobots need the ability to replicate in place quickly. The other exception is the need for self-replicating nanobot-based probes to explore planetary systems outside of our solar system.

Another good example of a useful ethical guideline is a ban on self-replicating physical entities that contain their own codes for self-replication. In what nanotechnologist Ralph Merkle calls the "broadcast architecture," such entities would have to obtain such codes from a centralized secure server, which would guard against undesirable replication.[36] The broadcast architecture is impossible in the biological world, so there's at least one way in which nanotechnology can be made safer than biotechnology. In other ways, nanotech is potentially more dangerous because nanobots can be physically stronger than protein-based entities and more intelligent.

As I described in chapter 5, we can apply a nanotechnology-based broadcast architecture to biology. A nanocomputer would augment or replace the nucleus in every cell and provide the DNA codes. A nanobot that incorporated molecular machinery similar to ribosomes (the molecules that interpret the base pairs in the mRNA outside the nucleus) would take the codes and produce the strings of amino acids. Since we could control the nanocomputer through wireless messages, we would be able to shut off unwanted replication, thereby eliminating cancer. We could produce special proteins as needed to combat disease. And we could correct the DNA errors and upgrade the DNA code. I comment further on the strengths and weaknesses of the broadcast architecture below.

Dealing with Abuse. Broad relinquishment is contrary to economic progress and ethically unjustified given the opportunity to alleviate disease, overcome poverty, and clean up the environment. As mentioned above, it would exacerbate the dangers. Regulations on safety—essentially fine-grained relinquishment—will remain appropriate.

However, we also need to streamline the regulatory process. Right now in the United States, we have a five- to ten-year delay on new health technologies for FDA approval (with comparable delays in other nations). The harm caused by holding up potential lifesaving treatments (for example, one million lives lost in the United States for each year we delay treatments for heart disease) is given very little weight against the possible risks of new therapies.

Other protections will need to include oversight by regulatory bodies, the development of technology-specific "immune" responses, and computer-assisted surveillance by law-enforcement organizations. Many people are not aware that our intelligence agencies already use advanced technologies such as automated keyword spotting to monitor a substantial flow of telephone, cable, satellite, and Internet conversations. As we go forward, balancing our cherished rights of privacy with our need to be protected from the malicious use of powerful twenty-first-century technologies will be one of many profound challenges. This is one reason such issues as an encryption "trapdoor" (in which law-enforcement authorities would have access to otherwise secure information) and the FBI's Carnivore e-mail-snooping system have been controversial.[37]

As a test case we can take a small measure of comfort from how we have dealt with one recent technological challenge. There exists today a new fully nonbiological self-replicating entity that didn't exist just a few decades ago: the computer virus. When this form of destructive intruder first appeared, strong concerns were voiced that as they became more sophisticated, software pathogens had the potential to destroy the computer-network medium in which they live. Yet the "immune system" that has evolved in response to this challenge has been largely effective. Although destructive self-replicating software entities do cause damage from time to time, the injury is but a small fraction of the benefit we receive from the computers and communication links that harbor them.

One might counter that computer viruses do not have the lethal potential of biological viruses or of destructive nanotechnology. This is not always the case; we rely on software to operate our 911 call centers, monitor patients in critical-care units, fly and land airplanes, guide intelligent weapons in our military campaigns, handle our financial transactions, operate our municipal utilities, and many other mission-critical tasks. To the extent that software viruses do not yet pose a lethal danger, however, this observation only strengthens my argument. The fact that computer viruses are not usually deadly to humans only means that more people are willing to create and release them. The vast majority of software-virus authors would not release viruses if they thought they would kill people. It also means that our response to the danger is that much less intense. Conversely, when it comes to self-replicating entities that are potentially lethal on a large scale, our response on all levels will be vastly more serious.

Although software pathogens remain a concern, the danger exists today mostly at a nuisance level. Keep in mind that our success in combating them has taken place in an industry in which there is no regulation and minimal certification for practitioners. The largely unregulated computer industry is also enormously productive. One could argue that it has contributed more to our technological and economic progress than any other enterprise in human history.

But the battle concerning software viruses and the panoply of software pathogens will never end. We are becoming increasingly reliant on mission-critical software systems, and the sophistication and potential destructiveness of self-replicating software weapons will continue to escalate. When we have software running in our brains and bodies and controlling the world's nanobot immune system, the stakes will be immeasurably greater.

The Threat from Fundamentalism. The world is struggling with an especially pernicious form of religious fundamentalism in the form of radical Islamic terrorism. Although it may appear that these terrorists have no program other than destruction, they do have an agenda that goes beyond literal interpretations of ancient scriptures: essentially, to turn the clock back on such modern ideas as democracy, women's rights, and education.

But religious extremism is not the only form of fundamentalism that represents a reactionary force. At the beginning of this chapter I quoted Patrick Moore, cofounder of Greenpeace, on his disillusionment with the movement he helped found. The issue that undermined Moore's support of Greenpeace was its total opposition to Golden Rice, a strain of rice genetically modified to contain high levels of beta-carotene, the precursor to vitamin A.[38] Hundreds of millions of people in Africa and Asia lack sufficient vitamin A, with half a million children going blind each year from the deficiency, and millions more contracting other related diseases. About seven ounces a day of Golden Rice would provide 100 percent of a child's vitamin A requirement. Extensive studies have shown that this grain, as well as many other genetically modified organisms (GMOs), is safe. For example, in 2001 the European Commission released eighty-one studies that concluded that GMOs have "not shown any new risks to human health or the environment, beyond the usual uncertainties of conventional plant breeding. Indeed, the use of more precise technology and the greater regulatory scrutiny probably make them even safer than conventional plants and foods."[39]

It is not my position that all GMOs are inherently safe; obviously safety testing of each product is needed. But the anti-GMO movement takes the position that every GMO is by its very nature hazardous, a view that has no scientific basis.

The availability of Golden Rice has been delayed by at least five years through the pressure of Greenpeace and other anti-GMO activists. Moore, noting that this delay will cause millions of additional children to go blind, quotes the grain's opponents as threatening "to rip the G.M. rice out of the fields if farmers dare to plant it." Similarly, African nations have been pressured to refuse GMO food aid and genetically modified seeds, thereby worsening conditions of famine.[40] Ultimately the demonstrated ability of technologies such as GMO to solve overwhelming problems will prevail, but the temporary delays caused by irrational opposition will nonetheless result in unnecessary suffering.

Certain segments of the environmental movement have become fundamentalist Luddites—"fundamentalist" because of their misguided attempt to preserve things as they are (or were); "Luddite" because of the reflexive stance against technological solutions to outstanding problems. Ironically it is GMO plants—many of which are designed to resist insects and other forms of blight and thereby require greatly reduced levels of chemicals, if any—that offer the best hope for reversing environmental assault from chemicals such as pesticides.

Actually my characterization of these groups as "fundamentalist Luddites" is redundant, because Ludditism is inherently fundamentalist. It reflects the idea that humanity will be better off without change, without progress. This brings us back to the idea of relinquishment, as the enthusiasm for relinquishing technology on a broad scale is coming from the same intellectual sources and activist groups that make up the Luddite segment of the environmental movement.

Fundamentalist Humanism. With G and N technologies now beginning to modify our bodies and brains, another form of opposition to progress has emerged in the form of "fundamentalist humanism": opposition to any change in the nature of what it means to be human (for example, changing our genes and taking other steps toward radical life extension). This effort, too, will ultimately fail, however, because the demand for therapies that can overcome the suffering, disease, and short lifespans inherent in our version 1.0 bodies will ultimately prove irresistible.

In the end, it is only technology—especially GNR—that will offer the leverage needed to overcome problems that human civilization has struggled with for many generations.

Development of Defensive Technologies and the Impact of Regulation

One of the reasons that calls for broad relinquishment have appeal is that they paint a picture of future dangers assuming they will be released in the context of today's unprepared world. The reality is that the sophistication and power of our defensive knowledge and technologies will grow along with the dangers. A phenomenon like gray goo (unrestrained nanobot replication) will be countered with "blue goo" ("police" nanobots that combat the "bad" nanobots). Obviously we cannot say with assurance that we will successfully avert all misuse. But the surest way to prevent development of effective defensive technologies would be to relinquish the pursuit of knowledge in a number of broad areas. We have been able to largely control harmful software-virus replication because the requisite knowledge is widely available to responsible practitioners. Attempts to restrict such knowledge would have given rise to a far less stable situation. Responses to new challenges would have been far slower, and it is likely that the balance would have shifted toward more destructive applications (such as self-modifying software viruses).

If we compare the success we have had in controlling engineered software viruses to the coming challenge of controlling engineered biological viruses, we are struck with one salient difference. As I noted above, the software industry is almost completely unregulated. The same is obviously not true for biotechnology. While a bioterrorist does not need to put his "inventions" through the FDA, we do require the scientists developing defensive technologies to follow existing regulations, which slow down the innovation process at every step. Moreover, under existing regulations and ethical standards, it is impossible to test defenses against bioterrorist agents. Extensive discussion is already under way to modify these regulations to allow for animal models and simulations to replace unfeasible human trials. This will be necessary, but I believe we will need to go beyond these steps to accelerate the development of vitally needed defensive technologies.

In terms of public policy the task at hand is to rapidly develop the defensive steps needed, which include ethical standards, legal standards, and defensive technologies themselves. It is quite clearly a race. As I noted, in the software field defensive technologies have responded quickly to innovations in the offensive ones. In the medical field, in contrast, extensive regulation slows down innovation, so we cannot have the same confidence with regard to the abuse of biotechnology. In the current environment, when one person dies in gene-therapy trials, research can be severely restricted.[41] There is a legitimate need to make biomedical research as safe as possible, but our balancing of risks is completely skewed. Millions of people desperately need the advances promised by gene therapy and other breakthrough biotechnology advances, but they appear to carry little political weight against a handful of well-publicized casualties from the inevitable risks of progress.

This risk-balancing equation will become even more stark when we consider the emerging dangers of bioengineered pathogens. What is needed is a change in public attitude in tolerance for necessary risk. Hastening defensive technologies is absolutely vital to our security. We need to streamline regulatory procedures to achieve this. At the same time we must greatly increase our investment explicitly in defensive technologies. In the biotechnology field this means the rapid development of antiviral medications. We will not have time to formulate specific countermeasures for each new challenge that comes along. We are close to developing more generalized antiviral technologies, such as RNA interference, and these need to be accelerated.

We're addressing biotechnology here because that is the immediate threshold and challenge that we now face. As the threshold for self-organizing nanotechnology approaches, we will then need to invest specifically in the development of defensive technologies in that area, including the creation of a technological immune system. Consider how our biological immune system works. When the body detects a pathogen the T cells and other immune-system cells self-replicate rapidly to combat the invader. A nanotechnology immune system would work similarly both in the human body and in the environment and would include nanobot sentinels that could detect rogue self-replicating nanobots. When a threat was detected, defensive nanobots capable of destroying the intruders would rapidly be created (eventually with self-replication) to provide an effective defensive force.

Bill Joy and other observers have pointed out that such an immune system would itself be a danger because of the potential of "autoimmune" reactions (that is, the immune-system nanobots attacking the world they are supposed to defend).[42] However this possibility is not a compelling reason to avoid the creation of an immune system. No one would argue that humans would be better off without an immune system because of the potential of developing autoimmune diseases. Although the immune system can itself present a danger, humans would not last more than a few weeks (barring extraordinary efforts at isolation) without one. And even so, the development of a technological immune system for nanotechnology will happen even without explicit efforts to create one. This has effectively happened with regard to software viruses, creating an immune system not through a formal grand-design project but rather through incremental responses to each new challenge and by developing heuristic algorithms for early detection. We can expect the same thing will happen as challenges from nanotechnology-based dangers emerge. The point for public policy will be to invest specifically in these defensive technologies.

It is premature today to develop specific defensive nanotechnologies, since we can now have only a general idea of what we are trying to defend against. However, fruitful dialogue and discussion on anticipating this issue are already taking place, and significantly expanded investment in these efforts is to be encouraged. As I mentioned above, the Foresight Institute, as one example, has devised a set of ethical standards and strategies for assuring the development of safe nanotechnology, based on guidelines for biotechnology.[43] When gene-splicing began in 1975 two biologists, Maxine Singer and Paul Berg, suggested a moratorium on the technology until safety concerns could be addressed. It seemed apparent that there was substantial risk if genes for poisons were introduced into pathogens, such as the common cold, that spread easily. After a ten-month moratorium guidelines were agreed to at the Asilomar conference, which included provisions for physical and biological containment, bans on particular types of experiments, and other stipulations. These biotechnology guidelines have been strictly followed, and there have not been reported accidents in the thirty-year history of the field.

More recently, the organization representing the world's organ transplantation surgeons has adopted a moratorium on the transplantation of vascularized animal organs into humans. This was done out of fear of the spread of long-dormant HIV-type xenoviruses from animals such as pigs or baboons into the human population. Unfortunately, such a moratorium can also slow down the availability of lifesaving xenografts (genetically modified animal organs that are accepted by the human immune system) to the millions of people who die each year from heart, kidney, and liver disease. Geoethicist Martine Rothblatt has proposed replacing this moratorium with a new set of ethical guidelines and regulations.[44]

In the case of nanotechnology, the ethics debate has started a couple of decades prior to the availability of the particularly dangerous applications. The most important provisions of the Foresight Institute guidelines include:

- "Artificial replicators must not be capable of replication in a natural, uncontrolled environment."
- "Evolution within the context of a self-replicating manufacturing system is discouraged."
- "MNT device designs should specifically limit proliferation and provide traceability of any replicating systems."

- "Distribution of molecular manufacturing *development* capability should be restricted whenever possible, to responsible actors that have agreed to use the Guidelines. No such restriction need apply to end products of the development process."

Other strategies that the Foresight Institute has proposed include:

- Replication should require materials not found in the natural environment.
- Manufacturing (replication) should be separated from the functionality of end products. Manufacturing devices can create end products but cannot replicate themselves, and end products should have no replication capabilities.
- Replication should require replication codes that are encrypted and time limited. The broadcast architecture mentioned earlier is an example of this recommendation.

These guidelines and strategies are likely to be effective for preventing accidental release of dangerous self-replicating nanotechnology entities. But dealing with the intentional design and release of such entities is a more complex and challenging problem. A sufficiently determined and destructive opponent could possibly defeat each of these layers of protections. Take, for example, the broadcast architecture. When properly designed, each entity is unable to replicate without first obtaining replication codes, which are not repeated from one replication generation to the next. However, a modification to such a design could bypass the destruction of the replication codes and thereby pass them on to the next generation. To counteract that possibility it has been recommended that the memory for the replication codes be limited to only a subset of the full code. However, this guideline could be defeated by expanding the size of the memory.

Another protection that has been suggested is to encrypt the codes and build in protections in the decryption systems, such as time-expiration limitations. However, we can see how easy it has been to defeat protections against unauthorized replications of intellectual property such as music files. Once replication codes and protective layers are stripped away, the information can be replicated without these restrictions.

This doesn't mean that protection is impossible. Rather, each level of protection will work only to a certain level of sophistication. The meta-lesson here is that we will need to place twenty-first-century society's highest priority on the continuing advance of defensive technologies, keeping them one or more steps ahead of the destructive technologies (or at least no more than a quick step behind).

Protection from "Unfriendly" Strong AI. Even as effective a mechanism as the broadcast architecture, however, won't serve as protection against abuses of strong AI. The barriers provided by the broadcast architecture rely on the lack of intelligence in nanoengineered entities. By definition, however, intelligent entities have the cleverness to easily overcome such barriers.

Eliezer Yudkowsky has extensively analyzed paradigms, architectures, and ethical rules that may help assure that once strong AI has the means of accessing and modifying its own design it remains friendly to biological humanity and supportive of its values. Given that self-improving strong AI cannot be recalled, Yudkowsky points out that we need to "get it right the first time," and that its initial design must have "zero nonrecoverable errors."[45]

Inherently there will be no absolute protection against strong AI. Although the argument is subtle I believe that maintaining an open free-market system for incremental scientific and technological progress, in which each step is subject to market acceptance, will provide the most constructive environment for technology to embody widespread human values. As I have pointed out, strong AI is emerging from many diverse efforts and will be deeply integrated into our civilization's infrastructure. Indeed, it will be intimately embedded in our bodies and brains. As such, it will reflect our values because it will be us. Attempts to control these technologies via secretive government programs, along with inevitable underground development, would only foster an unstable environment in which the dangerous applications would be likely to become dominant.

Decentralization. One profound trend already well under way that will provide greater stability is the movement from centralized technologies to distributed ones and from the real world to the virtual world discussed above. Centralized technologies involve an aggregation of resources such as people (for example, cities, buildings), energy (such as nuclear-power plants, liquid-natural-gas and oil tankers, energy pipelines), transportation (airplanes, trains), and other items. Centralized technologies are subject to disruption and disaster. They also tend to be inefficient, wasteful, and harmful to the environment.

Distributed technologies, on the other hand, tend to be flexible, efficient, and relatively benign in their environmental effects. The quintessential distributed technology is the Internet. The Internet has not been substantially disrupted to date, and as it continues to grow, its robustness and resilience continue to strengthen. If any hub or channel does go down, information simply routes around it.

Distributed Energy. In energy, we need to move away from the extremely concentrated and centralized installations on which we now depend. For example, one company is pioneering fuel cells that are microscopic, using MEMS technology.[46] They are manufactured like electronic chips but are actually energy-storage devices with an energy-to-size ratio significantly exceeding that of conventional technology. As I discussed earlier, nanoengineered solar panels will be able to meet our energy needs in a distributed, renewable, and clean fashion. Ultimately technology along these lines could power everything from our cell phones to our cars and homes. These types of decentralized energy technologies would not be subject to disaster or disruption.

As these technologies develop, our need for aggregating people in large buildings and cities will diminish, and people will spread out, living where they want and gathering together in virtual reality.

Civil Liberties in an Age of Asymmetric Warfare. The nature of terrorist attacks and the philosophies of the organizations behind them highlight how civil liberties can be at odds with legitimate state interests in surveillance and control. Our law-enforcement system—and indeed, much of our thinking about security—is based on the assumption that people are motivated to preserve their own lives and well-being. That logic underlies all our strategies, from protection at the local level to mutual assured destruction on the world stage. But a foe that values the destruction of both its enemy and itself is not amenable to this line of reasoning.

The implications of dealing with an enemy that does not value its own survival are deeply troublesome and have led to controversy that will only intensify as the stakes continue to escalate. For example, when the FBI identifies a likely terrorist cell, it will arrest the participants, even though there may be insufficient evidence to convict them of a crime and they may not yet even have committed a crime. Under the rules of engagement in our war on terrorism, the government continues to hold these individuals.

In a lead editorial, the *New York Times* objected to this policy, which it described as a "troubling provision."[47] The paper argued that the government should release these detainees because they have not yet committed a crime and should rearrest them only after they have done so. Of course by that time suspected terrorists might well be dead along with a large number of their victims. How can the authorities possibly break up a vast network of decentralized cells of suicide terrorists if they have to wait for each one to commit a crime?

On the other hand this very logic has been routinely used by tyrannical regimes to justify the waiving of the judicial protections we have come to cherish. It is likewise fair to argue that curtailing civil liberties in this way is exactly the aim of the terrorists, who despise our notions of freedoms and pluralism. However, I do not see the prospect of any technology "magic bullet" that would essentially change this dilemma.

The encryption trapdoor may be considered a technical innovation that the government has been proposing in an attempt to balance legitimate individual needs for privacy with the government's need for surveillance. Along with this type of technology we also need the requisite political innovation to provide for effective oversight, by both the judicial and legislative branches, of the executive branch's use of these trapdoors, to avoid the potential for abuse of power. The secretive nature of our opponents and their lack of respect for human life including their own will deeply test the foundations of our democratic traditions.

A Program for GNR Defense

> We come from goldfish, essentially, but that [doesn't] mean we turned around and killed all the goldfish. Maybe [the AIs] will feed us once a week. . . . If you had a machine with a 10 to the 18th power IQ over humans, wouldn't you want it to govern, or at least control your economy?
>
> —SETH SHOSTAK

How can we secure the profound benefits of GNR while ameliorating its perils? Here's a review of a suggested program for containing the GNR risks:

The most urgent recommendation is to *greatly increase our investment in defensive technologies.* Since we are already in the G era, *the bulk of this investment today should be in (biological) antiviral medications and treatments.* We have new tools that are well suited to this task. RNA interference, for example, can be used to block gene expression. Virtually all infections (as well as cancer) rely on gene expression at some point during their life cycles.

Efforts to anticipate the defensive technologies needed to safely guide N and R should also be supported, and these should be substantially increased as we get closer to the feasibility of molecular manufacturing and strong AI, respectively. A significant side benefit would be to accelerate effective treatments for infectious disease and cancer. I've testified before Congress on this issue, advocating the investment of tens of billions of dollars per year (less than 1 percent of the GDP) to address this new and underrecognized existential threat to humanity.[48]

- We need to streamline the regulatory process for genetic and medical technologies. The regulations do not impede the malevolent use of technology but significantly delay the needed defenses. As mentioned, we need to better balance the risks of new technology (for example, new medications) against the known harm of delay.

- A global program of confidential, random serum monitoring for unknown or evolving biological pathogens should be funded. Diagnostic tools exist to rapidly identify the existence of unknown protein or nucleic acid sequences. Intelligence is key to defense, and such a program could provide invaluable early warning of an impending epidemic. Such a "pathogen sentinel" program has been proposed for many years by public health authorities but has never received adequate funding.

- Well-defined and targeted temporary moratoriums, such as the one that occurred in the genetics field in 1975, may be needed from time to time. But such moratoriums are unlikely to be necessary with nanotechnology. Broad efforts at relinquishing major areas of technology serve only to continue vast human suffering by delaying the beneficial aspects of new technologies, and actually make the dangers worse.

- Efforts to define safety and ethical guidelines for nanotechnology should continue. Such guidelines will inevitably become more detailed and refined as we get closer to molecular manufacturing.

- To create the political support to fund the efforts suggested above, it is necessary to *raise public awareness of these dangers*. Because, of course, there exists the downside of raising alarm and generating uninformed backing for broad antitechnology mandates, we also need to create a public understanding of the profound benefits of continuing advances in technology.

- These risks cut across international boundaries—which is, of course, nothing new; biological viruses, software viruses, and missiles already cross such boundaries with impunity. *International cooperation* was vital to containing the SARS virus and will become increasingly vital in confronting future challenges. Worldwide organizations such as the World Health Organization, which helped coordinate the SARS response, need to be strengthened.

- A contentious contemporary political issue is the need for preemptive action to combat threats, such as terrorists with access to weapons of mass destruction or rogue nations that support such terrorists. Such measures will always be controversial, but the potential need for them is clear. A nuclear explosion can destroy a city in seconds. A self-replicating pathogen, whether biological or nanotechnology based, could destroy our civilization in a matter of days or weeks. We cannot always afford to wait for the massing of armies or other overt indications of ill intent before taking protective action.

- Intelligence agencies and policing authorities will have a vital role in forestalling the vast majority of potentially dangerous incidents. Their efforts need to involve the most powerful technologies available. For example, before this decade is over, devices the size of dust particles will be able to carry out reconnaissance missions. When we reach the 2020s and have software running in our bodies and brains, government authorities will have a legitimate need on occasion to monitor these software streams. The potential for abuse of such powers is obvious. We will need to achieve a middle road of preventing catastrophic events while preserving our privacy and liberty.

- The above approaches will be inadequate to deal with the danger from pathological R (strong AI). Our primary strategy in this area should be to optimize the likelihood that future nonbiological intelligence will reflect our values of liberty, tolerance, and respect for knowledge and diversity. The best way to accomplish this is to foster those values in our society today and going forward. If this sounds vague, it is. But there is no purely technical strategy that is workable in this area, because greater intelligence will always find a way to circumvent measures that are the product of a lesser intelligence. The nonbiological intelligence we are creating is and will be embedded in our societies and will reflect our values. The transbiological phase will involve nonbiological intelligence deeply integrated with biological intelligence. This will amplify our abilities, and our application of these greater intellectual powers will be governed by the values of its creators. The transbiological era will ultimately give way to the post-biological era, but it is to be hoped that our values will remain influential. This strategy is certainly not foolproof, but it is the primary means we have today to influence the future course of strong AI.

Technology will remain a double-edged sword. It represents vast power to be used for all humankind's purposes. GNR will provide the means to overcome age-old problems such as illness and poverty, but it will also empower destructive ideologies. We have no choice but to strengthen our defenses while we apply these quickening technologies to advance our human values, despite an apparent lack of consensus on what those values should be.

MOLLY 2004: *Okay, now run that stealthy scenario by me again—you know, the one where the bad nanobots spread quietly through the biomass to get themselves into position but don't actually expand to noticeably destroy anything until they're spread around the globe.*

RAY: *Well, the nanobots would spread at very low concentrations, say one carbon atom per 10^{15} in the biomass, so they would be seeded throughout the biomass. Thus, the speed of physical spread of the destructive nanobots would not be a limiting factor when they subsequently replicate in place. If they skipped the stealth phase and expanded instead from a single point, the spreading nanodisease would be noticed, and the spread around the world would be relatively slow.*

MOLLY 2004: *So how are we going to protect ourselves from that? By the time they start phase two, we've got only about ninety minutes, or much less if you want to avoid enormous damage.*

RAY: *Because of the nature of exponential growth, the bulk of the damage gets done in the last few minutes, but your point is well taken. Under any scenario, we won't have a chance without a nanotechnology immune system. Obviously, we can't wait until the beginning of a ninety-minute cycle of destruction to begin thinking about creating one. Such a system would be very comparable to our human immune system. How long would a biological human circa 2004 last without one?*

MOLLY 2004: *Not long, I suppose. How does this nano-immune system pick up these bad nanobots if they're only one in a thousand trillion?*

RAY: *We have the same issue with our biological immune system. Detection of even a single foreign protein triggers rapid action by biological antibody factories, so the immune system is there in force by the time a pathogen achieves a near critical level. We'll need a similar capability for the nano-immune system.*

CHARLES DARWIN: *Now tell me, do the immune-system nanobots have the ability to replicate?*

RAY: *They would need to be able to do this; otherwise they would not be able to keep pace with the replicating pathogenic nanobots. There have been proposals to seed the biomass with protective immune-system nanobots at a particular concentration, but as soon as the bad nanobots significantly exceeded this fixed concentration the immune system would lose. Robert Freitas proposes nonreplicating nanofactories able to turn out additional protective nanorobots when needed. I think this is likely to deal with threats for a while, but ultimately the defensive system will need the ability to replicate its immune capabilities in place to keep pace with emerging threats.*

CHARLES: *So aren't the immune-system nanobots entirely equivalent to the phase one malevolent nanobots? I mean seeding the biomass is the first phase of the stealth scenario.*

RAY: *But the immune-system nanobots are programmed to protect us, not destroy us.*

CHARLES: *I understand that software can be modified.*

RAY: *Hacked, you mean?*

CHARLES: *Yes, exactly. So if the immune-system software is modified by a hacker to simply turn on its self-replication ability without end—*

RAY: *—yes, well, we'll have to be careful about that, won't we?*

MOLLY 2004: *I'll say.*

RAY: *We have the same problem with our biological immune system. Our immune system is comparably powerful, and if it turns on us that's an autoimmune disease, which can be insidious. But there's still no alternative to having an immune system.*

MOLLY 2004: *So a software virus could turn the nanobot immune system into a stealth destroyer?*

RAY: *That's possible. It's fair to conclude that software security is going to be the decisive issue for many levels of the human-machine civilization. With everything becoming information, maintaining the software integrity of our defensive technologies will be critical to our survival. Even on an economic level, maintaining the business model that creates information will be critical to our well-being.*

MOLLY 2004: *This makes me feel rather helpless. I mean, with all these good and bad nanobots battling it out, I'll just be a hapless bystander.*

RAY: *That's hardly a new phenomenon. How much influence do you have in 2004 on the disposition of the tens of thousands of nuclear weapons in the world?*

MOLLY 2004: *At least I have a voice and a vote in elections that affect foreign-policy issues.*

RAY: *There's no reason for that to change. Providing for a reliable nanotechnology immune system will be one of the great political issues of the 2020s and 2030s.*

MOLLY 2004: *Then what about strong AI?*

RAY: *The good news is that it will protect us from malevolent nanotechnology because it will be smart enough to assist us in keeping our defensive technologies ahead of the destructive ones.*

NED LUDD: *Assuming it's on our side.*

RAY: *Indeed.*

CHAPTER NINE

Response to Critics

The human mind likes a strange idea as little as the body likes a strange protein and resists it with a similar energy.

—W. I. BEVERIDGE

If a . . . scientist says that something is possible he is almost certainly right, but if he says that it is impossible he is very probably wrong.

—ARTHUR C. CLARKE

A Panoply of Criticisms

In *The Age of Spiritual Machines*, I began to examine some of the accelerating trends that I have sought to explore in greater depth in this book. *ASM* inspired a broad variety of reactions, including extensive discussions of the profound, imminent changes it considered (for example, the promise-versus-peril debate prompted by Bill Joy's *Wired* story, "Why the Future Doesn't Need Us," as I reviewed in the previous chapter). The response also included attempts to argue on many levels why such transformative changes would not, could not, or should not happen. Here is a summary of the critiques I will be responding to in this chapter:

- The "criticism from Malthus": *It's a mistake to extrapolate exponential trends indefinitely, since they inevitably run out of resources to maintain the exponential growth. Moreover, we won't have enough energy to power the extraordinarily dense computational platforms forecast, and even if we did they would be as hot as the sun.* Exponential trends do reach an asymptote, but the matter and energy resources needed for computation and communication are so small per compute and per bit that these trends can continue to the point where nonbiological intelligence is trillions of trillions of times more powerful than biological intelligence. Reversible computing can reduce energy requirements, as well as heat dissipation, by many orders of magnitude. Even restricting computation to "cold" computers will achieve nonbiological computing platforms that vastly outperform biological intelligence.

- The "criticism from software": *We're making exponential gains in hardware, but software is stuck in the mud.* Although the doubling time for progress in software is longer than that for computational hardware, software is also accelerating in effectiveness, efficiency, and complexity. Many software applications, ranging from search engines to games, routinely use AI techniques that were only research projects a decade ago. Substantial gains have also been made in the overall complexity of software, in software productivity, and in the efficiency of software in solving key algorithmic problems. Moreover, we have an effective game plan to achieve the capabilities of human intelligence in a machine: reverse engineering the brain to capture its principles of operation and then implementing those principles in brain-capable computing platforms. Every aspect of brain reverse engineering is accelerating: the spatial and temporal resolution of brain scanning, knowledge about every level of the brain's operation, and efforts to realistically model and simulate neurons and brain regions.

- The "criticism from analog processing": *Digital computation is too rigid because digital bits are either on or off. Biological intelligence is mostly analog, so subtle gradations can be considered.* It's true that the human brain uses digital-controlled analog methods, but we can also use such methods in our machines. Moreover, digital computation can simulate analog transactions to any desired level of accuracy, whereas the converse statement is not true.

- The "criticism from the complexity of neural processing": *The information processes in the interneuronal connections (axons, dendrites, synapses) are far more complex than the simplistic models used in neural nets.* True, but brain-region simulations don't use such simplified models. We have achieved realistic mathematical models and computer simulations of neurons and interneuronal connections that do capture the nonlinearities and intricacies of their biological counterparts. Moreover, we have found that the complexity of processing brain regions is often simpler than the neurons they comprise. We already have effective models and simulations for several dozen regions of the human brain. The genome contains only about thirty to one hundred million bytes of design information when redundancy is considered, so the design information for the brain is of a manageable level.

- The "criticism from microtubules and quantum computing": *The microtubules in neurons are capable of quantum computing, and such quantum computing is a prerequisite for consciousness. To "upload" a personality, one would have to capture its precise quantum state.* No evidence exists to support either of these statements. Even if true, there is nothing that bars quantum computing from being carried out in nonbiological systems. We routinely use quantum effects in semiconductors (tunneling in transistors, for example), and machine-based quantum computing is also progressing. As for capturing a precise quantum state, I'm in a very different quantum state than I was before writing this sentence. So am I already a different person? Perhaps I am, but if one captured my state a minute ago, an upload based on that information would still successfully pass a "Ray Kurzweil" Turing test.

- The "criticism from the Church-Turing thesis": *We can show that there are broad classes of problems that cannot be solved by any Turing machine. It can also be shown that Turing machines can emulate any possible computer (that is, there exists a Turing machine that can solve any problem that any computer can solve), so this demonstrates a clear limitation on the problems that a computer can solve. Yet humans are capable of solving these problems, so machines will never emulate human intelligence.* Humans are no more capable of universally solving such "unsolvable" problems than machines. Humans can make educated guesses to solutions in certain instances, but machines can do the same thing and can often do so more quickly.

- The "criticism from failure rates": *Computer systems are showing alarming rates of catastrophic failure as their complexity increases. Thomas Ray writes that we are "pushing the limits of what we can effectively design and build through conventional approaches."* We have developed increasingly complex systems to manage a broad variety of mission-critical tasks, and failure rates in these systems are very low. However, imperfection is an inherent feature of any complex process, and that certainly includes human intelligence.

- The "criticism from 'lock-in' ": *The pervasive and complex support systems (and the huge investments in these systems) required by such fields as energy and transportation are blocking innovation, so this will prevent the kind of rapid change envisioned for the technologies underlying the Singularity.* It is specifically information processes that are growing exponentially in capability and price-performance. We have already seen rapid paradigm shifts in every aspect of information technology, unimpeded by any lock-in phenomenon (despite large infrastructure investments in such areas as the Internet and telecommunications). Even the energy and transportation sectors will witness revolutionary changes from new nanotechnology-based innovations.

- The "criticism from ontology": *John Searle describes several versions of his Chinese Room analogy. In one formulation a man follows a written program to answer questions in Chinese. The man appears to be answering questions competently in Chinese, but since he is just mechanically following a written program, he has no real understanding of Chinese and no real awareness of what he is doing. The "man" in the room doesn't understand anything, because, after all, "he is just a computer," according to Searle. So clearly, computers cannot understand what they are doing, since they are just following rules.* Searle's Chinese Room arguments are fundamentally tautological, as they just assume his conclusion that computers cannot possibly have any real understanding. Part of the philosophical sleight of hand in Searle's simple analogies is a matter of scale. He purports to describe a simple system and then asks the reader to consider how such a system could possibly have any real understanding. But the characterization itself is misleading. To be consistent with Searle's own assumptions the Chinese Room system that Searle describes would have to be as complex as a human brain and would, therefore, have as much understanding as a human brain. The man in the analogy would be acting as the central-processing unit, only a small part of the system. While the man may not see it, the understanding is distributed across the entire pattern of the program itself and the billions of notes he would have to make to follow the program. Consider that I understand English, but none of my neurons do. My understanding is represented in vast patterns of neurotransmitter strengths, synaptic clefts, and interneuronal connections.

- The "criticism from the rich-poor divide": *It's likely that through these technologies the rich may obtain certain opportunities that the rest of humankind does not have access to.* This, of course, would be nothing new, but I would point out that because of the ongoing exponential growth of price-performance, all of these technologies quickly become so inexpensive as to become almost free.

- The "criticism from the likelihood of government regulation": *Governmental regulation will slow down and stop the acceleration of technology.* Although the obstructive potential of regulation is an important concern, it has had as of yet little measurable effect on the trends discussed in this book. Absent a worldwide totalitarian state, the economic and other forces underlying technical progress will only grow with ongoing advances. Even controversial issues such as stem-cell research end up being like stones in a stream, the flow of progress rushing around them.

• The "criticism from theism": *According to William A. Dembski, "contemporary materialists such as Ray Kurzweil . . . see the motions and modifications of matter as sufficient to account for human mentality." But materialism is predictable, whereas reality is not. Predictability [is] materialism's main virtue . . . and hollowness [is] its main fault."* Complex systems of matter and energy are not predictable, since they are based on a vast number of unpredictable quantum events. Even if we accept a "hidden variables" interpretation of quantum mechanics (which says that quantum events only appear to be unpredictable but are based on undetectable hidden variables), the behavior of a complex system would still be unpredictable in practice. All of the trends show that we are clearly headed for nonbiological systems that are as complex as their biological counterparts. Such future systems will be no more "hollow" than humans and in many cases will be based on the reverse engineering of human intelligence. We don't need to go beyond the capabilities of patterns of matter and energy to account for the capabilities of human intelligence.

• The "criticism from holism": *To quote Michael Denton, organisms are "self-organizing, . . . self-referential, . . . self-replicating, . . . reciprocal, . . . self-formative, and . . . holistic." Such organic forms can be created only through biological processes, and such forms are "immutable, . . . impenetrable, and . . . fundamental realities of existence."*[1] It's true that biological design represents a profound set of principles. However, machines can use—and already are using—these same principles, and there is nothing that restricts nonbiological systems from harnessing the emergent properties of the patterns found in the biological world.

I've engaged in countless debates and dialogues responding to these challenges in a diverse variety of forums. One of my goals for this book is to provide a comprehensive response to the most important criticisms I have encountered. Most of my rejoinders to these critiques on feasibility and inevitability have been discussed throughout this book, but in this chapter I want to offer a detailed reply to several of the more interesting ones.

The Criticism from Incredulity

Perhaps the most candid criticism of the future I have envisioned here is simple disbelief that such profound changes could possibly occur. Chemist Richard Smalley, for example, dismisses the idea of nanobots being capable of performing missions in the human bloodstream as just "silly." But scientists' ethics call for caution in assessing the prospects for current work, and such reasonable prudence unfortunately often leads scientists to shy away from considering the power of generations of science and technology far beyond today's frontier. With the rate of paradigm shift occurring ever more quickly, this ingrained pessimism does not serve society's needs in assessing scientific capabilities in the decades ahead. Consider how incredible today's technology would seem to people even a century ago.

A related criticism is based on the notion that it is difficult to predict the future, and any number of bad predictions from other futurists in earlier eras can be cited to support this. Predicting which company or product will succeed is indeed very difficult, if not impossible. The same difficulty occurs in predicting which technical design or standard will prevail. (For example, how will the wireless-communication protocols WiMAX, CDMA, and 3G fare over the next several years?) However, as this book has extensively argued, we find remarkably precise and predictable exponential trends when assessing the overall effectiveness (as measured by price-performance, bandwidth, and other measures of capability) of information technologies. For example, the smooth exponential growth of the price-performance of computing dates back over a century. Given that the minimum amount of matter and energy required to compute or transmit a bit of information is known to be vanishingly small, we can confidently predict the continuation of these information-technology trends at least through this next century. Moreover, we can reliably predict the capabilities of these technologies at future points in time.

Consider that predicting the path of a single molecule in a gas is essentially impossible, but predicting certain properties of the entire gas (composed of a great many chaotically interacting molecules) can reliably be predicted through the laws of thermodynamics. Analogously, it is not possible to reliably predict the results of a specific project or company, but the overall capabilities of information technology (comprised of many chaotic activities) can nonetheless be dependably anticipated through the law of accelerating returns.

Many of the furious attempts to argue why machines—nonbiological systems—cannot ever possibly compare to humans appear to be fueled by this basic reaction of incredulity. The history of human thought is marked by many attempts to refuse to accept ideas that seem to threaten the accepted view that our species is special. Copernicus's insight that the Earth was not at the center of the universe was resisted, as was Darwin's that we were only slightly evolved from other primates. The notion that machines could match and even exceed human intelligence appears to challenge human status once again.

In my view there is something essentially special, after all, about human beings. We were the first species on Earth to combine a cognitive function and an effective opposable appendage (the thumb), so we were able to create technology that would extend our own horizons. No other species on Earth has accomplished this. (To be precise, we're the only surviving species in this ecological niche—others, such as the Neanderthals, did not survive.) And as I discussed in chapter 6, we have yet to discover any other such civilization in the universe.

The Criticism from Malthus

Exponential Trends Don't Last Forever. The classical metaphorical example of exponential trends hitting a wall is known as "rabbits in Australia." A species happening upon a hospitable new habitat will expand its numbers exponentially until its growth hits the limits of the ability of that environment to support it. Approaching this limit to exponential growth may even cause an overall reduction in numbers—for example, humans noticing a spreading pest may seek to eradicate it. Another common example is a microbe that may grow exponentially in an animal body until a limit is reached: the ability of that body to support it, the response of its immune system, or the death of the host.

Even the human population is now approaching a limit. Families in the more developed nations have mastered means of birth control and have set relatively high standards for the resources they wish to provide their children. As a result population expansion in the developed world has largely stopped. Meanwhile people in some (but not all) underdeveloped countries have continued to seek large families as a means of social security, hoping that at least one child will survive long enough to support them in old age. However, with the law of accelerating returns providing more widespread economic gains, the overall growth in human population is slowing.

So isn't there a comparable limit to the exponential trends that we are witnessing for information technologies?

The answer is yes, but not before the profound transformations described throughout this book take place. As I discussed in chapter 3, the amount of matter and energy required to compute or transmit one bit is vanishingly small. By using reversible logic gates, the input of energy is required only to transmit results and to correct errors. Otherwise, the heat released from each computation is immediately recycled to fuel the next computation.

As I discussed in chapter 5, nanotechnology-based designs for virtually all applications—computation, communication, manufacturing, and transportation—will require substantially less energy than they do today. Nanotechnology will also facilitate capturing renewable energy sources such as sunlight. We could meet all of our projected energy needs of thirty trillion watts in 2030 with solar power if we captured only 0.03 percent (three ten-thousandths) of the sun's energy as it hit the Earth. This will be feasible with extremely inexpensive, lightweight, and efficient nanoengineered solar panels together with nano–fuel cells to store and distribute the captured energy.

A Virtually Unlimited Limit. As I discussed in chapter 3 an optimally organized 2.2-pound computer using reversible logic gates has about 10^{25} atoms and can store about 10^{27} bits. Just considering electromagnetic interactions between the particles, there are at least 10^{15} state changes per bit per second that can be harnessed for computation, resulting in about 10^{42} calculations per second in the ultimate "cold" 2.2-pound computer. This is about 10^{16} times more powerful than all biological brains today. If we allow our ultimate computer to get hot, we can increase this further by as much as 10^8-fold. And we obviously won't restrict our computational resources to one kilogram of matter but will ultimately deploy a significant fraction of the matter and energy on the Earth and in the solar system and then spread out from there.

Specific paradigms do reach limits. We expect that Moore's Law (concerning the shrinking of the size of transistors on a flat integrated circuit) will hit a limit over the next two decades. The date for the demise of Moore's Law keeps getting pushed back. The first estimates predicted 2002, but now Intel says it won't take place until 2022. But as I discussed in chapter 2, every time a specific computing paradigm was seen to approach its limit, research interest and pressure increased to create the next paradigm. This has already happened four times in the century-long history of exponential growth in computation (from electromagnetic calculators to relay-based computers to vacuum tubes to discrete transistors to integrated circuits). We have already achieved many important milestones toward the next (sixth) paradigm of computing: three-dimensional self-organizing circuits at the molecular level. So the impending end of a given paradigm does not represent a true limit.

There are limits to the power of information technology, but these limits are vast. I estimated the capacity of the matter and energy in our solar system to support computation to be at least 10^{70} cps (see chapter 6). Given that there are at least 10^{20} stars in the universe, we get about 10^{90} cps for it, which matches Seth Lloyd's independent analysis. So yes, there are limits, but they're not very limiting.

The Criticism from Software

A common challenge to the feasibility of strong AI, and therefore the Singularity, begins by distinguishing between quantitative and qualitative trends. This argument acknowledges, in essence, that certain brute-force capabilities such as memory capacity, processor speed, and communications bandwidths are expanding exponentially but maintains that the software (that is, the methods and algorithms) are not.

This is the hardware-versus-software challenge, and it is a significant one. Virtual-reality pioneer Jaron Lanier, for example, characterizes my position and that of other so-called cybernetic totalists as, we'll just figure out the software in some unspecified way—a position he refers to as a software "deus ex machina."[2] This ignores, however, the specific and detailed scenario that I've described by which the software of intelligence will be achieved. The reverse engineering of the human brain, an undertaking that is much further along than Lanier and many other observers realize, will expand our AI toolkit to include the self-organizing methods underlying human intelligence. I'll return to this topic in a moment, but first let's address some other basic misconceptions about the so-called lack of progress in software.

Software Stability. Lanier calls software inherently "unwieldy" and "brittle" and has described at great length a variety of frustrations that he has encountered in using it. He writes that "getting computers to perform specific tasks of significant complexity in a reliable but modifiable way, without crashes or security breaches, is essentially impossible."[3] It is not my intention to defend all software, but it's not true that complex software is necessarily brittle and prone to catastrophic breakdown. Many examples of complex mission-critical software operate with very few, if any, breakdowns: for example, the sophisticated software programs that control an increasing percentage of airplane landings, monitor patients in critical-care facilities, guide intelligent weapons, control the investment of billions of dollars in automated pattern recognition-based hedge funds, and serve many other functions.[4] I am not aware of any airplane crashes that have been caused by failures of automated landing software; the same, however, cannot be said for human reliability.

Software Responsiveness. Lanier complains that "computer user interfaces tend to respond more slowly to user interface events, such as a key press, than they did fifteen years earlier. . . . What's gone wrong?"[5] I would invite Lanier to attempt using an old computer today. Even if we put aside the difficulty of setting one up (which is a different issue), he has forgotten just how unresponsive, unwieldy, and limited they were. Try getting some real work done to today's standards with twenty-year-old personal-computer software. It's simply not true to say that the old software was better in any qualitative or quantitative sense.

Although it's always possible to find poor-quality design, response delays, when they occur, are generally the result of new features and functions. If users were willing to freeze the functionality of their software, the ongoing exponential growth of computing speed and memory would quickly eliminate software-response delays. But the market demands ever-expanded capability. Twenty years ago there were no search engines or any other integration with the World Wide Web (indeed, there was no Web), only primitive language, formatting, and multimedia tools, and so on. So functionality always stays on the edge of what's feasible.

This romancing of software from years or decades ago is comparable to people's idyllic view of life hundreds of years ago, when people were "unencumbered" by the frustrations of working with machines. Life was unfettered, perhaps, but it was also short, labor-intensive, poverty filled, and disease and disaster prone.

Software Price-Performance. With regard to the price-performance of software, the comparisons in every area are dramatic. Consider the table on p. 103 on speech-recognition software. In 1985 five thousand dollars bought you a software package that provided a thousand-word vocabulary, did not offer continuous-speech capability, required three hours of training on your voice, and had relatively poor accuracy. In 2000 for only fifty dollars, you could purchase a software package with a hundred-thousand-word vocabulary that provided continuous-speech capability, required only five minutes of training on your voice, had dramatically improved accuracy, offered natural-language understanding (for editing commands and other purposes), and included many other features.[6]

Software Development Productivity. How about software development itself? I've been developing software myself for forty years, so I have some perspective on the topic. I estimate the doubling time of software development productivity to be approximately six years, which is slower than the doubling time for processor price-performance, which is approximately one year today. However, software productivity is nonetheless growing exponentially. The development tools, class libraries, and support systems available today are dramatically more effective than those of decades ago. In my current projects teams of just three or four people achieve in a few months objectives that are comparable to what twenty-five years ago required a team of a dozen or more people working for a year or more.

Software Complexity. Twenty years ago software programs typically consisted of thousands to tens of thousands of lines. Today, mainstream programs (for example, supply-channel control, factory automation, reservation systems, biochemical simulation) are measured in millions of lines or more. Software for major defense systems such as the Joint Strike Fighter contains tens of millions of lines.

Software to control software is itself rapidly increasing in complexity. IBM is pioneering the concept of autonomic computing, in which routine information-technology support functions will be automated.[7] These systems will be programmed with models of their own behavior and will be capable, according to IBM, of being "self-configuring, self-healing, self-optimizing, and self-protecting." The software to support autonomic computing will be measured in tens of millions of lines of code (with each line containing tens of bytes of information). So in terms of information complexity, software already exceeds the tens of millions of bytes of usable information in the human genome and its supporting molecules.

The amount of information contained in a program, however, is not the best measure of complexity. A software program may be long but may be bloated with useless information. Of course, the same can be said for the genome, which appears to be very inefficiently coded. Attempts have been made to formulate measures of software complexity—for example, the Cyclomatic Complexity Metric, developed by computer scientists Arthur Watson and Thomas McCabe at the National Institute of Standards and Technology.[8] This metric measures the complexity of program logic and takes into account the structure of branching and decision points. The anecdotal evidence strongly suggests rapidly increasing complexity if measured by these indexes, although there is insufficient data to track doubling times. However, the key point is that the most complex software systems in use in industry today have higher levels of complexity than software programs that are performing neuromorphic-based simulations of brain regions, as well as biochemical simulations of individual neurons. We can already handle levels of software complexity that exceed what is needed to model and simulate the parallel, self-organizing, fractal algorithms that we are discovering in the human brain.

Accelerating Algorithms. Dramatic improvements have taken place in the speed and efficiency of software algorithms (on constant hardware). Thus the price-performance of implementing a broad variety of methods to solve the basic mathematical functions that underlie programs like those used in signal processing, pattern recognition, and artificial intelligence has benefited from the acceleration of both hardware and software. These improvements vary depending on the problem, but are nonetheless pervasive.

For example, consider the processing of signals, which is a widespread and computationally intensive task for computers as well as for the human brain. Georgia Institute of Technology's Mark A. Richards and MIT's Gary A. Shaw have documented a broad trend toward greater signal-processing algorithm efficiency.[9] For example, to find patterns in signals it is often necessary to solve what are called partial differential equations. Algorithms expert Jon Bentley has shown a continual reduction in the number of computing operations required to solve this class of problem.[10] For example, from 1945 to 1985, for a representative application (finding an elliptic partial differential solution for a three-dimensional grid with sixty-four elements on each side), the number of operation counts has been reduced by a factor of three hundred thousand. This is a 38 percent increase in efficiency each year (not including hardware improvements).

Another example is the ability to send information on unconditioned phone lines, which has improved from 300 bits per second to 56,000 bps in twelve years, a 55 percent annual increase.[11] Some of this improvement was the result of improvements in hardware design, but most of it is a function of algorithmic innovation.

One of the key processing problems is converting a signal into its frequency components using Fourier transforms, which express signals as sums of sine waves. This method is used in the front end of computerized speech recognition and in many other applications. Human auditory perception also starts by breaking the speech signal into frequency components in the cochlea. The 1965 "radix-2 Cooley-Tukey algorithm" for a "fast Fourier transform" reduced the number of operations required for a 1,024-point Fourier transform by about two hundred.[12] An improved "radix-4" method further boosted the improvement to eight hundred. Recently "wavelet" transforms have been introduced, which are able to express arbitrary signals as sums of waveforms more complex than sine waves. These methods provide further dramatic increases in the efficiency of breaking down a signal into its key components.

The examples above are not anomalies; most computationally intensive "core" algorithms have undergone significant reductions in the number of operations required. Other examples include sorting, searching, autocorrelation (and other statistical methods), and information compression and decompression. Progress has also been made in parallelizing algorithms—that is, breaking a single method into multiple methods that can be performed simultaneously. As I discussed earlier, parallel processing inherently runs at a lower temperature. The brain uses massive parallel processing as one strategy to achieve more complex functions and faster reaction times, and we will need to utilize this approach in our machines to achieve optimal computational densities.

There is an inherent difference between the improvements in hardware price-performance and improvements in software efficiencies. Hardware improvements have been remarkably consistent and predictable. As we master each new level of speed and efficiency in hardware we gain powerful tools to continue to the next level of exponential improvement. Software improvements, on the other hand, are less predictable. Richards and Shaw call them "worm-holes in development time," because we can often achieve the equivalent of years of hardware improvement through a single algorithmic improvement. Note that we do not rely on ongoing progress in software efficiency, since we can count on the ongoing acceleration of hardware. Nonetheless, the benefits from algorithmic breakthroughs contribute significantly to achieving the overall computational power to emulate human intelligence, and they are likely to continue to accrue.

The Ultimate Source of Intelligent Algorithms. The most important point here is that there is a specific game plan for achieving human-level intelligence in a machine: reverse engineer the parallel, chaotic, self-organizing, and fractal methods used in the human brain and apply these methods to modern computational hardware. Having tracked the exponentially increasing knowledge about the human brain and its methods (see chapter 4), we can expect that within twenty years we will have detailed models and simulations of the several hundred information-processing organs we collectively call the human brain.

Understanding the principles of operation of human intelligence will add to our toolkit of AI algorithms. Many of these methods used extensively in our machine pattern-recognition systems exhibit subtle and complex behaviors that are not predictable by the designer. Self-organizing methods are not an easy shortcut to the creation of complex and intelligent behavior, but they are one important way the complexity of a system can be increased without incurring the brittleness of explicitly programmed logical systems.

As I discussed earlier, the human brain itself is created from a genome with only thirty to one hundred million bytes of useful, compressed information. How is it, then, that an organ with one hundred trillion connections can result from a genome that is so small? (I estimate that just the interconnection data alone needed to characterize the human brain is one million times greater than the information in the genome.)[13] The answer is that the genome specifies a set of processes, each of which utilizes chaotic methods (that is, initial randomness, then self-organization) to increase the amount of information represented. It is known, for example, that the wiring of the interconnections follows a plan that includes a great deal of randomness. As an individual encounters his environment the connections and the neurotransmitter-level patterns self-organize to better represent the world, but the initial design is specified by a program that is not extreme in its complexity.

It is not my position that we will program human intelligence link by link in a massive rule-based expert system. Nor do we expect the broad set of skills represented by human intelligence to emerge from a massive genetic algorithm. Lanier worries correctly that any such approach would inevitably get stuck in some local minima (a design that is better than designs that are very similar to it but that is not actually optimal). Lanier also interestingly points out, as does Richard Dawkins, that biological evolution "missed the wheel" (in that no organism evolved to have one). Actually, that's not entirely accurate—there are small wheel-like structures at the protein level, for example the ionic motor in the bacterial flagellum, which is used for transportation in a three-dimensional environment.[14] With larger organisms, wheels are not very useful, of course, without roads, which is why there are no biologically evolved wheels for two-dimensional surface transportation.[15] However, evolution did generate a species that created both wheels and roads, so it did succeed in creating a lot of wheels, albeit indirectly. There is nothing wrong with indirect methods; we use them in engineering all the time. Indeed, indirection is how evolution works (that is, the products of each stage create the next stage).

Brain reverse engineering is not limited to replicating each neuron. In chapter 5 we saw how substantial brain regions containing millions or billions of neurons could be modeled by implementing parallel algorithms that are functionally equivalent. The feasibility of such neuromorphic approaches has been demonstrated with models and simulations of a couple dozen regions. As I discussed, this often results in substantially reduced computational requirements, as shown by Lloyd Watts, Carver Mead, and others.

Lanier writes that "if there ever was a complex, chaotic phenomenon, we are it." I agree with that but don't see this as an obstacle. My own area of interest is chaotic computing, which is how we do pattern recognition, which in turn is the heart of human intelligence. Chaos is part of the process of pattern recognition—it drives the process—and there is no reason that we cannot harness these methods in our machines just as they are utilized in our brains.

Lanier writes that "evolution has evolved, introducing sex, for instance, but evolution has never found a way to be any speed but very slow." But Lanier's comment is only applicable to biological evolution, not technological evolution. That's precisely why we've moved beyond biological evolution. Lanier is ignoring the essential nature of an evolutionary process: it accelerates because each stage introduces more powerful methods for creating the next stage. We've gone from billions of years for the first steps of biological evolution (RNA) to the fast pace of technological evolution today. The World Wide Web emerged in only a few years, distinctly faster than, say, the Cambrian explosion. These phenomena are all part of the same evolutionary process, which started out slow, is now going relatively quickly, and within a few decades will go astonishingly fast.

Lanier writes that "the whole enterprise of Artificial Intelligence is based on an intellectual mistake." Until such time that computers at least match human intelligence in every dimension, it will always remain possible for skeptics to say the glass is half empty. Every new achievement of AI can be dismissed by pointing out other goals that have not yet been accomplished. Indeed, this is the frustration of the AI practitioner: once an AI goal is achieved, it is no longer considered as falling within the realm of AI and becomes instead just a useful general technique. AI is thus often regarded as the set of problems that have not yet been solved.

But machines are indeed growing in intelligence, and the range of tasks that they can accomplish—tasks that previously required intelligent human attention—is rapidly increasing. As we discussed in chapters 5 and 6 there are hundreds of examples of operational narrow AI today.

As one example of many, I pointed out in the sidebar "Deep Fritz Draws" on pp. 274–78 that computer chess software no longer relies just on computational brute force. In 2002 Deep Fritz, running on just eight personal computers, performed as well as IBM's Deep Blue in 1997 based on improvements in its pattern-recognition algorithms. We see many examples of this kind of qualitative improvement in software intelligence. However, until such time as the entire range of human intellectual capability is emulated, it will always be possible to minimize what machines are capable of doing.

Once we have achieved complete models of human intelligence, machines will be capable of combining the flexible, subtle human levels of pattern recognition with the natural advantages of machine intelligence, in speed, memory capacity, and, most important, the ability to quickly share knowledge and skills.

The Criticism from Analog Processing

Many critics, such as the zoologist and evolutionary-algorithm scientist Thomas Ray, charge theorists like me who postulate intelligent computers with an alleged "failure to consider the unique nature of the digital medium."[16]

First of all, my thesis includes the idea of combining analog and digital methods in the same way that the human brain does. For example, more advanced neural nets are already using highly detailed models of human neurons, including detailed nonlinear, analog activation functions. There's a significant efficiency advantage to emulating the brain's analog methods. Analog methods are also not the exclusive province of biological systems. We used to refer to "digital computers" to distinguish them from the more ubiquitous analog computers widely used during World War II. The work of Carver Mead has shown the ability of silicon circuits to implement digital-controlled analog circuits entirely analogous to, and indeed derived from, mammalian neuronal circuits. Analog methods are readily re-created by conventional transistors, which are essentially analog devices. It is only by adding the mechanism of comparing the transistor's output to a threshold that it is made into a digital device.

More important, there is nothing that analog methods can accomplish that digital methods are unable to accomplish just as well. Analog processes can be emulated with digital methods (by using floating point representations), whereas the reverse is not necessarily the case.

The Criticism from the Complexity of Neural Processing

Another common criticism is that the fine detail of the brain's biological design is simply too complex to be modeled and simulated using nonbiological technology. For example, Thomas Ray writes:

> The structure and function of the brain or its components cannot be separated. The circulatory system provides life support for the brain, but it also delivers hormones that are an integral part of the chemical information processing function of the brain. The membrane of a neuron is a

structural feature defining the limits and integrity of a neuron, but it is also the surface along which depolarization propagates signals. The structural and life-support functions cannot be separated from the handling of information.[17]

Ray goes on to describe several of the "broad spectrum of chemical communication mechanisms" that the brain exhibits.

In fact, all of these features can readily be modeled, and a great deal of progress has already been made in this endeavor. The intermediate language is mathematics, and translating the mathematical models into equivalent non-biological mechanisms (examples include computer simulations and circuits using transistors in their native analog mode) is a relatively straightforward process. The delivery of hormones by the circulatory system, for example, is an extremely low-bandwidth phenomenon, which is not difficult to model and replicate. The blood levels of specific hormones and other chemicals influence parameter levels that affect a great many synapses simultaneously.

Thomas Ray concludes that "a metallic computation system operates on fundamentally different dynamic properties and could never precisely and exactly 'copy' the function of a brain." Following closely the progress in the related fields of neurobiology, brain scanning, neuron and neural-region modeling, neuron-electronic communication, neural implants, and related endeavors, we find that our ability to replicate the salient functionality of biological information processing can meet any desired level of precision. In other words the copied functionality can be "close enough" for any conceivable purpose or goal, including satisfying a Turing-test judge. Moreover, we find that efficient implementations of the mathematical models require substantially less computational capacity than the theoretical potential of the biological neuron clusters being modeled. In chapter 4, I reviewed a number of brain-region models (Watts's auditory regions, the cerebellum, and others) that demonstrate this.

Brain Complexity. Thomas Ray also makes the point that we might have difficulty creating a system equivalent to "billions of lines of code," which is the level of complexity he attributes to the human brain. This figure, however, is highly inflated, for as we have seen our brains are created from a genome of only about thirty to one hundred million bytes of unique information (eight hundred million bytes without compression, but compression is clearly feasible given the massive redundancy), of which perhaps two thirds describe the principles of operation of the brain. It is self-organizing processes that incorporate significant elements of randomness (as well as exposure to the real world) that enable so relatively small an amount of design information to be expanded to the thousands of trillions of bytes of information represented in a mature human brain. Similarly, the task of creating human-level intelligence in a non-biological entity will involve creating not a massive expert system comprising billions of rules or lines of code but rather a learning, chaotic, self-organizing system, one that is ultimately biologically inspired.

Ray goes on to write, "The engineers among us might propose nano-molecular devices with fullerene switches, or even DNA-like computers. But I am sure they would never think of neurons. Neurons are astronomically large structures compared to the molecules we are starting with."

This is exactly my own point. The purpose of reverse engineering the human brain is not to copy the digestive or other unwieldy processes of biological neurons but rather to understand their key information-processing methods. The feasibility of doing this has already been demonstrated in dozens of contemporary projects. The complexity of the neuron clusters being emulated is scaling up by orders of magnitude, along with all of our other technological capabilities.

A Computer's Inherent Dualism. Neuroscientist Anthony Bell of Redwood Neuroscience Institute articulates two challenges to our ability to model and simulate the brain with computation. In the first he maintains that

> a computer is an intrinsically dualistic entity, with its physical set-up designed not to interfere with its logical set-up, which executes the computation. In empirical investigation, we find that the brain is not a dualistic entity. Computer and program may be two, but mind and brain are one. The brain is thus not a machine, meaning it is not a finite model (or computer) instantiated physically in such a way that the physical instantiation does not interfere with the execution of the model (or program).[18]

This argument is easily dispensed with. The ability to separate in a computer the program from the physical instantiation that performs the computation is an advantage, not a limitation. First of all, we do have electronic devices with dedicated circuitry in which the "computer and program" are not two, but one. Such devices are not programmable but are hardwired for one specific set of algorithms. Note that I am not just referring to computers with software (called "firmware") in read-only memory, as may be found in a cell phone or pocket computer. In such a system, the electronics and the software may still be considered dualistic even if the program cannot easily be modified.

I am referring instead to systems with dedicated logic that cannot be programmed at all—such as application-specific integrated circuits (used, for example, for image and signal processing). There is a cost efficiency in implementing algorithms in this way, and many electronic consumer products use such circuitry. Programmable computers cost more but provide the flexibility of allowing the software to be changed and upgraded. Programmable computers can emulate the functionality of any dedicated system, including the algorithms that we are discovering (through the efforts to reverse engineer the brain) for neural components, neurons, and brain regions.

There is no validity to calling a system in which the logical algorithm is inherently tied to its physical design "not a machine." If its principles of operation can be understood, modeled in mathematical terms, and then instantiated on another system (whether that other system is a machine with unchangeable dedicated logic or software on a programmable computer), then we can consider it to be a machine and certainly an entity whose capabilities can be re-created in a machine. As I discussed extensively in chapter 4, there are no barriers to our discovering the brain's principles of operation and successfully modeling and simulating them, from its molecular interactions upward.

Bell refers to a computer's "physical set-up [that is] designed not to interfere with its logical set-up," implying that the brain does not have this "limitation." He is correct that our thoughts do help create our brains, and as I pointed out earlier we can observe this phenomenon in dynamic brain scans. But we can readily model and simulate both the physical and logical aspects of the brain's plasticity in software. The fact that software in a computer is separate from its physical instantiation is an architectural advantage in that it allows the same software to be applied to ever-improving hardware. Computer software, like the brain's changing circuits, can also modify itself, as well as be upgraded.

Computer hardware can likewise be upgraded without requiring a change in software. It is the brain's relatively fixed architecture that is severely limited. Although the brain is able to create new connections and neurotransmitter patterns, it is restricted to chemical signaling more than one million times slower than electronics, to the limited number of interneuronal connections that can fit inside our skulls, and to having no ability to be upgraded, other than through the merger with nonbiological intelligence that I've been discussing.

Levels and Loops. Bell also comments on the apparent complexity of the brain:

> Molecular and biophysical processes control the sensitivity of neurons to incoming spikes (both synaptic efficiency and post-synaptic responsivity), the excitability of the neuron to produce spikes, the patterns of

spikes it can produce and the likelihood of new synapses forming (dynamic rewiring), to list only four of the most obvious interferences from the subneural level. Furthermore, transneural volume effects such as local electric fields and the transmembrane diffusion of nitric oxide have been seen to influence, respectively, coherent neural firing, and the delivery of energy (blood flow) to cells, the latter of which directly correlates with neural activity.

The list could go on. I believe that anyone who seriously studies neuromodulators, ion channels or synaptic mechanism and is honest, would have to reject the neuron level as a separate computing level, even while finding it to be a useful descriptive level.[19]

Although Bell makes the point here that the neuron is not the appropriate level at which to simulate the brain, his primary argument here is similar to that of Thomas Ray above: the brain is more complicated than simple logic gates.

He makes this explicit:

To argue that one piece of structured water or one quantum coherence is a necessary detail in the functional description of the brain would clearly be ludicrous. But if, in every cell, molecules derive systematic functionality from these submolecular processes, if these processes are used all the time, all over the brain, to reflect, record and propagate spatio-temporal correlations of molecular fluctuations, to enhance or diminish the probabilities and specificities of reactions, then we have a situation qualitatively different from the logic gate.

At one level he is disputing the simplistic models of neurons and interneuronal connections used in many neural-net projects. Brain-region simulations don't use these simplified models, however, but rather apply realistic mathematical models based on the results from brain reverse engineering.

The real point that Bell is making is that the brain is immensely complicated, with the consequent implication that it will therefore be very difficult to understand, model, and simulate its functionality. The primary problem with Bell's perspective is that he fails to account for the self-organizing, chaotic, and fractal nature of the brain's design. It's certainly true that the brain is complex, but a lot of the complication is more apparent than real. In other words, the principles of the design of the brain are simpler than they appear.

To understand this, let's first consider the fractal nature of the brain's organization, which I discussed in chapter 2. A fractal is a rule that is iteratively applied to create a pattern or design. The rule is often quite simple, but because of the iteration the resulting design can be remarkably complex. A famous example of this is the Mandelbrot set devised by mathematician Benoit Mandelbrot.[20] Visual images of the Mandelbrot set are remarkably complex, with endlessly complicated designs within designs. As we look at finer and finer detail in an image of the Mandelbrot set, the complexity never goes away, and we continue to see ever finer complication. Yet the formula underlying all of this complexity is amazingly simple: the Mandelbrot set is characterized by a single formula $Z = Z^2 + C$, in which Z is a "complex" (meaning two-dimensional) number and C is a constant. The formula is iteratively applied, and the resulting two-dimensional points are graphed to create the pattern.

Mandelbrot Set

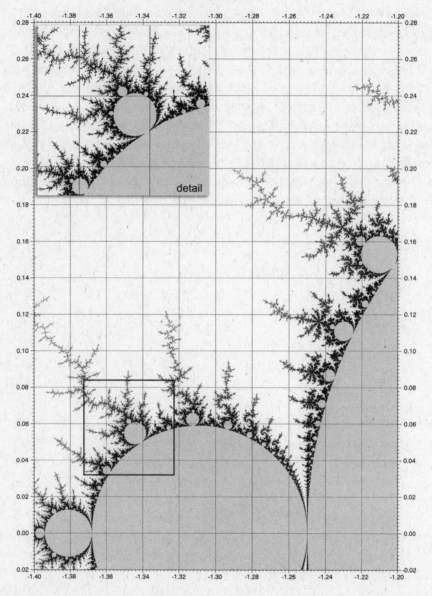

detail

The point here is that a simple design rule can create a lot of apparent complexity. Stephen Wolfram makes a similar point using simple rules on cellular automata (see chapter 2). This insight holds true for the brain's design. As I've discussed, the compressed genome is a relatively compact design, smaller than some contemporary software programs. As Bell points out, the actual implementation of the brain appears far more complex than this. Just as with the

Mandelbrot set, as we look at finer and finer features of the brain, we continue to see apparent complexity at each level. At a macro level the pattern of connections looks complicated, and at a micro level so does the design of a single portion of a neuron such as a dendrite. I've mentioned that it would take at least thousands of trillions of bytes to characterize the state of a human brain, but the design is only tens of millions of bytes. So the ratio of the apparent complexity of the brain to the design information is at least one hundred million to one. The brain's information starts out as largely random information, but as the brain interacts with a complex environment (that is, as the person learns and matures), that information becomes meaningful.

The actual design complexity is governed by the compressed information in the design (that is, the genome and supporting molecules), not by the patterns created through the iterative application of the design rules. I would agree that the roughly thirty to one hundred million bytes of information in the genome do not represent a simple design (certainly far more complex than the six characters in the definition of the Mandelbrot set), but it is a level of complexity that we can already manage with our technology. Many observers are confused by the apparent complexity in the brain's physical instantiation, failing to recognize that the fractal nature of the design means that the actual design information is far simpler than what we see in the brain.

I also mentioned in chapter 2 that the design information in the genome is a probabilistic fractal, meaning that the rules are applied with a certain amount of randomness each time a rule is iterated. There is, for example, very little information in the genome describing the wiring pattern for the cerebellum, which comprises more than half the neurons in the brain. A small number of genes describe the basic pattern of the four cell types in the cerebellum and then say in essence, "Repeat this pattern several billion times with some random variation in each repetition." The result may look very complicated, but the design information is relatively compact.

Bell is correct that trying to compare the brain's design to a conventional computer would be frustrating. The brain does not follow a typical top-down (modular) design. It uses its probabilistic fractal type of organization to create processes that are chaotic—that is, not fully predictable. There is a well-developed body of mathematics devoted to modeling and simulating chaotic systems, which are used to understand phenomena such as weather patterns and financial markets, that is also applicable to the brain.

Bell makes no mention of this approach. He argues why the brain is dramatically different from conventional logic gates and conventional software design, which leads to his unwarranted conclusion that the brain is not a machine and cannot be modeled by a machine. While he is correct that standard logic gates and the organization of conventional modular software are not the appropriate way to think about the brain, that does not mean that we are unable to simulate the brain on a computer. Because we can describe the brain's principles of operation in mathematical terms, and since we can model any mathematical process (including chaotic ones) on a computer, we are able to implement these types of simulations. Indeed, we're making solid and accelerating progress in doing so.

Despite his skepticism Bell expresses cautious confidence that we will understand our biology and brains well enough to improve on them. He writes: "Will there be a transhuman age? For this there is a strong biological precedent in the two major steps in biological evolution. The first, the incorporation into eukaryotic bacteria of prokaryotic symbiotes, and the second, the emergence of multicellular life-forms from colonies of eukaryotes. . . . I believe that something like [a transhumanist age] may happen."

The Criticism from Microtubules and Quantum Computing

> Quantum mechanics is mysterious, and consciousness is mysterious. Q.E.D.: Quantum mechanics and consciousness must be related.
>
> —CHRISTOF KOCH, MOCKING ROGER PENROSE'S THEORY OF QUANTUM
> COMPUTING IN NEURON TUBULES AS THE SOURCE OF HUMAN
> CONSCIOUSNESS[21]

Over the past decade Roger Penrose, a noted physicist and philosopher, in conjunction with Stuart Hameroff, an anesthesiologist, has suggested that fine structures in the neurons called microtubules perform an exotic form of computation called "quantum computing." As I discussed, quantum computing is computing using what are called qubits, which take on all possible combinations of solutions simultaneously. The method can be considered to be an extreme form of parallel processing (because every combination of values of the qubits is tested simultaneously). Penrose suggests that the microtubules and their quantum-computing capabilities complicate the concept of re-creating neurons and reinstantiating mind files.[22] He also hypothesizes that the brain's quantum computing is responsible for consciousness and that systems, biological or otherwise, cannot be conscious without quantum computing.

Although some scientists have claimed to detect quantum wave collapse (resolution of ambiguous quantum properties such as position, spin, and velocity) in the brain, no one has suggested that human capabilities actually require a capacity for quantum computing. Physicist Seth Lloyd said:

> I think that it is incorrect that microtubules perform computing tasks in the brain, in the way that [Penrose] and Hameroff have proposed. The brain is a hot, wet place. It is not a very favorable environment for exploiting quantum coherence. The kinds of superpositions and assembly/disassembly of microtubules for which they search do not seem to exhibit quantum entanglement.... The brain clearly isn't a classical, digital computer by any means. But my guess is that it performs most of its tasks in a "classical" manner. If you were to take a large enough computer, and model all of the neurons, dendrites, synapses, and such, [then] you could probably get the thing to do most of the tasks that brains perform. I don't think that the brain is exploiting any quantum dynamics to perform tasks.[23]

Anthony Bell also remarks that "there is no evidence that large-scale macroscopic quantum coherences, such as those in superfluids and superconductors, occur in the brain."[24]

However, even if the brain does do quantum computing, this does not significantly change the outlook for human-level computing (and beyond), nor does it suggest that brain uploading is infeasible. First of all, if the brain does do quantum computing this would only verify that quantum computing is feasible. There would be nothing in such a finding to suggest that quantum computing is restricted to biological mechanisms. Biological quantum-computing mechanisms, if they exist, could be replicated. Indeed, recent experiments with small-scale quantum computers appear to be successful. Even the conventional transistor relies on the quantum effect of electron tunneling.

Penrose's position has been interpreted to imply that it is impossible to perfectly replicate a set of quantum states, so therefore perfect downloading is impossible. Well, how perfect does a download have to be? If we develop downloading technology to the point where the "copies" are as close to the original as the original person is to him- or herself over the course of one minute, that would be good enough for any conceivable purpose yet would not require copying quantum states. As the technology improves, the accuracy of the copy could become as close as the original to within ever briefer periods of time (one second, one millisecond, one microsecond).

When it was pointed out to Penrose that neurons (and even neural connections) were too big for quantum computing, he came up with the tubule theory as a possible mechanism for neural quantum computing. If one is searching for barriers to replicating brain function it is an ingenious theory, but it fails to introduce any genuine barriers. However, there is little evidence to suggest that microtubules, which provide structural integrity to the neural cells, perform quantum computing and that this capability contributes to the thinking process. Even generous models of human knowledge and potential are more than accounted for by current estimates of brain size, based on contemporary models of neuron functioning that do not include microtubule-based quantum computing. Recent experiments showing that hybrid biological/ nonbiological networks perform similarly to all-biological networks, while not definitive, are strongly suggestive that our microtubuleless models of neuron functioning are adequate. Lloyd Watts's software simulation of his intricate model of human auditory processing uses orders of magnitude less computation than the networks of neurons he is simulating, and again there is no suggestion that quantum computing is needed. I reviewed other ongoing efforts to model and simulate brain regions in chapter 4, while in chapter 3 I discussed estimates of the amount of computation necessary to simulate all regions of the brain based on functionally equivalent simulations of different regions. None of these analyses demonstrates the necessity for quantum computing in order to achieve human-level performance.

Some detailed models of neurons (in particular those by Penrose and Hameroff) do assign a role to the microtubules in the functioning and growth of dendrites and axons. However, successful neuromorphic models of neural regions do not appear to require microtubule components. For neuron models that do consider microtubules, results appear to be satisfactory by modeling their overall chaotic behavior without modeling each microtubule filament individually. However, even if the Penrose-Hameroff tubules are an important factor, accounting for them doesn't change the projections I have discussed above to any significant degree. According to my model of computational growth, if the tubules multiplied neuron complexity by even a factor of one thousand (and keep in mind that our current tubuleless neuron models are already complex, including on the order of one thousand connections per neuron, multiple nonlinearities, and other details), this would delay our reaching brain capacity by only about nine years. If we're off by a factor of one million, that's still a delay of only seventeen years. A factor of a billion is around twenty-four years (recall that computation is growing by a double exponential).[25]

The Criticism from the Church-Turing Thesis

Early in the twentieth century mathematicians Alfred North Whitehead and Bertrand Russell published their seminal work, *Principia Mathematica*, which sought to determine axioms that could serve as the basis for all of mathematics.[26] However, they were unable to prove conclusively that an axiomatic system that can generate the natural numbers (the positive integers or counting numbers) would not give rise to contradictions. It was assumed that such a proof would be found sooner or later, but in the 1930s a young Czech mathematician, Kurt Gödel, stunned the mathematical world by proving that within such a system there inevitably exist propositions that can be neither proved nor disproved. It was later shown that such unprovable propositions are as common as provable ones. Gödel's incompleteness theorem, which is fundamentally a proof demonstrating that there are definite limits to what logic, mathematics, and by extension computation can do, has been called the most important in all mathematics, and its implications are still being debated.[27]

A similar conclusion was reached by Alan Turing in the context of understanding the nature of computation. When in 1936 Turing presented the Turing machine (described in chapter 2) as a theoretical model of a computer, which continues today to form the basis of modern computational theory, he reported an unexpected discovery similar to Gödel's.[28] In his paper that year he described the concept of unsolvable problems—that is, problems that are well defined, with unique answers that can be shown to exist, but that we can also show can never be computed by a Turing machine.

The fact that there are problems that cannot be solved by this particular theoretical machine may not seem particularly startling until you consider the other conclusion of Turing's paper: that the Turing machine can model any computational process. Turing showed that there are as many unsolvable problems as solvable ones, the number of each being the lowest order of infinity, the so-called countable infinity (that is, counting the number of integers). Turing also demonstrated that the problem of determining the truth or falsity of any logical proposition in an arbitrary system of logic powerful enough to represent the natural numbers was one example of an unsolved problem, a result similar to Gödel's. (In other words, there is no procedure guaranteed to answer this question for all such propositions.)

Around the same time Alonzo Church, an American mathematician and philosopher, published a theorem that examined a similar question in the context of arithmetic. Church independently came to the same conclusion as Turing.[29] Taken together, the works of Turing, Church, and Gödel were the first formal proofs that there are definite limits to what logic, mathematics, and computation can do.

In addition, Church and Turing also advanced, independently, an assertion that has become known as the Church-Turing thesis. This thesis has both weak and strong interpretations. The weak interpretation is that if a problem that can be presented to a Turing machine is not solvable by one, then it is not solvable by any machine. This conclusion follows from Turing's demonstration that the Turing machine could model any algorithmic process. It is only a small step from there to describe the behavior of a machine as following an algorithm.

The strong interpretation is that problems that are not solvable on a Turing machine cannot be solved by human thought, either. The basis of this thesis is that human thought is performed by the human brain (with some influence by the body), that the human brain (and body) comprises matter and energy, that matter and energy follow natural laws, that these laws are describable in mathematical terms, and that mathematics can be simulated to any degree of precision by algorithms. Therefore there exist algorithms that can simulate human thought. The strong version of the Church-Turing thesis postulates an essential equivalence between what a human can think or know and what is computable.

It is important to note that although the existence of Turing's unsolvable problems is a mathematical certainty, the Church-Turing thesis is not a mathematical proposition at all. It is, rather, a conjecture that, in various disguises, is at the heart of some of our most profound debates in the philosophy of mind.[30]

The criticism of strong AI based on the Church-Turing thesis argues the following: since there are clear limitations to the types of problems that a computer can solve, yet humans are capable of solving these problems, machines will never emulate the full range of human intelligence. This conclusion, however, is not warranted. Humans are no more capable of universally solving such "unsolvable" problems than machines are. We can make educated guesses to solutions in certain instances and can apply heuristic methods (procedures that attempt to solve problems but that are not guaranteed to work) that succeed on occasion. But both these approaches are also algorithmically based processes, which means that machines are also capable of doing them. Indeed, machines can often search for solutions with far greater speed and thoroughness than humans can.

The strong formulation of the Church-Turing thesis implies that biological brains and machines are equally subject to the laws of physics, and therefore mathematics can model and simulate them equally. We've already demonstrated the ability to model and simulate the function of neurons, so why not a system of a hundred billion neurons? Such a system would display the same complexity and lack of predictability as human intelligence. Indeed, we already have computer algorithms (for example, genetic algorithms) with results that are complex and unpredictable and that provide intelligent solutions to problems. If anything, the Church-Turing thesis implies that brains and machines are essentially equivalent.

To see machines' ability to use heuristic methods, consider one of the most interesting of the unsolvable problems, the "busy beaver" problem, formulated by Tibor Rado in 1962.[31] Each Turing machine has a certain number of states that its internal program can be in, which correspond to the number of steps in its internal program. There are a number of different 4-state Turing machines that are possible, a certain number of 5-state machines, and so on. In the "busy beaver" problem, given a positive integer n, we construct all the Turing machines that have n states. The number of such machines will always be finite. Next we eliminate those n-state machines that get into an infinite loop (that is, never halt). Finally, we select the machine (one that does halt) that writes the largest number of 1s on its tape. The number of 1s that this Turing machine writes is called the busy beaver of n. Rado showed that there is no algorithm—that is, no Turing machine—that can compute this function for all ns. The crux of the problem is sorting out those n-state machines that get into infinite loops. If we program a Turing machine to generate and simulate all possible n-state Turing machines, this simulator *itself* gets into an infinite loop when it attempts to simulate one of the n-state machines that gets into an infinite loop.

Despite its status as an unsolvable problem (and one of the most famous), we can determine the busy-beaver function for some ns. (Interestingly, it is also an unsolvable problem to separate those ns for which we can determine the busy beaver of n from those for which we cannot.) For example, the busy beaver of 6 is easily determined to be 35. With seven states, a Turing machine can multiply, so the busy beaver of 7 is much bigger: 22,961. With eight states, a Turing machine can compute exponentials, so the busy beaver of 8 is even bigger: approximately 10^{43}. We can see that this is an "intelligent" function, in that it requires greater intelligence to solve for larger ns.

By the time we get to 10, a Turing machine can perform types of calculations that are impossible for a human to follow (without help from a computer). So we were able to determine the busy beaver of 10 only with a computer's assistance. The answer requires an exotic notation to write down, in which we have a stack of exponents, the height of which is determined by another stack of exponents, the height of which is determined by another stack of exponents, and so on. Because a computer can keep track of such complex numbers, whereas the human brain cannot, it appears that computers will prove more capable of solving unsolvable problems than humans will.

The Criticism from Failure Rates

Jaron Lanier, Thomas Ray, and other observers all cite high failure rates of technology as a barrier to its continued exponential growth. For example, Ray writes:

> The most complex of our creations are showing alarming failure rates. Orbiting satellites and telescopes, space shuttles, interplanetary probes, the Pentium chip, computer operating systems, all seem to be pushing the limits of what we can effectively design and build through conventional approaches. . . . Our most complex software (operating systems and telecommunications control systems) already contains tens of millions of lines of code. At present it seems unlikely that we can produce and manage software with hundreds of millions or billions of lines of code.[32]

First, we might ask what alarming failure rates Ray is referring to. As mentioned earlier, computerized systems of significant sophistication routinely fly and land our airplanes automatically and monitor intensive care units in hospitals, yet almost never malfunction. If alarming failure rates are of concern, they're more often attributable to human error. Ray alludes to problems with Intel microprocessor chips, but these problems have been extremely subtle, have caused almost no repercussions, and have quickly been rectified.

The complexity of computerized systems has indeed been scaling up, as we have seen, and moreover the cutting edge of our efforts to emulate human intelligence will utilize the self-organizing paradigms that we find in the human brain. As we continue our progress in reverse engineering the human brain, we will add new self-organizing methods to our pattern recognition and AI toolkit. As I have discussed, self-organizing methods help to alleviate the need for unmanageable levels of complexity. As I pointed out earlier, we will not need systems with "billions of lines of code" to emulate human intelligence.

It is also important to point out that imperfection is an inherent feature of any complex process, and that certainly includes human intelligence.

The Criticism from "Lock-In"

Jaron Lanier and other critics have cited the prospect of a "lock-in," a situation in which old technologies resist displacement because of the large investment in the infrastructure supporting them. They argue that pervasive and complex support systems have blocked innovation in such fields as transportation, which have not seen the rapid development that we've seen in computation.[33]

The concept of lock-in is not the primary obstacle to advancing transportation. If the existence of a complex support system necessarily caused lock-in, then why don't we see this phenomenon affecting the expansion of every aspect of the Internet? After all, the Internet certainly requires an enormous and complex infrastructure. Because it is specifically the processing and movement of information that is growing exponentially, however, one reason that an area such as transportation has reached a plateau (that is, resting at the top of an S-curve) is that many if not most of its purposes have been satisfied by exponentially growing communication technologies. My own organization, for example, has colleagues in different parts of the country, and most of our needs that in times past would have required a person or a package to be transported can be met through the increasingly viable virtual meetings (and electronic distribution of documents and other intellectual creations) made possible by a panoply of communication technologies, some of which Lanier himself is working to advance. More important, we will see advances in transportation facilitated by the nanotechnology-based energy technologies I discussed in chapter 5. However, with increasingly realistic, high-resolution full-immersion forms of virtual reality continuing to emerge, our needs to be together will increasingly be met through computation and communication.

As I discussed in chapter 5, the full advent of MNT-based manufacturing will bring the law of accelerating returns to such areas as energy and transportation. Once we can create virtually any physical product from information and very inexpensive raw materials, these traditionally slow-moving industries will see the same kind of annual doubling of price-performance and capacity that we see in information technologies. Energy and transportation will effectively become information technologies.

We will see the advent of nanotechnology-based solar panels that are efficient, lightweight, and inexpensive, as well as comparably powerful fuel cells and other technologies to store and distribute that energy. Inexpensive energy will in turn transform transportation. Energy obtained from nanoengineered solar cells and other renewable technologies and stored in nanoengineered fuel cells will provide clean and inexpensive energy for every type of transportation. In addition, we will be able to manufacture devices—including flying machines of varying sizes—for almost no cost, other than the cost of the design (which needs to be amortized only once). It will be feasible, therefore, to build inexpensive small flying devices that can transport a package directly to your destination in a matter of hours without going through intermediaries such as shipping companies. Larger but still inexpensive vehicles will be able to fly people from place to place with nanoengineered microwings.

Information technologies are already deeply influential in every industry. With the full realization of the GNR revolutions in a few decades, every area of human endeavor will essentially comprise information technologies and thus will directly benefit from the law of accelerating returns.

The Criticism from Ontology: Can a Computer Be Conscious?

> Because we do not understand the brain very well we are constantly tempted to use the latest technology as a model for trying to understand it. In my childhood we were always assured that the brain was a telephone switchboard. ("What else could it be?") I was amused to see that Sherrington, the great British neuroscientist, thought that the brain worked like a telegraph system. Freud often compared the brain to hydraulic and electromagnetic systems. Leibniz compared it to a mill, and I am told some of the ancient Greeks thought the brain functions like a catapult. At present, obviously, the metaphor is the digital computer.
>
> —JOHN R. SEARLE, "MINDS, BRAINS, AND SCIENCE"

Can a computer—a nonbiological intelligence—be conscious? We have first, of course, to agree on what the question means. As I discussed earlier, there are conflicting perspectives on what may at first appear to be a straightforward issue. Regardless of how we attempt to define the concept, however, we must acknowledge that consciousness is widely regarded as a crucial, if not essential, attribute of being human.[34]

John Searle, distinguished philosopher at the University of California at Berkeley, is popular among his followers for what they believe is a staunch defense of the deep mystery of human consciousness against trivialization by strong-AI "reductionists" like Ray Kurzweil. And even though I have always found Searle's logic in his celebrated Chinese Room argument to be tautological, I had expected an elevating treatise on the paradoxes of consciousness. Thus it is with some surprise that I find Searle writing statements such as, "human brains cause consciousness by a series of specific neurobiological processes in the brain";

"The essential thing is to recognize that consciousness is a biological process like digestion, lactation, photosynthesis, or mitosis";

"The brain is a machine, a biological machine to be sure, but a machine all the same. So the first step is to figure out how the brain does it and then build an artificial machine that has an equally effective mechanism for causing consciousness"; and

"We know that brains cause consciousness with specific biological mechanisms."[35]

So who is being the reductionist here? Searle apparently expects that we can measure the subjectivity of another entity as readily as we measure the oxygen output of photosynthesis.

Searle writes that I "frequently cite IBM's Deep Blue as evidence of superior intelligence in the computer." Of course, the opposite is the case: I cite Deep Blue not to belabor the issue of chess but rather to examine the clear contrast it illustrates between the human and contemporary machine approaches to the game. As I pointed out earlier, however, the pattern-recognition ability of chess programs is increasing, so chess machines are beginning to combine the analytical strength of traditional machine intelligence with more humanlike pattern recognition. The human paradigm (of self-organizing chaotic processes) offers profound advantages: we can recognize and respond to extremely subtle patterns. But we can build machines with the same abilities. That, indeed, has been my own area of technical interest.

Searle is best known for his Chinese Room analogy and has presented various formulations of it over twenty years. One of the more complete descriptions of it appears in his 1992 book, *The Rediscovery of the Mind:*

I believe the best-known argument against strong AI was my Chinese room argument . . . that showed that a system could instantiate a program so as to give a perfect simulation of some human cognitive capacity, such as the capacity to understand Chinese, even though that system had no understanding of Chinese whatever. Simply imagine that someone who understands no Chinese is locked in a room with a lot of Chinese symbols and a computer program for answering questions in Chinese. The input to the system consists in Chinese symbols in the form of questions; the output of the system consists in Chinese symbols in answer to the questions. We might suppose that the program is so good that the answers to the questions are indistinguishable from those of a native Chinese speaker. But all the same, neither the person inside nor any other part of the system literally understands Chinese; and because the programmed computer has nothing that this system does not have, the programmed computer, qua computer, does not understand Chinese either. Because the program is purely formal or syntactical and because minds have mental or semantic contents, any attempt to produce a mind purely with computer programs leaves out the essential features of the mind.[36]

Searle's descriptions illustrate a failure to evaluate the essence of either brain processes or the nonbiological processes that could replicate them. He starts with the assumption that the "man" in the room doesn't understand anything because, after all, "he is just a computer," thereby illuminating his own bias. Not surprisingly Searle then concludes that the computer (as implemented by the man) doesn't understand. Searle combines this tautology with a basic contradiction: the computer doesn't understand Chinese, yet (according to Searle) can convincingly answer questions in Chinese. But if an entity—biological or otherwise—really doesn't understand human language, it will quickly be unmasked by a competent interlocutor. In addition, for the program to respond convincingly, it would have to be as complex as a human brain. The observers would long be dead while the man in the room spends millions of years following a program many millions of pages long.

Most important, the man is acting only as the central processing unit, a small part of a system. While the man may not see it, the understanding is distributed across the entire pattern of the program itself and the billions of notes he would have to make to follow the program. *I understand English, but none of my neurons do.* My understanding is represented in vast patterns of neurotransmitter strengths, synaptic clefts, and interneuronal connections. Searle fails to account for the significance of distributed patterns of information and their emergent properties.

A failure to see that computing processes are capable of being—just like the human brain—chaotic, unpredictable, messy, tentative, and emergent is behind much of the criticism of the prospect of intelligent machines that we hear from Searle and other essentially materialist philosophers. Inevitably Searle comes back to a criticism of "symbolic" computing: that orderly sequential symbolic processes cannot re-create true thinking. I think that's correct (depending, of course, on what level we are modeling an intelligent process), but the manipulation of symbols (in the sense that Searle implies) is not the only way to build machines, or computers.

So-called computers (and part of the problem is the word "computer," because machines can do more than "compute") are not limited to symbolic processing. Nonbiological entities can also use the emergent self-organizing paradigm, which is a trend well under way and one that will become even more important over the next several decades. Computers do not have to use only 0 and 1, nor do they have to be all digital. Even if a computer is all digital, digital algorithms can simulate analog processes to any degree of precision (or lack of precision). Machines can be massively parallel. And machines can use chaotic emergent techniques just as the brain does.

The primary computing techniques that we have used in pattern-recognition systems do not use symbol manipulation but rather self-organizing methods such as those described in chapter 5 (neural nets, Markov models, genetic algorithms, and more complex paradigms based on brain reverse engineering). A machine that could really do what Searle describes in the Chinese Room argument would not merely be manipulating language symbols, because that approach doesn't work. This is at the heart of the philosophical sleight of hand underlying the Chinese Room. The nature of computing is not limited to manipulating logical symbols. Something is going on in the human brain, and there is nothing that prevents these biological processes from being reverse engineered and replicated in nonbiological entities.

Adherents appear to believe that Searle's Chinese Room argument demonstrates that machines (that is, nonbiological entities) can never truly understand anything of significance, such as Chinese. First, it is important to recognize that for this system—the person and the computer—to, as Searle puts it, "give a perfect simulation of some human cognitive capacity, such as the capacity to understand Chinese," and to convincingly answer questions in Chinese, it must essentially pass a Chinese Turing test. Keep in mind that we are not talking about answering questions from a fixed list of stock questions (because that's a trivial task) but answering any unanticipated question or sequence of questions from a knowledgeable human interrogator.

Now, the human in the Chinese Room has little or no significance. He is just feeding things into the computer and mechanically transmitting its output (or, alternatively, just following the rules in the program). And neither the computer nor the human needs to be in a room. Interpreting Searle's description to imply that the man himself is implementing the program does not change anything other than to make the system far slower than real time and extremely error prone. *Both the human and the room are irrelevant.* The only thing that is significant is the computer (either an electronic computer or the computer comprising the man following the program).

For the computer to really perform this "perfect simulation," it would indeed have to understand Chinese. According to the very premise it has "the capacity to understand Chinese," so it is then entirely contradictory to say that "the programmed computer . . . does not understand Chinese."

A computer and computer program *as we know them today* could not successfully perform the described task. So if we are to understand the computer to be like today's computers, then it cannot fulfill the premise. The only way that it could do so would be if it had the depth and complexity of a human. Turing's brilliant insight in proposing his test was that convincingly answering any possible sequence of questions from an intelligent human questioner in a human language really probes all of human intelligence. A computer that is capable of accomplishing this—a computer that will exist a few decades from now—will need to be of human complexity or greater and will indeed understand Chinese in a deep way, because otherwise it would never be convincing in its claim to do so.

Merely stating, then, that the computer "does not literally understand Chinese" does not make sense, for it contradicts the entire premise of the argument. To claim that the computer is not conscious is not a compelling contention, either. To be consistent with some of Searle's other statements, we have to conclude that we really don't know if it is conscious or not. With regard to relatively simple machines, including today's computers, while we can't state for certain that these entities are not conscious, their behavior, including their inner workings, doesn't give us that impression. But that will not be true for a computer that can really do what is needed in the Chinese Room. Such a machine will at least *seem* conscious, even if we cannot say definitively whether it is or not. But just declaring that it is obvious that the computer (or the entire system of the computer, person, and room) is not conscious is far from a compelling argument.

In the quote above Searle states that "the program is purely formal or syntactical." But as I pointed out earlier, that is a bad assumption, based on Searle's failure to account for the requirements of such a technology. This assumption is behind much of Searle's criticism of AI. A program that is purely formal or syntactical will not be able to understand Chinese, and it won't "give a perfect simulation of some human cognitive capacity."

But again, we don't have to build our machines that way. We can build them in the same fashion that nature built the human brain: using chaotic emergent methods that are massively parallel. Furthermore, there is nothing inherent in the concept of a machine that restricts its expertise to the level of syntax alone and prevents it from mastering semantics. Indeed, if the machine inherent in Searle's conception of the Chinese Room had not mastered semantics, it would not be able to convincingly answer questions in Chinese and thus would contradict Searle's own premise.

In chapter 4 I discussed the ongoing effort to reverse engineer the human brain and to apply these methods to computing platforms of sufficient power. So, like a human brain, if we teach a computer Chinese, it will understand Chinese. This may seem to be an obvious statement, but it is one with which Searle takes issue. To use his own terminology, I am not talking about a simulation per se but rather a duplication of the causal powers of the massive neuron cluster that constitutes the brain, at least those causal powers salient and relevant to thinking.

Will such a copy be conscious? I don't think the Chinese Room tells us anything about this question.

It is also important to point out that Searle's Chinese Room argument can be applied to the human brain itself. Although it is clearly not his intent, his line of reasoning implies that the human brain has no understanding. He writes: "The computer . . . succeeds by manipulating formal symbols. The symbols themselves are quite meaningless: they have only the meaning we have attached to them. The computer knows nothing of this, it just shuffles the symbols." Searle acknowledges that biological neurons are machines, so if we simply substitute the phrase "human brain" for "computer" and "neurotransmitter concentrations and related mechanisms" for "formal symbols," we get:

> The [human brain] . . . succeeds by manipulating [neurotransmitter concentrations and related mechanisms]. The [neurotransmitter concentrations and related mechanisms] themselves are quite meaningless: they have only the meaning we have attached to them. The [human brain] knows nothing of this, it just shuffles the [neurotransmitter concentrations and related mechanisms].

Of course, neurotransmitter concentrations and other neural details (for example, interneuronal connection and neurotransmitter patterns) have no meaning in and of themselves. The meaning and understanding that emerge in the human brain are exactly that: an *emergent* property of its complex patterns of activity. The same is true for machines. Although "shuffling symbols" does not have meaning in and of itself, the emergent patterns have the same potential role in nonbiological systems as they do in biological systems such as the brain. Hans Moravec has written, "Searle is looking for understanding in the wrong places. . . . [He] seemingly cannot accept that real meaning can exist in mere patterns."[37]

Let's address a second version of the Chinese Room. In this conception the room does not include a computer or a man simulating a computer but has a room full of people manipulating slips of paper with Chinese symbols on them—essentially, a lot of people simulating a computer. This system would convincingly answer questions in Chinese, but none of the participants would know Chinese, nor could we say that the whole system really knows Chinese— at least not in a conscious way. Searle then essentially ridicules the idea that this "system" could be conscious. What are we to consider conscious, he asks: the slips of paper? The room?

One of the problems with this version of the Chinese Room argument is that it does not come remotely close to really solving the specific problem of answering questions in Chinese. Instead it is really a description of a machine-like process that uses the equivalent of a table lookup, with perhaps some straightforward logical manipulations, to answer questions. It would be able to answer a limited number of canned questions, but if it were to answer *any* arbitrary question that it might be asked, it would really have to understand Chinese in the same way that a Chinese-speaking person does. Again, it is essentially being asked to pass a Chinese Turing test, and as such, would have to be as clever, and about as complex, as a human brain. Straightforward table lookup algorithms are simply not going to achieve that.

If we want to re-create a brain that understands Chinese using people as little cogs in the re-creation, we would really need billions of people simulating the processes in a human brain (essentially the people would be simulating a computer, which would be simulating human brain methods). This would require a rather large room, indeed. And even if extremely efficiently organized, this system would run many thousands of times slower than the Chinese-speaking brain it is attempting to re-create.

Now, it's true that none of these billions of people would need to know anything about Chinese, and none of them would necessarily know what is going on in this elaborate system. But that's equally true of the neural connections in a real human brain. None of the hundred trillion connections in my brain knows anything about this book I am writing, nor do any of them know English, nor any of the other things that I know. None of them is conscious of this chapter, nor of any of the things I am conscious of. Probably none of them is conscious at all. But the entire system of them—that is, Ray Kurzweil—is conscious. At least I'm claiming that I'm conscious (and so far, these claims have not been challenged).

So if we scale up Searle's Chinese Room to be the rather massive "room" it needs to be, who's to say that the entire system of billions of people simulating a brain that knows Chinese isn't conscious? Certainly it would be correct to say that such a system knows Chinese. And we can't say that it is not conscious any more than we can say that about any other brain process. We can't know the subjective experience of another entity (and in at least some of Searle's other writings, he appears to acknowledge this limitation). And this massive multi-billion-person "room" is an entity. And perhaps it is conscious. Searle is just declaring ipso facto that it isn't conscious and that this conclusion is obvious. It may seem that way when you call it a room and talk about a limited number of people manipulating a small number of slips of paper. But as I said, such a system doesn't remotely work.

Another key to the philosophical confusion implicit in the Chinese Room argument is specifically related to the complexity and scale of the system. Searle says that whereas he cannot prove that his typewriter or tape recorder is not conscious, he feels it is obvious that they are not. Why is this so obvious? At least one reason is because a typewriter and a tape recorder are relatively simple entities.

But the existence or absence of consciousness is not so obvious in a system that is as complex as the human brain—indeed, one that may be a direct copy of the organization and "causal powers" of a real human brain. If such a "system" acts human and knows Chinese in a human way, is it conscious? Now the answer is no longer so obvious. What Searle is saying in the Chinese Room argument is that we take a simple "machine" and then consider how absurd it is to consider such a simple machine to be conscious. The fallacy has everything to do with the scale and complexity of the system. Complexity alone does not necessarily give us consciousness, but the Chinese Room tells us nothing about whether or not such a system is conscious.

Kurzweil's Chinese Room. I have my own conception of the Chinese Room—call it Ray Kurzweil's Chinese Room.

In my thought experiment there is a human in a room. The room has decorations from the Ming dynasty, including a pedestal on which sits a mechanical typewriter. The typewriter has been modified so that its keys are marked with Chinese symbols instead of English letters. And the mechanical linkages have been cleverly altered so that when the human types in a question in Chinese, the typewriter does not type the question but instead types the answer to the question. Now, the person receives questions in Chinese characters and dutifully presses the appropriate keys on the typewriter. The typewriter types out not the question, but the appropriate answer. The human then passes the answer outside the room.

So here we have a room with a human in it who appears from the outside to know Chinese yet clearly does not. And clearly the typewriter does not know Chinese, either. It is just an ordinary typewriter with its mechanical linkages modified. So despite the fact that the man in the room can answer questions in Chinese, who or what can we say truly knows Chinese? The decorations?

Now, you might have some objections to my Chinese Room.

You might point out that the decorations don't seem to have any significance.

Yes, that's true. Neither does the pedestal. The same can be said for the human and for the room.

You might also point out that the premise is absurd. Just changing the mechanical linkages in a mechanical typewriter could not possibly enable it to convincingly answer questions in Chinese (not to mention the fact that we can't fit the thousands of Chinese-character symbols on the keys of a typewriter).

Yes, that's a valid objection, as well. The only difference between my Chinese Room conception and the several proposed by Searle is that it is patently obvious in my conception that it couldn't possibly work and is by its very nature absurd. That may not be quite as apparent to many readers or listeners with regard to the Searle Chinese Rooms. However, it is equally the case.

And yet we can make my conception work, just as we can make Searle's conceptions work. All you have to do is to make the typewriter linkages as complex as a human brain. And that's theoretically (if not practically) possible. But the phrase "typewriter linkages" does not suggest such vast complexity. The same is true of Searle's description of a person manipulating slips of paper or following a book of rules or a computer program. These are all equally misleading conceptions.

Searle writes: "Actual human brains cause consciousness by a series of specific neurobiological processes in the brain." However, he has yet to provide any basis for such a startling view. To illuminate Searle's perspective, I quote from a letter he sent me:

> *It may turn out that rather simple organisms like termites or snails are conscious. . . . The essential thing is to recognize that consciousness is a biological process like digestion, lactation, photosynthesis, or mitosis, and you should look for its specific biology as you look for the specific biology of these other processes.*[38]

I replied:

> *Yes, it is true that consciousness emerges from the biological process(es) of the brain and body, but there is at least one difference. If I ask the question, "does a particular entity emit carbon dioxide," I can answer that question*

through clear objective measurement. If I ask the question, "is this entity conscious," I may be able to provide inferential arguments—possibly strong and convincing ones—but not clear objective measurement.

With regard to the snail, I wrote:

Now when you say that a snail may be conscious, I think what you are saying is the following: that we may discover a certain neurophysiological basis for consciousness (call it "x") in humans such that when this basis was present humans were conscious, and when it was not present humans were not conscious. So we would presumably have an objectively measurable basis for consciousness. And then if we found that in a snail, we could conclude that it was conscious. But this inferential conclusion is just a strong suggestion, it is not a proof of subjective experience on the snail's part. It may be that humans are conscious because they have "x" as well as some other quality that essentially all humans share, call this "y." The "y" may have to do with a human's level of complexity or something having to do with the way we are organized, or with the quantum properties of our microtubules (although this may be part of "x"), or something else entirely. The snail has "x" but doesn't have "y" and so it may not be conscious.

How would one settle such an argument? You obviously can't ask the snail. Even if we could imagine a way to pose the question, and it answered yes, that still wouldn't prove that it was conscious. You can't tell from its fairly simple and more-or-less predictable behavior. Pointing out that it has "x" may be a good argument, and many people may be convinced by it. But it's just an argument—not a direct measurement of the snail's subjective experience. Once again, objective measurement is incompatible with the very concept of subjective experience.

Many such arguments are taking place today—though not so much about snails as about higher-level animals. It is apparent to me that dogs and cats are conscious (and Searle has said that he acknowledges this as well). But not all humans accept this. I can imagine scientific ways of strengthening the argument by pointing out many similarities between these animals and humans, but again these are just arguments, not scientific proof.

Searle expects to find some clear biological "cause" of consciousness, and he seems unable to acknowledge that either understanding or consciousness may emerge from an overall pattern of activity. Other philosophers, such as Daniel Dennett, have articulated such "pattern emergent" theories of consciousness. But whether it is "caused" by a specific biological process or by a pattern of activity, Searle provides no foundation for how we would measure or detect consciousness. Finding a neurological correlate of consciousness in humans does not prove that consciousness is necessarily present in other entities with the same correlate, nor does it prove that the absence of such a correlate indicates the absence of consciousness. Such inferential arguments necessarily stop short of direct measurement. In this way, consciousness differs from objectively measurable processes such as lactation and photosynthesis.

As I discussed in chapter 4, we have discovered a biological feature unique to humans and a few other primates: the spindle cells. And these cells with their deep branching structures do appear to be heavily involved with our conscious responses, especially emotional ones. Is the spindle cell structure the neuro-physiological basis "x" for human consciousness? What sort of experiment could possibly prove that? Cats and dogs don't have spindle cells. Does that prove that they have no conscious experience?

Searle writes: "It is out of the question, for purely neurobiological reasons, to suppose that the chair or the computer is conscious." I agree that chairs don't seem to be conscious, but as for computers of the future that have the same complexity, depth, subtlety, and capabilities as humans, I don't think we can rule out this possibility. Searle just assumes that they are not, and that it is "out of the question" to suppose otherwise. There is really nothing more of a substantive nature to Searle's "arguments" than this tautology.

Now, part of the appeal of Searle's stance against the possibility of a computer's being conscious is that the computers we know today just don't seem to be conscious. Their behavior is brittle and formulaic, even if they are occasionally unpredictable. But as I pointed out above, computers today are on the order of one million times simpler than the human brain, which is at least one reason they don't share all of the endearing qualities of human thought. But that disparity is rapidly shrinking and will ultimately reverse itself in a couple of decades. The early twenty-first-century machines I am talking about in this book will appear and act very differently than the relatively simple computers of today.

Searle articulates the view that nonbiological entities are capable of only manipulating logical symbols and he appears to be unaware of other paradigms. It is true that manipulating symbols is largely how rule-based expert systems and game-playing programs work. But the current trend is in a different direction, toward self-organizing chaotic systems that employ biologically inspired methods, including processes derived directly from the reverse engineering of the hundreds of neuron clusters we call the human brain.

Searle acknowledges that biological neurons are machines—indeed, that the entire brain is a machine. As I discussed in chapter 4, we have already re-created in an extremely detailed way the "causal powers" of individual neurons as well as those of substantial neuron clusters. There is no conceptual barrier to scaling these efforts up to the entire human brain.

The Criticism from the Rich-Poor Divide

Another concern expressed by Jaron Lanier and others is the "terrifying" possibility that through these technologies the rich may gain certain advantages and opportunities to which the rest of humankind does not have access.[39] Such inequality, of course, would be nothing new, but with regard to this issue the law of accelerating returns has an important and beneficial impact. Because of the ongoing exponential growth of price-performance, all of these technologies quickly become so inexpensive as to become almost free.

Look at the extraordinary amount of high-quality information available at no cost on the Web today that did not exist at all just a few years ago. And if one wants to point out that only a fraction of the world today has Web access, keep in mind that the explosion of the Web is still in its infancy, and access is growing exponentially. Even in the poorest countries of Africa, Web access is expanding rapidly.

Each example of information technology starts out with early-adoption versions that do not work very well and that are unaffordable except by the elite. Subsequently the technology works a bit better and becomes merely expensive. Then it works quite well and becomes inexpensive. Finally it works extremely well and is almost free. The cell phone, for example, is somewhere between these last two stages. Consider that a decade ago if a character in a movie took out a portable telephone, this was an indication that this person must be very wealthy, powerful, or both. Yet there are societies around the world in which the majority of the population were farming with their hands two decades ago and now have thriving information-based economies with widespread use of cell phones (for example, Asian societies, including rural areas of China). This lag from very expensive early adopters to very inexpensive, ubiquitous adoption now takes about a decade. But in keeping with the doubling of the paradigm-shift rate each decade, this lag will be only five years a decade from now. In twenty years, the lag will be only two to three years (see chapter 2).

The rich-poor divide remains a critical issue, and at each point in time there is more that can and should be done. It is tragic, for example, that the developed nations were not more proactive in sharing AIDS drugs with poor countries in Africa and elsewhere, with millions of lives lost as a result. But the exponential improvement in the price-performance of information technologies is rapidly mitigating this divide. Drugs are essentially an information technology, and we see the same doubling of price-performance each year as we do with other forms of information technology such as computers, communications, and DNA base-pair sequencing. AIDS drugs started out not working very well and costing tens of thousands of dollars per patient per year. Today these drugs work reasonably well and are approaching one hundred dollars per patient per year in poor countries such as those in Africa.

In chapter 2 I cited the World Bank report for 2004 of higher economic growth in the developing world (over 6 percent) compared to the world average (of 4 percent), and an overall reduction in poverty (for example, a reduction of 43 percent in extreme poverty in the East Asian and Pacific region since 1990). Moreover, economist Xavier Sala-i-Martin examined eight measures of global inequality among individuals, and found that all were declining over the past quarter century.[40]

The Criticism from the Likelihood of Government Regulation

These guys talking here act as though the government is not part of their lives. They may wish it weren't, but it is. As we approach the issues they debated here today, they had better believe that those issues will be debated by the whole country. The majority of Americans will not simply sit still while some elite strips off their personalities and uploads themselves into their cyberspace paradise. They will have something to say about that. There will be vehement debate about that in this country.

—LEON FUERTH, FORMER NATIONAL SECURITY ADVISER TO VICE PRESIDENT AL GORE, AT THE 2002 FORESIGHT CONFERENCE

Human life without death would be something other than human; consciousness of mortality gives rise to our deepest longings and greatest accomplishments.

—LEON KASS, CHAIR OF THE PRESIDENTIAL COMMISSION ON BIOETHICS, 2003

The criticism concerning governmental control is that regulation will slow down and stop the acceleration of technology. Although regulation is a vital issue, it has actually had no measurable effect on the trends discussed in this book, which have occurred with extensive regulation in place. Short of a worldwide totalitarian state, the economic and other forces underlying technical progress will only grow with ongoing advances.

Consider the issue of stem-cell research, which has been especially controversial, and for which the U.S. government is restricting its funding. Stem-cell research is only one of numerous ideas concerned with controlling and influencing the information processes underlying biology that are being pursued as part of the biotechnology revolution. Even within the field of cell therapies the controversy over embryonic stem-cell research has served only to accelerate other ways of accomplishing the same goal. For example, transdifferentiation (converting one type of cell such as a skin cell into other types of cells) has moved ahead quickly.

As I reported in chapter 5, scientists have recently demonstrated the ability to reprogram skin cells into several other cell types. This approach represents the holy grail of cell therapies in that it promises an unlimited supply of differentiated cells with the patient's own DNA. It also allows cells to be selected without DNA errors and will ultimately be able to provide extended telomere strings (to make the cells more youthful). Even embryonic stem-cell research itself has moved ahead, for example, with projects like Harvard's major new research center and California's successful three-billion-dollar bond initiative to support such work.

Although the restrictions on stem-cell research are unfortunate, it is hard to say that cell-therapy research, let alone the broad field of biotechnology, has been affected to a significant degree.

Some governmental restrictions reflect the perspective of fundamentalist humanism, which I addressed in the previous chapter. For example, the Council of Europe proclaimed that "human rights imply the right to inherit a genetic pattern that has not been artificially changed."[41] Perhaps the most interesting aspect of the council's edict is its posing a restriction as a right. In the same spirit, I assume the council would advocate the human right not to be cured from natural disease by unnatural means, just as activists "protected" starving African nations from the indignity of consuming bioengineered crops.[42]

Ultimately the benefits of technical progress overwhelm such reflexive anti-technology sentiments. The majority of crops in the United States are already GMOs, while Asian nations are aggressively adopting the technology to feed their large populations, and even Europe is now beginning to approve GMO foods. The issue is important because unnecessary restrictions, although temporary, can result in exacerbated suffering of millions of people. But technical progress is advancing on thousands of fronts, fueled by irresistible economic gains and profound improvements in human health and well-being.

Leon Fuerth's observation quoted above reveals an inherent misconception about information technologies. Information technologies are not available only to an elite. As discussed, desirable information technologies rapidly become ubiquitous and almost free. It is only when they don't work very well (that is, in an early stage of development) that they are expensive and restricted to an elite.

Early in the second decade of this century, the Web will provide full immersion visual-auditory virtual reality with images written directly to our retinas from our eyeglasses and lenses and very high-bandwidth wireless Internet access woven in our clothing. These capabilities will not be restricted just to the privileged. Just like cell phones, by the time they work well they will be everywhere.

In the 2020s we will routinely have nanobots in our bloodstream keeping us healthy and augmenting our mental capabilities. By the time these work well they will be inexpensive and widely used. As I discussed above, reducing the lag between early and late adoption of information technologies will itself accelerate from the current ten-year period to only a couple of years two decades from now. Once nonbiological intelligence gets a foothold in our brains, it will at least double in capability each year, as is the nature of information technology. Thus it will not take long for the nonbiological portion of our intelligence to predominate. This will not be a luxury reserved for the rich, any more than search engines are today. And to the extent that there will be a debate about the desirability of such augmentation, it's easy to predict who will win, since those with enhanced intelligence will be far better debaters.

The Unbearable Slowness of Social Institutions. MIT senior research scientist Joel Cutcher-Gershenfeld writes: "Just looking back over the course of the past century and a half, there have been a succession of political regimes where each was the solution to an earlier dilemma, but created new dilemmas in the subsequent era. For example, Tammany Hall and the political patron model were a vast improvement over the dominant system based on landed gentry—many more people were included in the political process. Yet, problems emerged with patronage, which led to the civil service model—a strong solution to the preceding problem by introducing the meritocracy. Then, of course, civil service became the barrier to innovation and we move to reinventing government. And the story continues."[43] Gershenfeld is pointing out that social institutions even when innovative in their day become "a drag on innovation."

First I would point out that the conservatism of social institutions is not a new phenomenon. It is part of the evolutionary process of innovation, and the law of accelerating returns has always operated in this context. Second, innovation has a way of working around the limits imposed by institutions. The advent of decentralized technology empowers the individual to bypass all kinds of restrictions, and does represent a primary means for social change to accelerate. As one of many examples, the entire thicket of communications regulations is in the process of being bypassed by emerging point-to-point techniques such as voice over Internet protocol (VOIP).

Virtual reality will represent another means of hastening social change. People will ultimately be able to have relationships and engage in activities in immersive and highly realistic virtual-reality environments that they would not be able or willing to do in real reality.

As technology becomes more sophisticated it increasingly takes on traditional human capabilities and requires less adaptation. You had to be technically adept to use early personal computers, whereas using computerized systems today, such as cell phones, music players, and Web browsers, requires much less technical ability. In the second decade of this century, we will routinely be interacting with virtual humans that, although not yet Turing-test capable, will have sufficient natural language understanding to act as our personal assistants for a wide range of tasks.

There has always been a mix of early and late adopters of new paradigms. We still have people today who want to live as we did in the seventh century. This does not restrain the early adopters from establishing new attitudes and social conventions, for example new Web-based communities. A few hundred years ago, only a handful of people such as Leonardo da Vinci and Newton were exploring new ways of understanding and relating to the world. Today, the worldwide community that participates in and contributes to the social innovation of adopting and adapting to new technological innovation is a substantial portion of the population, another reflection of the law of accelerating returns.

The Criticism from Theism

Another common objection explicitly goes beyond science to maintain that there is a spiritual level that accounts for human capabilities and that is not penetrable by objective means. William A. Dembski, a distinguished philosopher and mathematician, decries the outlook of such thinkers as Marvin Minsky, Daniel Dennett, Patricia Churchland, and Ray Kurzweil, whom he calls "contemporary materialists" who "see the motions and modifications of matter as sufficient to account for human mentality."[44]

Dembski ascribes "predictability [as] materialism's main virtue" and cites "hollowness [as] its main fault." He goes on to say that "humans have aspirations. We long for freedom, immortality, and the beatific vision. We are restless until we find our rest in God. The problem for the materialist, however, is that these aspirations cannot be redeemed in the coin of matter." He concludes that humans cannot be mere machines because of "the strict absence of extra-material factors from such systems."

I would prefer that we call Dembski's concept of materialism "capability materialism," or better yet "capability patternism." Capability materialism/patternism is based on the observation that biological neurons and their interconnections are made up of sustainable patterns of matter and energy. It also holds that their methods can be described, understood, and modeled with either replicas or functionally equivalent re-creations. I use the word "capability" because it encompasses all of the rich, subtle, and diverse ways in which humans interact with the world, not just those narrower skills that one might label as intellectual. Indeed, our ability to understand and respond to emotions is at least as complex and diverse as our ability to process intellectual issues.

John Searle, for example, acknowledges that human neurons are biological machines. Few serious observers have postulated capabilities or reactions of human neurons that require Dembski's "extra-material factors." Relying on the patterns of matter and energy in the human body and brain to explain its behavior and proficiencies need not diminish our wonderment at its remarkable qualities. Dembski has an outdated understanding of the concept of "machine."

Dembski also writes that "unlike brains, computers are neat and precise. . . . [C]omputers operate deterministically." This statement and others reveal a view of machines, or entities made up of patterns of matter and energy ("material" entities), that is limited to the literally simpleminded mechanisms of nineteenth-century automatons. These devices, with their hundreds and even thousands of parts, were quite predictable and certainly not capable of longings for freedom and other such endearing qualities of the human entity. The same observations largely hold true for today's machines, with their billions of parts. But the same cannot necessarily be said for machines with *millions of billions* of interacting "parts," entities with the complexity of the human brain and body.

Moreover it is incorrect to say that materialism is predictable. Even today's computer programs routinely use simulated randomness. If one needs truly random events in a process, there are devices that can provide this as well. Fundamentally, everything we perceive in the material world is the result of many trillions of quantum events, each of which displays a profound and irreducible quantum randomness at the core of physical reality (or so it seems—the scientific jury is still out on the true nature of the apparent randomness underlying quantum events). The material world—at both the macro and micro levels—is anything but predictable.

Although many computer programs do operate the way Dembski describes, the predominant techniques in my own field of pattern recognition use biology-inspired chaotic-computing methods. In these systems the unpredictable interaction of millions of processes, many of which contain random and unpredictable elements, provide unexpected yet appropriate answers to subtle questions of recognition. The bulk of human intelligence consists of just these sorts of pattern-recognition processes.

As for our responses to emotions and our highest aspirations, these are properly regarded as emergent properties—profound ones to be sure but nonetheless emergent patterns that result from the interaction of the human brain with its complex environment. The complexity and capacity of nonbiological entities is increasing exponentially and will match biological systems including the human brain (along with the rest of the nervous system and the endocrine system) within a couple of decades. Indeed, many of the designs of future machines will be biologically inspired—that is, derivative of biological designs. (This is already true of many contemporary systems.) It is my thesis that by sharing the complexity as well as the actual patterns of human brains, these future nonbiological entities will display the intelligence and emotionally rich reactions (such as "aspirations") of humans.

Will such a nonbiological entity be conscious? Searle claims that we can (at least in theory) readily resolve this question by ascertaining if it has the correct "specific neurobiological processes." It is my view that many humans, ultimately the vast majority of humans, will come to believe that such human-derived but nonetheless nonbiological intelligent entities are conscious, but that's a political and psychological prediction, not a scientific or philosophical judgment. My bottom line: I agree with Dembski that this is not a scientific question, because it cannot be resolved through objective observation. Some observers say that if it's not a scientific question, it's not an important or even a real question. My view (and I'm sure Dembski agrees) is that precisely because the question is not scientific, it is a philosophical one—indeed, the fundamental philosophical question.

Dembski writes: "We need to transcend ourselves to find ourselves. Now the motions and modifications of matter offer no opportunity for transcending ourselves. . . . Freud . . . Marx . . . Nietzsche, . . . each regarded the hope for transcendence as a delusion." This view of transcendence as an ultimate goal is reasonably stated. But I disagree that the material world offers no "opportunity for transcending." The material world inherently evolves, and each stage transcends the stage before it. As I discussed in chapter 7, evolution moves toward greater complexity, greater elegance, greater knowledge, greater intelligence, greater beauty, greater creativity, greater love. And God has been called all these things, only without any limitation: infinite knowledge, infinite intelligence, infinite beauty, infinite creativity, and infinite love. Evolution does not achieve an infinite level, but as it explodes exponentially it certainly moves in that direction. So evolution moves inexorably toward our conception of God, albeit never reaching this ideal.

Dembski continues:

A machine is fully determined by the constitution, dynamics, and interrelationships of its physical parts. . . . "[M]achines" stresses the strict absence of extra-material factors. . . . The replacement principle is relevant to this discussion because it implies that machines have no substantive history. . . . But a machine, properly speaking, has no history. Its history is a superfluous rider—an addendum that could easily have been different without altering the machine. . . . For a machine, all that is is what it is at this moment. . . . Machines access or fail to access items in storage. . . . Mutatis mutandis, items that represent counterfactual occurrences (i.e., things that never happened) but which are accessible can be, as far as the machine is concerned, just as though they did happen.

It need hardly be stressed that the whole point of this book is that many of our dearly held assumptions about the nature of machines and indeed of our own human nature will be called into question in the next several decades. Dembski's conception of "history" is just another aspect of our humanity that necessarily derives from the richness, depth, and complexity of being human. Conversely, not having a history in the Dembski sense is just another attribute of the simplicity of the machines that we have known up to this time. It is precisely my thesis that machines of the 2030s and beyond will be of such great complexity and richness of organization that their behavior will evidence emotional reactions, aspirations, and, yes, history. So Dembski is merely describing today's limited machines and just assuming that these limitations are inherent, a line of argument equivalent to stating that "today's machines are not as capable as humans, therefore machines will never reach this level of performance." Dembski is just assuming his conclusion.

Dembski's view of the ability of machines to understand their own history is limited to their "accessing" items in storage. Future machines, however, will possess not only a record of their own history but an ability to understand that history and to reflect insightfully upon it. As for "items that represent counterfactual occurrences," surely the same can be said for our human memories.

Dembski's lengthy discussion of spirituality is summed up thus:

> But how can a machine be aware of God's presence? Recall that machines are entirely defined by the constitution, dynamics, and interrelationships among their physical parts. It follows that God cannot make his presence known to a machine by acting upon it and thereby changing its state. Indeed, the moment God acts upon a machine to change its state, it no longer properly is a machine, for an aspect of the machine now transcends its physical constituents. It follows that awareness of God's presence by a machine must be independent of any action by God to change the state of the machine. How then does the machine come to awareness of God's presence? The awareness must be self-induced. Machine spirituality is the spirituality of self-realization, not the spirituality of an active God who freely gives himself in self-revelation and thereby transforms the beings with which he is in communion. For Kurzweil to modify "machine" with the adjective "spiritual" therefore entails an impoverished view of spirituality.

Dembski states that an entity (for example, a person) cannot be aware of God's presence without God's acting upon her, yet God cannot act upon a machine, so therefore a machine cannot be aware of God's presence. Such reasoning is entirely tautological and humancentric. God communes only with humans, and only biological ones at that. I have no problem with Dembski's subscribing to this as a personal belief, but he fails to make the "strong case" that he promises, that "humans are not machines—period." As with Searle, Dembski just assumes his conclusion.

Like Searle, Dembski cannot seem to grasp the concept of the emergent properties of complex distributed patterns. He writes:

> Anger presumably is correlated with certain localized brain excitations. But localized brain excitations hardly explain anger any better than overt behaviors associated with anger, like shouting obscenities. Localized brain excitations may be reliably correlated with anger, but what accounts for one person interpreting a comment as an insult and experiencing anger, and another person interpreting that same comment as a

joke and experiencing laughter? A full materialist account of mind needs to understand localized brain excitations in terms of other localized brain excitations. Instead we find localized brain excitations (representing, say, anger) having to be explained in terms of semantic contents (representing, say, insults). But this mixture of brain excitations and semantic contents hardly constitutes a materialist account of mind or intelligent agency.

Dembski assumes that anger is correlated with a "localized brain excitation," but anger is almost certainly the reflection of complex distributed patterns of activity in the brain. Even if there is a localized neural correlate associated with anger, it nonetheless results from multifaceted and interacting patterns. Dembski's question as to why different people react differently to similar situations hardly requires us to resort to his extramaterial factors for an explanation. The brains and experiences of different people are clearly not the same, and these differences are well explained by differences in their physical brains resulting from varying genes and experiences.

Dembski's resolution of the ontological problem is that the ultimate basis of what exists is what he calls the "real world of things" that are not reducible to material stuff. Dembski does not list what "things" we might consider as fundamental, but presumably human minds would be on the list, as might be other things, such as money and chairs. There may be a small congruence of our views in this regard. I regard Dembski's "things" as patterns. Money, for example, is a vast and persisting pattern of agreements, understandings, and expectations. "Ray Kurzweil" is perhaps not so vast a pattern but thus far is also persisting. Dembski apparently regards patterns as ephemeral and not substantial, but I have a profound respect for the power and endurance of patterns. It is not unreasonable to regard patterns as a fundamental ontological reality. We are unable to really touch matter and energy directly, but we do directly experience the patterns underlying Dembski's "things." Fundamental to this thesis is that as we apply our intelligence, and the extension of our intelligence called technology, to understanding the powerful patterns in our world (for example, human intelligence), we can re-create—and extend!—these patterns in other substrates. The patterns are more important than the materials that embody them.

Finally, if Dembski's intelligence-enhancing extramaterial stuff really exists, then I'd like to know where I can get some.

The Criticism from Holism

Another common criticism says the following: machines are organized as rigidly structured hierarchies of modules, whereas biology is based on holistically organized elements in which every element affects every other. The unique capabilities of biology (such as human intelligence) can result only from this type of holistic design. Furthermore, only biological systems can use this design principle.

Michael Denton, a biologist at the University of Otago in New Zealand, points out the apparent differences between the design principles of biological entities and those of the machines he has known. Denton eloquently describes organisms as "self-organizing, . . . self-referential, . . . self-replicating, . . . reciprocal, . . . self-formative, and . . . holistic."[45] He then makes the unsupported leap—a leap of faith, one might say—that such organic forms can be created only through biological processes and that such forms are "immutable, . . . impenetrable, and . . . fundamental" realities of existence.

I do share Denton's "awestruck" sense of "wonderment" at the beauty, intricacy, strangeness, and interrelatedness of organic systems, ranging from the "eerie other-worldly . . . impression" left by asymmetric protein shapes to the extraordinary complexity of higher-order organs such as the human brain. Further, I agree with Denton that biological design represents a profound set of principles. However, it is precisely my thesis, which neither Denton nor other critics from the holistic school acknowledge or respond to, that machines (that is, entities derivative of human-directed design) can access—and already are using—these same principles. This has been the thrust of my own work and represents the wave of the future. Emulating the ideas of nature is the most effective way to harness the enormous powers that future technology will make available.

Biological systems are not completely holistic, and contemporary machines are not completely modular; both exist on a continuum. We can identify units of functionality in natural systems even at the molecular level, and discernible mechanisms of action are even more evident at the higher level of organs and brain regions. The process of understanding the functionality and information transformations performed in specific brain regions is well under way, as we discussed in chapter 4.

It is misleading to suggest that every aspect of the human brain interacts with every other aspect and that it is therefore impossible to understand its methods. Researchers have already identified and modeled the transformations of information in several dozen of its regions. Conversely there are numerous examples of contemporary machines that were not designed in a modular fashion, and in which many of the design aspects are deeply interconnected, such as the examples of genetic algorithms described in chapter 5. Denton writes:

> Today almost all professional biologists have adopted the mechanistic/reductionist approach and assume that the basic parts of an organism (like the cogs of a watch) are the primary essential things, that a living organism (like a watch) is no more than the sum of its parts, and that it is the parts that determine the properties of the whole and that (like a watch) a complete description of all the properties of an organism may be had by characterizing its parts in isolation.

Denton, too, is ignoring here the ability of complex processes to exhibit emergent properties that go beyond "its parts in isolation." He appears to recognize this potential in nature when he writes: "In a very real sense organic forms . . . represent genuinely emergent realities." However, it is hardly necessary to resort to Denton's "vitalistic model" to explain emergent realities. Emergent properties derive from the power of patterns, and nothing restricts patterns and their emergent properties to natural systems.

Denton appears to acknowledge the feasibility of emulating the ways of nature when he writes:

> Success in engineering new organic forms from proteins up to organisms will therefore require a completely novel approach, a sort of designing from "the top down." Because the parts of organic wholes only exist in the whole, organic wholes cannot be specified bit by bit and built up from a set of relatively independent modules; consequently the entire undivided unity must be specified together *in toto.*

Here Denton provides sound advice and describes an approach to engineering that I and other researchers use routinely in the areas of pattern recognition, complexity (chaos) theory, and self-organizing systems. Denton appears to be unaware of these methodologies, however, and after describing examples of bottom-up, component-driven engineering and their limitations concludes with no justification that there is an unbridgeable chasm between the two design philosophies. The bridge is, in fact, already under construction.

As I discussed in chapter 5, we can create our own "eerie other-worldly" but effective designs through applied evolution. I described how to apply the principles of evolution to creating intelligent designs through genetic algorithms.

In my own experience with this approach, the results are well represented by Denton's description of organic molecules in the "apparent illogic of the design and the lack of any obvious modularity or regularity, . . . the sheer chaos of the arrangement, . . . [and the] non-mechanical impression."

Genetic algorithms and other bottom-up self-organizing design methodologies (such as neural nets, Markov models, and others that we discussed in chapter 5) incorporate an unpredictable element, so that the results of such systems are different every time the process is run. Despite the common wisdom that machines are deterministic and therefore predictable, there are numerous readily available sources of randomness available to machines. Contemporary theories of quantum mechanics postulate a profound randomness at the core of existence. According to certain theories of quantum mechanics, what appears to be the deterministic behavior of systems at a macro level is simply the result of overwhelming statistical preponderances based on enormous numbers of fundamentally unpredictable events. Moreover, the work of Stephen Wolfram and others has demonstrated that even a system that is in theory fully deterministic can nonetheless produce effectively random and, most important, entirely unpredictable results.

Genetic algorithms and similar self-organizing approaches give rise to designs that could not have been arrived at through a modular component-driven approach. The "strangeness, . . . [the] chaos, . . . the dynamic interaction" of parts to the whole that Denton attributes exclusively to organic structures describe very well the qualities of the results of these human-initiated chaotic processes.

In my own work with genetic algorithms I have examined the process by which such an algorithm gradually improves a design. A genetic algorithm does not accomplish its design achievements through designing individual subsystems one at a time but effects an incremental "all at once" approach, making many small distributed changes throughout the design that progressively improve the overall fit or "power" of the solution. The solution itself emerges gradually and unfolds from simplicity to complexity. While the solutions it produces are often asymmetric and ungainly but effective, just as in nature, they can also appear elegant and even beautiful.

Denton is correct in observing that most contemporary machines, such as today's conventional computers, are designed using the modular approach. There are certain significant engineering advantages to this traditional technique. For example, computers have much more accurate memories than humans and can perform logical transformations far more effectively than unaided human intelligence. Most important, computers can share their memories and patterns instantly. The chaotic nonmodular approach of nature also has clear advantages that Denton well articulates, as evidenced by the deep powers of human pattern recognition. But it is a wholly unjustified leap to say that because of the current (and diminishing!) limitations of human-directed technology that biological systems are inherently, even ontologically, a world apart.

The exquisite designs of nature (the eye, for example) have benefited from a profound evolutionary process. Our most complex genetic algorithms today incorporate genetic codes of tens of thousands of bits, whereas biological entities such as humans are characterized by genetic codes of billions of bits (only tens of millions of bytes with compression).

However, as is the case with all information-based technology, the complexity of genetic algorithms and other nature-inspired methods is increasing exponentially. If we examine the rate at which this complexity is increasing, we find that they will match the complexity of human intelligence within about two decades, which is consistent with my estimates drawn from direct trends in hardware and software.

Denton points out we have not yet succeeded in folding proteins in three dimensions, "even one consisting of only 100 components." However, it is only in the recent few years that we have had the tools even to visualize these three-dimensional patterns. Moreover, modeling the interatomic forces will require on the order of one hundred thousand billion (10^{14}) calculations per second. In late 2004 IBM introduced a version of its Blue Gene/L supercomputer with a capability of seventy teraflops (nearly 10^{14} cps), which, as the name suggests, is expected to provide the ability to simulate protein folding.

We have already succeeded in cutting, splicing, and rearranging genetic codes and harnessing nature's own biochemical factories to produce enzymes and other complex biological substances. It is true that most contemporary work of this type is done in two dimensions, but the requisite computational resources to visualize and model the far more complex three-dimensional patterns found in nature are not far from realization.

In discussions of the protein issue with Denton himself, he acknowledged that the problem would eventually be solved, estimating that it was perhaps a decade away. The fact that a certain technical feat has not *yet* been accomplished is not a strong argument that it never will be.

Denton writes:

From knowledge of the genes of an organism it is impossible to predict the encoded organic forms. Neither the properties nor structure of individual proteins nor those of any higher order forms—such as ribosomes and whole cells—can be inferred even from the most exhaustive analysis of the genes and their primary products, linear sequences of amino acids.

Although Denton's observation above is essentially correct, it basically points out that the genome is only part of the overall system. The DNA code is not the whole story, and the rest of the molecular support system is required for the system to work and for it to be understood. We also need the design of the ribosome and other molecules that make the DNA machinery function. However, adding these designs does not significantly change the amount of design information in biology.

But re-creating the massively parallel, digitally controlled analog, hologramlike, self-organizing, and chaotic processes of the human brain does not require us to fold proteins. As discussed in chapter 4 there are dozens of contemporary projects that have succeeded in creating detailed re-creations of neurological systems. These include neural implants that successfully function inside people's brains without folding any proteins. However, while I understand Denton's argument about proteins to be evidence regarding the holistic ways of nature, as I have pointed out there are no essential barriers to our emulating these ways in our technology, and we are already well down this path.

In summary, Denton is far too quick to conclude that complex systems of matter and energy in the physical world are incapable of exhibiting the "emergent . . . vital characteristics of organisms such as self-replication, 'morphing,' self-regeneration, self-assembly and the holistic order of biological design" and that, therefore, "organisms and machines belong to different categories of being." Dembski and Denton share the same limited view of machines as entities that can be designed and constructed only in a modular way. We can build and already are building "machines" that have powers far greater than the sum of their parts by combining the self-organizing design principles of the natural world with the accelerating powers of our human-initiated technology. It will be a formidable combination.

Epilogue

I do not know what I may appear to the world, but to myself I seem to have been only like a boy playing on the seashore, and diverting myself in now and then finding a smoother pebble or a prettier shell than ordinary, whilst the great ocean of truth lay undiscovered before me.

—Isaac Newton[1]

The meaning of life is creative love. Not love as an inner feeling, as a private sentimental emotion, but love as a dynamic power moving out into the world and doing something original.

—Tom Morris, *If Aristotle Ran General Motors*

No exponential is forever . . . but we can delay "forever."

—Gordon E. Moore, 2004

How Singular? How singular is the Singularity? Will it happen in an instant? Let's consider again the derivation of the word. In mathematics a singularity is a value that is beyond any limit—in essence, infinity. (Formally the value of a function that contains such a singularity is said to be undefined at the singularity point, but we can show that the value of the function at nearby points exceeds any specific finite value).[2]

The Singularity, as we have discussed it in this book, does not achieve infinite levels of computation, memory, or any other measurable attribute. But it certainly achieves vast levels of all of these qualities, including intelligence. With the reverse engineering of the human brain we will be able to apply the parallel, self-organizing, chaotic algorithms of human intelligence to enormously powerful computational substrates. This intelligence will then be in a position to improve its own design, both hardware and software, in a rapidly accelerating iterative process.

But there still appears to be a limit. The capacity of the universe to support intelligence appears to be only about 10^{90} calculations per second, as I discussed in chapter 6. There are theories such as the holographic universe that suggest the possibility of higher numbers (such as 10^{120}), but these levels are all decidedly finite.

Of course, the capabilities of such an intelligence may appear infinite for all practical purposes to our current level of intelligence. A universe saturated with intelligence at 10^{90} cps would be one trillion trillion trillion trillion trillion times more powerful than all biological human brains on Earth today.[3] Even a one-kilogram "cold" computer has a peak potential of 10^{42} cps, as I reviewed in chapter 3, which is ten thousand trillion (10^{16}) times more powerful than all biological human brains.[4]

Given the power of exponential notation, we can easily conjure up bigger numbers, even if we lack the imagination to contemplate all of their implications. We can imagine the possibility of our future intelligence spreading into other universes. Such a scenario is conceivable given our current understanding of cosmology, although speculative. This could potentially allow our future intelligence to go beyond any limits. If we gained the ability to create and colonize other universes (and if there is a way to do this, the vast intelligence of our future civilization is likely to be able to harness it), our intelligence would ultimately be capable of exceeding any specific finite level. That's exactly what we can say for singularities in mathematical functions.

How does our use of "singularity" in human history compare to its use in physics? The word was borrowed from mathematics by physics, which has always shown a penchant for anthropomorphic terms (such as "charm" and "strange" for names of quarks). In physics "singularity" theoretically refers to a point of zero size with infinite density of mass and therefore infinite gravity. But because of quantum uncertainty there is no actual point of infinite density, and indeed quantum mechanics disallows infinite values.

Just like the Singularity as I have discussed it in this book, a singularity in physics denotes unimaginably large values. And the area of interest in physics is not actually zero in size but rather is an event horizon around the theoretical singularity point inside a black hole (which is not even black). Inside the event horizon particles and energy, such as light, cannot escape because gravity is too strong. Thus from outside the event horizon, we cannot see easily inside the event horizon with certainty.

However, there does appear to be a way to see inside a black hole, because black holes give off a shower of particles. Particle-antiparticle pairs are created near the event horizon (as happens everywhere in space), and for some of these pairs, one of the pair is pulled into the black hole while the other manages to escape. These escaping particles form a glow called Hawking radiation, named after its discoverer, Stephen Hawking. The current thinking is that this radiation does reflect (in a coded fashion, and as a result of a form of quantum entanglement with the particles inside) what is happening inside the black hole. Hawking initially resisted this explanation but now appears to agree.

So, we find our use of the term "Singularity" in this book to be no less appropriate than the deployment of this term by the physics community. Just as we find it hard to see beyond the event horizon of a black hole, we also find it difficult to see beyond the event horizon of the historical Singularity. How can we, with our brains each limited to 10^{16} to 10^{19} cps, imagine what our future civilization in 2099 with its 10^{60} cps will be capable of thinking and doing?

Nevertheless, just as we can draw conclusions about the nature of black holes through our conceptual thinking, despite never having actually been inside one, our thinking today is powerful enough to have meaningful insights into the implications of the Singularity. That's what I've tried to do in this book.

Human Centrality. A common view is that science has consistently been correcting our overly inflated view of our own significance. Stephen Jay Gould said, "The most important scientific revolutions all include, as their only common feature, the dethronement of human arrogance from one pedestal after another of previous convictions about our centrality in the cosmos."[5]

But it turns out that we are central, after all. Our ability to create models—virtual realities—in our brains, combined with our modest-looking thumbs, has been sufficient to usher in another form of evolution: technology. That development enabled the persistence of the accelerating pace that started with biological evolution. It will continue until the entire universe is at our fingertips.

Resources and Contact Information

Singularity.com

New developments in the diverse fields discussed in this book are accumulating at an accelerating pace. To help you keep pace, I invite you to visit Singularity. com, where you will find

- Recent news stories
- A compilation of thousands of relevant news stories going back to 2001 from KurzweilAI.net (see below)
- Hundreds of articles on related topics from KurzweilAI.net
- Research links
- Data and citation for all graphs
- Material about this book
- Excerpts from this book
- Online endnotes

KurzweilAI.net

You are also invited to visit our award-winning Web site, KurzweilAI.net, which includes over six hundred articles by over one hundred "big thinkers" (many of whom are cited in this book), thousands of news articles, listings of events, and other features. Over the past six months, we have had more than one million readers. Memes on KurzweilAI.net include:

- The Singularity
- Will Machines Become Conscious?
- Living Forever
- How to Build a Brain
- Virtual Realities

- Nanotechnology
- Dangerous Futures
- Visions of the Future
- Point/Counterpoint

You can sign up for our free (daily or weekly) e-newsletter by putting your e-mail address in the simple one-line form on the KurzweilAI.net home page. We do not share your e-mail address with anyone.

Fantastic-Voyage.net and RayandTerry.com

For those of you who would like to optimize your health today, and to maximize your prospects of living long enough to actually witness and experience the Singularity, visit Fantastic-Voyage.net and RayandTerry.com. I developed these sites with Terry Grossman, M.D., my health collaborator and coauthor of *Fantastic Voyage: Live Long Enough to Live Forever*. These sites contain extensive information about improving your health with today's knowledge so that you can be in good health and spirits when the biotechnology and nanotechnology revolutions are fully mature.

Contacting the Author

Ray Kurzweil can be reached at ray@singularity.com.

APPENDIX

The Law of Accelerating Returns Revisited

The following analysis provides the basis of understanding evolutionary change as a doubly exponential phenomenon (that is, exponential growth in which the rate of exponential growth—the exponent—is itself growing exponentially). I will describe here the growth of computational power, although the formulas are similar for other aspects of evolution, especially information-based processes and technologies, including our knowledge of human intelligence, which is a primary source of the software of intelligence.

We are concerned with three variables:

V: Velocity (that is, power) of computation (measured in calculations per second per unit cost)

W: World knowledge as it pertains to designing and building computational devices

t: Time

As a first-order analysis, we observe that computer power is a linear function of W. We also note that W is cumulative. This is based on the observation that relevant technology algorithms are accumulated in an incremental way. In the case of the human brain, for example, evolutionary psychologists argue that the brain is a massively modular intelligence system, evolved over time in an incremental manner. Also, in this simple model, the instantaneous increment to knowledge is proportional to computational power. These observations lead to the conclusion that computational power grows exponentially over time.

In other words, computer power is a linear function of the knowledge of how to build computers. This is actually a conservative assumption. In general, innovations improve V by a multiple, not in an additive way. Independent innovations (each representing a linear increment to knowledge) multiply one another's effects. For example, a circuit advance such as CMOS (complementary metal oxide semiconductor), a more efficient IC wiring methodology, a processor innovation such as pipelining, or an algorithmic improvement such as the fast Fourier transform, all increase V by independent multiples.

As noted, our initial observations are:

The velocity of computation is proportional to world knowledge:

(1) $V = c_1 W$

The rate of change of world knowledge is proportional to the velocity of computation:

(2) $\dfrac{dW}{dt} = c_2 V$

Substituting (1) into (2) gives:

(3) $\dfrac{dW}{dt} = c_1 c_2 W$

The solution to this is:

(4) $W = W_0 e^{c_1 c_2 t}$

and W grows exponentially with time (e is the base of the natural logarithms).

The data that I've gathered shows that there is exponential growth in the rate of (exponent for) exponential growth (we doubled computer power every three years early in the twentieth century and every two years in the middle of the century, and are doubling it every one year now). The exponentially growing power of technology results in exponential growth of the economy. This can be observed going back at least a century. Interestingly, recessions, including the Great Depression, can be modeled as a fairly weak cycle on top of the underlying exponential growth. In each case, the economy "snaps back" to where it would have been had the recession/depression never existed in the first place. We can see even more rapid exponential growth in specific industries tied to the exponentially growing technologies, such as the computer industry.

If we factor in the exponentially growing resources for computation, we can see the source for the second level of exponential growth.

Once again we have:

(5) $V = c_1 W$

But now we include the fact that the resources deployed for computation, N, are also growing exponentially:

(6) $N = c_3 e^{c_4 t}$

The rate of change of world knowledge is now proportional to the product of the velocity of computation and the deployed resources:

(7) $\dfrac{dW}{dt} = c_2 NV$

Substituting (5) and (6) into (7) we get:

(8) $\dfrac{dW}{dt} = c_1 c_2 c_3 e^{c_4 t} W$

The solution to this is:

(9) $W = W_0 \exp\left(\dfrac{c_1 c_2 c_3}{c_4} e^{c_4 t} \right)$

and world knowledge accumulates at a double exponential rate.

Now let's consider some real-world data. In chapter 3, I estimated the computational capacity of the human brain, based on the requirements for functional simulation of all brain regions, to be approximately 10^{16} cps. Simulating the salient nonlinearities in every neuron and interneuronal connection would require a higher level of computing: 10^{11} neurons times an average 10^3 connections per neuron (with the calculations taking place primarily in the connections) times 10^2 transactions per second times 10^3 calculations per transaction—a total of about 10^{19} cps. The analysis below assumes the level for functional simulation (10^{16} cps).

Analysis Three

Considering the data for actual calculating devices and computers during the twentieth century:

Let S = cps/$1K: calculations per second for $1,000.

Twentieth-century computing data matches:

$$S = 10^{\left[6.00 \times \left[\left(\frac{20.40}{6.00}\right)^{\left[\frac{Year - 1900}{100}\right]}\right] - 11.00\right]}$$

We can determine the growth rate, G, over a period of time:

$$G = 10^{\left(\frac{\log(Sc) - \log(Sp)}{Yc - Yp}\right)}$$

where Sc is cps/$1K for current year, Sp is cps/$1K of previous year, Yc is current year, and Yp is previous year.

Human brain = 10^{16} calculations per second.

Human race = 10 billion (10^{10}) human brains = 10^{26} calculations per second.

We achieve one human brain capability (10^{16} cps) for $1,000 around the year 2023.

We achieve one human brain capability (10^{16} cps) for one cent around the year 2037.

We achieve one human race capability (10^{26} cps) for $1,000 around the year 2049.

If we factor in the exponentially growing economy, particularly with regard to the resources available for computation (already about one trillion dollars per year), we can see that nonbiological intelligence will be billions of times more powerful than biological intelligence before the middle of the century.

We can derive the double exponential growth in another way. I noted above that the rate of adding knowledge (dW/dt) was at least proportional to the knowledge at each point in time. This is clearly conservative given that many innovations (increments to knowledge) have a multiplicative rather than additive impact on the ongoing rate.

However, if we have an exponential growth rate of the form:

$$(10) \quad \frac{dW}{dt} = C^w$$

where $C > 1$, this has the solution:

$$(11) \quad W = \frac{1}{\ln C} \ln \left(\frac{1}{1 - t \ln C} \right)$$

which has a slow logarithmic growth while $t < 1/\ln C$ but then explodes close to the singularity at $t = 1/\ln C$.

Even the modest $dW/dt = W^2$ results in a singularity.

Indeed any formula with a power law growth rate of the form:

$$(12) \quad \frac{dW}{dt} = W^a$$

where $a > 1$, leads to a solution with a singularity:

$$(13) \quad W = W_o \, \frac{1}{(T - t)^{\frac{1}{a-1}}}$$

at the time T. The higher the value of a, the closer the singularity.

My view is that it is hard to imagine infinite knowledge, given apparently finite resources of matter and energy, and the trends to date match a double exponential process. The additional term (to W) appears to be of the form $W \times \log(W)$. This term describes a network effect. If we have a network such as the Internet, its effect or value can reasonably be shown to be proportional to $n \times \log(n)$ where n is the number of nodes. Each node (each user) benefits, so this accounts for the n multiplier. The value to each user (to each node) $= \log(n)$. Bob Metcalfe (inventor of Ethernet) has postulated the value of a network of n nodes $= c \times n^2$, but this is overstated. If the Internet doubles in size, its value to me does increase but it does not double. It can be shown that a reasonable estimate is that a network's value to each user is proportional to the log of the size of the network. Thus, its overall value is proportional to $n \times \log(n)$.

If the growth rate instead includes a logarithmic network effect, we get an equation for the rate of change that is given by:

(14) $\dfrac{dW}{dt} = W + W \ln W$

The solution to this is a double exponential, which we have seen before in the data:

(15) $W = \exp(e^t)$

Notes

Prologue: The Power of Ideas

1. My mother is a talented artist specializing in watercolor paintings. My father was a noted musician, conductor of the Bell Symphony, founder and former chairman of the Queensborough College Music Department.

2. The Tom Swift Jr. series, which was launched in 1954 by Grosset and Dunlap and written by a series of authors under the pseudonym Victor Appleton, continued until 1971. The teenage Tom Swift, along with his pal Bud Barclay, raced around the universe exploring strange places, conquering bad guys, and using exotic gadgets such as house-sized spacecraft, a space station, a flying lab, a cycloplane, an electric hydrolung, a diving seacopter, and a repellatron (which repelled things; underwater, for example, it would repel water, thus forming a bubble in which the boys could live).

   The first nine books in the series are *Tom Swift and His Flying Lab* (1954), *Tom Swift and His Jetmarine* (1954), *Tom Swift and His Rocket Ship* (1954), *Tom Swift and His Giant Robot* (1954), *Tom Swift and His Atomic Earth Blaster* (1954), *Tom Swift and His Outpost in Space* (1955), *Tom Swift and His Diving Seacopter* (1956), *Tom Swift in the Caves of Nuclear Fire* (1956), and *Tom Swift on the Phantom Satellite* (1956).

3. The program was called Select. Students filled out a three-hundred-item questionnaire. The computer software, which contained a database of about two million pieces of information on three thousand colleges, selected six to fifteen schools that matched the student's interests, background, and academic standing. We processed about ten thousand students on our own and then sold the program to the publishing company Harcourt, Brace, and World.

4. *The Age of Intelligent Machines*, published in 1990 by MIT Press, was named Best Computer Science Book by the Association of American Publishers. The book explores the development of artificial intelligence and predicts a range of philosophic, social, and economic impacts of intelligent machines. The narrative is complemented by twenty-three articles on AI from thinkers such as Sherry Turkle, Douglas Hofstadter, Marvin Minsky, Seymour Papert, and George Gilder. For the entire text of the book, see http://www.KurzweilAI.net/aim.

5. Key measures of capability (such as price-performance, bandwidth, and capacity) increase by multiples (that is, the measures are multiplied by a factor for each increment of time) rather than being added to linearly.

6. Douglas R. Hofstadter, *Gödel, Escher, Bach: An Eternal Golden Braid* (New York: Basic Books, 1979).

Chapter One: The Six Epochs

1. According to the Transtopia site (http://transtopia.org/faq.html#1.11), "Singularitarian" was "originally defined by Mark Plus ('91) to mean 'one who believes the concept of a Singularity.' " Another definition of this term is " 'Singularity activist' or 'friend of the Singularity'; that is, one who acts so as to bring about a Singularity [Mark Plus, 1991; *Singularitarian Principles,* Eliezer Yudkowsky, 2000]." There is not universal agreement on this definition, and many Transhumanists are still Singularitarians in the original sense—that is, "believers in the Singularity concept" rather than "activists" or "friends."

Eliezer S. Yudkowsky, in *The Singularitarian Principles*, version 1.0.2 (January 1, 2000), http://yudkowsky.net/sing/principles.ext.html, proposed an alternate definition: "A Singularitarian is someone who believes that technologically creating a greater-than-human intelligence is desirable, and who works to that end. A Singularitarian is friend, advocate, defender, and agent of the future known as the Singularity."

My view: one can advance the Singularity and in particular make it more likely to represent a constructive advance of knowledge in many ways and in many spheres of human discourse—for example, advancing democracy, combating totalitarian and fundamentalist belief systems and ideologies, and creating knowledge in all of its diverse forms: music, art, literature, science, and technology. I regard a Singularitarian as someone who understands the transformations that are coming in this century and who has reflected on their implications for his or her own life.

2. We will examine the doubling rates of computation in the next chapter. Although the number of transistors per unit cost has doubled every two years, transistors have been getting progressively faster, and there have been many other levels of innovation and improvement. The overall power of computation per unit cost has recently been doubling every year. In particular, the amount of computation (in computations per second) that can be brought to bear to a computer chess machine doubled every year during the 1990s.

3. John von Neumann, paraphrased by Stanislaw Ulam, "Tribute to John von Neumann," *Bulletin of the American Mathematical Society* 64.3, pt. 2 (May 1958): 1–49. Von Neumann (1903–1957) was born in Budapest into a Jewish banking family and came to Princeton University to teach mathematics in 1930. In 1933 he became one of the six original professors in the new Institute for Advanced Study

in Princeton, where he stayed until the end of his life. His interests were far ranging: he was the primary force in defining the new field of quantum mechanics; along with coauthor Oskar Morgenstern, he wrote *Theory of Games and Economic Behavior*, a text that transformed the study of economics; and he made significant contributions to the logical design of early computers, including building MANIAC (Mathematical Analyzer, Numeral Integrator, and Computer) in the late 1930s.

Here is how Oskar Morgenstern described von Neumann in the obituary "John von Neumann, 1903–1957," in the *Economic Journal* (March 1958: 174): "Von Neumann exercised an unusually large influence upon the thought of other men in his personal relations. . . . His stupendous knowledge, the immediate response, the unparalleled intuition held visitors in awe. He would often solve their problems before they had finished stating them. His mind was so unique that some people have asked themselves—they too eminent scientists—whether he did not represent a new stage in human mental development."

4. See notes 20 and 21 in chapter 2.

5. The conference was held February 19–21, 2003, in Monterey, California. Among the topics covered were stem-cell research, biotechnology, nanotechnology, cloning, and genetically modified food. For a list of books recommended by conference speakers, see http://www.thefutureoflife.com/books.htm.

6. The Internet, as measured by the number of nodes (servers), was doubling every year during the 1980s but was only tens of thousands of nodes in 1985. This grew to tens of millions of nodes by 1995. By January 2003, the Internet Software Consortium (http://www.isc.org/ds/host-count-history.html) counted 172 million Web hosts, which are the servers hosting Web sites. That number represents only a subset of the total number of nodes.

7. At the broadest level, the anthropic principle states that the fundamental constants of physics must be compatible with our existence; if they were not, we would not be here to observe them. One of the catalysts for the development of the principle is the study of constants, such as the gravitational constant and the electromagnetic-coupling constant. If the values of these constants were to stray beyond a very narrow range, intelligent life would not be possible in our universe. For example, if the electromagnetic-coupling constant were stronger, there would be no bonding between electrons and other atoms. If it were weaker, electrons could not be held in orbit. In other words, if this single constant strayed outside an extremely narrow range, molecules would not form. Our universe, then, appears to proponents of the anthropic principle to be fine-tuned for the evolution of intelligent life. (Detractors such as Victor Stenger claim the fine-tuning is not so fine after all; there are compensatory mechanisms that would support a wider window for life to form under different conditions.)

The anthropic principle comes up again in the context of contemporary cosmology theories that posit multiple universes (see notes 8 and 9, below), each with its own set of laws. Only in a universe in which the laws allowed thinking beings to exist could we be here asking these questions.

One of the seminal texts in the discussion is John Barrow and Frank Tipler, *The Anthropic Cosmological Principle* (New York: Oxford University Press, 1988). See also Steven Weinberg, "A Designer Universe?" at http://www.physlink.com/Education/essay_weinberg.cfm.

8. According to some cosmological theories, there were multiple big bangs, not one, leading to multiple universes (parallel multiverses or "bubbles"). Different physical constants and forces apply in the different bubbles; conditions in some (or at least one) of these bubbles support carbon-based life. See Max Tegmark, "Parallel Universes," *Scientific American* (May 2003): 41–53; Martin Rees, "Exploring Our Universe and Others," *Scientific American* (December 1999): 78–83; Andrei Linde, "The Self-Reproducing Inflationary Universe," *Scientific American* (November 1994): 48–55.

9. The "many worlds" or multiverse theory as an interpretation of quantum mechanics was developed to solve a problem presented by quantum mechanics and then has been combined with the anthropic principle. As summarized by Quentin Smith:

> A serious difficulty associated with the conventional or Copenhagen interpretation of quantum mechanics is that it cannot be applied to the general relativity space-time geometry of a closed universe. A quantum state of such a universe is describable as a wave function with varying spatial-temporal amplitude; the probability of the state of the universe being found at any given point is the square of the amplitude of the wave function at that point. In order for the universe to make the transition from the superposition of many points of varying probabilities to one of these points—the one in which it actually is—a measuring apparatus must be introduced that collapses the wave function and determines the universe to be at that point. But this is impossible, for there is nothing outside the universe, no external measuring apparatus, that can collapse the wave function.
>
> A possible solution is to develop an interpretation of quantum mechanics that does not rely on the notion of external observation or measurement that is central to the Copenhagen interpretation. A quantum mechanics can be formulated that is internal to a closed system.
>
> It is such an interpretation that Hugh Everett developed in his 1957 paper, "Relative State Formulation of Quantum Mechanics." Each point in the superposition represented by the wave function is regarded as actually containing one state of the observer (or measuring apparatus) and one state of the system being observed. Thus "with each succeeding observation (or interaction), the observer state 'branches' into a number of different states. Each branch represents a different outcome of the measurement and the corresponding eigenstate for the object-system state. All branches exist simultaneously in the superposition after any given sequence of observations."

Each branch is causally independent of each other branch, and consequently no observer will ever be aware of any "splitting" process. The world will seem to each observer as it does in fact seem.

Applied to the universe as a whole, this means that the universe is regularly dividing into numerous different and causally independent branches, consequent upon the measurement-like interactions among its various parts. Each branch can be regarded as a separate world, with each world constantly splitting into further worlds.

Given that these branches—the set of universes—will include ones both suitable and unsuitable for life, Smith continues, "At this point it can be stated how the strong anthropic principle in combination with the many-worlds interpretation of quantum mechanics can be used in an attempt to resolve the apparent problem mentioned at the beginning of this essay. The seemingly problematic fact that a world with intelligent life is actual, rather than one of the many lifeless worlds, is found not to be a fact at all. If worlds with life and without life are both actual, then it is not surprising that this world is actual but is something to be expected."

Quentin Smith, "The Anthropic Principle and Many-Worlds Cosmologies," *Australasian Journal of Philosophy* 63.3 (September 1985), available at http://www.qsmithwmu.com/the_anthropic_principle_and_many-worlds_cosmologies.htm.

10. See chapter 4 for a complete discussion of the brain's self-organizing principles and the relationship of this principle of operation to pattern recognition.

11. With a "linear" plot (where all graph divisions are equal), it would be impossible to visualize all of the data (such as billions of years) in a limited space (such as a page of this book). A logarithmic ("log") plot solves that by plotting the order of magnitude of the values rather than the actual values, allowing you to see a wider range of data.

12. Theodore Modis, professor at DUXX, Graduate School in Business Leadership in Monterrey, Mexico, attempted to develop a "precise mathematical law that governs the evolution of change and complexity in the Universe." To research the pattern and history of these changes, he required an analytic data set of significant events where the events equate to major change. He did not want to rely solely on his own list, because of selection bias. Instead, he compiled thirteen multiple independent lists of major events in the history of biology and technology from these sources:

Carl Sagan, *The Dragons of Eden: Speculations on the Evolution of Human Intelligence* (New York: Ballantine Books, 1989). Exact dates provided by Modis.

American Museum of Natural History. Exact dates provided by Modis.

The data set "important events in the history of life" in the *Encyclopaedia Britannica*.

Educational Resources in Astronomy and Planetary Science (ERAPS), University of Arizona, http://ethel.as.arizona.edu/~collins/astro/subjects/evolve-26.html.

Paul D. Boyer, biochemist, winner of the 1997 Nobel Prize, private communication. Exact dates provided by Modis.

J. D. Barrow and J. Silk, "The Structure of the Early Universe," *Scientific American* 242.4 (April 1980): 118–28.

J. Heidmann, *Cosmic Odyssey: Observatoir de Paris,* trans. Simon Mitton (Cambridge, U.K.: Cambridge University Press, 1989).

J. W. Schopf, ed., *Major Events in the History of Life,* symposium convened by the IGPP Center for the Study of Evolution and the Origin of Life, 1991 (Boston: Jones and Bartlett, 1991).

Phillip Tobias, "Major Events in the History of Mankind," chap. 6 in Schopf, *Major Events in the History of Life.*

David Nelson, "Lecture on Molecular Evolution I," http://drnelson.utmem.edu/evolution.html, and "Lecture Notes for Evolution II," http://drnelson.utmem.edu/evolution2.html.

G. Burenhult, ed., *The First Humans: Human Origins and History to 10,000 BC* (San Francisco: HarperSanFrancisco, 1993).

D. Johanson and B. Edgar, *From Lucy to Language* (New York: Simon & Schuster, 1996).

R. Coren, *The Evolutionary Trajectory: The Growth of Information in the History and Future of Earth,* World Futures General Evolution Studies (Amsterdam: Gordon and Breach, 1998).

These lists date from the 1980s and 1990s, with most covering the known history of the universe, while three focus on the narrower period of hominoid evolution. The dates used by some of the older lists are imprecise, but it is the events themselves, and the relative locations of these events in history, that are of primary interest.

Modis then combined these lists to find clusters of major events, his "canonical milestones." This resulted in 28 canonical milestones from the 203 milestone events in the lists. Modis also used another independent list by Coren as a check to see if it corroborated his methods. See T. Modis, "Forecasting the Growth of Complexity and Change," *Technological Forecasting and Social Change* 69.4 (2002); http://ourworld.compuserve.com/homepages/tmodis/TedWEB.htm.

13. Modis notes that errors can arise from variations in the size of lists and from variations in dates assigned to events (see T. Modis, "The Limits of Complexity and Change," *The Futurist* [May–June 2003], http://ourworld.compuserve.com/homepages/tmodis/Futurist.pdf). So he used clusters of dates to define his canonical milestones. A milestone represents an average, with known errors assumed to be the standard deviation. For events without multiple sources, he "arbitrarily assign[ed] the average error as error." Modis also points out other sources of error—cases where precise dates are unknown or where there is the possibility of inappropriate assumption of equal importance for each data point—which are not caught in the standard deviation.

Note that Modis's date of 54.6 million years ago for the dinosaur extinction is not far enough back.

14. Typical interneuronal reset times are on the order of five milliseconds, which allows for two hundred digital-controlled analog transactions per second. Even accounting for multiple nonlinearities in neuronal information processing, this is on the order of a million times slower than contemporary electronic circuits, which can switch in less than one nanosecond (see the analysis of computational capacity in chapter 2).

15. A new analysis by Los Alamos National Lab researchers of the relative concentrations of radioactive isotopes in the world's only known natural nuclear reactor (at Oklo in Gabon, West Africa) has found a decrease in the fine-structure constant, or alpha (the speed of light is inversely proportional to alpha), over two billion years. That translates into a small increase in the speed of light, although this finding clearly needs to be confirmed. See "Speed of Light May Have Changed Recently," *New Scientist*, June 30, 2004, http://www.newscientist.com/news/news.jsp?id=ns99996092. See also http://www.sciencedaily.com/releases/2005/05/050512120842.htm.

16. Stephen Hawking declared at a scientific conference in Dublin on July 21, 2004, that he had been wrong in a controversial assertion he made thirty years ago about black holes. He had said information about what had been swallowed by a black hole could never be retrieved from it. This would have been a violation of quantum theory, which says that information is preserved. "I'm sorry to disappoint science fiction fans, but if information is preserved there is no possibility of using black holes to travel to other universes," he said. "If you jump into a black hole, your mass energy will be returned to our universe, but in a mangled form, which contains the information about what you were like, but in an unrecognizable state." See Dennis Overbye, "About Those Fearsome Black Holes? Never Mind," *New York Times*, July 22, 2004.

17. An event horizon is the outer boundary, or perimeter, of a spherical region surrounding the singularity (the black hole's center, characterized by infinite density and pressure). Inside the event horizon, the effects of gravity are so strong that not even light can escape, although there is radiation emerging from the surface owing to quantum effects that cause particle-antiparticle pairs to form, with one of the pair being pulled into the black hole and the other being emitted as radiation (so-called Hawking radiation). This is the reason why these regions are called "black holes," a term invented by Professor John Wheeler. Although black holes were originally predicted by German astrophysicist Kurt Schwarzschild in 1916 based on Einstein's theory of general relativity, their existence at the centers of galaxies has only recently been experimentally demonstrated. For further reading, see Kimberly Weaver, "The Galactic Odd Couple," http://www.scientificamerican.com, June 10, 2003; Jean-Pierre Lasota, "Unmasking Black Holes," *Scientific American* (May 1999): 41–47; Stephen Hawking, *A Brief History of Time: From the Big Bang to Black Holes* (New York: Bantam, 1988).

18. Joel Smoller and Blake Temple, "Shock-Wave Cosmology Inside a Black Hole," *Proceedings of the National Academy of Sciences* 100.20 (September 30, 2003): 11216–18.

19. Vernor Vinge, "First Word," *Omni* (January 1983): 10.

20. Ray Kurzweil, *The Age of Intelligent Machines* (Cambridge, Mass.: MIT Press, 1989).

21. Hans Moravec, *Mind Children: The Future of Robot and Human Intelligence* (Cambridge, Mass.: Harvard University Press, 1988).

22. Vernor Vinge, "The Coming Technological Singularity: How to Survive in the Post-Human Era," VISION-21 Symposium, sponsored by the NASA Lewis Research Center and the Ohio Aerospace Institute, March 1993. The text is available at http://www.KurzweilAI.net/vingesing.

23. Ray Kurzweil, *The Age of Spiritual Machines: When Computers Exceed Human Intelligence* (New York: Viking, 1999).

24. Hans Moravec, *Robot: Mere Machine to Transcendent Mind* (New York: Oxford University Press, 1999).

25. Damien Broderick, two works: *The Spike: Accelerating into the Unimaginable Future* (Sydney, Australia: Reed Books, 1997) and *The Spike: How Our Lives Are Being Transformed by Rapidly Advancing Technologies*, rev. ed. (New York: Tor/Forge, 2001).

26. One of John Smart's overviews, "What Is the Singularity," can be found at http://www.KurzweilAI.net/meme/frame.html?main=/articles/art0133.html; for a collection of John Smart's writings on technology acceleration, the Singularity, and related issues, see http://www.singularitywatch.com and http://www.Accelerating.org.

    John Smart runs the "Accelerating Change" conference, which covers issues related to "artificial intelligence and intelligence amplification." See http://www.accelerating.org/ac2005/index.html.

27. An emulation of the human brain running on an electronic system would run much faster than our biological brains. Although human brains benefit from massive parallelism (on the order of one hundred trillion interneuronal connections, all potentially operating simultaneously), the reset time of the connections is extremely slow compared to contemporary electronics.

28. See notes 20 and 21 in chapter 2.

29. See the appendix, "The Law of Accelerating Returns Revisited," for a mathematical analysis of the exponential growth of information technology as it applies to the price-performance of computation.

30. In a 1950 paper published in *Mind: A Quarterly Review of Psychology and Philosophy*, the computer theoretician Alan Turing posed the famous questions "Can a machine think? If a computer could think, how could we tell?" The answer to the second question is the Turing test. As the test is currently defined, an expert committee interrogates a remote correspondent on a wide range of topics such as love,

current events, mathematics, philosophy, and the correspondent's personal history to determine whether the correspondent is a computer or a human. The Turing test is intended as a measure of *human* intelligence; failure to pass the test does not imply a lack of intelligence. Turing's original article can be found at http://www.abelard.org/turpap/turpap.htm; see also the *Stanford Encyclopedia of Philosophy,* http://plato.stanford.edu/entries/turing-test, for a discussion of the test.

There is no set of tricks or algorithms that would allow a machine to pass a properly designed Turing test without actually possessing intelligence at a fully human level. Also see Ray Kurzweil, "A Wager on the Turing Test: Why I Think I Will Win," http://www.KurzweilAI.net/turingwin.

31. See John H. Byrne, "Propagation of the Action Potential," *Neuroscience Online,* https://oac22.hsc.uth.tmc.edu/courses/nba/s1/i3-1.html: "The propagation velocity of the action potentials in nerves can vary from 100 meters per second (580 miles per hour) to less than a tenth of a meter per second (0.6 miles per hour)."

Also see Kenneth R. Koehler, "The Action Potential," http://www.rwc.uc.edu/koehler/biophys/4d.html: "The speed of propagation for mammalian motor neurons is 10–120 m/s, while for nonmyelinated sensory neurons it's about 5–25 m/s (nonmyelinated neurons fire in a continuous fashion, without the jumps; ion leakage allows effectively complete circuits but slows the rate of propagation)."

32. A 2002 study published in *Science* highlighted the role of the beta-catenin protein in the horizontal expansion of the cerebral cortex in humans. This protein plays a key role in the folding and grooving of the surface of the cerebral cortex; it is this folding, in fact, that increases the surface area of this part of the brain and makes room for more neurons. Mice that overproduced the protein developed wrinkled, folded cerebral cortexes with substantially more surface area than the smooth, flat cerebral cortexes of control mice. Anjen Chenn and Christopher Walsh, "Regulation of Cerebral Cortical Size by Control of Cell Cycle Exit in Neural Precursors," *Science* 297 (July 2002): 365–69.

A 2003 comparison of cerebral-cortex gene-expression profiles for humans, chimpanzees, and rhesus macaques showed a difference of expression in only ninety-one genes associated with brain organization and cognition. The study authors were surprised to find that 90 percent of these differences involved up-regulation (higher activity). See M. Cacares et al., "Elevated Gene Expression Levels Distinguish Human from Non-human Primate Brains," *Proceedings of the National Academy of Sciences* 100.22 (October 28, 2003): 13030–35.

However, University of California–Irvine College of Medicine researchers have found that gray matter in specific regions in the brain is more related to IQ than is overall brain size and that only about 6 percent of all the gray matter in the brain appears related to IQ. The study also discovered that because these regions related to intelligence are located throughout the brain, a single "intelligence center," such as the frontal lobe, is unlikely. See "Human Intelligence Determined by Volume and Location of Gray Matter Tissue in Brain," University of California–Irvine news release (July 19, 2004), http://today.uci.edu/news/release_detail.asp?key=1187.

A 2004 study found that human nervous system genes displayed accelerated evolution compared with nonhuman primates and that all primates had accelerated evolution compared with other mammals. Steve Dorus et al., "Accelerated Evolution of Nervous System Genes in the Origin of *Homo sapiens*," *Cell* 119 (December 29, 2004): 1027–40. In describing this finding, the lead researcher, Bruce Lahn, states, "Humans evolved their cognitive abilities not due to a few accidental mutations, but rather from an enormous number of mutations acquired through exceptionally intense selection favoring more complex cognitive abilities." Catherine Gianaro, *University of Chicago Chronicle* 24.7 (January 6, 2005).

A single mutation to the muscle fiber gene MYH16 has been proposed as one change allowing humans to have much larger brains. The mutation made ancestral humans' jaws weaker, so that humans did not require the brain-size limiting muscle anchors found in other great apes. Stedman et al., "Myosin Gene Mutation Correlates with Anatomical Changes in the Human Lineage," *Nature* 428 (March 25, 2004): 415–18.

33. Robert A. Freitas Jr., "Exploratory Design in Medical Nanotechnology: A Mechanical Artificial Red Cell," *Artificial Cells, Blood Substitutes, and Immobil. Biotech.* 26 (1998): 411–30; http://www.foresight.org/Nanomedicine/Respirocytes.html; see also the Nanomedicine Art Gallery images (http://www.foresight.org/Nanomedicine/Gallery/Species/Respirocytes.html) and award-winning animation (http://www.phleschbubble.com/album/beyondhuman/respirocyte01.htm) of the respirocytes.

34. Foglets are the conception of the nanotechnology pioneer and Rutgers professor J. Storrs Hall. Here is a snippet of his description: "Nanotechnology is based on the concept of tiny, self-replicating robots. The Utility Fog is a very simple extension of the idea: Suppose, instead of building the object you want atom by atom, the tiny robots [foglets] linked their arms together to form a solid mass in the shape of the object you wanted? Then, when you got tired of that avant-garde coffee table, the robots could simply shift around a little and you'd have an elegant Queen Anne piece instead." J. Storrs Hall, "What I Want to Be When I Grow Up, Is a Cloud," *Extropy*, Quarters 3 and 4, 1994. Published on KurzweilAI.net July 6, 2001: http://www.KurzweilAI.net/foglets. See also J. Storrs Hall, "Utility Fog: The Stuff That Dreams Are Made Of," in *Nanotechnology: Molecular Speculations on Global Abundance*, B. C. Crandall, ed. (Cambridge, Mass.: MIT Press, 1996). Published on KurzweilAI.net July 5, 2001: http://www.KurzweilAI.net/utilityfog.

35. Sherry Turkle, ed., "Evocative Objects: Things We Think With," forthcoming.

36. See the "Exponential Growth of Computing" figure in chapter 2 (p. 70). Projecting the double exponential growth of the price-performance of computation to the end of the twenty-first century, one thousand dollars' worth of computation will provide 10^{60} calculations per second (cps). As we will discuss in chapter 2, three different analyses of the amount of computing required to functionally emulate

the human brain result in an estimate of 10^{15} cps. A more conservative estimate, which assumes that it will be necessary to simulate all of the nonlinearities in every synapse and dendrite, results in an estimate of 10^{19} cps for neuromorphic emulation of the human brain. Even taking the more conservative figure, we get a figure of 10^{29} cps for the approximately 10^{10} humans. Thus, the 10^{60} cps that can be purchased for one thousand dollars circa 2099 will represent 10^{31} (ten million trillion trillion) human civilizations.

37. The invention of the power loom and the other textile automation machines of the early eighteenth century destroyed the livelihoods of the cottage industry of English weavers, who had passed down stable family businesses for hundreds of years. Economic power passed from the weaving families to the owners of the machines. As legend has it, a young and feebleminded boy named Ned Ludd broke two textile factory machines out of sheer clumsiness. From that point on, whenever factory equipment was found to have mysteriously been damaged, anyone suspected of foul play would say, "But Ned Ludd did it." In 1812 the desperate weavers formed a secret society, an urban guerrilla army. They made threats and demands of factory owners, many of whom complied. When asked who their leader was, they replied, "Why, General Ned Ludd, of course." Although the Luddites, as they became known, initially directed most of their violence against the machines, a series of bloody engagements erupted later that year. The tolerance of the Tory government for the Luddites ended, and the movement dissolved with the imprisonment and hanging of prominent members. Although they failed to create a sustained and viable movement, the Luddites have remained a powerful symbol of opposition to automation and technology.

38. See note 34 above.

Chapter Two: A Theory of Technology Evolution: The Law of Accelerating Returns

1. John Smart, Abstract to "Understanding Evolutionary Development: A Challenge for Futurists," presentation to World Futurist Society annual meeting, Washington, D.C., August 3, 2004.

2. That epochal events in evolution represent increases in complexity is Theodore Modis's view. See Theodore Modis, "Forecasting the Growth of Complexity and Change," *Technological Forecasting and Social Change* 69.4 (2002), http://ourworld. compuserve.com/homepages/tmodis/TedWEB.htm.

3. Compressing files is a key aspect of both data transmission (such as a music or text file over the Internet) and data storage. The smaller the file is, the less time it will take to transmit and the less space it will require. The mathematician Claude Shannon, often called the father of information theory, defined the basic theory of data compression in his paper "A Mathematical Theory of Communication," *The Bell System Technical Journal* 27 (July–October 1948): 379–423, 623–56. Data

compression is possible because of factors such as redundancy (repetition) and probability of appearance of character combinations in data. For example, silence in an audio file could be replaced by a value that indicates the duration of the silence, and letter combinations in a text file could be replaced with coded identifiers in the compressed file.

Redundancy can be removed by lossless compression, as Shannon explained, which means there is no loss of information. There is a limit to lossless compression, defined by what Shannon called the entropy rate (compression increases the "entropy" of the data, which is the amount of actual information in it as opposed to predetermined and thus predictable data structures). Data compression removes redundancy from data; lossless compression does it without losing data (meaning that the exact original data can be restored). Alternatively, lossy compression, which is used for graphics files or streaming video and audio files, does result in information loss, though that loss is often imperceptible to our senses.

Most data-compression techniques use a code, which is a mapping of the basic units (or symbols) in the source to a code alphabet. For example, all the spaces in a text file could be replaced by a single code word and the number of spaces. A compression algorithm is used to set up the mapping and then create a new file using the code alphabet; the compressed file will be smaller than the original and thus easier to transmit or store. Here are some of the categories into which common lossless-compression techniques fall:

- Run-length compression, which replaces repeating characters with a code and a value representing the number of repetitions of that character (examples: PackBits and PCX).
- Minimum redundancy coding or simple entropy coding, which assigns codes on the basis of probability, with the most frequent symbols receiving the shortest codes (examples: Huffman coding and arithmetic coding).
- Dictionary coders, which use a dynamically updated symbol dictionary to represent patterns (examples: Lempel-Ziv, Lempel-Ziv-Welch, and DEFLATE).
- Block-sorting compression, which reorganizes characters rather than using a code alphabet; run-length compression can then be used to compress the repeating strings (example: Burrows-Wheeler transform).
- Prediction by partial mapping, which uses a set of symbols in the uncompressed file to predict how often the next symbol in the file appears.

4. Murray Gell-Mann, "What Is Complexity?" in *Complexity*, vol. 1 (New York: John Wiley and Sons, 1995).

5. The human genetic code has approximately six billion (about 10^{10}) bits, not considering the possibility of compression. So the 10^{27} bits that theoretically can be stored in a one-kilogram rock is greater than the genetic code by a factor of 10^{17}. See note 57 below for a discussion of genome compression.

6. Of course, a human, who is also composed of an enormous number of particles, contains an amount of information comparable to a rock of similar weight when

we consider the properties of all the particles. As with the rock, the bulk of this information is not needed to characterize the state of the person. On the other hand, much more information is needed to characterize a person than a rock.

7. See note 175 in chapter 5 for an algorithmic description of genetic algorithms.

8. Humans, chimpanzees, gorillas, and orangutans are all included in the scientific classification of hominids (family *Hominidae*). The human lineage is thought to have diverged from its great ape relatives five to seven million years ago. The human genus *Homo* within the *Hominidae* includes extinct species such as *H. erectus* as well as modern man (*H. sapiens*).

   In chimpanzee hands, the fingers are much longer and less straight than in humans, and the thumb is shorter, weaker, and not as mobile. Chimps can flail with a stick but tend to lose their grip. They cannot pinch hard because their thumbs do not overlap their index fingers. In the modern human, the thumb is longer, and the fingers rotate toward a central axis, so you can touch all the tips of your fingers to the tip of your thumb, a quality that is called full opposability. These and other changes gave humans two new grips: the precision and power grips. Even prehominoid hominids such as the *Australopithecine* from Ethiopia called Lucy, who is thought to have lived around three million years ago, could throw rocks with speed and accuracy. Since then, scientists claim, continual improvements in the hand's capacity to throw and club, along with associated changes in other parts of the body, have resulted in distinct advantages over other animals of similar size and weight. See Richard Young, "Evolution of the Human Hand: The Role of Throwing and Clubbing," *Journal of Anatomy* 202 (2003): 165–74; Frank Wilson, *The Hand: How Its Use Shapes the Brain, Language, and Human Culture* (New York: Pantheon, 1998).

9. The Santa Fe Institute has played a pioneering role in developing concepts and technology related to complexity and emergent systems. One of the principal developers of paradigms associated with chaos and complexity is Stuart Kauffman. Kauffman's *At Home in the Universe: The Search for the Laws of Self-Organization and Complexity* (Oxford: Oxford University Press, 1995) looks "at the forces for order that lie at the edge of chaos."

   In his book *Evolution of Complexity by Means of Natural Selection* (Princeton: Princeton University Press, 1988), John Tyler Bonner asks the questions "How is it that an egg turns into an elaborate adult? How is it that a bacterium, given many millions of years, could have evolved into an elephant?"

   John Holland is another leading thinker from the Santa Fe Institute in the emerging field of complexity. His book *Hidden Order: How Adaptation Builds Complexity* (Reading, Mass.: Addison-Wesley, 1996) includes a series of lectures that he presented at the Santa Fe Institute in 1994. See also John H. Holland, *Emergence: From Chaos to Order* (Reading, Mass.: Addison-Wesley, 1998) and Mitchell Waldrop, *Complexity: The Emerging Science at the Edge of Order and Chaos* (New York: Simon & Schuster, 1992).

10. The second law of thermodynamics explains why there is no such thing as a

perfect engine that uses all the heat (energy) produced by burning fuel to do work: some heat will inevitably be lost to the environment. This same principle of nature holds that heat will flow from a hot pan to cold air rather than in reverse. It also posits that closed ("isolated") systems will spontaneously become more disordered over time—that is, they tend to move from order to disorder. Molecules in ice chips, for example, are limited in their possible arrangements. So a cup of ice chips has less entropy (disorder) than the cup of water the ice chips become when left at room temperature. There are many more possible molecular arrangements in the glass of water than in the ice; greater freedom of movement equals higher entropy. Another way to think of entropy is as multiplicity. The more ways that a state could be achieved, the higher the multiplicity. Thus, for example, a jumbled pile of bricks has a higher multiplicity (and higher entropy) than a neat stack.

11. Max More articulates the view that "advancing technologies are combining and cross-fertilizing to accelerate progress even faster." Max More, "Track 7 Tech Vectors to Take Advantage of Technological Acceleration," *ManyWorlds*, August 1, 2003.

12. For more information, see J. J. Emerson et al., "Extensive Gene Traffic on the Mammalian X Chromosome," *Science* 303.5657 (January 23, 2004): 537–40, http://www3.uta.edu/faculty/betran/science2004.pdf; Nicholas Wade, "Y Chromosome Depends on Itself to Survive," *New York Times*, June 19, 2003; and Bruce T. Lahn and David C. Page, "Four Evolutionary Strata on the Human X Chromosome," *Science* 286.5441 (October 29, 1999): 964–67, http://inside.wi.mit.edu/page/Site/Page%20PDFs/Lahn_and_Page_strata_1999.pdf.

   Interestingly, the second X chromosome in girls is turned off in a process called X inactivation so that the genes on only one X chromosome are expressed. Research has shown that the X chromosome from the father is turned off in some cells and the X chromosome from the mother in other cells.

13. Human Genome Project, "Insights Learned from the Sequence," http://www.ornl.gov/sci/techresources/Human_Genome/project/journals/insights.html. Even though the human genome has been sequenced, most of it does not code for proteins (the so-called junk DNA), so researchers are still debating how many genes will be identified among the three billion base pairs in human DNA. Current estimates suggest less than thirty thousand, though during the Human Genome Project estimates ranged as high as one hundred thousand. See "How Many Genes Are in the Human Genome?" (http://www.ornl.gov/sci/techresources/Human_Genome/faq/genenumber.shtml) and Elizabeth Pennisi, "A Low Number Wins the GeneSweep Pool," *Science* 300.5625 (June 6, 2003): 1484.

14. Niles Eldredge and the late Stephen Jay Gould proposed this theory in 1972 (N. Eldredge and S. J. Gould, "Punctuated Equilibria: An Alternative to Phyletic Gradualism," in T. J. M. Schopf, ed., *Models in Paleobiology* [San Francisco: Freeman, Cooper], pp. 82–115). It has sparked heated discussions among paleontolo-

gists and evolutionary biologists ever since, though it has gradually gained acceptance. According to this theory, millions of years may pass with species in relative
stability. This stasis is then followed by a burst of change, resulting in new species
and the extinction of old (called a "turnover pulse" by Elisabeth Vrba). The effect
is ecosystemwide, affecting many unrelated species. Eldredge and Gould's proposed pattern required a new perspective: "For no bias can be more constricting
than invisibility—and stasis, inevitably read as absence of evolution, had always
been treated as a non-subject. How odd, though, to define the most common of
all palaeontological phenomena as beyond interest or notice!" S. J. Gould and
N. Eldredge, "Punctuated Equilibrium Comes of Age," *Nature* 366 (November 18, 1993): 223–27.

See also K. Sneppen et al., "Evolution As a Self-Organized Critical Phenomenon," *Proceedings of the National Academy of Sciences* 92.11 (May 23, 1995):
5209–13; Elisabeth S. Vrba, "Environment and Evolution: Alternative Causes of
the Temporal Distribution of Evolutionary Events," *South African Journal of Science* 81 (1985): 229–36.

15. As I will discuss in chapter 6, if the speed of light is not a fundamental limit to
rapid transmission of information to remote portions of the universe, then intelligence and computation will continue to expand exponentially until they saturate
the potential of matter and energy to support computation throughout the entire
universe.

16. Biological evolution continues to be of relevance to humans, however, in that disease processes such as cancer and viral diseases use evolution against us (that is,
cancer cells and viruses evolve to counteract specific countermeasures such as
chemotherapy drugs and antiviral medications respectively). But we can use our
human intelligence to outwit the intelligence of biological evolution by attacking
disease processes at sufficiently fundamental levels and by using "cocktail"
approaches that attack a disease in several orthogonal (independent) ways at once.

17. Andrew Odlyzko, "Internet Pricing and the History of Communications," AT&T
Labs Research, revised version February 8, 2001, http://www.dtc.umn.edu/
~odlyzko/doc/history.communications1b.pdf.

18. Cellular Telecommunications and Internet Association, Semi-Annual Wireless
Industry Survey, June 2004, http://www.ctia.org/research_statistics/index.cfm/
AID/10030.

19. Electricity, telephone, radio, television, mobile phones: FCC, www.fcc.gov/
Bureaus/Common_Carrier/Notices/2000/fc00057a.xls. Home computers and
Internet use: Eric C. Newburger, U.S. Census Bureau, "Home Computers and Internet Use in the United States: August 2000" (September 2001), http://www.census.
gov/prod/2001pubs/p23-207.pdf. See also "The Millennium Notebook," *News-
week*, April 13, 1998, p. 14.

20. The paradigm-shift rate, as measured by the amount of time required to adopt
new communications technologies, is currently doubling (that is, the amount of

time for mass adoption—defined as being used by a quarter of the U.S. population—is being cut in half) every nine years. See also note 21.

21. The "Mass Use of Inventions" chart in this chapter on p. 50 shows that the time required for adoption by 25 percent of the U.S. population steadily declined over the past 130 years. For the telephone, 35 years were required compared to 31 for the radio—a reduction of 11 percent, or 0.58 percent per year in the 21 years between these two inventions. The time required to adopt an invention dropped 0.60 percent per year between the radio and television, 1.0 percent per year between television and the PC, 2.6 percent per year between the PC and the mobile phone, and 7.4 percent per year between the mobile phone and the World Wide Web. Mass adoption of the radio beginning in 1897 required 31 years, while the Web required a mere 7 years after it was introduced in 1991—a reduction of 77 percent over 94 years, or an average rate of 1.6 percent reduction in adoption time per year. Extrapolating this rate for the entire twentieth century results in an overall reduction of 79 percent for the century. At the current rate of reducing adoption time of 7.4 percent each year, it would take only 20 years at today's rate of progress to achieve the same reduction of 79 percent that was achieved in the twentieth century. At this rate, the paradigm-shift rate doubles (that is, adoption times are reduced by 50 percent) in about 9 years. Over the twenty-first century, eleven doublings of the rate will result in multiplying the rate by 2^{11}, to about 2,000 times the rate in 2000. The increase in rate will actually be greater than this because the current rate will continue to increase as it steadily did over the twentieth century.

22. Data from 1967–1999, Intel data, see Gordon E. Moore, "Our Revolution," http://www.sia-online.org/downloads/Moore.pdf. Data from 2000–2016, International Technology Roadmap for Semiconductors (ITRS) 2002 Update and 2004 Update, http://public.itrs.net/Files/2002Update/2002Update.pdf and http://www.itrs.net/Common/2004Update/2004_00_Overview.pdf.

23. The ITRS DRAM cost is the cost per bit (packaged microcents) at production. Data from 1971–2000: VLSI Research Inc. Data from 2001–2002: ITRS, 2002 Update, Table 7a, Cost-Near-Term Years, p. 172. Data from 2003–2018: ITRS, 2004 Update, Tables 7a and 7b, Cost-Near-Term Years, pp. 20–21.

24. Intel and Dataquest reports (December 2002), see Gordon E. Moore, "Our Revolution," http://www.sia-online.org/downloads/Moore.pdf.

25. Randall Goodall, D. Fandel, and H. Huffet, "Long-Term Productivity Mechanisms of the Semiconductor Industry," Ninth International Symposium on Silicon Materials Science and Technology, May 12–17, 2002, Philadelphia, sponsored by the Electrochemical Society (ECS) and International Sematech.

26. Data from 1976–1999: E. R. Berndt, E. R. Dulberger, and N. J. Rappaport, "Price and Quality of Desktop and Mobile Personal Computers: A Quarter Century of History," July 17, 2000, http://www.nber.org/~confer/2000/si2000/berndt.pdf. Data from 2001–2016: ITRS, 2002 Update, On-Chip Local Clock in Table 4c: Performance and Package Chips: Frequency On-Chip Wiring Levels—Near-Term Years, p. 167.

27. See note 26 for clock speed (cycle times) and note 24 for cost per transistor.

28. Intel transistors on microprocessors: *Microprocessor Quick Reference Guide,* Intel Research, http://www.intel.com/pressroom/kits/quickrefyr.htm. See also Silicon Research Areas, Intel Research, http://www.intel.com/research/silicon/mooreslaw.htm.

29. Data from Intel Corporation. See also Gordon Moore, "No Exponential Is Forever . . . but We Can Delay 'Forever,' " presented at the International Solid State Circuits Conference (ISSCC), February 10, 2003, ftp://download.intel.com/research/silicon/Gordon_Moore_ISSCC_021003.pdf.

30. Steve Cullen, "Semiconductor Industry Outlook," InStat/MDR, report no. IN0401550SI, April 2004, http://www.instat.com/abstract.asp?id=68&SKU=IN0401550SI.

31. World Semiconductor Trade Statistics, http://wsts.www5.kcom.at.

32. Bureau of Economic Analysis, U.S. Department of Commerce, http://www.bea.gov/bea/dn/home/gdp.htm.

33. See notes 22–24 and 26–30.

34. International Technology Roadmap for Semiconductors, 2002 update, International Sematech.

35. "25 Years of Computer History," http://www.compros.com/timeline.html; Linley Gwennap, "Birth of a Chip," *BYTE* (December 1996), http://www.byte.com/art/9612/sec6/art2.htm; "The CDC 6000 Series Computer," http://www.moorecad.com/standardpascal/cdc6400.html; "A Chronology of Computer History," http://www.cyberstreet.com/hcs/museum/chron.htm; Mark Brader, "A Chronology of Digital Computing Machines (to 1952)," http://www.davros.org/misc/chronology.html; Karl Kempf, "Electronic Computers Within the Ordnance Corps," November 1961, http://ftp.arl.mil/~mike/comphist/61ordnance/index.html; Ken Polsson, "Chronology of Personal Computers," http://www.islandnet.com/~kpolsson/comphist; "The History of Computing at Los Alamos," http://bang.lanl.gov/video/sunedu/computer/comphist.html (requires password); the Machine Room, http://www.machine-room.org; Mind Machine Web Museum, http://www.userwww.sfsu.edu/~hl/mmm.html; Hans Moravec, computer data, http://www.frc.ri.cmu.edu/~hpm/book97/ch3/processor.list; "PC Magazine Online: Fifteen Years of PC Magazine," http://www.pcmag.com/article2/0,1759,23390,00.asp; Stan Augarten, *Bit by Bit: An Illustrated History of Computers* (New York: Ticknor and Fields, 1984); International Association of Electrical and Electronics Engineers (IEEE), *Annals of the History of the Computer* 9.2 (1987): 150–53 and 16.3 (1994): 20; Hans Moravec, *Mind Children: The Future of Robot and Human Intelligence* (Cambridge, Mass.: Harvard University Press, 1988); René Moreau, *The Computer Comes of Age* (Cambridge, Mass.: MIT Press, 1984).

36. The plots in this chapter labeled "Logarithmic Plot" are technically semilogarithmic plots in that one axis (time) is on a linear scale, and the other axis is on a logarithmic scale. However, I am calling these plots "logarithmic plots" for simplicity.

37. See the appendix, "The Law of Accelerating Returns Revisited," which provides a

mathematical derivation of why there are two levels of exponential growth (that is, exponential growth over time in which the rate of the exponential growth—the exponent—is itself growing exponentially over time) in computational power as measured by MIPS per unit cost.

38. Hans Moravec, "When Will Computer Hardware Match the Human Brain?" *Journal of Evolution and Technology* 1 (1998), http://www.jetpress.org/volume1/moravec.pdf.

39. See note 35 above.

40. Achieving the first MIPS per $1,000 took from 1900 to 1990. We're now doubling the number of MIPS per $1,000 in about 400 days. Because current price-performance is about 2,000 MIPS per $1,000, we are adding price-performance at the rate of 5 MIPS per day, or 1 MIPS about every 5 hours.

41. "IBM Details Blue Gene Supercomputer," *CNET News,* May 8, 2003, http://news.com.com/2100-1008_3-1000421.html.

42. See Alfred North Whitehead, *An Introduction to Mathematics* (London: Williams and Norgate, 1911), which he wrote at the same time he and Bertrand Russell were working on their seminal three-volume *Principia Mathematica*.

43. While originally projected to take fifteen years, "the Human Genome Project was finished two and a half years ahead of time and, at $2.7 billion in FY 1991 dollars, significantly under original spending projections": http://www.ornl.gov/sci/tech resources/Human_Genome/project/50yr/press4_2003.shtml.

44. Human Genome Project Information, http://www.ornl.gov/sci/techresources/Human_Genome/project/privatesector.shtml; Stanford Genome Technology Center, http://sequence-www.stanford.edu/group/techdev/auto.html; National Human Genome Research Institute, http://www.genome.gov; Tabitha Powledge, "How Many Genomes Are Enough?" *Scientist,* November 17, 2003, http://www.biomedcentral.com/news/20031117/07.

45. Data from National Center for Biotechnology Information, "GenBank Statistics," revised May 4, 2004, http://www.ncbi.nlm.nih.gov/Genbank/genbankstats.html.

46. Severe acute respiratory syndrome (SARS) was sequenced within thirty-one days of the virus being identified by the British Columbia Cancer Agency and the American Centers for Disease Control. The sequencing from the two centers differed by only ten base pairs out of twenty-nine thousand. This work identified SARS as a coronavirus. Dr. Julie Gerberding, director of the CDC, called the quick sequencing "a scientific achievement that I don't think has been paralleled in our history." See K. Philipkoski, "SARS Gene Sequence Unveiled," *Wired News,* April 15, 2003, http://www.wired.com/news/medtech/0,1286,58481,00.html?tw= wn_story_related.

In contrast, the efforts to sequence HIV began in the 1980s. HIV 1 and HIV 2 were completely sequenced in 2003 and 2002 respectively. National Center for Biotechnology Information, http://www.ncbi.nlm.nih.gov/genomes/framik.cgi? db=genome&gi=12171; HIV Sequence Database maintained by the Los Alamos National Laboratory, http://www.hiv.lanl.gov/content/hiv-db/HTML/outline.html.

47. Mark Brader, "A Chronology of Digital Computing Machines (to 1952),"
http://www.davros.org/misc/chronology.html; Richard E. Matick, *Computer Storage Systems and Technology* (New York: John Wiley and Sons, 1977); University of
Cambridge Computer Laboratory, EDSAC99, http://www.cl.cam.ac.uk/UoCCL/
misc/EDSAC99/statistics.html; Mary Bellis, "Inventors of the Modern Computer:
The History of the UNIVAC Computer—J. Presper Eckert and John Mauchly,"
http://inventors.about.com/library/weekly/aa062398.htm; "Initial Date of Operation of Computing Systems in the USA (1950–1958)," compiled from 1968 OECD
data, http://members.iinet.net.au/~dgreen/timeline.html; Douglas Jones, "Frequently Asked Questions about the DEC PDP-8 computer," ftp://rtfm.mit.edu/
pub/usenet/alt.sys.pdp8/PDP-8_Frequently_Asked_Questions_%28posted_
every_other_month%29; *Programmed Data Processor-1 Handbook*, Digital Equipment Corporation (1960–1963), http://www.dbit.com/~greeng3/pdp1/pdp1.html
#INTRODUCTION; John Walker, "Typical UNIVAC® 1108 Prices: 1968,"
http://www.fourmilab.ch/documents/univac/config1108.html; Jack Harper, "LISP
1.5 for the Univac 1100 Mainframe," http://www.frobenius.com/univac.htm;
Wikipedia, "Data General Nova," http://www.answers.com/topic/data-general-
nova; Darren Brewer, "Chronology of Personal Computers 1972–1974," http://
uk.geocities.com/magoos_universe/comp1972.htm; www.pricewatch.com; http://
www.jc-news.com/parse.cgi?news/pricewatch/raw/pw-010702; http://www.jc-news.
com/parse.cgi?news/pricewatch/raw/pw-020624; http://www.pricewatch.com
(11/17/04); http://sharkyextreme.com/guides/WMPG/article.php/10706_2227191_2;
Byte advertisements, September 1975–March 1998; *PC Computing* advertisements, March 1977–April 2000.

48. Seagate, "Products," http://www.seagate.com/cda/products/discsales/index; *Byte*
advertisements, 1977–1998; *PC Computing* advertisements, March 1999; Editors of
Time-Life Books, *Understanding Computers: Memory and Storage*, rev. ed. (New
York: Warner Books, 1990); "Historical Notes about the Cost of Hard Drive Storage
Space," http://www.alts.net/ns1625/winchest.html; "IBM 305 RAMAC Computer
with Disk Drive," http://www.cedmagic.com/history/ibm-305-ramac.html; John C.
McCallum, "Disk Drive Prices (1955–2004)," http://www.jcmit.com/diskprice.htm.

49. James DeRose, *The Wireless Data Handbook* (St. Johnsbury, Vt.: Quantrum,
1996); First Mile Wireless, http://www.firstmilewireless.com/; J. B. Miles, "Wireless
LANs," *Government Computer News* 18.28 (April 30, 1999), http://www.gcn.com/
vol18_no28/guide/514-1.html; *Wireless Week* (April 14, 1997), http://www.
wirelessweek.com/toc/4%2F14%2F1997; Office of Technology Assessment, "Wireless Technologies and the National Information Infrastructure," September 1995,
http://infoventures.com/emf/federal/ota/ota95-tc.html; Signal Lake, "Broadband
Wireless Network Economics Update," January 14, 2003, http://www.signallake.
com/publications/broadbandupdate.pdf; BridgeWave Communications communication, http://www.bridgewave.com/050604.htm.

50. Internet Software Consortium (http://www.isc.org), ISC Domain Survey: Number of Internet Hosts, http://www.isc.org/ds/host-count-history.html.

51. Ibid.

52. Average traffic on Internet backbones in the U.S. during December of each year is used to estimate traffic for the year. A. M. Odlyzko, "Internet Traffic Growth: Sources and Implications," *Optical Transmission Systems and Equipment for WDM Networking II*, B. B. Dingel, W. Weiershausen, A. K. Dutta, and K.-I. Sato, eds., *Proc. SPIE* (The International Society for Optical Engineering) 5247 (2003): 1–15, http://www.dtc.umn.edu/~odlyzko/doc/oft.internet.growth.pdf; data for 2003–2004 values: e-mail correspondence with A. M. Odlyzko.

53. Dave Kristula, "The History of the Internet" (March 1997, update August 2001), http://www.davesite.com/webstation/net-history.shtml; Robert Zakon, "Hobbes' Internet Timeline v8.0," http://www.zakon.org/robert/internet/timeline; *Converge Network Digest*, December 5, 2002, http://www.convergedigest.com/Daily/daily. asp?vn=v9n229&fecha=December%2005,%202002; V. Cerf, "Cerf's Up," 2004, http://global.mci.com/de/resources/cerfs_up/.

54. H. C. Nathanson et al., "The Resonant Gate Transistor," *IEEE Transactions on Electron Devices* 14.3 (March 1967): 117–33; Larry J. Hornbeck, "128 x 128 Deformable Mirror Device," *IEEE Transactions on Electron Devices* 30.5 (April 1983): 539–43; J. Storrs Hall, "Nanocomputers and Reversible Logic," *Nanotechnology* 5 (July 1994): 157–67; V. V. Aristov et al., "A New Approach to Fabrication of Nanostructures," *Nanotechnology* 6 (April 1995): 35–39; C. Montemagno et al., "Constructing Biological Motor Powered Nanomechanical Devices," *Nanotechnology* 10 (1999): 225–31, http://www.foresight.org/Conferences/MNT6/Papers/Montemagno/; Celeste Biever, "Tiny 'Elevator' Most Complex Nanomachine Yet," *NewScientist.com News Service*, March 18, 2004, http://www.newscientist.com/article.ns?id=dn4794.

55. ETC Group, "From Genomes to Atoms: The Big Down," p. 39, http://www.etcgroup.org/documents/TheBigDown.pdf.

56. Ibid., p. 41.

57. Although it is not possible to determine precisely the information content in the genome, because of the repeated base pairs it is clearly much less than the total uncompressed data. Here are two approaches to estimating the compressed information content of the genome, both of which demonstrate that a range of thirty to one hundred million bytes is conservatively high.

   1. In terms of the uncompressed data, there are three billion DNA rungs in the human genetic code, each coding two bits (since there are four possibilities for each DNA base pair). Thus, the human genome is about 800 million bytes uncompressed. The noncoding DNA used to be called "junk DNA," but it is now clear that it plays an important role in gene expression. However, it is very inefficiently coded. For one thing, there are massive redundancies (for example, the sequence called "ALU" is repeated hundreds of thousands of times), which compression algorithms can take advantage of.

With the recent explosion of genetic data banks, there is a great deal of interest in compressing genetic data. Recent work on applying standard data compression algorithms to genetic data indicates that reducing the data by 90 percent (for bit-perfect compression) is feasible: Hisahiko Sato et al., "DNA Data Compression in the Post Genome Era," *Genome Informatics* 12 (2001): 512–14, http://www.jsbi.org/journal/GIW01/GIW01P130.pdf.

Thus we can compress the genome to about 80 million bytes without loss of information (meaning we can perfectly reconstruct the full 800-million-byte uncompressed genome).

Now consider that more than 98 percent of the genome does not code for proteins. Even after standard data compression (which eliminates redundancies and uses a dictionary lookup for common sequences), the algorithmic content of the noncoding regions appears to be rather low, meaning that it is likely that we could code an algorithm that would perform the same function with fewer bits. However, since we are still early in the process of reverse engineering the genome, we cannot make a reliable estimate of this further decrease based on a functionally equivalent algorithm. I am using, therefore, a range of 30 to 100 million bytes of compressed information in the genome. The top part of this range assumes only data compression and no algorithmic simplification.

Only a portion (although the majority) of this information characterizes the design of the brain.

2. Another line of reasoning is as follows. Though the human genome contains around 3 billion bases, only a small percentage, as mentioned above, codes for proteins. By current estimates, there are 26,000 genes that code for proteins. If we assume those genes average 3,000 bases of useful data, those equal only approximately 78 million bases. A base of DNA requires only two bits, which translate to about 20 million bytes (78 million bases divided by four). In the protein-coding sequence of a gene, each "word" (codon) of three DNA bases translates into one amino acid. There are, therefore, 4^3 (64) possible codon codes, each consisting of three DNA bases. There are, however, only 20 amino acids used plus a stop codon (null amino acid) out of the 64. The rest of the 43 codes are used as synonyms of the 21 useful ones. Whereas 6 bits are required to code for 64 possible combinations, only about 4.4 ($\log_2$ 21) bits are required to code for 21 possibilities, a savings of 1.6 out of 6 bits (about 27 percent), bringing us down to about 15 million bytes. In addition, some standard compression based on repeating sequences is feasible here, although much less compression is possible on this protein-coding portion of the DNA than in the so-called junk DNA, which has massive redundancies. So this will bring the

figure probably below 12 million bytes. However, now we have to add information for the noncoding portion of the DNA that controls gene expression. Although this portion of the DNA comprises the bulk of the genome, it appears to have a low level of information content and is replete with massive redundancies. Estimating that it matches the approximately 12 million bytes of protein-coding DNA, we again come to approximately 24 million bytes. From this perspective, an estimate of 30 to 100 million bytes is conservatively high.

58. Continuous values can be represented by floating-point numbers to any desired degree of accuracy. A floating-point number consists of two sequences of bits. One "exponent" sequence represents a power of 2. The "base" sequence represents a fraction of 1. By increasing the number of bits in the base, any desired degree of accuracy can be achieved.

59. Stephen Wolfram, *A New Kind of Science* (Champaign, Ill.: Wolfram Media, 2002).

60. Early work on a digital theory of physics was also presented by Frederick W. Kantor, *Information Mechanics* (New York: John Wiley and Sons, 1977). Links to several of Kantor's papers can be found at http://w3.execnet.com/kantor/pm00.htm (1997); http://w3.execnet.com/kantor/1b2p.htm (1989); and http://w3.execnet.com/kantor/ipoim.htm (1982). Also see at http://www.kx.com/listbox/k/msg05621.html.

61. Konrad Zuse, "Rechnender Raum," *Elektronische Datenverarbeitung*, 1967, vol. 8, pp. 336–44. Konrad Zuse's book on a cellular automaton–based universe was published two years later: *Rechnender Raum, Schriften zur Datenverarbeitung* (Braunschweig, Germany: Friedrich Vieweg & Sohn, 1969). English translation: *Calculating Space*, MIT Technical Translation AZT-70-164-GEMIT, February 1970. MIT Project MAC, Cambridge, MA 02139. PDF.

62. Edward Fredkin quoted in Robert Wright, "Did the Universe Just Happen?" *Atlantic Monthly*, April 1988, 29–44, http://digitalphysics.org/Publications/Wri88a/html.

63. Ibid.

64. Many of Fredkin's results come from studying his own model of computation, which explicitly reflects a number of fundamental principles of physics. See the classic article Edward Fredkin and Tommaso Toffoli, "Conservative Logic," *International Journal of Theoretical Physics* 21.3–4 (1982): 219–53, http://www.digitalphilosophy.org/download_documents/ConservativeLogic.pdf. Also, a set of concerns about the physics of computation analytically similar to those of Fredkin's may be found in Norman Margolus, "Physics and Computation," Ph.D. thesis, MIT/LCS/TR-415, MIT Laboratory for Computer Science, 1988.

65. I discussed Norbert Wiener and Ed Fredkin's view of information as the fundamental building block for physics and other levels of reality in my 1990 book, *The Age of Intelligent Machines*.

   The complexity of casting all of physics in terms of computational transformations proved to be an immensely challenging project, but Fredkin has continued his efforts. Wolfram has devoted a considerable portion of his work over the past decade to this notion, apparently with only limited communication with some of the others in the physics community who are also pursuing the idea. Wolfram's stated goal "is not to present a specific ultimate model for physics," but in his "Note for Physicists" (which essentially equates to a grand challenge), Wolfram describes the "features that [he] believe[s] such a model will have" (*A New Kind of Science*, pp. 1043–65, http://www.wolframscience.com/nksonline/page-1043c-text).

In *The Age of Intelligent Machines*, I discuss "the question of whether the ultimate nature of reality is analog or digital" and point out that "as we delve deeper and deeper into both natural and artificial processes, we find the nature of the process often alternates between analog and digital representations of information." As an illustration, I discussed sound. In our brains, music is represented as the digital firing of neurons in the cochlea, representing different frequency bands. In the air and in the wires leading to loudspeakers, it is an analog phenomenon. The representation of sound on a compact disc is digital, which is interpreted by digital circuits. But the digital circuits consist of thresholded transistors, which are analog amplifiers. As amplifiers, the transistors manipulate individual electrons, which can be counted and are, therefore, digital, but at a deeper level electrons are subject to analog quantum-field equations. At a yet deeper level, Fredkin and now Wolfram are theorizing a digital (computational) basis to these continuous equations.

It should be further noted that if someone actually does succeed in establishing such a digital theory of physics, we would then be tempted to examine what sorts of deeper mechanisms are actually implementing the computations and links of the cellular automata. Perhaps underlying the cellular automata that run the universe are yet more basic analog phenomena, which, like transistors, are subject to thresholds that enable them to perform digital transactions. Thus, establishing a digital basis for physics will not settle the philosophical debate as to whether reality is ultimately digital or analog. Nonetheless, establishing a viable computational model of physics would be a major accomplishment.

So how likely is this? We can easily establish an existence proof that a digital model of physics is feasible, in that continuous equations can always be expressed to any desired level of accuracy in the form of discrete transformations on discrete changes in value. That is, after all, the basis for the fundamental theorem of calculus. However, expressing continuous formulas in this way is an inherent complication and would violate Einstein's dictum to express things "as simply as possible, but no simpler." So the real question is whether we can express the basic relationships that we are aware of in more elegant terms, using cellular-automata algorithms. One test of a new theory of physics is whether it is capable of making verifiable predictions. In at least one important way, that might be a difficult challenge for a cellular automata–based theory because lack of predictability is one of the fundamental features of cellular automata.

Wolfram starts by describing the universe as a large network of nodes. The nodes do not exist in "space," but rather space, as we perceive it, is an illusion created by the smooth transition of phenomena through the network of nodes. One can easily imagine building such a network to represent "naive" (Newtonian) physics by simply building a three-dimensional network to any desired degree of granularity. Phenomena such as "particles" and "waves" that appear to move through space would be represented by "cellular gliders," which are patterns that are advanced through the network for each cycle of computation. Fans of the game *Life* (which is based on cellular automata) will recognize the common phenomenon of gliders and the diversity of patterns that can move smoothly through a cellular-automaton network. The speed of light, then, is the result of the clock speed of the celestial computer, since gliders can advance only one cell per computational cycle.

Einstein's general relativity, which describes gravity as perturbations in space itself, as if our three-dimensional world were curved in some unseen fourth dimension, is also straightforward to represent in this scheme. We can imagine a four-dimensional network and can represent apparent curvatures in space in the same way that one represents normal curvatures in three-dimensional space. Alternatively, the network can become denser in certain regions to represent the equivalent of such curvature.

A cellular-automata conception proves useful in explaining the apparent increase in entropy (disorder) that is implied by the second law of thermodynamics. We have to assume that the cellular-automata rule underlying the universe is a class 4 rule (see main text)—otherwise the universe would be a dull place indeed. Wolfram's primary observation that a class 4 cellular automaton quickly produces apparent randomness (despite its determinate process) is consistent with the tendency toward randomness that we see in Brownian motion and that is implied by the second law.

Special relativity is more difficult. There is an easy mapping from the Newtonian model to the cellular network. But the Newtonian model breaks down in special relativity. In the Newtonian world, if a train is going eighty miles per hour and you drive along it on a parallel road at sixty miles per hour, the train will appear to pull away from you at twenty miles per hour. But in the world of special relativity, if you leave Earth at three quarters of the speed of light, light will still appear to you to move away from you at the full speed of light. In accordance with this apparently paradoxical perspective, both the size and subjective passage of time for two observers will vary depending on their relative speed. Thus, our fixed mapping of space and nodes becomes considerably more complex. Essentially, each observer needs his or her own network. However, in considering special relativity, we can essentially apply the same conversion to our "Newtonian" network as we do to Newtonian space. However, it is not clear that we are achieving greater simplicity in representing special relativity in this way.

A cellular-node representation of reality may have its greatest benefit in understanding some aspects of the phenomenon of quantum mechanics. It could provide an explanation for the apparent randomness that we find in quantum phenomena. Consider, for example, the sudden and apparently random creation of particle-antiparticle pairs. The randomness could be the same sort of randomness that we see in class 4 cellular automata. Although predetermined, the behavior of class 4 automata cannot be anticipated (other than by running the cellular automata) and is effectively random.

This is not a new view. It's equivalent to the "hidden variables" formulation of quantum mechanics, which states that there are some variables that we cannot otherwise access that control what appears to be random behavior that we can observe. The hidden-variables conception of quantum mechanics is not inconsistent with the formulas for quantum mechanics. It is possible but is not popular with quantum physicists because it requires a large number of assumptions to work out in a very particular way. However, I do not view this as a good argument against it. The existence of our universe is itself very unlikely and requires many assumptions to all work out in a very precise way. Yet here we are.

A bigger question is, How could a hidden-variables theory be tested? If based on cellular-automata-like processes, the hidden variables would be inherently unpredictable, even if deterministic. We would have to find some other way to "unhide" the hidden variables.

Wolfram's network conception of the universe provides a potential perspective on the phenomenon of quantum entanglement and the collapse of the wave function. The collapse of the wave function, which renders apparently ambiguous properties of a particle (for example, its location) retroactively determined, can be viewed from the cellular-network perspective as the interaction of the observed phenomenon with the observer itself. As observers, we are not outside the network but exist inside it. We know from cellular mechanics that two entities cannot interact without both being changed, which suggests a basis for wave-function collapse.

Wolfram writes, "If the universe is a network, then it can in a sense easily contain threads that continue to connect particles even when the particles get far apart in terms of ordinary space." This could provide an explanation for recent dramatic experiments showing nonlocality of action in which two "quantum entangled" particles appear to continue to act in concert with each other even though separated by large distances. Einstein called this "spooky action at a distance" and rejected it, although recent experiments appear to confirm it.

Some phenomena fit more neatly into this cellular automata–network conception than others. Some of the suggestions appear elegant, but as Wolfram's "Note for Physicists" makes clear, the task of translating all of physics into a consistent cellular-automata–based system is daunting indeed.

Extending his discussion to philosophy, Wolfram "explains" the apparent phenomenon of free will as decisions that are determined but unpredictable. Since there is no way to predict the outcome of a cellular process without actually running the process, and since no simulator could possibly run faster than the universe itself, there is therefore no way to reliably predict human decisions. So even though our decisions are determined, there is no way to preidentify what they will be. However, this is not a fully satisfactory examination of the concept. This observation concerning the lack of predictability can be made for the outcome of most physical processes—such as where a piece of dust will fall on the ground. This view thereby equates human free will with the random descent of a piece of dust. Indeed, that appears to be Wolfram's view when he states that the process in the human brain is "computationally equivalent" to those taking place in processes such as fluid turbulence.

Some of the phenomena in nature (for example, clouds, coastlines) are characterized by repetitive simple processes such as cellular automata and fractals, but intelligent patterns (such as the human brain) require an evolutionary process (or alternatively, the reverse engineering of the results of such a process). Intelligence is the inspired product of evolution and is also, in my view, the most powerful "force" in the world, ultimately transcending the powers of mindless natural forces.

In summary, Wolfram's sweeping and ambitious treatise paints a compelling but ultimately overstated and incomplete picture. Wolfram joins a growing community of voices that maintain that patterns of information, rather than matter and energy, represent the more fundamental building blocks of reality. Wolfram has added to our knowledge of how patterns of information create the world we experience, and I look forward to a period of collaboration between Wolfram and his colleagues so that we can build a more robust vision of the ubiquitous role of algorithms in the world.

The lack of predictability of class 4 cellular automata underlies at least some of the apparent complexity of biological systems and does represent one of the important biological paradigms that we can seek to emulate in our technology. It does not explain all of biology. It remains at least possible, however, that such methods can explain all of physics. If Wolfram, or anyone else for that matter, succeeds in formulating physics in terms of cellular-automata operations and their patterns, Wolfram's book will have earned its title. In any event, I believe the book to be an important work of ontology.

66. Rule 110 states that a cell becomes white if its previous color was, and its two neighbors are, all black or all white, or if its previous color was white and the two neighbors are black and white, respectively; otherwise, the cell becomes black.

67. Wolfram, New Kind of Science, p. 4, http://www.wolframscience.com/nksonline/page-4-text.

68. Note that certain interpretations of quantum mechanics imply that the world is not based on deterministic rules and that there is an inherent quantum randomness to every interaction at the (small) quantum scale of physical reality.

69. As discussed in note 57 above, the uncompressed genome has about six billion bits of information (order of magnitude = 10^{10} bits), and the compressed genome is about 30 to 100 million bytes. Some of this design information applies, of course, to other organs. Even assuming all of 100 million bytes applies to the brain, we get a conservatively high figure of 10^9 bits for the design of the brain in the genome. In chapter 3, I discuss an estimate for "human memory on the level of individual interneuronal connections," including "the connection patterns and neurotransmitter concentrations" of 10^{18} (billion billion) bits in a mature brain. This is about a billion (10^9) times more information than that in the genome which describes the brain's design. This increase comes about from the self-organization of the brain as it interacts with the person's environment.

70. See the sections "Disdisorder" and "The Law of Increasing Entropy Versus the Growth of Order" in my book *The Age of Spiritual Machines: When Computers Exceed Human Intelligence* (New York: Viking, 1999), pp. 30–33.

71. A universal computer can accept as input the definition of any other computer and then simulate that other computer. This does not address the speed of simulation, which might be relatively slow.

72. C. Geoffrey Woods, "Crossing the Midline," *Science* 304.5676 (June 4, 2004): 1455–56; Stephen Matthews, "Early Programming of the Hypothalamo-Pituitary-Adrenal Axis," *Trends in Endocrinology and Metabolism* 13.9 (November 1, 2002): 373–80; Justin Crowley and Lawrence Katz, "Early Development of Ocular Dominance Columns," *Science* 290.5495 (November 17, 2000): 1321–24; Anna Penn et al., "Competition in the Retinogeniculate Patterning Driven by Spontaneous Activity," *Science* 279.5359 (March 27, 1998): 2108–12.

73. The seven commands of a Turing machine are: (1) Read Tape, (2) Move Tape Left, (3) Move Tape Right, (4) Write 0 on the Tape, (5) Write 1 on the Tape, (6) Jump to Another Command, and (7) Halt.

74. In what is perhaps the most impressive analysis in his book, Wolfram shows how a Turing machine with only two states and five possible colors can be a universal Turing machine. For forty years, we've thought that a universal Turing machine had to be more complex than this. Also impressive is Wolfram's demonstration that rule 110 is capable of universal computation, given the right software. Of course, universal computation by itself cannot perform useful tasks without appropriate software.

75. The "nor" gate transforms two inputs into one output. The output of "nor" is true if and only if neither A *nor* B is true.

76. See the section "A *nor* B: The Basis of Intelligence?" in *The Age of Intelligent Machines* (Cambridge, Mass.: MIT Press, 1990), pp. 152–57, http://www.KurzweilAI.net/meme/frame.html?m=12.

77. United Nations Economic and Social Commission for Asia and the Pacific, "Regional Road Map Towards an Information Society in Asia and the Pacific," ST/ESCAP/2283, http://www.unescap.org/publications/detail.asp?id=771; Eco-

nomic and Social Commission for Western Asia, "Regional Profile of the Information Society in Western Asia," October 8, 2003, http://www.escwa.org.lb/information/publications/ictd/docs/ictd-03-11-e.pdf; John Enger, "Asia in the Global Information Economy: The Rise of Region-States, The Role of Telecommunications," presentation at the International Conference on Satellite and Cable Television in Chinese and Asian Regions, Communication Arts Research Institute, Fu Jen Catholic University, June 4–6, 1996.

78. See "The 3 by 5 Initiative," Fact Sheet 274, December 2003, http://www.who.int/mediacentre/factsheets/2003/fs274/en/print.html.

79. Technology investments accounted for 76 percent of 1998 venture-capital investments ($10.1 billion) (PricewaterhouseCoopers news release, "Venture Capital Investments Rise 24 Percent and Set Record at $14.7 Billion, Pricewaterhouse-Coopers Finds," February 16, 1999). In 1999, technology-based companies cornered 90 percent of venture-capital investments ($32 billion) (PricewaterhouseCoopers news release, "Venture Funding Explosion Continues: Annual and Quarterly Investment Records Smashed, According to PricewaterhouseCoopers Money Tree National Survey," February 14, 2000). Venture-capital levels certainly dropped during the high-tech recession; but in just the second quarter of 2003, software companies alone attracted close to $1 billion (PricewaterhouseCoopers news release, "Venture Capital Investments Stabilize in Q2 2003," July 29, 2003). In 1974 in all U.S. manufacturing industries forty-two firms received a total of $26.4 million in venture-capital disbursements (in 1974 dollars, or $81 million in 1992 dollars). Samuel Kortum and Josh Lerner, "Assessing the Contribution of Venture Capital to Innovation," *RAND Journal of Economics* 31.4 (Winter 2000): 674–92, http://econ.bu.edu/kortum/rje_Winter'00_Kortum.pdf. As Paul Gompers and Josh Lerner say, "Inflows to venture capital funds have expanded from virtually zero in the mid-1970s. . . ." Gompers and Lerner, *The Venture Capital Cycle*, (Cambridge, Mass.: MIT Press, 1999). See also Paul Gompers, "Venture Capital," in B. Espen Eckbo, ed., *Handbook of Corporate Finance: Empirical Corporate Finance*, in the Handbooks in Finance series (Holland: Elsevier, forthcoming), chapter 11, 2005, http://mba.tuck.dartmouth.edu/pages/faculty/espen.eckbo/PDFs/Handbookpdf/CH11-VentureCapital.pdf.

80. An account of how "new economy" technologies are making important transformations to "old economy" industries: Jonathan Rauch, "The New Old Economy: Oil, Computers, and the Reinvention of the Earth," *Atlantic Monthly*, January 3, 2001.

81. U.S. Department of Commerce, Bureau of Economic Analysis (http://www.bea.doc.gov), use the following site and select Table 1.1.6: http://www.bea.doc.gov/bea/dn/nipaweb/SelectTable.asp?Selected=N.

82. U.S. Department of Commerce, Bureau of Economic Analysis, http://www.bea.doc.gov. Data for 1920–1999: Population Estimates Program, Population Division, U.S. Census Bureau, "Historical National Population Estimates: July 1,

1900 to July 1, 1999," http://www.census.gov/popest/archives/1990s/popclockest.txt; data for 2000–2004: http://www.census.gov/popest/states/tables/NST-EST2004-01.pdf.

83. "The Global Economy: From Recovery to Expansion," Results from *Global Economic Prospects 2005: Trade, Regionalism and Prosperity* (World Bank, 2004), http://globaloutlook.worldbank.org/globaloutlook/outside/globalgrowth.aspx; "World Bank: 2004 Economic Growth Lifts Millions from Poverty," *Voice of America News*, http://www.voanews.com/english/2004-11-17-voa41.cfm.

84. Mark Bils and Peter Klenow, "The Acceleration in Variety Growth," *American Economic Review* 91.2 (May 2001): 274–80, http://www.klenow.com/Acceleration.pdf.

85. See notes 84, 86, and 87.

86. U.S. Department of Labor, Bureau of Labor Statistics, news report, June 3, 2004. You can generate productivity reports at http://www.bls.gov/bls/productivity.htm.

87. Bureau of Labor Statistics, Major Sector Multifactor Productivity Index, Manufacturing Sector: Output per Hour All Persons (1996 = 100), http://data.bls.gov/PDQ/outside.jsp?survey=mp (Requires JavaScript: select "Manufacturing," "Output Per Hour All Persons," and starting year 1949), or http://data.bls.gov/cgi-bin/srgate (use series "MPU300001," "All Years," and Format 2).

88. George M. Scalise, Semiconductor Industry Association, in "Luncheon Address: The Industry Perspective on Semiconductors," *2004 Productivity and Cyclicality in Semiconductors: Trends, Implications, and Questions—Report of a Symposium (2004)* (National Academies Press, 2004), p. 40, http://www.nap.edu/openbook/0309092744/html/index.html.

89. Data from Kurzweil Applied Intelligence, now part of ScanSoft (formerly Kurzweil Computer Products).

90. eMarketer, "E-Business in 2003: How the Internet Is Transforming Companies, Industries, and the Economy—a Review in Numbers," February 2003; "US B2C E-Commerce to Top $90 Billion in 2003," April 30, 2003, http://www.emarketer.com/Article.aspx?1002207; and "Worldwide B2B E-Commerce to Surpass $1 Trillion By Year's End," March 19, 2003, http://www.emarketer.com/Article.aspx?1002125.

91. The patents used in this chart are, as described by the U.S. Patent and Trademark Office, "patents for inventions," also known as "utility" patents. The U.S. Patent and Trademark Office, Table of Annual U.S. Patent Activity, http://www.uspto.gov/web/offices/ac/ido/oeip/taf/h_counts.htm.

92. The doubling time for IT's share of the economy is twenty-three years. U.S. Department of Commerce, Economics and Statistics Administration, "The Emerging Digital Economy," figure 2, http://www.technology.gov/digeconomy/emerging.htm.

93. The doubling time for U.S. education expenditures per capita is twenty-three years. National Center for Education Statistics, Digest of Education Statistics, 2002, http://nces.ed.gov/pubs2003/digest02/tables/dt030.asp.

94. The United Nations estimated that the total global equity market capitalization in

2000 was thirty-seven trillion dollars. United Nations, "Global Finance Profile," *Report of the High-Level Panel of Financing for Development*, June 2001, http://www.un.org/reports/financing/profile.htm.

If our perception of future growth rates were to increase (compared to current expectations) by an annual compounded rate of as little as 2 percent, and considering an annual discount rate (for discounting future values today) of 6 percent, then considering the increased present value resulting from only twenty years of compounded and discounted future (additional) growth, present values should triple. As the subsequent dialogue points out, this analysis does not take into consideration the likely increase in the discount rate that would result from such a perception of increased future growth.

Chapter Three: Achieving the Computational Capacity of the Human Brain

1. Gordon E. Moore, "Cramming More Components onto Integrated Circuits," *Electronics* 38.8 (April 19, 1965): 114–17, ftp://download.intel.com/research/silicon/moorespaper.pdf.

2. Moore's initial projection in this 1965 paper was that the number of components would double every year. In 1975 this was revised to every two years. However, this more than doubles price-performance every two years because smaller components run faster (because the electronics have less distance to travel). So overall price-performance (for the cost of each transistor cycle) has been coming down by half about every thirteen months.

3. Paolo Gargini quoted in Ann Steffora Mutschler, "Moore's Law Here to Stay," ElectronicsWeekly.com, July 14, 2004, http://www.electronicsweekly.co.uk/articles/article.asp?liArticleID=36829. See also Tom Krazit, "Intel Prepares for Next 20 Years of Chip Making," *Computerworld*, October 25, 2004, http://www.computerworld.com/hardwaretopics/hardware/story/0,10801,96917,00.html.

4. Michael Kanellos, " 'High-rise' Chips Sneak on Market," CNET News.com, July 13, 2004, http://zdnet.com.com/2100-1103-5267738.html.

5. Benjamin Fulford, "Chipmakers Are Running Out of Room: The Answer Might Lie in 3-D," Forbes.com, July 22, 2002, http://www.forbes.com/forbes/2002/0722/173_print.html.

6. NTT news release, "Three-Dimensional Nanofabrication Using Electron Beam Lithography," February 2, 2004, http://www.ntt.co.jp/news/news04e/0402/040202.html.

7. László Forró and Christian Schönenberger, "Carbon Nanotubes, Materials for the Future," *Europhysics News* 32.3 (2001), http://www.europhysicsnews.com/full/09/article3/article3.html. Also see http://www.research.ibm.com/nanoscience/nanotubes.html for an overview of nanotubes.

8. Michael Bernstein, American Chemical Society news release, "High-Speed Nanotube Transistors Could Lead to Better Cell Phones, Faster Computers," April 27, 2004, http://www.eurekalert.org/pub_releases/2004-04/acs-nt042704.php.

9. I estimate a nanotube-based transistor and supporting circuitry and connections require approximately a ten-nanometer cube (the transistor itself will be a fraction of this), or 10^3 cubic nanometers. This is conservative, since single-walled nanotubes are only one nanometer in diameter. One inch = 2.54 centimeters = 2.54×10^7 nanometers. Thus, a 1-inch cube = $2.54^3 \times 10^{21} = 1.6 \times 10^{22}$ cubic nanometers. So a one-inch cube could provide 1.6×10^{19} transistors. With each computer requiring approximately 10^7 transistors (which is a much more complex apparatus than that comprising the calculations in a human interneuronal connection), we can support about 10^{12} (one trillion) parallel computers.

A nanotube transistor–based computer at 10^{12} calculations per second (based on Burke's estimate) gives us a speed estimate of 10^{24} cps for the one-inch cube of nanotube circuitry. Also see Bernstein, "High-Speed Nanotube Transistors."

With an estimate of 10^{16} cps for functional emulation of the human brain (see discussion later in this chapter), this gives us about 100 million (10^8) human-brain equivalents. If we use the more conservative 10^{19} cps estimate needed for neuromorphic simulation (simulating every nonlinearity in every neural component; see subsequent discussion in this chapter), a one-inch cube of nanotube circuitry would provide only one hundred thousand human-brain equivalents.

10. "Only four years ago did we measure for the first time any electronic transport through a nanotube. Now, we are exploring what can be done and what cannot in terms of single-molecule devices. The next step will be to think about how to combine these elements into complex circuits," says one of the authors, Cees Dekker, of Henk W. Ch. Postma et al., "Carbon Nanotube Single-Electron Transistors at Room Temperature," *Science* 293.5527 (July 6, 2001): 76–129, described in the American Association for the Advancement of Science news release, "Nano-transistor Switches with Just One Electron May Be Ideal for Molecular Computers, *Science* Study Shows," http://www.eurekalert.org/pub_releases/2001-07/aaft-nsw062901.php.

11. The IBM researchers solved a problem in nanotube fabrication. When carbon soot is heated to create the tubes, a large number of unusable metallic tubes are created along with the semiconductor tubes suitable for transistors. The team included both types of nanotubes in a circuit and then used electrical pulses to shatter the undesirable ones—a far more efficient approach than cherry-picking the desirable tubes with an atomic-force microscope. Mark K. Anderson, "Mega Steps Toward the Nanochip," *Wired News,* April 27, 2001, at http://www.wired.com/news/technology/0,1282,43324,00.html, referring to Philip G. Collins, Michael S. Arnold, and Phaedon Avouris, "Engineering Carbon Nanotubes and Nanotube Circuits Using Electrical Breakdown," *Science* 292.5517 (April 27, 2001): 706–9.

12. "A carbon nanotube, which looks like rolled chicken wire when examined at the atomic level, is tens of thousands of times thinner than a human hair, yet remarkably strong." University of California at Berkeley press release, "Researchers Create First Ever Integrated Silicon Circuit with Nanotube Transistors," January 5, 2004, http://www.berkeley.edu/news/media/releases/2004/01/05_nano.shtml, referring to Yu-Chih Tseng et al., "Monolithic Integration of Carbon Nanotube Devices

with Silicon MOS Technology," *Nano Letters* 4.1 (2004): 123–27, http://pubs.acs. org/cgi-bin/sample.cgi/nalefd/2004/4/i01/pdf/nl0349707.pdf.

13. R. Colin Johnson, "IBM Nanotubes May Enable Molecular-Scale Chips," *EETimes*, April 26, 2001, http://eetimes.com/article/showArticle.jhtml?articleId=10807704.

14. Avi Aviram and Mark A. Ratner, "Molecular Rectifiers," *Chemical Physics Letters* (November 15, 1974): 277–83, referred to in Charles M. Lieber, "The Incredible Shrinking Circuit," *Scientific American* (September 2001), at http://www.sciam. com and http://www-mcg.uni-r.de/downloads/lieber.pdf. The single-molecule rectifier described in Aviram and Ratner could pass current preferentially in either direction.

15. Will Knight, "Single Atom Memory Device Stores Data," NewScientist.com, September 10, 2002, http://www.newscientist.com/news/news.jsp?id=ns99992775, referring to R. Bennewitz et al., "Atomic Scale Memory at a Silicon Surface," *Nanotechnology* 13 (July 4, 2002): 499–502.

16. Their transistor is made from indium phosphide and indium gallium arsenide. University of Illinois at Urbana-Champaign news release, "Illinois Researchers Create World's Fastest Transistor—Again," http://www.eurekalert.org/pub_ releases/2003-11/uoia-irc110703.php.

17. Michael R. Diehl et al., "Self-Assembled Deterministic Carbon Nanotube Wiring Networks," *Angewandte Chemie International Edition* 41.2 (2002): 353–56; C. P. Collier et al., "Electronically Configurable Molecular-Based Logic Gates," *Science* 285.5426 (July 1999): 391–94. See http://www.its.caltech.edu/~heathgrp/ papers/Paperfiles/2002/diehlangchemint.pdf and http://www.cs.duke.edu/~thl/ papers/Heath.Switch.pdf.

18. The "rosette nanotubes" designed by the Purdue team contain carbon, nitrogen, hydrogen, and oxygen. The rosettes self-assemble because their interiors are hydrophobic and their exteriors are hydrophilic; therefore, to protect their insides from water, the rosettes stack into nanotubes. "The physical and chemical properties of our rosette nanotubes can now be modified almost at will through a novel dial-in approach," according to lead researcher Hicham Fenniri. R. Colin Johnson, "Purdue Researchers Build Made-to-Order Nanotubes," *EETimes*, October 24, 2002, http://www.eetimes.com/article/showArticle.jhtml?articleId=18307660; H. Fenniri et al., "Entropically Driven Self-Assembly of Multichannel Rosette Nanotubes," *Proceedings of the National Academy of Sciences* 99, suppl. 2 (April 30, 2002): 6487–92; Purdue news release, "Adaptable Nanotubes Make Way for Custom-Built Structures, Wires," http://news.uns.purdue.edu/UNS/html4ever/ 020311.Fenniri.scaffold.html.

Similar work has been done by scientists in the Netherlands: Gaia Vince, "Nano-Transistor Self-Assembles Using Biology," NewScientist.com, November 20, 2003, http://www.newscientist.com/news/news.jsp?id=ns99994406.

19. Liz Kalaugher, "Lithography Makes a Connection for Nanowire Devices," June 9, 2004, http://www.nanotechweb.org/articles/news/3/6/6/1, referring to Song Jin et

al., "Scalable Interconnection and Integration of Nanowire Devices Without Registration," *Nano Letters* 4.5 (2004): 915–19.

20. Chao Li et al., "Multilevel Memory Based on Molecular Devices," *Applied Physics Letters* 84.11 (March 15, 2004): 1949–51. Also see http://www.technology review.com/articles/rnb_051304.asp?p=1. See also http://nanolab.usc.edu/PDF% 5CAPL84-1949.pdf.

21. Gary Stix, "Nano Patterning," *Scientific American* (February 9, 2004), http://www. sciam.com/print_version.cfm?articleID=000170D6-C99F-101E-861F83414B7F 0000; Michael Kanellos, "IBM Gets Chip Circuits to Draw Themselves," CNET News.com, http://zdnet.com.com/2100-1103-5114066.html. See also http://www. nanopolis.net/news_ind.php?type_id=3.

22. IBM is working on chips that automatically reconfigure as needed, such as by adding memory or accelerators. "In the future, the chip you have may not be the chip you bought," said Bernard Meyerson, chief technologist, IBM Systems and Technology Group. IBM press release, "IBM Plans Industry's First Openly Customizable Microprocessor," http://www.ibm.com/investor/press/mar-2004/ 31-03-04-1.phtml.

23. BBC News, " 'Nanowire' Breakthrough Hailed," April 1, 2003, http://news.bbc. co.uk/1/hi/sci/tech/2906621.stm. Published article is Thomas Scheibel et al., "Conducting Nanowires Built by Controlled Self-Assembly of Amyloid Fibers and Selective Metal Deposition," *Proceedings of the National Academy of Sciences* 100.8 (April 15, 2003): 4527–32, published online April 2, 2003, http://www.pnas.org/ cgi/content/full/100/8/4527.

24. Duke University press release, "Duke Scientists 'Program' DNA Molecules to Self Assemble into Patterned Nanostructures," http://www.eurekalert.org/pub_releases/ 2003-09/du-ds092403.php, referring to Hao Yan et al., "DNA-Templated Self-Assembly of Protein Arrays and Highly Conductive Nanowires," *Science* 301.5641 (September 26, 2003): 1882–84. See also http://www.phy.duke.edu/~gleb/Pdf_ FILES/DNA_science.pdf.

25. Ibid.

26. Here is an example of the procedure to solve what's called the traveling-salesperson problem. We try to find an optimal route for a hypothetical traveler among multiple cities without having to visit a city more than once. Only certain city pairs are connected by routes, so finding the right path is not straightforward.

   To solve the traveling-salesperson problem, mathematician Leonard Adleman of the University of Southern California performed the following steps:

   1. Generate a small strand of DNA with a unique code for each city.
   2. Replicate each such strand (one for each city) trillions of times using PCR.
   3. Next, put the pools of DNA (one for each city) together in a test tube. This step uses DNA's affinity to link strands together. Longer strands will form automatically. Each such strand represents a possible route of multiple cities. The small strands representing each city link up with each other in a random fashion, so

there is no mathematical certainty that a linked strand representing the correct answer (sequence of cities) will be formed. However, the number of strands is so vast that it is virtually certain that at least one strand—and probably millions—will be formed that represents the correct answer.

The next steps use specially designed enzymes to eliminate the trillions of strands that represent wrong answers, leaving only the strands representing the correct answer:

4. Use molecules called "primers" to destroy those DNA strands that do not start with the start city, as well as those that do not end with the end city; then replicate the surviving strands, using PCR.

5. Use an enzyme reaction to eliminate those DNA strands that represent a travel path greater than the total number of cities.

6. Use an enzyme reaction to destroy those strands that do not include city 1. Repeat for each of the cities.

7. Now, each of the surviving strands represents the correct answer. Replicate these surviving strands (using PCR) until there are billions of such strands.

8. Using a technique called electrophoresis, read out the DNA sequence of these correct strands (as a group). The readout looks like a set of distinct lines, which specifies the correct sequence of cities.

See L. M. Adleman, "Molecular Computation of Solutions to Combinatorial Problems," *Science* 266 (1994): 1021–24.

27. Charles Choi, "DNA Computer Sets Guinness Record," http://www.upi.com/view.cfm?StoryID=20030224-045551-7398r. See also Y. Benenson et al., "DNA Molecule Provides a Computing Machine with Both Data and Fuel," *Proceedings of the National Academy of Sciences* 100.5 (March 4, 2003): 2191–96, available at http://www.pubmedcentral.nih.gov/articlerender.fcgi?tool=pubmed&pubmedid=12601148; Y. Benenson et al., "An Autonomous Molecular Computer for Logical Control of Gene Expression," *Nature* 429.6990 (May 27, 2004): 423–29 (published online, April 28, 2004), available at http://www.wisdom.weizmann.ac.il/~udi/ShapiroNature2004.pdf.

28. Stanford University news release, " 'Spintronics' Could Enable a New Generation of Electronic Devices, Physicists Say," http://www.eurekalert.org/pub_releases/2003-08/su-ce080803.php, referring to Shuichi Murakami, Naoto Nagaosa, and Shou-Cheng Zhang, "Dissipationless Quantum Spin Current at Room Temperature," *Science* 301.5638 (September 5, 2003): 1348–51.

29. Celeste Biever, "Silicon-Based Magnets Boost Spintronics," NewScientist.com, March 22, 2004, http://www.newscientist.com/news/news.jsp?id=ns99994801, referring to Steve Pearton, "Silicon-Based Spintronics," *Nature Materials* 3.4 (April 2004): 203–4.

30. Will Knight, "Digital Image Stored in Single Molecule," NewScientist.com, December 1, 2002, http://www.newscientist.com/news/news.jsp?id=ns99993129,

referring to Anatoly K. Khitrin, Vladimir L. Ermakov, and B. M. Fung, "Nuclear Magnetic Resonance Molecular Photography," *Journal of Chemical Physics* 117.15 (October 15, 2002): 6903–6.

31. Reuters, "Processing at the Speed of Light," *Wired News,* http://www.wired.com/news/technology/0,1282,61009,00.html.

32. To date, the largest number to be factored is one of 512 bits, according to RSA Security.

33. Stephan Gulde et al., "Implementation of the Deutsch-Jozsa Algorithm on an Ion-Trap Quantum Computer," *Nature* 421 (January 2, 2003): 48–50. See http://heart-c704.uibk.ac.at/Papers/Nature03–Gulde.pdf.

34. Since we are currently doubling the price-performance of computation each year, a factor of a thousand requires ten doublings, or ten years. But we are also (slowly) decreasing the doubling time itself, so the actual figure is eight years.

35. Each subsequent thousandfold increase is itself occurring at a slightly faster rate. See the previous note.

36. Hans Moravec, "Rise of the Robots," *Scientific American* (December 1999): 124–35, http://www.sciam.com and http://www.frc.ri.cmu.edu/~hpm/project.archive/robot.papers/1999/SciAm.scan.html. Moravec is a professor at the Robotics Institute at Carnegie Mellon University. His Mobile Robot Laboratory explores how to use cameras, sonars, and other sensors to give robots 3-D spatial awareness. In the 1990s, he described a succession of robot generations that would "essentially [be] our off-spring, by unconventional means. Ultimately, I think they're on their own and they'll do things that we can't imagine or understand— you know, just the way children do" (Nova Online interview with Hans Moravec, October 1997, http://www.pbs.org/wgbh/nova/robots/moravec.html). His books *Mind Children: The Future of Robot and Human Intelligence* and *Robot: Mere Machine to Transcendent Mind* explore the capabilities of the current and future robot generations.

   Disclosure: The author is an investor in and on the board of directors of Moravec's robotics company, Seegrid.

37. Although instructions per second as used by Moravec and calculations per second are slightly different concepts, these are close enough for the purposes of these order-of-magnitude estimates. Moravec developed the mathematical techniques for his robot vision independent of biological models, but similarities (between Moravec's algorithms and those performed biologically) were noted after the fact. Functionally, Moravec's computations re-create what is accomplished in these neural regions, so computational estimates based on Moravec's algorithms are appropriate in determining what is required to achieve functionally equivalent transformations.

38. Lloyd Watts, "Event-Driven Simulation of Networks of Spiking Neurons," seventh Neural Information Processing Systems Foundation Conference, 1993; Lloyd Watts, "The Mode-Coupling Liouville-Green Approximation for a Two-Dimensional Cochlear Model," *Journal of the Acoustical Society of America* 108.5 (November

2000): 2266–71. Watts is the founder of Audience, Inc., which is devoted to applying functional simulation of regions of the human auditory system to applications in sound processing, including creating a way of preprocessing sound for automated speech-recognition systems. For more information, see http://www.lloydwatts.com/neuroscience.shtml.

Disclosure: The author is an adviser to Audience.

39. U.S. Patent Application 20030095667, U.S. Patent and Trademark Office, May 22, 2003.

40. The Medtronic MiniMed closed-loop artificial pancreas currently in human clinical trials is returning encouraging results. The company has announced that the device should be on the market within the next five years. Medtronic news release, "Medtronic Supports Juvenile Diabetes Research Foundation's Recognition of Artificial Pancreas as a Potential 'Cure' for Diabetes," March 23, 2004, http://www.medtronic.com/newsroom/news_2004323a.html. Such devices require a glucose sensor, an insulin pump, and an automated feedback mechanism to monitor insulin levels (International Hospital Federation, "Progress in Artificial Pancreas Development for Treating Diabetes," http://www.hospitalmanagement.net/informer/technology/tech10). Roche is also in the race to produce an artificial pancreas by 2007. See http://www.roche.com/pages/downloads/science/pdf/rtdcmannh02-6.pdf.

41. A number of models and simulations have been created based on analyses of individual neurons and interneuronal connections. Tomaso Poggio writes, "One view of the neuron is that it is more like a chip with thousands of logical-gates-equivalents rather than a single threshold element." Tomaso Poggio, private communication to Ray Kurzweil, January 2005.

See also T. Poggio and C. Koch, "Synapses That Compute Motion," *Scientific American* 256 (1987): 46–52.

C. Koch and T. Poggio, "Biophysics of Computational Systems: Neurons, Synapses, and Membranes," in *Synaptic Function*, G. M. Edelman, W. E. Gall, and W. M. Cowan, eds. (New York: John Wiley and Sons, 1987), pp. 637–97.

Another set of detailed neuron-level models and simulations is being created at the University of Pennsylvania's Neuroengineering Research Lab based on reverse engineering brain function at the neuron level. Dr. Leif Finkel, head of the laboratory, says, "Right now we're building a cellular-level model of a small piece of visual cortex. It's a very detailed computer simulation which reflects with some accuracy at least the basic operations of real neurons. [My colleague Kwabena Boahen] has a chip that accurately models the retina and produces output spikes that closely match real retinae." See http://nanodot.org/article.pl?sid=01/12/18/1552221.

Reviews of these and other models and simulations at the neuron level indicate that an estimate of 10^3 calculations per neural transaction (a single transaction involving signal transmission and reset on a single dendrite) is a reasonable upper bound. Most simulations use considerably less than this.

42. Plans for Blue Gene/L, the second generation of Blue Gene computers, were announced in late 2001. The new supercomputer, planned to be fifteen times faster than today's supercomputers and one twentieth the size, is being built jointly by the National Nuclear Security Agency's Lawrence Livermore National Laboratory and IBM. In 2002, IBM announced that open-source Linux had been chosen as the operating system for the new supercomputers. By July 2003, the innovative processor chips for the supercomputer, which are complete systems on chips, were in production. "Blue Gene/L is a poster child for what is possible with the system-on-a-chip concept. More than 90 percent of this chip was built from standard blocks in our technology library," according to Paul Coteus, one of the managers of the project (Timothy Morgan, "IBM's Blue Gene/L Shows Off Minimalist Server Design," *The Four Hundred,* http://www.midrangeserver.com/tfh/tfh120103-story05.html). By June 2004, the Blue Gene/L prototype systems appeared for the first time on the list of top ten supercomputers. IBM press release, "IBM Surges Past HP to Lead in Global Supercomputing," http://www.research.ibm.com/bluegene.

43. This type of network is also called peer-to-peer, many-to-many, and "multihop." In it, nodes in the network can be connected to all the other nodes or to a subset, and there are multiple paths through meshed nodes to each destination. These networks are highly adaptable and self-organizing. "The signature of a mesh network is that there is no central orchestrating device. Instead, each node is outfitted with radio communications gear and acts as a relay point for other nodes." Sebastian Rupley, "Wireless: Mesh Networks," *PC Magazine,* July 1, 2003, http://www.pcmag.com/article2/0,1759,1139094,00.asp; Robert Poor, "Wireless Mesh Networks," Sensors Online, February 2003, http://www.sensorsmag.com/articles/0203/38/main.shtml; Tomas Krag and Sebastian Büettrich, "Wireless Mesh Networking," O'Reilly Wireless DevCenter, January 22, 2004, http://www.oreillynet.com/pub/a/wireless/2004/01/22/wirelessmesh.html.

44. Carver Mead, founder of more than twenty-five companies and holder of more than fifty patents, is pioneering the new field of neuromorphic electronic systems, circuits modeled on the brain and nervous system. See Carver A. Mead, "Neuromorphic Electronic Systems," *IEEE Proceedings* 78.10 (October 1990): 1629–36. His work led to the computer touch pad and the cochlear chip used in digital hearing aids. His 1999 start-up company Foveon makes analog image-sensors that imitate the properties of film.

45. Edward Fredkin, "A Physicist's Model of Computation," Proceedings of the Twenty-sixth Recontre de Moriond, Texts of Fundamental Symmetries (1991): 283–97, http://digitalphilosophy.org/physicists_model.htm.

46. Gene Frantz, "Digital Signal Processing Trends," *IEEE Micro* 20.6 (November/December 2000): 52–59, http://csdl.computer.org/comp/mags/mi/2000/06/m6052abs.htm.

47. In 2004 Intel announced a "right hand turn" switch toward dual-core (more than one processor on a chip) architecture after reaching a "thermal wall" (or "power

wall") caused by too much heat from ever-faster single processors: http://www. intel.com/employee/retiree/circuit/righthandturn.htm.

48. R. Landauer, "Irreversibility and Heat Generation in the Computing Process," *IBM Journal of Research Development* 5 (1961): 183–91, http://www.research.ibm. com/journal/rd/053/ibmrd0503C.pdf.

49. Charles H. Bennett, "Logical Reversibility of Computation," *IBM Journal of Research Development* 17 (1973): 525–32, http://www.research.ibm.com/journal/ rd/176/ibmrd1706G.pdf; Charles H. Bennett, "The Thermodynamics of Computation—a Review," *International Journal of Theoretical Physics* 21 (1982): 905–40; Charles H. Bennett, "Demons, Engines, and the Second Law," *Scientific American* 257 (November 1987): 108–16.

50. Edward Fredkin and Tommaso Toffoli, "Conservative Logic," *International Journal of Theoretical Physics* 21 (1982): 219–53, http://digitalphilosophy.org/download_ documents/ConservativeLogic.pdf. Edward Fredkin, "A Physicist's Model of Computation," Proceedings of the Twenty-sixth Recontre de Moriond, Tests of Fundamental Symmetries (1991): 283–97, http://www.digitalphilosophy.org/ physicists_model.htm.

51. Knight, "Digital Image Stored in Single Molecule," referring to Khitrin et al., "Nuclear Magnetic Resonance Molecular Photography"; see note 30 above.

52. Ten billion (10^{10}) humans at 10^{19} cps each is 10^{29} cps for all human brains; 10^{42} cps is greater than this by ten trillion (10^{13}).

53. Fredkin, "Physicist's Model of Computation"; see notes 45 and 50 above.

54. Two such gates are the Interaction Gate, a two-input, four-output universal, reversible-logic gate

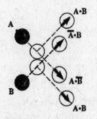

and the Feynman Gate, a two-input, three-output reversible, universal-logic gate.

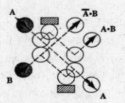

Both images are from ibid., p. 7.

55. Ibid., p. 8.

56. C. L. Seitz et al., "Hot-Clock nMOS," *Proceedings of the 1985 Chapel Hill Conference on VLSI* (Rockville, Md.: Computer Science Press, 1985), pp. 1–17, http://caltechcstr. library.caltech.edu/archive/00000365; Ralph C. Merkle, "Reversible Electronic Logic Using Switches," *Nanotechnology* 4 (1993): 21–40; S. G. Younis and T. F. Knight, "Practical Implementation of Charge Recovering Asymptotic Zero Power CMOS," *Proceedings of the 1993 Symposium on Integrated Systems* (Cambridge, Mass.: MIT Press, 1993), pp. 234–50.

57. Hiawatha Bray, "Your Next Battery," *Boston Globe,* November 24, 2003, http://www.boston.com/business/technology/articles/2003/11/24/your_next_battery.

58. Seth Lloyd, "Ultimate Physical Limits to Computation," *Nature* 406 (2000): 1047–54.

    Early work on the limits of computation was done by Hans J. Bremermann in 1962: Hans J. Bremermann, "Optimization Through Evolution and Recombination," in M. C. Yovits, C. T. Jacobi, C. D. Goldstein, eds., *Self-Organizing Systems* (Washington, D.C.: Spartan Books, 1962), pp. 93–106.

    In 1984 Robert A. Freitas Jr. built on Bremermann's work in Robert A. Freitas Jr., "Xenopsychology," *Analog* 104 (April 1984): 41–53, http://www.rfreitas. com/Astro/Xenopsychology.htm#SentienceQuotient.

59. $\pi \times$ maximum energy (10^{17} kg $\times$ meter ^{2}/second^2) / (6.6×10^{-34}) joule-seconds = $\sim 5 \times 10^{50}$ operations/second.

60. 5×10^{50} cps is equivalent to 5×10^{21} (5 billion trillion) human civilizations (each requiring 10^{29} cps).

61. Ten billion (10^{10}) humans at 10^{16} cps each is 10^{26} cps for human civilization. So 5×10^{50} cps is equivalent to 5×10^{24} (5 trillion trillion) human civilizations.

62. This estimate makes the conservative assumption that we've had ten billion humans for the past ten thousand years, which is obviously not the case. The actual number of humans has been increasing gradually over the past to reach about 6.1 billion in 2000. There are 3×10^7 seconds in a year, and 3×10^{11} seconds in ten thousand years. So, using the estimate of 10^{26} cps for human civilization, human thought over ten thousand years is equivalent to certainly no more than 3×10^{37} calculations. The ultimate laptop performs 5×10^{50} calculations in one second. So simulating ten thousand years of ten billion humans' thoughts would take it about 10^{-13} seconds, which is one ten-thousandth of a nanosecond.

63. Anders Sandberg, "The Physics of the Information Processing Superobjects: Daily Life Among the Jupiter Brains," *Journal of Evolution & Technology* 5 (December 22, 1999), http://www.transhumanist.com/volume5/Brains2.pdf.

64. See note 62 above; 10^{42} cps is a factor of 10^{-8} less than 10^{50} cps, so one ten-thousandth of a nanosecond becomes 10 microseconds.

65. See http://e-drexler.com/p/04/04/0330drexPubs.html for a list of Drexler's publications and patents.

66. At the rate of $10^{12} and 10^{26} cps per thousand dollars (10^3), we get 10^{35} cps per

year in the mid-2040s. The ratio of this to the 10^{26} cps for all of the biological thinking in human civilization is 10^9 (one billion).

67. In 1984 Robert A. Freitas proposed a logarithmic scale of "sentience quotient" (SQ) based on the computational capacity of a system. In a scale that ranges from -70 to 50, human brains come out at 13. The Cray 1 supercomputer comes out at 9. Freitas's sentience quotient is based on the amount of computation per unit mass. A very fast computer with a simple algorithm would come out with a high SQ. The measure I describe for computation in this section builds on Freitas's SQ and attempts to take into consideration the usefulness of the computation. So if a simpler computation is equivalent to the one actually being run, then we base the computational efficiency on the equivalent (simpler) computation. Also in my measure, the computation needs to be "useful." Robert A. Freitas Jr., "Xenopsychology," *Analog* 104 (April 1984): 41–53, http://www.rfreitas.com/Astro/Xeno psychology.htm#SentienceQuotient.

68. As an interesting aside, engravings on the side of small rocks did in fact represent an early form of computer storage. One of the earliest forms of written language, cuneiform, which was developed in Mesopotamia circa 3000 B.C., used pictorial markings on stones to store information. Agricultural records were maintained as cuneiform markings on stones placed in trays, and organized in rows and columns. These marked stones were essentially the first spreadsheet. One such cuneiform stone record is a prized artifact in my collection of historical computers.

69. One thousand (10^3) bits is less than the theoretical capacity of the atoms in the stone to store information (estimated at 10^{27} bits) by a factor of 10^{-24}.

70. 1 cps (10^0 cps) is less than the theoretical computing capacity of the atoms in the stone (estimated at 10^{42} cps) by a factor of 10^{-42}.

71. Edgar Buckingham, "Jet Propulsion for Airplanes," NACA report no. 159, in *Ninth Annual Report of NACA-1923* (Washington, D.C.: NACA, 1924), pp. 75–90. See http://naca.larc.nasa.gov/reports/1924/naca-report-159/.

72. Belle Dumé, "Microscopy Moves to the Picoscale," *PhysicsWeb*, June 10, 2004, http://physicsweb.org/article/news/8/6/6, referring to Stefan Hembacher, Franz J. Giessibl, and Jochen Mannhart, "Force Microscopy with Light-Atom Probes," *Science* 305.5682 (July 16, 2004): 380–83. This new "higher harmonic" force microscope, developed by University of Augsburg physicists, uses a single carbon atom as a probe and has a resolution that is at least three times better than that of traditional scanning tunneling microscopes. How it works: as the tungsten tip of the probe is made to oscillate at subnanometer amplitudes, the interaction between the tip atom and the carbon atom produces higher harmonic components in the underlying sinusoidal-wave pattern. The scientists measured these signals to obtain an ultrahigh-resolution image of the tip atom that showed features just 77 picometers (thousandths of a nanometer) across.

73. Henry Fountain, "New Detector May Test Heisenberg's Uncertainty Principle," *New York Times*, July 22, 2003.

74. Mitch Jacoby, "Electron Moves in Attoseconds," *Chemical and Engineering News*

82.25 (June 21, 2004): 5, referring to Peter Abbamonte et al., "Imaging Density Disturbances in Water with a 41.3-Attosecond Time Resolution," *Physical Review Letters* 92.23 (June 11, 2004): 237–401.

75. S. K. Lamoreaux and J. R. Torgerson, "Neutron Moderation in the Oklo Natural Reactor and the Time Variation of Alpha," *Physical Review* D 69 (2004): 121701–6, http://scitation.aip.org/getabs/servlet/GetabsServlet?prog=normal&id=PRVDAQ 000069000012121701000001&idtype=cvips&gifs=yes; Eugenie S. Reich, "Speed of Light May Have Changed Recently," *New Scientist,* June 30, 2004, http://www. newscientist.com/news/news.jsp?id=ns99996092.

76. Charles Choi, "Computer Program to Send Data Back in Time," UPI, October 1, 2002, http://www.upi.com/view.cfm?StoryID=20021001-125805-3380r; Todd Brun, "Computers with Closed Timelike Curves Can Solve Hard Problems," *Foundation of Physics Letters* 16 (2003): 245–53. Electronic edition, September 11, 2002, http://arxiv.org/PS_cache/gr-qc/pdf/0209/0209061.pdf.

Chapter Four: Achieving the Software of Human Intelligence: How to Reverse Engineer the Human Brain

1. Lloyd Watts, "Visualizing Complexity in the Brain," in D. Fogel and C. Robinson, eds., *Computational Intelligence: The Experts Speak* (Piscataway, N.J.: IEEE Press/Wiley, 2003), http://www.lloydwatts.com/wcci.pdf.

2. J. G. Taylor, B. Horwitz, and K. J. Friston, "The Global Brain: Imaging and Modeling," *Neural Networks* 13, special issue (2000): 827.

3. Neil A. Busis, "Neurosciences on the Internet," http://www.neuroguide.com; "Neuroscientists Have Better Tools on the Brain," *Bio IT Bulletin,* http://www.bio-it world.com/news/041503_report2345.html; "Brain Projects to Reap Dividends for Neurotech Firms," *Neurotech Reports,* http://www.neurotechreports.com/pages/ brainprojects.html.

4. Robert A. Freitas Jr., *Nanomedicine,* vol. 1, *Basic Capabilities,* section 4.8.6, "Noninvasive Neuroelectric Monitoring" (Georgetown, Tex.: Landes Bioscience, 1999), pp. 115–16, http://www.nanomedicine.com/NMI/4.8.6.htm.

5. Chapter 3 analyzed this issue; see the section "The Computational Capacity of the Human Brain."

6. Speech-recognition research and development, Kurzweil Applied Intelligence, which I founded in 1982, now part of ScanSoft (formerly Kurzweil Computer Products).

7. Lloyd Watts, U.S. Patent Application, U.S. Patent and Trademark Office, 20030095667, May 22, 2003, "Computation of Multi-sensor Time Delays." Abstract: "Determining a time delay between a first signal received at a first sensor and a second signal received at a second sensor is described. The first signal is analyzed to derive a plurality of first signal channels at different frequencies and the second signal is analyzed to derive a plurality of second signal channels at different frequencies. A first feature is detected that occurs at a first time in one of the first

signal channels. A second feature is detected that occurs at a second time in one of the second signal channels. The first feature is matched with the second feature and the first time is compared to the second time to determine the time delay." See also Nabil H. Farhat, U.S. Patent Application 20040073415, U.S. Patent and Trademark Office, April 15, 2004, "Dynamical Brain Model for Use in Data Processing Applications."

8. I estimate the compressed genome at about thirty to one hundred million bytes (see note 57 for chapter 2); this is smaller than the object code for Microsoft Word and much smaller than the source code. See Word 2003 system requirements, October 20, 2003, http://www.microsoft.com/office/word/prodinfo/sysreq.mspx.

9. Wikipedia, http://en.wikipedia.org/wiki/Epigenetics.

10. See note 57 in chapter 2 for an analysis of the information content in the genome, which I estimate to be 30 to 100 million bytes, therefore less than 10^9 bits. See the section "Human Memory Capacity" in chapter 3 (p. 126) for my analysis of the information in a human brain, estimated at 10^{18} bits.

11. Marie Gustafsson and Christian Balkenius, "Using Semantic Web Techniques for Validation of Cognitive Models against Neuroscientific Data," AILS 04 Workshop, SAIS/SSLS Workshop (Swedish Artificial Intelligence Society; Swedish Society for Learning Systems), April 15–16, 2004, Lund, Sweden, www.lucs.lu.se/People/Christian.Balkenius/PDF/Gustafsson.Balkenius.2004.pdf.

12. See discussion in chapter 3. In one useful reference, when modeling neuron by neuron, Tomaso Poggio and Christof Koch describe the neuron as similar to a chip with thousands of logical gates. See T. Poggio and C. Koch, "Synapses That Compute Motion," *Scientific American* 256 (1987): 46–52. Also C. Koch and T. Poggio, "Biophysics of Computational Systems: Neurons, Synapses, and Membranes," in *Synaptic Function*, G. M. Edelman, W. E. Gall, and W. M. Cowan, eds. (New York: John Wiley and Sons, 1987), pp. 637–97.

13. On Mead, see http://www.technology.gov/Medal/2002/bios/Carver_A._Mead.pdf. Carver Mead, *Analog VLSI and Neural Systems* (Reading, Mass.: Addison-Wesley, 1986).

14. See note 172 in chapter 5 for an algorithmic description of a self-organizing neural net and note 175 in chapter 5 for a description of a self-organizing genetic algorithm.

15. See Gary Dudley et al., "Autonomic Self-Healing Systems in a Cross-Product IT Environment," proceedings of the IEEE International Conference on Autonomic Computing, New York City, May 17–19, 2004, http://csdl.computer.org/comp/proceedings/icac/2004/2114/00/21140312.pdf; "About IBM Autonomic Computing," http://www-3.ibm.com/autonomic/about.shtml; and Ric Telford, "The Autonomic Computing Architecture," April 14, 2004, http://www.dcs.st-andrews.ac.uk/undergrad/current/dates/disclec/2003–2/RicTelfordDistinguished2.pdf.

16. Christine A. Skarda and Walter J. Freeman, "Chaos and the New Science of the Brain," *Concepts in Neuroscience* 1.2 (1990): 275–85.

17. C. Geoffrey Woods, "Crossing the Midline," *Science* 304.5676 (June 4, 2004): 1455–56; Stephen Matthews, "Early Programming of the Hypothalamo-Pituitary-Adrenal Axis," *Trends in Endocrinology and Metabolism* 13.9 (November 1, 2002): 373–80; Justin Crowley and Lawrence Katz, "Early Development of Ocular Dominance Columns," *Science* 290.5495 (November 17, 2000): 1321–24; Anna Penn et al., "Competition in the Retinogeniculate Patterning Driven by Spontaneous Activity," *Science* 279.5359 (March 27, 1998): 2108–12; M. V. Johnston et al., "Sculpting the Developing Brain," *Advances in Pediatrics* 48 (2001): 1–38; P. La Cerra and R. Bingham, "The Adaptive Nature of the Human Neurocognitive Architecture: An Alternative Model," *Proceedings of the National Academy of Sciences* 95 (September 15, 1998): 11290–94.

18. Neural nets are simplified models of neurons that can self-organize and solve problems. See note 172 in chapter 5 for an algorithmic description of neural nets. Genetic algorithms are models of evolution using sexual reproduction with controlled mutation rates. See note 175 in chapter 5 for a detailed description of genetic algorithms. Markov models are products of a mathematical technique that are similar in some respects to neural nets.

19. Aristotle, *The Works of Aristotle*, trans. W. D. Ross (Oxford: Clarendon Press, 1908–1952 (see, in particular, *Physics*); see also http://www.encyclopedia.com/html/section/aristotl_philosophy.asp.

20. E. D. Adrian, *The Basis of Sensation: The Action of Sense Organs* (London: Christophers, 1928).

21. A. L. Hodgkin and A. F. Huxley, "Action Potentials Recorded from Inside a Nerve Fibre," *Nature* 144 (1939): 710–12.

22. A. L. Hodgkin and A. F. Huxley, "A Quantitative Description of Membrane Current and Its Application to Conduction and Excitation in Nerve," *Journal of Physiology* 117 (1952): 500–544.

23. W. S. McCulloch and W. Pitts, "A Logical Calculus of the Ideas Immanent in Nervous Activity," *Bulletin of Mathematical Biophysics* 5 (1943): 115–33. This seminal paper is a difficult one to understand. For a clear introduction and explanation, see "A Computer Model of the Neuron," the Mind Project, Illinois State University, http://www.mind.ilstu.edu/curriculum/perception/mpneuron1.html.

24. See note 172 in chapter 5 for an algorithmic description of neural nets.

25. E. Salinas and P. Thier, "Gain Modulation: A Major Computational Principle of the Central Nervous System," *Neuron* 27 (2000): 15–21.

26. K. M. O'Craven and R. L. Savoy, "Voluntary Attention Can Modulate fMRI Activity in Human MT/MST," *Investigational Ophthalmological Vision Science* 36 (1995): S856 (supp.).

27. Marvin Minsky and Seymour Papert, *Perceptrons* (Cambridge, Mass.: MIT Press, 1969).

28. Frank Rosenblatt, Cornell Aeronautical Laboratory, "The Perceptron: A Probabilistic Model for Information Storage and Organization in the Brain," *Psycholog-*

ical Review 65.6 (1958): 386–408; see Wikipedia, http://en.wikipedia.org/wiki/Perceptron.

29. O. Sporns, G. Tononi, and G. M. Edelman, "Connectivity and Complexity: The Relationship Between Neuroanatomy and Brain Dynamics," *Neural Networks* 13.8–9 (2000): 909–22.

30. R. H. Hahnloser et al., "Digital Selection and Analogue Amplification Coexist in a Cortex-Inspired Silicon Circuit," *Nature* 405.6789 (June 22, 2000): 947–51; "MIT and Bell Labs Researchers Create Electronic Circuit That Mimics the Brain's Circuitry," *MIT News*, June 21, 2000, http://web.mit.edu/newsoffice/nr/2000/machinebrain.html.

31. Manuel Trajtenberg, *Economic Analysis of Product Innovation: The Case of CT Scanners* (Cambridge, Mass.: Harvard University Press, 1990); Michael H. Friebe, Ph.D., president, CEO, NEUROMED GmbH; P-M. L. Robitaille, A. M. Abduljalil, and A. Kangarlu, "Ultra High Resolution Imaging of the Human Head at 8 Tesla: 2K x 2K for Y2K," *Journal of Computer Assisted Tomography* 24.1 (January–February 2000): 2–8.

32. Seong-Gi Kim, "Progress in Understanding Functional Imaging Signals," *Proceedings of the National Academy of Sciences* 100.7 (April 1, 2003): 3550–52, http://www.pnas.org/cgi/content/full/100/7/3550. See also Seong-Gi Kim et al., "Localized Cerebral Blood Flow Response at Submillimeter Columnar Resolution," *Proceedings of the National Academy of Sciences* 98.19 (September 11, 2001): 10904–9, http://www.pnas.org/cgi/content/abstract/98/19/10904.

33. K. K. Kwong et al., "Dynamic Magnetic Resonance Imaging of Human Brain Activity During Primary Sensory Stimulation," *Proceedings of the National Academy of Sciences* 89.12 (June 15, 1992): 5675–79.

34. C. S. Roy and C. S. Sherrington, "On the Regulation of the Blood Supply of the Brain," *Journal of Physiology* 11 (1890): 85–105.

35. M. I. Posner et al., "Localization of Cognitive Operations in the Human Brain," *Science* 240.4859 (June 17, 1988): 1627–31.

36. F. M. Mottaghy et al., "Facilitation of Picture Naming after Repetitive Transcranial Magnetic Stimulation," *Neurology* 53.8 (November 10, 1999): 1806–12.

37. Daithí Ó hAnluain, "TMS: Twilight Zone Science?" *Wired News*, April 18, 2002, http://wired.com/news/medtech/0,1286,51699,00.html.

38. Lawrence Osborne, "Savant for a Day," *New York Times Magazine*, June 22, 2003, available at http://www.wireheading.com/brainstim/savant.html.

39. Bruce H. McCormick, "Brain Tissue Scanner Enables Brain Microstructure Surveys," *Neurocomputing* 44–46 (2002): 1113–18; Bruce H. McCormick, "Design of a Brain Tissue Scanner," *Neurocomputing* 26–27 (1999): 1025–32; Bruce H. McCormick, "Development of the Brain Tissue Scanner," *Brain Networks Laboratory Technical Report*, Texas A&M University Department of Computer Science, College Station, Tex., March 18, 2002, http://research.cs.tamu.edu/bnl/pubs/McC02.pdf.

40. Leif Finkel et al., "Meso-scale Optical Brain Imaging of Perceptual Learning," University of Pennsylvania grant 2000–01737 (2000).

41. E. Callaway and R. Yuste, "Stimulating Neurons with Light," *Current Opinions in Neurobiology* 12.5 (October 2002): 587–92.

42. B. L. Sabatini and K. Svoboda, "Analysis of Calcium Channels in Single Spines Using Optical Fluctuation Analysis," *Nature* 408.6812 (November 30, 2000): 589–93.

43. John Whitfield, "Lasers Operate Inside Single Cells," *News@nature.com*, October 6, 2003, http://www.nature.com/nsu/030929/030929-12.html (subscription required). Mazur's lab: http://mazur-www.harvard.edu/research/. Jason M. Samonds and A. B. Bonds, "From Another Angle: Differences in Cortical Coding Between Fine and Coarse Discrimination of Orientation," *Journal of Neurophysiology* 91 (2004): 1193–1202.

44. Robert A. Freitas Jr., *Nanomedicine*, vol. 2A, *Biocompatibility*, section 15.6.2, "Bloodstream Intrusiveness" (Georgetown, Tex.: Landes Bioscience, 2003), pp. 157–59, http://www.nanomedicine.com/NMIIA/15.6.2.htm.

45. Robert A. Freitas Jr., *Nanomedicine*, vol. 1, *Basic Capabilities*, section 7.3, "Communication Networks" (Georgetown, Tex.: Landes Bioscience, 1999), pp. 186–88, http://www.nanomedicine.com/NMI/7.3.htm.

46. Robert A. Freitas Jr., *Nanomedicine*, vol. 1, *Basic Capabilities*, section 9.4.4.3, "Intercellular Passage" (Georgetown, Tex.: Landes Bioscience, 1999), pp. 320–21, http://www.nanomedicine.com/NMI/9.4.4.3.htm#p2.

47. Keith L. Black, M.D., and Nagendra S. Ningaraj, "Modulation of Brain Tumor Capillaries for Enhanced Drug Delivery Selectively to Brain Tumor," *Cancer Control* 11.3 (May/June 2004): 165–73, http://www.moffitt.usf.edu/pubs/ccj/v11n3/pdf/165.pdf.

48. Robert A. Freitas Jr., *Nanomedicine*, vol. 1, *Basic Capabilities*, section 4.1, "Nanosensor Technology" (Georgetown, Tex.: Landes Bioscience, 1999), p. 93, http://www.nanomedicine.com/NMI/4.1.htm.

49. Conference on Advanced Nanotechnology (http://www.foresight.org/Conferences/AdvNano2004/index.html), NanoBioTech Congress and Exhibition (http://www.nanobiotec.de/), NanoBusiness Trends in Nanotechnology (http://www.nanoevent.com/), and NSTI Nanotechnology Conference and Trade Show (http://www.nsti.org/events.html).

50. Peter D. Kramer, *Listening to Prozac* (New York: Viking, 1993).

51. LeDoux's research is on the brain regions that deal with threatening stimuli, of which the central player is the amygdala, an almond-shaped region of neurons located at the base of the brain. The amygdala stores memories of threatening stimuli and controls responses having to do with fear.

     MIT brain researcher Tomaso Poggio points out that "synaptic plasticity is one hardware substratum for learning but it may be important to emphasize that learning is much more than memory." See T. Poggio and E. Bizzi, "Generalization in Vision and Motor Control," *Nature* 431 (2004): 768–74. See also E. Benson, "The Synaptic Self," *APA Online*, November 2002, http://www.apa.org/monitor/nov02/synaptic.html.

52. Anthony J. Bell, "Levels and Loops: The Future of Artificial Intelligence and Neuroscience," *Philosophical Transactions of the Royal Society of London B* 354.1352 (December 29, 1999): 2013–20, http://www.cnl.salk.edu/~tony/ptrsl.pdf.

53. Peter Dayan and Larry Abbott, *Theoretical Neuroscience: Computational and Mathematical Modeling of Neural Systems* (Cambridge, Mass.: MIT Press, 2001).

54. D. O. Hebb, *The Organization of Behavior: A Neuropsychological Theory* (New York: Wiley, 1949).

55. Michael Domjan and Barbara Burkhard, *The Principles of Learning and Behavior*, 3d ed. (Pacific Grove, Calif.: Brooks/Cole, 1993).

56. J. Quintana and J. M. Fuster, "From Perception to Action: Temporal Integrative Functions of Prefrontal and Parietal Neurons," *Cerebral Cortex* 9.3 (April–May 1999): 213–21; W. F. Asaad, G. Rainer, and E. K. Miller, "Neural Activity in the Primate Prefrontal Cortex During Associative Learning," *Neuron* 21.6 (December 1998): 1399–1407.

57. G. G. Turrigiano et al., "Activity-Dependent Scaling of Quantal Amplitude in Neocortical Neurons," *Nature* 391.6670 (February 26, 1998): 892–96; R. J. O'Brien et al., "Activity-Dependent Modulation of Synaptic AMPA Receptor Accumulation," *Neuron* 21.5 (November 1998): 1067–78.

58. From "A New Window to View How Experiences Rewire the Brain," Howard Hughes Medical Institute (December 19, 2002), http://www.hhmi.org/news/svoboda2.html. See also J. T. Trachtenberg et al., "Long-Term in Vivo Imaging of Experience-Dependent Synaptic Plasticity in Adult Cortex," *Nature* 420.6917 (December 2002): 788–94, http://cpmcnet.columbia.edu/dept/physio/physio2/Trachtenberg_NATURE.pdf; and Karen Zita and Karel Svoboda, "Activity-Dependent Synaptogenesis in the Adult Mammalian Cortex," *Neuron* 35.6 (September 2002): 1015–17, http://svobodalab.cshl.edu/reprints/2414zito02neur.pdf.

59. See http://whyfiles.org/184make_memory/4.html. For more information on neuronal spines and memory, see J. Grutzendler et al., "Long-Term Dendritic Spine Stability in the Adult Cortex," *Nature* 420.6917 (Dec. 19–26, 2002): 812–16.

60. S. R. Young and E. W. Rubel, "Embryogenesis of Arborization Pattern and Typography of Individual Axons in N. Laminaris of the Chicken Brain Stem," *Journal of Comparative Neurology* 254.4 (December 22, 1986): 425–59.

61. Scott Makeig, "Swartz Center for Computational Neuroscience Vision Overview," http://www.sccn.ucsd.edu/VisionOverview.html.

62. D. H. Hubel and T. N. Wiesel, "Binocular Interaction in Striate Cortex of Kittens Reared with Artificial Squint," *Journal of Neurophysiology* 28.6 (November 1965): 1041–59.

63. Jeffrey M. Schwartz and Sharon Begley, *The Mind and the Brain: Neuroplasticity and the Power of Mental Force* (New York: Regan Books, 2002). See also C. Xerri, M. Merzenich et al., "The Plasticity of Primary Somatosensory Cortex Paralleling

Sensorimotor Skill Recovery from Stroke in Adult Monkeys," *The Journal of Neurophysiology,* 79.4 (April 1980): 2119–48. See also S. Begley, "Survival of the Busiest," *Wall Street Journal,* October 11, 2002, http://webreprints.djreprints.com/606120211414.html.

64. Paula Tallal et al., "Language Comprehension in Language-Learning Impaired Children Improved with Acoustically Modified Speech," *Science* 271 (January 5, 1996): 81–84. Paula Tallal is Board of Governors Professor of Neuroscience and codirector of the CMBN (Center for Molecular and Behavioral Neuroscience) at Rutgers University, and cofounder and director of SCIL (Scientific Learning Corporation); see http://www.cmbn.rutgers.edu/faculty/tallal.html. See also Paula Tallal, "Language Learning Impairment: Integrating Research and Remediation," *New Horizons for Learning* 4.4 (August–September 1998), http://www.newhorizons.org/neuro/tallal.htm; A. Pascual-Leone, "The Brain That Plays Music and Is Changed by It," *Annals of the New York Academy of Sciences* 930 (June 2001): 315–29. See also note 63 above.

65. F. A. Wilson, S. P. Scalaidhe, and P. S. Goldman-Rakic, "Dissociation of Object and Spatial Processing Domains in Primate Prefrontal Cortex." *Science* 260.5116 (June 25, 1993): 1955–58.

66. C. Buechel, J. T. Coull, and K. J. Friston, "The Predictive Value of Changes in Effective Connectivity for Human Learning," *Science* 283.5407 (March 5, 1999): 1538–41.

67. They produced dramatic images of brain cells forming temporary and permanent connections in response to various stimuli, illustrating structural changes between neurons that, many scientists have long believed, take place when we store memories. "Pictures Reveal How Nerve Cells Form Connections to Store Short- and Long-Term Memories in Brain," University of California, San Diego, November 29, 2001, http://ucsdnews.ucsd.edu/newsrel/science/mccell.htm; M. A. Colicos et al., "Remodeling of Synaptic Action Induced by Photoconductive Stimulation," *Cell* 107.5 (November 30, 2001): 605–16. Video link: http://www.qflux.net/NeuroStim01.rm, Neural Silicon Interface—Quantum Flux.

68. S. Lowel and W. Singer, "Selection of Intrinsic Horizontal Connections in the Visual Cortex by Correlated Neuronal Activity," *Science* 255.5041 (January 10, 1992): 209–12.

69. K. Si et al., "A Neuronal Isoform of CPEB Regulates Local Protein Synthesis and Stabilizes Synapse-Specific Long-Term Facilitation in Aplysia," *Cell* 115.7 (December 26, 2003): 893–904; K. Si, S. Lindquist, and E. R. Kandel, "A Neuronal Isoform of the Aplysia CPEB Has Prion-Like Properties," *Cell* 115.7 (December 26, 2003): 879–91. These researchers have found that CPEB may help form and preserve long-term memories by undergoing shape changes in synapses similar to deformations of prions (protein fragments implicated in mad-cow disease and other neurologic illnesses). The study suggests that this protein does its good work while in a prion state, contradicting a widely held belief that a protein that has prion

activity is toxic or at least doesn't function properly. This prion mechanism may also have roles in areas such as cancer maintenance and organ development, suspects Eric R. Kandel, University Professor of physiology and cell biophysics, psychiatry, biochemistry, and molecular biophysics at Columbia University and winner of a 2000 Nobel Prize for Medicine. See Whitehead Institute press release, http://www.wi.mit.edu/nap/features/nap_feature_memory.html.

70. M. C. Anderson et al., "Neural Systems Underlying the Suppression of Unwanted Memories," *Science* 303.5655 (January 9, 2004): 232–35. The findings could encourage the development of new ways for people to overcome traumatizing memories. Keay Davidson, "Study Suggests Brain Is Built to Forget: MRIs in Stanford Experiments Indicate Active Suppression of Unneeded Memories," *San Francisco Chronicle,* January 9, 2004, http://www.sfgate.com/cgi-bin/article.cgi?file=/c/a/2004/01/09/FORGET.TMP&type=science.

71. Dieter C. Lie et al., "Neurogenesis in the Adult Brain: New Strategies for CNS Diseases," *Annual Review of Pharmacology and Toxicology* 44 (2004): 399–421.

72. H. van Praag, G. Kempermann, and F. H. Gage, "Running Increases Cell Proliferation and Neurogenesis in the Adult Mouse Dentate Gyrus," *Nature Neuroscience* 2.3 (March 1999): 266–70.

73. Minsky and Papert, *Perceptrons.*

74. Ray Kurzweil, *The Age of Spiritual Machines* (New York: Viking, 1999), p. 79.

75. Basis functions are nonlinear functions that can be combined linearly (by adding together multiple weighted-basis functions) to approximate any nonlinear function. Pouget and Snyder, "Computational Approaches to Sensorimotor Transformations," *Nature Neuroscience* 3.11 Supplement (November 2000): 1192–98.

76. T. Poggio, "A Theory of How the Brain Might Work," in *Proceedings of Cold Spring Harbor Symposia on Quantitative Biology* 4 (Cold Spring Harbor, N.Y.: Cold Spring Harbor Laboratory Press, 1990), 899–910. Also see T. Poggio and E. Bizzi, "Generalization in Vision and Motor Control," *Nature* 431 (2004): 768–74.

77. R. Llinas and J. P. Welsh, "On the Cerebellum and Motor Learning," *Current Opinion in Neurobiology* 3.6 (December 1993): 958–65; E. Courchesne and G. Allen, "Prediction and Preparation, Fundamental Functions of the Cerebellum," *Learning and Memory* 4.1 (May–June 1997): 1–35; J. M. Bower, "Control of Sensory Data Acquisition," *International Review of Neurobiology* 41 (1997): 489–513.

78. J. Voogd and M. Glickstein, "The Anatomy of the Cerebellum," *Trends in Neuroscience* 21.9 (September 1998): 370–75; John C. Eccles, Masao Ito, and János Szentágothai, *The Cerebellum as a Neuronal Machine* (New York: Springer-Verlag, 1967); Masao Ito, *The Cerebellum and Neural Control* (New York: Raven, 1984).

79. N. Bernstein, *The Coordination and Regulation of Movements* (New York: Pergamon Press, 1967).

80. U.S. Office of Naval Research press release, "Boneless, Brainy, and Ancient," September 26, 2001, http://www.eurekalert.org/pub_releases/2001-11/oonr-bba112601.php; the octopus arm "could very well be the basis of next-generation robotic arms for undersea, space, as well as terrestrial applications."

81. S. Grossberg and R. W. Paine, "A Neural Model of Cortico-Cerebellar Interactions During Attentive Imitation and Predictive Learning of Sequential Handwriting Movements," *Neural Networks* 13.8–9 (October–November 2000): 999–1046.

82. Voogd and Glickstein, "Anatomy of the Cerebellum"; Eccles, Ito, and Szentágothai, *Cerebellum as a Neuronal Machine*; Ito, *Cerebellum and Neural Control*; R. Llinas, in *Handbook of Physiology*, vol. 2, *The Nervous System,* ed. V. B. Brooks (Bethesda, Md.: American Physiological Society, 1981), pp. 831–976.

83. J. L. Raymond, S. G. Lisberger, and M. D. Mauk, "The Cerebellum: A Neuronal Learning Machine?" *Science* 272.5265 (May 24, 1996): 1126–31; J. J. Kim and R. F. Thompson, "Cerebellar Circuits and Synaptic Mechanisms Involved in Classical Eyeblink Conditioning," *Trends in Neuroscience* 20.4 (April 1997): 177–81.

84. The simulation included 10,000 granule cells, 900 Golgi cells, 500 mossy fiber cells, 20 Purkinje cells, and 6 nucleus cells.

85. J. F. Medina et al., "Timing Mechanisms in the Cerebellum: Testing Predictions of a Large-Scale Computer Simulation," *Journal of Neuroscience* 20.14 (July 15, 2000): 5516–25; Dean Buonomano and Michael Mauk, "Neural Network Model of the Cerebellum: Temporal Discrimination and the Timing of Motor Reponses," *Neural Computation* 6.1 (1994): 38–55.

86. Medina et al., "Timing Mechanisms in the Cerebellum."

87. Carver Mead, *Analog VLSI and Neural Systems* (Boston: Addison-Wesley Longman, 1989).

88. Lloyd Watts, "Visualizing Complexity in the Brain," in *Computational Intelligence: The Experts Speak*, D. Fogel and C. Robinson, eds. (Hoboken, N.J.: IEEE Press/ Wiley, 2003), pp. 45–56, http://www.lloydwatts.com/wcci.pdf.

89. Ibid.

90. See http://www.lloydwatts.com/neuroscience.shtml. NanoComputer Dream Team, "The Law of Accelerating Returns, Part II," http://nanocomputer.org/index.cfm? content=90&Menu=19.

91. See http://info.med.yale.edu/bbs/faculty/she_go.html.

92. Gordon M. Shepherd, ed., *The Synaptic Organization of the Brain*, 4th ed. (New York: Oxford University Press, 1998), p. vi.

93. E. Young, "Cochlear Nucleus," in ibid., pp. 121–58.

94. Tom Yin, "Neural Mechanisms of Encoding Binaural Localization Cues in the Auditory Brainstem," in D. Oertel, R. Fay, and A. Popper, eds., *Integrative Functions in the Mammalian Auditory Pathway* (New York: Springer-Verlag, 2002), pp. 99–159.

95. John Casseday, Thane Fremouw, and Ellen Covey, "The Inferior Colliculus: A Hub for the Central Auditory System," in Oertel, Fay, and Popper, *Integrative Functions in the Mammalian Auditory Pathway*, pp. 238–318.

96. Diagram by Lloyd Watts, http://www.lloydwatts.com/neuroscience.shtml, adapted from E. Young, "Cochlear Nucleus" in G. Shepherd, ed., *The Synaptic Organization of the Brain*, 4th ed. (New York: Oxford University Press, 2003 [first published 1998]), pp. 121–58; D. Oertel in D. Oertel, R. Fay, and A. Popper, eds., *Integrative Functions in the Mammalian Auditory Pathway* (New York: Springer-Verlag, 2002), pp. 1–5;

John Casseday, T. Fremouw, and E. Covey, "Inferior Colliculus" in ibid.; J. LeDoux, *The Emotional Brain* (New York: Simon & Schuster, 1997); J. Rauschecker and B. Tian, "Mechanisms and Streams for Processing of 'What' and 'Where' in Auditory Cortex," *Proceedings of the National Academy of Sciences* 97.22: 11800–11806.

Brain regions modeled:

Cochlea: Sense organ of hearing. Thirty thousand fibers convert motion of the stapes into spectrotemporal representations of sound.

MC: Multipolar cells. Measure spectral energy.

GBC: Globular bushy cells. Relay spikes from the auditory nerve to the lateral superior olivary complex (includes LSO and MSO). Encoding of timing and amplitude of signals for binaural comparison of level.

SBC: Spherical bushy cells. Provide temporal sharpening of time of arrival, as a preprocessor for interaural time-difference calculation (difference in time of arrival between the two ears, used to tell where a sound is coming from).

OC: Octopus cells. Detection of transients.

DCN: Dorsal cochlear nucleus. Detection of spectral edges and calibrating for noise levels.

VNTB: Ventral nucleus of the trapezoid body. Feedback signals to modulate outer hair-cell function in the cochlea.

VNLL, PON: Ventral nucleus of the lateral lemniscus; peri-olivary nuclei: processing transients from the OC.

MSO: Medial superior olive. Computing interaural time difference.

LSO: Lateral superior olive. Also involved in computing interaural level difference.

ICC: Central nucleus of the inferior colliculus. The site of major integration of multiple representations of sound.

ICx: Exterior nucleus of the inferior colliculus. Further refinement of sound localization.

SC: Superior colliculus. Location of auditory/visual merging.

MGB: Medial geniculate body. The auditory portion of the thalamus.

LS: Limbic system. Comprising many structures associated with emotion, memory, territory, et cetera.

AC: Auditory cortex.

97. M. S. Humayun et al., "Human Neural Retinal Transplantation," *Investigative Ophthalmology and Visual Science* 41.10 (September 2000): 3100–3106.

98. Information Science and Technology Colloquium Series, May 23, 2001, http://isandtcolloq.gsfc.nasa.gov/spring2001/speakers/poggio.html.

99. Kah-Kay Sung and Tomaso Poggio, "Example-Based Learning for View-Based Human Face Detection," *IEEE Transactions on Pattern Analysis and Machine Intelligence* 20.1 (1998): 39–51, http://portal.acm.org/citation.cfm?id=275345&dl=ACM&coll=GUIDE.

100. Maximilian Riesenhuber and Tomaso Poggio, "A Note on Object Class Representa-

tion and Categorical Perception," Center for Biological and Computational Learning, MIT, AI Memo 1679 (1999), ftp://publications.ai.mit.edu/ai-publications/pdf/AIM-1679.pdf.

101. K. Tanaka, "Inferotemporal Cortex and Object Vision," *Annual Review of Neuroscience* 19 (1996): 109–39; Anuj Mohan, "Object Detection in Images by Components," Center for Biological and Computational Learning, MIT, AI Memo 1664 (1999), http://citeseer.ist.psu.edu/cache/papers/cs/12185/ftp:zSzzSzpublications.ai.mit.eduzSzai-publicationszSz1500–1999zSzAIM-1664.pdf/mohan99object.pdf; Anuj Mohan, Constantine Papageorgiou, and Tomaso Poggio, "Example-Based Object Detection in Images by Components," *IEEE Transactions on Pattern Analysis and Machine Intelligence* 23.4 (April 2001), http://cbcl.mit.edu/projects/cbcl/publications/ps/mohan-ieee.pdf; B. Heisele, T. Poggio, and M. Pontil, "Face Detection in Still Gray Images," Artificial Intelligence Laboratory, MIT, Technical Report AI Memo 1687 (2000). Also see Bernd Heisele, Thomas Serre, and Stanley Bilesch, "Component-Based Approach to Face Detection," Artificial Intelligence Laboratory and the Center for Biological and Computational Learning, MIT (2001), http://www.ai.mit.edu/research/abstracts/abstracts2001/vision-applied-to-people/03heisele2.pdf.

102. D. Van Essen and J. Gallant, "Neural Mechanisms of Form and Motion Processing in the Primate Visual System," *Neuron* 13.1 (July 1994): 1–10.

103. Shimon Ullman, *High-Level Vision: Object Recognition and Visual Cognition* (Cambridge, Mass.: MIT Press, 1996); D. Mumford, "On the Computational Architecture of the Neocortex. II. The Role of Corticocortical Loops," *Biological Cybernetics* 66.3 (1992): 241–51; R. Rao and D. Ballard, "Dynamic Model of Visual Recognition Predicts Neural Response Properties in the Visual Cortex," *Neural Computation* 9.4 (May 15, 1997): 721–63.

104. B. Roska and F. Werblin, "Vertical Interactions Across Ten Parallel, Stacked Representations in the Mammalian Retina," *Nature* 410.6828 (March 29, 2001): 583–87; University of California, Berkeley, news release, "Eye Strips Images of All but Bare Essentials Before Sending Visual Information to Brain, UC Berkeley Research Shows," March 28, 2001, www.berkeley.edu/news/media/releases/2001/03/28_wers1.html.

105. Hans Moravec and Scott Friedman have founded a robotics company called Seegrid based on Moravec's research. See www.Seegrid.com.

106. M. A. Mahowald and C. Mead, "The Silicon Retina," *Scientific American* 264.5 (May 1991): 76–82.

107. Specifically, a low-pass filter is applied to one receptor (such as a photoreceptor). This is multiplied by the signal of the neighboring receptor. If this is done in both directions and the results of each operation subtracted from zero, we get an output that reflects the direction of movement.

108. On Berger, see http://www.usc.edu/dept/engineering/CNE/faculty/Berger.html.

109. "The World's First Brain Prosthesis," *New Scientist* 177.2386 (March 15, 2003): 4, http://www.newscientist.com/news/news.jsp?id=ns99993488.

110. Charles Choi, "Brain-Mimicking Circuits to Run Navy Robot," UPI, June 7, 2004, http://www.upi.com/view.cfm?StoryID=20040606-103352-6086r.

111. Giacomo Rizzolatti et al., "Functional Organization of Inferior Area 6 in the Macaque Monkey. II. Area F5 and the Control of Distal Movements," *Experimental Brain Research* 71.3 (1998): 491–507.

112. M. A. Arbib, "The Mirror System, Imitation, and the Evolution of Language," in Kerstin Dautenhahn and Chrystopher L. Nehaniv, eds., *Imitation in Animals and Artifacts* (Cambridge, Mass.: MIT Press, 2002).

113. Marc D. Hauser, Noam Chomsky, and W. Tecumseh Fitch, "The Faculty of Language: What Is It, Who Has It, and How Did It Evolve?" *Science* 298 (November 2002): 1569–79, www.wjh.harvard.edu/~mnkylab/publications/languagespeech/Hauser,Chomsky,Fitch.pdf.

114. Daniel C. Dennett, *Freedom Evolves* (New York: Viking, 2003).

115. See Sandra Blakeslee, "Humanity? Maybe It's All in the Wiring," *New York Times,* December 11, 2003, http://www.nytimes.com/2003/12/09/science/09BRAI.html?ex=1386306000&en=294f5e91dd262a1a&ei=5007&partner=USERLAND.

116. Antonio R. Damasio, *Descartes' Error: Emotion, Reason and the Human Brain* (New York: Putnam, 1994).

117. M. P. Maher et al., "Microstructures for Studies of Cultured Neural Networks," *Medical and Biological Engineering and Computing* 37.1 (January 1999): 110–18; John Wright et al., "Towards a Functional MEMS Neurowell by Physiological Experimentation," *Technical Digest,* ASME, 1996 International Mechanical Engineering Congress and Exposition, Atlanta, November 1996, DSC (Dynamic Systems and Control Division), vol. 59, pp. 333–38.

118. W. French Anderson, "Genetics and Human Malleability," *Hastings Center Report* 23.20 (January/February 1990): 1.

119. Ray Kurzweil, "A Wager on the Turing Test: Why I Think I Will Win," KurzweilAI.net, April 9, 2002, http://www.KurzweilAI.net/meme/frame.html?main=/articles/art0374.html.

120. Robert A. Freitas Jr. proposes a future nanotechnology-based brain-uploading system that would effectively be instantaneous. According to Freitas (personal communication, January 2005), "An in vivo fiber network as proposed in http://www.nanomedicine.com/NMI/7.3.1.htm can handle 10^{18} bits/sec of data traffic, capacious enough for real-time brain-state monitoring. The fiber network has a 30 cm^3 volume and generates 4–6 watts waste heat, both small enough for safe installation in a 1400 cm^3 25-watt human brain. Signals travel at most a few meters at nearly the speed of light, so transit time from signal origination at neuron sites inside the brain to the external computer system mediating the upload are ~0.00001 msec which is considerably less than the minimum ~5 msec neuron discharge cycle time. Neuron-monitoring chemical sensors located on average ~2 microns apart can capture relevant chemical events occurring within a ~5 msec time window, since this is the approximate diffusion time for, say, a small neuropeptide across a 2-micron distance (http://www.nanomedicine.com/NMI/

Tables/3.4.jpg). Thus human brain state monitoring can probably be instantaneous, at least on the timescale of human neural response, in the sense of 'nothing of significance was missed.' "

121. M. C. Diamond et al., "On the Brain of a Scientist: Albert Einstein," *Experimental Neurology* 88 (1985): 198–204.

Chapter Five: GNR: Three Overlapping Revolutions

1. Samuel Butler (1835–1902), "Darwin Among the Machines," *Christ Church Press,* June 13, 1863 (republished by Festing Jones in 1912 in *The Notebooks of Samuel Butler*).

2. Peter Weibel, "Virtual Worlds: The Emperor's New Bodies," in *Ars Electronica: Facing the Future*, ed. Timothy Druckery (Cambridge, Mass.: MIT Press, 1999), pp. 207–23; available online at http://www.aec.at/en/archiv_files/19902/E1990b_009.pdf.

3. James Watson and Francis Crick, "Molecular Structure of Nucleic Acids: A Structure for Deoxyribose Nucleic Acid," *Nature* 171.4356 (April 23, 1953): 737–38, http://www.nature.com/nature/dna50/watsoncrick.pdf.

4. Robert Waterston quoted in "Scientists Reveal Complete Sequence of Human Genome," CBC News, April 14, 2003, http://www.cbc.ca/story/science/national/2003/04/14/genome030414.html.

5. See chapter 2, note 57.

6. The original reports of Crick and Watson, which still make compelling reading today, may be found in James A. Peters, ed., *Classic Papers in Genetics* (Englewood Cliffs, N.J.: Prentice-Hall, 1959). An exciting account of the successes and failures that led to the double helix is given in J. D. Watson, *The Double Helix: A Personal Account of the Discovery of the Structure of DNA* (New York: Atheneum, 1968). Nature.com has a collection of Crick's papers available online at http://www.nature.com/nature/focus/crick/index.html.

7. Morislav Radman and Richard Wagner, "The High Fidelity of DNA Duplication," *Scientific American* 259.2 (August 1988): 40–46.

8. The structure and behavior of DNA and RNA are described in Gary Felsenfeld, "DNA," and James Darnell, "RNA," both in *Scientific American* 253.4 (October 1985), p. 58–67 and 68–78 respectively.

9. Mark A. Jobling and Chris Tyler-Smith, "The Human Y Chromosome: An Evolutionary Marker Comes of Age," *Nature Reviews Genetics* 4 (August 2003): 598–612; Helen Skaletsky et al., "The Male-Specific Region of the Human Y Chromosome Is a Mosaic of Discrete Sequence Classes," *Nature* 423 (June 19, 2003): 825–37.

10. Misformed proteins are perhaps the most dangerous toxin of all. Research suggests that misfolded proteins may be at the heart of numerous disease processes in the body. Such diverse diseases as Alzheimer's disease, Parkinson's disease, the human form of mad-cow disease, cystic fibrosis, cataracts, and diabetes are all

thought to result from the inability of the body to adequately eliminate misfolded proteins.

Protein molecules perform the lion's share of cellular work. Proteins are made within each cell according to DNA blueprints. They begin as long strings of amino acids, which must then be folded into precise three-dimensional configurations in order to function as enzymes, transport proteins, et cetera. Heavy-metal toxins interfere with normal function of these enzymes, further exacerbating the problem. There are also genetic mutations that predispose individuals to misformed-protein buildup.

When protofibrils begin to stick together, they form filaments, fibrils, and ultimately larger globular structures called amyloid plaque. Until recently these accumulations of insoluble plaque were regarded as the pathologic agents for these diseases, but it is now known that the protofibrils themselves are the real problem. The speed with which a protofibril is turned into insoluble amyloid plaque is inversely related to disease progression. This explains why some individuals are found to have extensive accumulation of plaque in their brains but no evidence of Alzheimer's disease, while others have little visible plaque yet extensive manifestations of the disease. Some people form amyloid plaque quickly, which protects them from further protofibril damage. Other individuals turn protofibrils into amyloid plaque less rapidly, allowing more extensive damage. These people also have little visible amyloid plaque. See Per Hammarström, Frank Schneider, and Jeffrey W. Kelly, "*Trans*-Suppression of Misfolding in an Amyloid Disease," *Science* 293.5539 (September 28, 2001): 2459–62.

11. A fascinating account of the new biology is given in Horace F. Judson, *The Eighth Day of Creation: The Makers of the Revolution in Biology* (Woodbury, N.Y.: CSHL Press, 1996).

12. Raymond Kurzweil and Terry Grossman, M.D., *Fantastic Voyage: Live Long Enough to Live Forever* (New York: Rodale, 2004). See http://www.Fantastic-Voyage.net and http://www.RayandTerry.com.

13. Raymond Kurzweil, *The 10% Solution for a Healthy Life: How to Eliminate Virtually All Risk of Heart Disease and Cancer* (New York: Crown Books, 1993).

14. Kurzweil and Grossman, *Fantastic Voyage.* "Ray & Terry's Longevity Program" is articulated throughout the book.

15. The test for "biological age," called the H-scan test, includes tests for auditory-reaction time, highest audible pitch, vibrotactile sensitivity, visual-reaction time, muscle-movement time, lung (forced expiratory) volume, visual-reaction time with decision, muscle-movement time with decision, memory (length of sequence), alternative button-tapping time, and visual accommodation. The author had this test done at Frontier Medical Institute (Grossman's health and longevity clinic), http://www.FMIClinic.com. For information on the H-scan test, see Diagnostic and Lab Testing, Longevity Institute, Dallas, http://www.lidhealth.com/diagnostic.html.

16. Kurzweil and Grossman, *Fantastic Voyage*, chapter 10: "Ray's Personal Program."
17. Ibid.
18. Aubrey D. N. J. de Grey, "The Foreseeability of Real Anti-Aging Medicine: Focusing the Debate," *Experimental Gerontology* 38.9 (September 2003): 927–34; Aubrey D. N. J. de Grey, "An Engineer's Approach to the Development of Real Anti-Aging Medicine," *Science of Aging, Knowledge, Environment* 1 (2003): Aubrey D. N. J. de Grey et al., "Is Human Aging Still Mysterious Enough to Be Left Only to Scientists?" *BioEssays* 24.7 (July 2002): 667–76.
19. Aubrey D. N. J. de Grey, ed., *Strategies for Engineered Negligible Senescence: Why Genuine Control of Aging May Be Foreseeable*, Annals of the New York Academy of Sciences, vol. 1019 (New York: New York Academy of Sciences, June 2004).
20. In addition to providing the functions of different types of cells, two other reasons for cells to control the expression of genes are environmental cues and developmental processes. Even simple organisms such as bacteria can turn on and off the synthesis of proteins depending on environmental cues. *E. coli,* for example, can turn off the synthesis of proteins that allow it to control the level of nitrogen gas from the air when there are other, less energy-intensive sources of nitrogen in its environment. A recent study of 1,800 strawberry genes found that the expression of 200 of those genes varied during different stages of development. E. Marshall, "An Array of Uses: Expression Patterns in Strawberries, Ebola, TB, and Mouse Cells," *Science* 286.5439 (1999): 445.
21. Along with a protein-encoding region, genes include regulatory sequences called promoters and enhancers that control where and when that gene is expressed. Promoters of genes that encode proteins are typically located immediately "upstream" on the DNA. An enhancer activates the use of a promoter, thereby controlling the rate of gene expression. Most genes require enhancers to be expressed. Enhancers have been called "the major determinant of differential transcription in space (cell type) and time"; and any given gene can have several different enhancer sites linked to it (S. F. Gilbert, *Developmental Biology*, 6th ed. [Sunderland, Mass.: Sinauer Associates, 2000]; available online at www.ncbi.nlm.nih.gov/books/bv.fcgi?call=bv.View..ShowSection&rid=.0BpKYEB-SPfx18nm8Q OxH).

   By binding to enhancer or promoter regions, transcription factors start or repress the expression of a gene. New knowledge of transcription factors has transformed our understanding of gene expression. Per Gilbert in the chapter "The Genetic Core of Development: Differential Gene Expression": "The gene itself is no longer seen as an independent entity controlling the synthesis of proteins. Rather, the gene both directs and is directed by protein synthesis. Natalie Anger (1992) has written, 'A series of discoveries suggests that DNA is more like a certain type of politician, surrounded by a flock of protein handlers and advisors that must vigorously massage it, twist it and, on occasion, reinvent it before the grand blueprint of the body can make any sense at all.' "

22. Bob Holmes, "Gene Therapy May Switch Off Huntington's," March 13, 2003, http://www.newscientist.com/news/news.jsp?id=ns99993493. "Emerging as a powerful tool for reverse genetic analysis, RNAi is rapidly being applied to study the function of many genes associated with human disease, in particular those associated with oncogenesis and infectious disease." J. C. Cheng, T. B. Moore, and K. M. Sakamoto, "RNA Interference and Human Disease," *Molecular Genetics and Metabolism* 80.1–2 (October 2003): 121–28. RNAi is a "potent and highly sequence-specific mechanism." L. Zhang, D. K. Fogg, and D. M. Waisman, "RNA Interference-Mediated Silencing of the S100A10 Gene Attenuates Plasmin Generation and Invasiveness of Colo 222 Colorectal Cancer Cells," *Journal of Biological Chemistry* 279.3 (January 16, 2004): 2053–62.

23. Each chip contains synthetic oligonucleotides that replicate sequences that identify specific genes. "To determine which genes have been expressed in a sample, researchers isolate messenger RNA from test samples, convert it to complementary DNA (cDNA), tag it with fluorescent dye, and run the sample over the wafer. Each tagged cDNA will stick to an oligo with a matching sequence, lighting up a spot on the wafer where the sequence is known. An automated scanner then determines which oligos have bound, and hence which genes were expressed. . . ." E. Marshall, "Do-It-Yourself Gene Watching," *Science* 286.5439 (October 15, 1999): 444–47.

24. Ibid.

25. J. Rosamond and A. Allsop, "Harnessing the Power of the Genome in the Search for New Antibiotics," *Science* 287.5460 (March 17, 2000): 1973–76.

26. T. R. Golub et al., "Molecular Classification of Cancer: Class Discovery and Class Prediction by Gene Expression Monitoring," *Science* 286.5439 (October 15, 1999): 531–37.

27. Ibid., as reported in A. Berns, "Cancer: Gene Expression in Diagnosis," *Nature* 403 (February 3, 2000): 491–92. In another study, 1 percent of the genes studied showed reduced expression in aged muscles. These genes produced proteins associated with energy production and cell building, so a reduction makes sense given the weakening associated with age. Genes with increased expression produced stress proteins, which are used to repair damaged DNA or proteins. J. Marx, "Chipping Away at the Causes of Aging," *Science* 287.5462 (March 31, 2000): 2390.

As another example, liver metastases are a common cause of colorectal cancer. These metastases respond differently to treatment depending on their genetic profile. Expression profiling is an excellent way to determine an appropriate mode of treatment. J. C. Sung et al., "Genetic Heterogeneity of Colorectal Cancer Liver Metastases," *Journal of Surgical Research* 114.2 (October 2003): 251.

As a final example, researchers have had difficulty analyzing the Reed-Sternberg cell of Hodgkin's disease because of its extreme rarity in diseased tissue. Expression profiling is now providing a clue regarding the heritage of this cell. J. Cossman et al., "Reed-Sternberg Cell Genome Expression Supports a B-Cell Lineage," *Blood* 94.2 (July 15, 1999): 411–16.

28. T. Ueland et al., "Growth Hormone Substitution Increases Gene Expression of Members of the IGF Family in Cortical Bone from Women with Adult Onset Growth Hormone Deficiency—Relationship with Bone Turn-Over," *Bone* 33.4 (October 2003): 638–45.

29. R. Lovett, "Toxicologists Brace for Genomics Revolution," *Science* 289.5479 (July 28, 2000): 536–37.

30. Gene transfer to somatic cells affects a subset of cells in the body for a period of time. It is theoretically possible also to alter genetic information in egg and sperm (germ-line) cells, for the purpose of passing on those changes to the next generations. Such therapy poses many ethical concerns and has not yet been attempted. "Gene Therapy," Wikipedia, http://en.wikipedia.org/wiki/Gene_therapy.

31. Genes encode proteins, which perform vital functions in the human body. Abnormal or mutated genes encode proteins that are unable to perform those functions, resulting in genetic disorders and diseases. The goal of gene therapy is to replace the defective genes so that normal proteins are produced. This can be done in a number of ways, but the most typical way is to insert a therapeutic replacement gene into the patient's target cells using a carrier molecule called a vector. "Currently, the most common vector is a virus that has been genetically altered to carry normal human DNA. Viruses have evolved a way of encapsulating and delivering their genes to human cells in a pathogenic manner. Scientists have tried to take advantage of this capability and manipulate the virus genome to remove the disease-causing genes and insert therapeutic genes" (Human Genome Project, "Gene Therapy," http://www.ornl.gov/TechResources/Human_Genome/medicine/gene therapy.html). See the Human Genome Project site for more information about gene therapy and links. Gene therapy is an important enough area of research that there are currently six scientific peer-reviewed gene-therapy journals and four professional associations dedicated to this topic.

32. K. R. Smith, "Gene Transfer in Higher Animals: Theoretical Considerations and Key Concepts," *Journal of Biotechnology* 99.1 (October 9, 2002): 1–22.

33. Anil Ananthaswamy, "Undercover Genes Slip into the Brain," March 20, 2003, http://www.newscientist.com/news/news.jsp?id=ns99993520.

34. A. E. Trezise et al., "In Vivo Gene Expression: DNA Electrotransfer," *Current Opinion in Molecular Therapeutics* 5.4 (August 2003): 397–404.

35. Sylvia Westphal, "DNA Nanoballs Boost Gene Therapy," May 12, 2002, http://www.newscientist.com/news/news.jsp?id=ns99992257.

36. L. Wu, M. Johnson, and M. Sato, "Transcriptionally Targeted Gene Therapy to Detect and Treat Cancer," *Trends in Molecular Medicine* 9.10 (October 2003): 421–29.

37. S. Westphal, "Virus Synthesized in a Fortnight," November 14, 2003, http://www.newscientist.com/news/news.jsp?id=ns99994383.

38. G. Chiesa, "Recombinant Apolipoprotein A-I(Milano) Infusion into Rabbit Carotid Artery Rapidly Removes Lipid from Fatty Streaks," *Circulation Research* 90.9 (May 17, 2002): 974–80; P. K. Shah et al., "High-Dose Recombinant

Apolipoprotein A-I(Milano) Mobilizes Tissue Cholesterol and Rapidly Reduces Plaque Lipid and Macrophage Content in Apolipoprotein e-Deficient Mice," *Circulation* 103.25 (June 26, 2001): 3047–50.

39. S. E. Nissen et al., "Effect of Recombinant Apo A-I Milano on Coronary Atherosclerosis in Patients with Acute Coronary Syndromes: A Randomized Controlled Trial," *JAMA* 290.17 (November 5, 2003): 2292–2300.

40. A recent phase 2 study reported "markedly increased HDL cholesterol levels and also decreased LDL cholesterol levels." M. E. Brousseau et al., "Effects of an Inhibitor of Cholesteryl Ester Transfer Protein on HDL Cholesterol," *New England Journal of Medicine* 350.15 (April 8, 2004): 1505–15, http://content.nejm. org/cgi/content/abstract/350/15/1505. Global phase 3 trials began in late 2003. Information on Torcetrapib is available on the Pfizer site: http://www.pfizer.com/ are/investors_reports/annual_2003/review/p2003ar14_15.htm.

41. O. J. Finn, "Cancer Vaccines: Between the Idea and the Reality," *Nature Reviews: Immunology* 3.8 (August 2003): 630–41; R. C. Kennedy and M. H. Shearer, "A Role for Antibodies in Tumor Immunity," *International Reviews of Immunology* 22.2 (March–April 2003): 141–72.

42. T. F. Greten and E. M. Jaffee, "Cancer Vaccines," *Journal of Clinical Oncology* 17.3 (March 1999): 1047–60.

43. "Cancer 'Vaccine' Results Encouraging," BBCNews, January 8, 2001, http://news. bbc.co.uk/2/hi/health/1102618.stm, reporting on research by E. M. Jaffee et al., "Novel Allogeneic Granulocyte-Macrophage Colony-Stimulating Factor-Secreting Tumor Vaccine for Pancreatic Cancer: A Phase I Trial of Safety and Immune Activation," *Journal of Clinical Oncology* 19.1 (January 1, 2001): 145–56.

44. John Travis, "Fused Cells Hold Promise of Cancer Vaccines," March 4, 2000, http://www.sciencenews.org/articles/20000304/fob3.asp, referring to D. W. Kufe, "Smallpox, Polio and Now a Cancer Vaccine?" *Nature Medicine* 6 (March 2000): 252–53.

45. J. D. Lewis, B. D. Reilly, and R. K. Bright, "Tumor-Associated Antigens: From Discovery to Immunity," *International Reviews of Immunology* 22.2 (March–April 2003): 81–112.

46. T. Boehm et al., "Antiangiogenic Therapy of Experimental Cancer Does Not Induce Acquired Drug Resistance," *Nature* 390.6658 (November 27, 1997): 404–7.

47. Angiogenesis Foundation, "Understanding Angiogenesis," http://www.angio.org/ understanding/content_understanding.html; L. K. Lassiter and M. A. Carducci, "Endothelin Receptor Antagonists in the Treatment of Prostate Cancer," *Seminars in Oncology* 30.5 (October 2003): 678–88. For an explanation of the process, see the National Cancer Institute Web site, "Understanding Angiogenesis," http:// press2.nci.nih.gov/sciencebehind/angiogenesis/angio02.htm.

48. I. B. Roninson, "Tumor Cell Senescence in Cancer Treatment," *Cancer Research* 63.11 (June 1, 2003): 2705–15; B. R. Davies et al., "Immortalization of Human Ovarian Surface Epithelium with Telomerase and Temperature-Sensitive

SV40 Large T Antigen," *Experimental Cell Research* 288.2 (August 15, 2003): 390–402.

49. See also R. C. Woodruff and J. N. Thompson Jr., "The Role of Somatic and Germline Mutations in Aging and a Mutation Interaction Model of Aging," *Journal of Anti-Aging Medicine* 6.1 (Spring 2003): 29–39. See also notes 18 and 19.

50. Aubrey D. N. J. de Grey, "The Reductive Hotspot Hypothesis of Mammalian Aging: Membrane Metabolism Magnifies Mutant Mitochondrial Mischief," *European Journal of Biochemistry* 269.8 (April 2002): 2003–9; P. F. Chinnery et al., "Accumulation of Mitochondrial DNA Mutations in Ageing, Cancer, and Mitochondrial Disease: Is There a Common Mechanism?" *Lancet* 360.9342 (October 26, 2002): 1323–25; A. D. de Grey, "Mitochondrial Gene Therapy: An Arena for the Biomedical Use of Inteins," *Trends in Biotechnology* 18.9 (September 2000): 394–99.

51. "The notion of 'vaccinating' individuals against a neurodegenerative disorder such as Alzheimer's disease is a marked departure from classical thinking about mechanism and treatment, and yet therapeutic vaccines for both Alzheimer's disease and multiple sclerosis have been validated in animal models and are in the clinic. Such approaches, however, have the potential to induce unwanted inflammatory responses as well as to provide benefit" (H. L. Weiner and D. J. Selkoe, "Inflammation and Therapeutic Vaccination in CNS Diseases," *Nature* 420.6917 [December 19–26, 2002]: 879–84). These researchers showed that a vaccine in the form of nose drops could slow the brain deterioration of Alzheimer's. H. L. Weiner et al., "Nasal Administration of Amyloid-beta Peptide Decreases Cerebral Amyloid Burden in a Mouse Model of Alzheimer's Disease," *Annals of Neurology* 48.4 (October 2000): 567–79.

52. S. Vasan, P. Foiles, and H. Founds, "Therapeutic Potential of Breakers of Advanced Glycation End Product-Protein Crosslinks," *Archives of Biochemistry and Biophysics* 419.1 (November 1, 2003): 89–96; D. A. Kass, "Getting Better Without AGE: New Insights into the Diabetic Heart," *Circulation Research* 92.7 (April 18, 2003): 704–6.

53. S. Graham, "Methuselah Worm Remains Energetic for Life," October 27, 2003, www.sciam.com/article.cfm?chanID=sa003&articleID=000C601F-8711-1F99-86FB83414B7F0156.

54. Ron Weiss's home page at Princeton University (http://www.princeton.edu/~rweiss) lists his publications, such as "Genetic Circuit Building Blocks for Cellular Computation, Communications, and Signal Processing," *Natural Computing, an International Journal* 2.1 (January 2003): 47–84.

55. S. L. Garfinkel, "Biological Computing," *Technology Review* (May–June 2000), http://static.highbeam.com/t/technologyreview/may012000/biologicalcomputing.

56. Ibid. See also the list of current research on the MIT Media Lab Web site, http://www.media.mit.edu/research/index.html.

57. Here is one possible explanation: "In mammals, female embryos have two X-chromosomes and males have one. During early development in females, one

of the X's and most of its genes are normally silenced or inactivated. That way, the amount of gene expression in males and females is the same. But in cloned animals, one X-chromosome is already inactivated in the donated nucleus. It must be reprogrammed and then later inactivated again, which introduces the possibility of errors." CBC News online staff, "Genetic Defects May Explain Cloning Failures," May 27, 2002, http://www.cbc.ca/stories/2002/05/27/cloning_errors020527. That story reports on F. Xue et al., "Aberrant Patterns of X Chromosome Inactivation in Bovine Clones," *Nature Genetics* 31.2 (June 2002): 216–20.

58. Rick Weiss, "Clone Defects Point to Need for 2 Genetic Parents," *Washington Post,* May 10, 1999, http://www.gene.ch/genet/1999/Jun/msg00004.html.

59. A. Baguisi et al., "Production of Goats by Somatic Cell Nuclear Transfer," *Nature Biotechnology* 5 (May 1999): 456–61. For more information on the partnership between Genzyme Transgenics Corporation, Louisiana State University, and Tufts University School of Medicine that produced this work, see the April 27, 1999, press release, "Genzyme Transgenics Corporation Announces First Successful Cloning of Transgenic Goat," http://www.transgenics.com/pressreleases/pr042799.html.

60. Luba Vangelova, "True or False? Extinction Is Forever," *Smithsonian*, June 2003, http://www.smithsonianmag.com/smithsonian/issues03/jun03/phenomena.html.

61. J. B. Gurdon and A. Colman, "The Future of Cloning," *Nature* 402.6763 (December 16, 1999): 743–46; Gregory Stock and John Campbell, eds., *Engineering the Human Germline: An Exploration of the Science and Ethics of Altering the Genes We Pass to Our Children* (New York: Oxford University Press, 2000).

62. As the Scripps Research Institute points out, "The ability to dedifferentiate or reverse lineage-committed cells to multipotent progenitor cells might overcome many of the obstacles associated with using ESCs and adult stem cells in clinical applications (inefficient differentiation, rejection of allogenic cells, efficient isolation and expansion, etc.). With an efficient dedifferentiation process, it is conceivable that healthy, abundant and easily accessible adult cells could be used to generate different types of functional cells for the repair of damaged tissues and organs" (http://www.scripps.edu/chem/ding/sciences.htm).

> The direct conversion of one differentiated cell type into another—a process referred to as transdifferentiation—would be beneficial for producing isogenic [patient's own] cells to replace sick or damaged cells or tissue. Adult stem cells display a broader differentiation potential than anticipated and might contribute to tissues other than those in which they reside. As such, they could be worthy therapeutic agents. Recent advances in transdifferentiation involve nuclear transplantation, manipulation of cell culture conditions, induction of ectopic gene expression and uptake of molecules from cellular extracts. These approaches open the doors to new avenues for engineering isogenic replacement cells. To avoid unpredictable tissue transformation, nuclear reprogramming requires controlled and heritable epigenetic modifications. Considerable efforts remain to unravel the molecular processes

underlying nuclear reprogramming and evaluate stability of the changes in reprogrammed cells.

Quoted from P. Collas and Anne-Mari Håkelien, "Teaching Cells New Tricks," *Trends in Biotechnology* 21.8 (August 2003): 354–61; P. Collas, "Nuclear Reprogramming in Cell-Free Extracts," *Philosophical Transactions of the Royal Society of London, B* 358.1436 (August 29, 2003): 1389–95.

63. Researchers have converted human liver cells to pancreas cells in the laboratory: Jonathan Slack et al., "Experimental Conversion of Liver to Pancreas," *Current Biology* 13.2 (January 2003): 105–15. Researchers reprogrammed cells to behave like other cells using cell extracts; for example, skin cells were reprogrammed to exhibit T-cell characteristics. Anne-Mari Håkelien et al., "Reprogramming Fibroblasts to Express T-Cell Functions Using Cell Extracts," *Nature Biotechnology* 20.5 (May 2002): 460–66; Anne-Mari Håkelien and P. Collas, "Novel Approaches to Transdifferentiation," *Cloning Stem Cells* 4.4 (2002): 379–87. See also David Tosh and Jonathan M. W. Slack, "How Cells Change Their Phenotype," *Nature Reviews Molecular Cell Biology* 3.3 (March 2002): 187–94.

64. See the description of transcription factors in note 21, above.

65. R. P. Lanza et al., "Extension of Cell Life-Span and Telomere Length in Animals Cloned from Senescent Somatic Cells," *Science* 288.5466 (April 28, 2000): 665–69. See also J. C. Ameisen, "On the Origin, Evolution, and Nature of Programmed Cell Death: A Timeline of Four Billion Years," *Cell Death and Differentiation* 9.4 (April 2002): 367–93; Mary-Ellen Shay, "Transplantation Without a Donor," *Dream: The Magazine of Possibilities* (Children's Hospital, Boston), Fall 2001.

66. In 2000 the Immune Tolerance Network (http://www.immunetolerance.org), a project of the National Institutes of Health (NIH) and the Juvenile Diabetes Foundation, announced a multicenter clinical trial to assess the effectiveness of islet transplantation.

According to a clinical-trial research summary (James Shapiro, "Campath-1H and One-Year Temporary Sirolimus Maintenance Monotherapy in Clinical Islet Transplantation," http://www.immunetolerance.org/public/clinical/islet/trials/shapiro2.html), "This therapy is not suitable for all patients with Type I diabetes, even if there were no limitation in islet supply, because of the potential long-term risks of cancer, life-threatening infections and drug side-effects related to the antirejection therapy. If tolerance [indefinite graft function without a need for long-term drugs to prevent rejection] could be achieved at minimal up-front risk, then islet transplant could be used safely earlier in the course of diabetes, and eventually in children at the time of diagnosis."

67. "Lab Grown Steaks Nearing Menu," http://www.newscientist.com/news/news.jsp?id=ns99993208, includes discussion of technical issues.

68. The halving time for feature sizes is five years in each dimension. See discussion in chapter 2.

69. An analysis by Robert A. Freitas Jr. indicates that replacing 10 percent of a person's

red blood cells with robotic respirocytes would enable holding one's breath for about four hours, which is about 240 times longer than one minute (about the length of time feasible with all biological red blood cells). Since this increase derives from replacing only 10 percent of the red blood cells, the respirocytes are thousands of times more effective.

70. Nanotechnology is "thorough, inexpensive control of the structure of matter based on molecule-by-molecule control of products and byproducts; the products and processes of molecular manufacturing, including molecular machinery" (Eric Drexler and Chris Peterson, *Unbounding the Future: The Nanotechnology Revolution* [New York: William Morrow, 1991]). According to the authors:

> Technology has been moving toward greater control of the structure of matter for millennia. . . . [P]ast advanced technologies—microwave tubes, lasers, superconductors, satellites, robots, and the like—have come trickling out of factories, at first with high price tags and narrow applications. Molecular manufacturing, though, will be more like computers: a flexible technology with a huge range of applications. And molecular manufacturing won't come trickling out of conventional factories as computers did; it will replace factories and replace or upgrade their products. This is something new and basic, not just another twentieth-century gadget. It will arise out of twentieth-century trends in science, but it will break the trend-lines in technology, economics, and environmental affairs. [chap. 1]

Drexler and Peterson outline the possible scope of the effects of the revolution: efficient solar cells "as cheap as newspaper and as tough as asphalt," molecular mechanisms that can kill cold viruses in six hours before biodegrading, immune machines that destroy malignant cells in the body at the push of a button, pocket supercomputers, the end of the use of fossil fuels, space travel, and restoration of lost species. Also see E. Drexler, *Engines of Creation* (New York: Anchor Books, 1986). The Foresight Institute has a useful list of nanotechnology FAQs (http://www.foresight.org/NanoRev/FIFAQ1.html) and other information. Other Web resources include the National Nanotechnology Initiative (http://www.nano.gov), http://nanotechweb.org, Dr. Ralph Merkle's nanotechnology page (http://www.zyvex.com/nano), and *Nanotechnology*, an online journal (http://www.iop.org/EJ/journal/0957-4484). Extensive material on nanotechnology can be found on the author's Web site at http://www.kurzweilAI.net/meme/frame.html?m=18.

71. Richard P. Feynman, "There's Plenty of Room at the Bottom," American Physical Society annual meeting, Pasadena, California, 1959; transcript at http://www.zyvex.com/nanotech/feynman.html.

72. John von Neumann, *Theory of Self-Reproducing Automata*, A. W. Burks, ed. (Urbana: University of Illinois Press, 1966).

73. The most comprehensive survey of kinematic machine replication is Robert A.

Freitas Jr. and Ralph C. Merkle, *Kinematic Self-Replicating Machines* (Georgetown, Tex.: Landes Bioscience, 2004), http://www.MolecularAssembler.com/KSRM.htm.

74. K. Eric Drexler, *Engines of Creation,* and K. Eric Drexler, *Nanosystems: Molecular Machinery, Manufacturing, and Computation* (New York: Wiley Interscience, 1992).

75. See the discussion of nanotube circuitry in chapter 3, including the analysis of the potential of nanotube circuitry in note 9 of that chapter.

76. K. Eric Drexler and Richard E. Smalley, "Nanotechnology: Drexler and Smalley Make the Case for and Against 'Molecular Assemblers,' " *Chemical and Engineering News*, November 30, 2003, http://pubs.acs.org/cen/coverstory/8148/8148counter point.html.

77. Ralph C. Merkle, "A Proposed 'Metabolism' for a Hydrocarbon Assembler," *Nanotechnology* 8 (December 1997): 149–62, http://www.iop.org/EJ/abstract/0957-4484/8/4/001 or http://www.zyvex.com/nanotech/hydroCarbonMetabolism.html. See also Ralph C. Merkle, "Binding Sites for Use in a Simple Assembler," *Nanotechnology* 8 (1997): 23–28, http://www.zyvex.com/nanotech/bindingSites.html; Ralph C. Merkle, "A New Family of Six Degree of Freedom Positional Devices," *Nanotechnology* 8 (1997): 47–52, http://www.zyvex.com/nanotech/6dof.html; Ralph C. Merkle, "Casing an Assembler," *Nanotechnology* 10 (1999): 315–22, http://www.zyvex.com/nanotech/casing; Robert A. Freitas Jr., "A Simple Tool for Positional Diamond Mechanosynthesis, and Its Method of Manufacture," U.S. Provisional Patent Application No. 60/543,802, filed February 11, 2004, process described in lecture at http://www.MolecularAssembler.com/Papers/PathDiam MolMfg.htm; Ralph C. Merkle and Robert A. Freitas Jr., "Theoretical Analysis of a Carbon-Carbon Dimer Placement Tool for Diamond Mechanosynthesis," *Journal of Nanoscience and Nanotechnology* 3 (August 2003): 319–24, http://www. rfreitas.com/Nano/JNNDimerTool.pdf; Robert A. Freitas Jr. and Ralph C. Merkle, "Merkle-Freitas Hydrocarbon Molecular Assembler," in *Kinematic Self-Replicating Machines,* section 4.11.3 (Georgetown, Tex.: Landes Bioscience, 2004), pp. 130–35, http://www.MolecularAssembler.com/KSRM/4.11.3.htm.

78. Robert A. Freitas Jr., *Nanomedicine,* vol. 1, *Basic Capabilities,* section 6.3.4.5, "Chemoelectric Cells" (Georgetown, Tex.: Landes Bioscience, 1999), pp. 152–54, http://www.nanomedicine.com/NMI/6.3.4.5.htm; Robert A. Freitas Jr., *Nanomedicine,* vol. 1, *Basic Capabilities,* section 6.3.4.4, "Glucose Engines" (Georgetown, Tex.: Landes Bioscience, 1999), pp. 149–52, http://www.nanomedicine.com/ NMI/6.3.4.4.htm; K. Eric Drexler, *Nanosystems: Molecular Machinery, Manufacturing, and Computation,* section 16.3.2, "Acoustic Power and Control" (New York: Wiley Interscience, 1992), pp. 472–76. See also Robert A. Freitas Jr. and Ralph C. Merkle, *Kinematic Self-Replicating Machines,* appendix B.4, "Acoustic Transducer for Power and Control" (Georgetown, Tex.: Landes Bioscience, 2004), pp. 225–33, http://www.MolecularAssembler.com/KSRM/AppB.4.htm.

79. The most comprehensive survey of these proposals may be found in Robert A.

Freitas Jr. and Ralph C. Merkle, *Kinematic Self-Replicating Machines*, chapter 4, "Microscale and Molecular Kinematic Machine Replicators" (Georgetown, Tex.: Landes Bioscience, 2004), pp. 89–144, http://www.MolecularAssembler.com/KSRM/4.htm.

80. Drexler, *Nanosystems*, p. 441.

81. The most comprehensive survey of these proposals may be found in Robert A. Freitas Jr. and Ralph C. Merkle, *Kinematic Self-Replicating Machines*, chapter 4, "Microscale and Molecular Kinematic Machine Replicators" (Georgetown, Tex.: Landes Bioscience, 2004), pp. 89–144, http://www.MolecularAssembler.com/KSRM/4.htm.

82. T. R. Kelly, H. De Silva, and R. A. Silva, "Unidirectional Rotary Motion in a Molecular System," *Nature* 401.6749 (September 9, 1999): 150–52.

83. Carlo Montemagno and George Bachand, "Constructing Nanomechanical Devices Powered by Biomolecular Motors," *Nanotechnology* 10 (1999): 225–31; George D. Bachand and Carlo D. Montemagno, "Constructing Organic/Inorganic NEMS Devices Powered by Biomolecular Motors," *Biomedical Microdevices* 2.3 (June 2000): 179–84.

84. N. Koumura et al., "Light-Driven Monodirectional Molecular Rotor," *Nature* 401.6749 (September 9, 1999): 152–55.

85. Berkeley Lab, "A Conveyor Belt for the Nano-Age," April 28, 2004, http://www.lbl.gov/Science-Articles/Archive/MSD-conveyor-belt-for-nanoage.html.

86. "Study: Self-Replicating Nanomachines Feasible," June 2, 2004, http://www.smalltimes.com/document_display.cfm?section_id=53&document_id=8007, reporting on Tihamer Toth-Fejel, "Modeling Kinematic Cellular Automata," April 30, 2004, http://www.niac.usra.edu/files/studies/final_report/pdf/883Toth-Fejel.pdf.

87. W. U. Dittmer, A. Reuter, and F. C. Simmel, "A DNA-Based Machine That Can Cyclically Bind and Release Thrombin," *Angewandte Chemie International Edition* 43 (2004): 3550–53.

88. Shiping Liao and Nadrian C. Seeman, "Translation of DNA Signals into Polymer Assembly Instructions," *Science* 306 (December 17, 2004): 2072–74, http://www.sciencemag.org/cgi/reprint/306/5704/2072.pdf.

89. Scripps Research Institute, "Nano-origami," February 11, 2004, http://www.eurekalert.org/pub_releases/2004-02/sri-n021004.php.

90. Jenny Hogan, "DNA Robot Takes Its First Steps," May 6, 2004, http://www.newscientist.com/news/news.jsp?id=ns99994958, reporting on Nadrian Seeman and William Sherman, "A Precisely Controlled DNA Biped Walking Device," *Nano Letters* 4.7 (July 2004): 1203–7.

91. Helen Pearson, "Construction Bugs Find Tiny Work," *Nature News,* July 11, 2003, http://www.nature.com/news/2003/030707/full/030707-9.html.

92. Richard E. Smalley, "Nanofallacies: Of Chemistry, Love and Nanobots," *Scientific American* 285.3 (September 2001): 76–77; subscription required for this link: http://www.sciamdigital.com/browse.cfm?sequencenameCHAR=item2&methodnameCHAR=resource_getitembrowse&interfacenameCHAR=browse.cfm&ISSU

EID_CHAR=6A628AB3-17A5-4374-B100-3185A0CCC86&ARTICLEID_CHAR=
F90C4210-C153-4B2F-83A1-28F2012B637&sc=I100322.

93. See the bibliography of references in notes 108 and 109 below. See also Drexler, *Nanosystems,* for his proposal. For sample confirmations, see Xiao Yan Chang, Martin Perry, James Peploski, Donald L. Thompson, and Lionel M. Raff, "Theoretical Studies of Hydrogen-Abstraction Reactions from Diamond and Diamondlike Surfaces," *Journal of Chemical Physics* 99 (September 15, 1993): 4748–58. See also L. J. Lauhon and W. Ho, "Inducing and Observing the Abstraction of a Single Hydrogen Atom in Bimolecular Reaction with a Scanning Tunneling Microscope," *Journal of Physical Chemistry* 105 (2000): 3987–92; G. Allis and K. Eric Drexler, "Design and Analysis of a Molecular Tool for Carbon Transfer in Mechanosynthesis," *Journal of Computational and Theoretical Nanoscience* 2.1 (March–April 2005, in press).

94. Lea Winerman, "How to Grab an Atom," *Physical Review Focus,* May 2, 2003, http://focus.aps.org/story/v11/st19, reporting on Noriaki Oyabu, "Mechanical Vertical Manipulation of Selected Single Atoms by Soft Nanoindentation Using a Near Contact Atomic Force Microscope," *Physical Review Letters* 90.17 (May 2, 2003): 176102.

95. Robert A. Freitas Jr., "Technical Bibliography for Research on Positional Mechanosynthesis," Foresight Institute Web site, December 16, 2003, http://foresight.org/stage2/mechsynthbib.html.

96. See equation and explanation on p. 3 of Ralph C. Merkle, "That's Impossible! How Good Scientists Reach Bad Conclusions," http://www.zyvex.com/nanotech/impossible.html.

97. "Thus ΔX_C is just ~5% of the typical atomic electron cloud diameter of ~0.3 nm, imposing only a modest additional constraint on the fabrication and stability of nanomechanical structures. (Even in most liquids at their boiling points, each molecule is free to move only ~0.07 nm from its average position.)" Robert A. Freitas Jr., *Nanomedicine,* vol. 1, *Basic Capabilities,* section 2.1, "Is Molecular Manufacturing Possible?" (Georgetown, Tex.: Landes Bioscience, 1999), p. 39, http://www.nanomedicine.com/NMI/2.1.htm#p9.

98. Robert A. Freitas Jr., *Nanomedicine,* vol. 1, *Basic Capabilities,* section 6.3.4.5, "Chemoelectric Cells" (Georgetown, Tex.: Landes Bioscience, 1999), pp. 152–54, http://www.nanomedicine.com/NMI/6.3.4.5.htm.

99. Montemagno and Bachand, "Constructing Nanomechanical Devices Powered by Biomolecular Motors."

100. Open letter from Foresight chairman K. Eric Drexler to Nobel laureate Richard Smalley, http://www.foresight.org/NanoRev/Letter.html, and reprinted here: http://www.KurzweilAI.net/meme/frame.html?main=/articles/art0560.html. The full story can be found at Ray Kurzweil, "The Drexler-Smalley Debate on Molecular Assembly," http://www.KurzweilAI.net/meme/frame.html?main=/articles/art0604.html.

101. K. Eric Drexler and Richard E. Smalley, "Nanotechnology: Drexler and Smalley

Make the Case for and Against 'Molecular Assemblers,' " *Chemical & Engineering News* 81.48 (Dec. 1, 2003): 37–42, http://pubs.acs.org/cen/coverstory/8148/8148 counterpoint.html.

102. A. Zaks and A. M. Klibanov, "Enzymatic Catalysis in Organic Media at 100 Degrees C," *Science* 224.4654 (June 15, 1984): 1249–51.

103. Patrick Bailey, "Unraveling the Big Debate About Small Machines," *BetterHumans,* August 16, 2004, http://www.betterhumans.com/Features/Reports/report.aspx? articleID=2004-08-16-1.

104. Charles B. Musgrave et al., "Theoretical Studies of a Hydrogen Abstraction Tool for Nanotechnology," *Nanotechnology* 2 (October 1991): 187–95; Michael Page and Donald W. Brenner, "Hydrogen Abstraction from a Diamond Surface: *Ab initio* Quantum Chemical Study with Constrained Isobutane as a Model," *Journal of the American Chemical Society* 113.9 (1991): 3270–74; Xiao Yan Chang, Martin Perry, James Peploski, Donald L. Thompson, and Lionel M. Raff, "Theoretical Studies of Hydrogen-Abstraction Reactions from Diamond and Diamond-like Surfaces," *Journal of Chemical Physics* 99 (September 15, 1993): 4748–58; J. W. Lyding, K. Hess, G. C. Abeln, et al., "UHV-STM Nanofabrication and Hydrogen/ Deuterium Desorption from Silicon Surfaces: Implications for CMOS Technology," *Applied Surface Science* 132 (1998): 221; http://www.hersam-group.north western.edu/publications.html; E. T. Foley et al., "Cryogenic UHV-STM Study of Hydrogen and Deuterium Desorption from Silicon(100)," *Physical Review Letters* 80 (1998): 1336–39, http://prola.aps.org/abstract/PRL/v80/i6/p1336_1; L. J. Lauhon and W. Ho, "Inducing and Observing the Abstraction of a Single Hydrogen Atom in Bimolecular Reaction with a Scanning Tunneling Microscope," *Journal of Physical Chemistry* 105 (2000): 3987–92.

105. Stephen P. Walch and Ralph C. Merkle, "Theoretical Studies of Diamond Mechanosynthesis Reactions," *Nanotechnology* 9 (September 1998): 285–96; Fedor N. Dzegilenko, Deepak Srivastava, and Subhash Saini, "Simulations of Carbon Nanotube Tip Assisted Mechano-Chemical Reactions on a Diamond Surface," *Nanotechnology* 9 (December 1998): 325–30; Ralph C. Merkle and Robert A. Freitas Jr., "Theoretical Analysis of a Carbon-Carbon Dimer Placement Tool for Diamond Mechanosynthesis," *Journal of Nanoscience and Nanotechnology* 3 (August 2003): 319–24, http://www.rfreitas.com/Nano/DimerTool.htm; Jingping Peng, Robert A. Freitas Jr., and Ralph C. Merkle, "Theoretical Analysis of Diamond Mechano-Synthesis. Part I. Stability of C_2 Mediated Growth of Nanocrystalline Diamond C(110) Surface," *Journal of Computational and Theoretical Nanoscience* 1 (March 2004): 62–70, http://www.molecularassembler.com/JCTNPengMar04.pdf; David J. Mann, Jingping Peng, Robert A. Freitas Jr., and Ralph C. Merkle, "Theoretical Analysis of Diamond MechanoSynthesis. Part II. C_2 Mediated Growth of Diamond C(110) Surface via Si/Ge-Triadamantane Dimer Placement Tools," *Journal of Computational and Theoretical Nanoscience* 1 (March 2004), 71–80, http://www. molecularassembler.com/JCTNMannMar04.pdf.

106. The analysis of the hydrogen abstraction tool and carbon deposition tools has involved many people, including: Donald W. Brenner, Tahir Cagin, Richard J. Colton, K. Eric Drexler, Fedor N. Dzegilenko, Robert A. Freitas Jr., William A. Goddard III, J. A. Harrison, Charles B. Musgrave, Ralph C. Merkle, Michael Page, Jason K. Perry, Subhash Saini, O. A. Shenderova, Susan B. Sinnott, Deepak Srivastava, Stephen P. Walch, and Carter T. White.

107. Ralph C. Merkle, "A Proposed 'Metabolism' for a Hydrocarbon Assembler," *Nanotechnology* 8 (December 1997): 149–62, http://www.iop.org/EJ/abstract/0957-4484/8/4/001 or http://www.zyvex.com/nanotech/hydroCarbonMetabolism.html.

108. A useful bibliography of references: Robert A. Freitas Jr., "Technical Bibliography for Research on Positional Mechanosynthesis," Foresight Institute Web site, December 16, 2003, http://foresight.org/stage2/mechsynthbib.html; Wilson Ho and Hyojune Lee, "Single Bond Formation and Characterization with a Scanning Tunneling Microscope," *Science* 286.5445 (November 26, 1999): 1719–22, http://www.physics.uci.edu/~wilsonho/stm-iets.html; K. Eric Drexler, *Nanosystems,* chapter 8; Ralph Merkle, "Proposed 'Metabolism' for a Hydrocarbon Assembler"; Musgrave et al., "Theoretical Studies of a Hydrogen Abstraction Tool for Nanotechnology"; Michael Page and Donald W. Brenner, "Hydrogen Abstraction from a Diamond Surface: *Ab initio* Quantum Chemical Study with Constrained Isobutane as a Model," *Journal of the American Chemical Society* 113.9 (1991): 3270–74; D. W. Brenner et al., "Simulated Engineering of Nanostructures," *Nanotechnology* 7 (September 1996): 161–67, http://www.zyvex.com/nanotech/nano4/brennerPaper.pdf; S. P. Walch, W. A. Goddard III, and Ralph Merkle, "Theoretical Studies of Reactions on Diamond Surfaces," Fifth Foresight Conference on Molecular Nanotechnology, 1997, http://www.foresight.org/Conferences/MNT05/Abstracts/Walcabst.html; Stephen P. Walch and Ralph C. Merkle, "Theoretical Studies of Diamond Mechanosynthesis Reactions," *Nanotechnology* 9 (September 1998): 285–96; Fedor N. Dzegilenko, Deepak Srivastava, and Subhash Saini, "Simulations of Carbon Nanotube Tip Assisted Mechano-Chemical Reactions on a Diamond Surface," *Nanotechnology* 9 (December 1998): 325–30; J. W. Lyding et al., "UHV-STM Nanofabrication and Hydrogen/Deuterium Desorption from Silicon Surfaces: Implications for CMOS Technology," *Applied Surface Science* 132 (1998): 221, http://www.hersam-group.northwestern.edu/publications.html; E. T. Foley et al., "Cryogenic UHV-STM Study of Hydrogen and Deuterium Desorption from Silicon(100)," *Physical Review Letters* 80 (1998): 1336–39, http://prola.aps.org/abstract/PRL/v80/i6/p1336_1; M. C. Hersam, G. C. Abeln, and J. W. Lyding, "An Approach for Efficiently Locating and Electrically Contacting Nanostructures Fabricated via UHV-STM Lithography on Si(100)," *Microelectronic Engineering* 47 (1999): 235–37; L. J. Lauhon and W. Ho, "Inducing and Observing the Abstraction of a Single Hydrogen Atom in Bimolecular Reaction with a Scanning Tunneling Microscope," *Journal of Physical Chemistry* 105 (2000): 3987–92, http://www.physics.uci.edu/~wilsonho/stm-iets.html.

109. Eric Drexler, "Drexler Counters," first published on KurzweilAI.net on November 1, 2003: http://www.KurzweilAI.net/meme/frame.html?main=/articles/art0606. html. See also K. Eric Drexler, *Nanosystems: Molecular Machinery, Manufacturing, and Computation* (New York: Wiley Interscience, 1992), chapter 8; Ralph C. Merkle, "Foresight Debate with *Scientific American*" (1995), http://www.foresight.org/ SciAmDebate/SciAmResponse.html; Wilson Ho and Hyojune Lee, "Single Bond Formation and Characterization with a Scanning Tunneling Microscope," *Science* 286. 5445 (November 26, 1999): 1719–22, http://www.physics.uci.edu/~wilsonho/ stm-iets.html; K. Eric Drexler, David Forrest, Robert A. Freitas Jr., J. Storrs Hall, Neil Jacobstein, Tom McKendree, Ralph Merkle, and Christine Peterson, "On Physics, Fundamentals, and Nanorobots: A Rebuttal to Smalley's Assertion that Self-Replicating Mechanical Nanorobots Are Simply Not Possible: A Debate About Assemblers" (2001), http://www.imm.org/SciAmDebate2/smalley.html.

110. See http://pubs.acs.org/cen/coverstory/8148/8148counterpoint.html; http://www. kurzweilAI.net/meme/frame.html?main=/articles/art0604.html?.

111. D. Maysinger et al., "Block Copolymers Modify the Internalization of Micelle-Incorporated Probes into Neural Cells," *Biochimica et Biophysica Acta* 1539.3 (June 20, 2001): 205–17; R. Savic et al., "Micellar Nanocontainers Distribute to Defined Cytoplasmic Organelles," *Science* 300.5619 (April 25, 2003): 615–18.

112. T. Yamada et al., "Nanoparticles for the Delivery of Genes and Drugs to Human Hepatocytes," *Nature Biotechnology* 21.8 (August 2003): 885–90. Published electronically June 29, 2003. Abstract: http://www.nature.com/cgi-taf/DynaPage. taf?file=/nbt/journal/v21/n8/abs/nbt843.html. Short press release from *Nature*: http://www.nature.com/nbt/press_release/nbt0803.html.

113. Richards Grayson et al., "A BioMEMS Review: MEMS Technology for Physiologically Integrated Devices," *IEEE Proceedings* 92 (2004): 6–21; Richards Grayson et al., "Molecular Release from a Polymeric Microreservoir Device: Influence of Chemistry, Polymer Swelling, and Loading on Device Performance," *Journal of Biomedical Materials Research* 69A.3 (June 1, 2004): 502–12.

114. D. Patrick O'Neal et al., "Photo-thermal Tumor Ablation in Mice Using Near Infrared-Absorbing Nanoparticles," *Cancer Letters* 209.2 (June 25, 2004): 171–76.

115. International Energy Agency, from an R. E. Smalley presentation, "Nanotechnology, the S&T Workforce, Energy & Prosperity," p. 12, presented at PCAST (President's Council of Advisors on Science and Technology), Washington, D.C., March 3, 2003, http://www.ostp.gov/PCAST/PCAST%203-3-03%20R%20Smalley%20 Slides.pdf; also at http://cohesion.rice.edu/NaturalSciences/Smalley/emplibrary/ PCAST%20March%203,%202003.ppt.

116. Smalley, "Nanotechnology, the S&T Workforce, Energy & Prosperity."

117. "FutureGen—A Sequestration and Hydrogen Research Initiative," U.S. Department of Energy, Office of Fossil Energy, February 2003, http://www.fossil.energy. gov/programs/powersystems/futuregen/futuregen_factsheet.pdf.

118. Drexler, *Nanosystems*, pp. 428, 433.

119. Barnaby J. Feder, "Scientist at Work/Richard Smalley: Small Thoughts for a Global

Grid," *New York Times*, September 2, 2003; the following link requires subscription or purchase: http://query.nytimes.com/gst/abstract.html?res=F30C17FC3D5C0C 718CDDA00894DB404482.

120. International Energy Agency, from Smalley, "Nanotechnology, the S&T Workforce, Energy & Prosperity," p. 12.

121. American Council for the United Nations University, Millennium Project Global Challenge 13: http://www.acunu.org/millennium/ch-13.html.

122. "Wireless Transmission in Earth's Energy Future," Environment News Service, November 19, 2002, reporting on Jerome C. Glenn and Theodore J. Gordon in "2002 State of the Future," American Council for the United Nations University (August 2002).

123. Disclosure: the author is an adviser to and investor in this company.

124. "NEC Unveils Methanol-Fueled Laptop," Associated Press, June 30, 2003, http://www.siliconvalley.com/mld/siliconvalley/news/6203790.htm, reporting on NEC press release, "NEC Unveils Notebook PC with Built-In Fuel Cell," June 30, 2003, http://www.nec.co.jp/press/en/0306/3002.html.

125. Tony Smith, "Toshiba Boffins Prep Laptop Fuel Cell," *The Register*, March 5, 2003, http://www.theregister.co.uk/2003/03/05/toshiba_boffins_prep_laptop_fuel; Yoshiko Hara, "Toshiba Develops Matchbox-Sized Fuel Cell for Mobile Phones," *EE Times*, June 24, 2004, http://www.eet.com/article/showArticle.jhtml?article Id=22101804, reporting on Toshiba press release, "Toshiba Announces World's Smallest Direct Methanol Fuel Cell with Energy Output of 100 Milliwats," http://www.toshiba.com/taec/press/dmfc_04_222.shtml.

126. Karen Lurie, "Hydrogen Cars," *ScienceCentral News*, May 13, 2004, http://www. sciencentral.com/articles/view.php3?language=english&type=article&article_id= 218392247.

127. Louise Knapp, "Booze to Fuel Gadget Batteries," *Wired News*, April 2, 2003, http://www.wired.com/news/gizmos/0,1452,58119,00.html, and St. Louis University press release, "Powered by Your Liquor Cabinet, New Biofuel Cell Could Replace Rechargeable Batteries," March 24, 2003, http://www.slu.edu/readstory/ newsinfo/2474, reporting on Nick Akers and Shelley Minteer, "Towards the Development of a Membrane Electrode Assembly," presented at the American Chemical Society national meeting, Anaheim, Calif. (2003).

128. "Biofuel Cell Runs on Metabolic Energy to Power Medical Implants," *Nature Online*, November 12, 2002, http://www.nature.com/news/2002/021111/full/ 021111-1.html, reporting on N. Mano, F. Mao, and A. Heller, "A Miniature Biofuel Cell Operating in a Physiological Buffer," *Journal of the American Chemical Society* 124 (2002): 12962–63.

129. "Power from Blood Could Lead to 'Human Batteries,' " *FairfaxDigital*, August 4, 2003, http://www.smh.com.au/articles/2003/08/03/1059849278131.html?oneclick=true. Read more about the microbial fuel cells here: http://www.geobacter.org/research/ microbial/. Matsuhiko Nishizawa's BioMEMs laboratory diagrams a micro-biofuel cell: http://www.biomems.mech.tohoku.ac.jp/research_e.html. This short article

describes work on an implantable, nontoxic power source that now can produce 0.2 watts: http://www.iol.co.za/index.php?set_id=1&click_id=31&art_id=qw111596760 144B215.

130. Mike Martin, "Pace-Setting Nanotubes May Power Micro-Devices," *NewsFactor*, February 27, 2003, http://physics.iisc.ernet.in/~asood/Pace-Setting%20Nano tubes%20May%20Power%20Micro-Devices.htm.

131. "Finally, it is possible to derive a limit to the total planetary active nanorobot mass by considering the global energy balance. Total solar insolation received at the Earth's surface is ~1.75 × 10^{17} watts (I_{Earth} ~ 1370 W/m² ± 0.4% at normal incidence)." Robert A. Freitas Jr., *Nanomedicine*, vol. 1, *Basic Capabilities*, section 6.5.7, "Global Hypsithermal Limit" (Georgetown, Tex.: Landes Bioscience, 1999), pp. 175–76, http://www.nanomedicine.com/NMI/6.5.7.htm#p1.

132. This assumes 10 billion (10^{10}) persons, a power density for nanorobots of around 10^7 watts per cubic meter, a nanorobot size of one cubic micron, and a power draw of about 10 picowatts (10^{-11} watts) per nanorobot. The hypsithermal limit of 10^{16} watts implies about 10 kilograms of nanorobots per person, or 10^{16} nanorobots per person. Robert A. Freitas Jr., *Nanomedicine*, vol. 1, *Basic Capabilities*, section 6.5.7 "Global Hypsithermal Limit" (Georgetown, Tex.: Landes Bioscience, 1999), pp. 175–76, http://www.nanomedicine.com/NMI/6.5.7.htm#p4.

133. Alternatively, nanotechnology can be designed to be extremely energy efficient in the first place so that energy recapture would be unnecessary, and infeasible because there would be relatively little heat dissipation to recapture. In a private communication (January 2005), Robert A. Freitas Jr. writes: "Drexler (*Nanosystems*: 396) claims that energy dissipation may in theory be as low as E_{diss} ~ 0.1 MJ/kg 'if one assumes the development of a set of mechanochemical processes capable of transforming feedstock molecules into complex product structures using only reliable, nearly reversible steps.' 0.1 MJ/kg of diamond corresponds roughly to the minimum thermal noise at room temperature (e.g., kT ~ 4 zJ/atom at 298 K)."

134. Alexis De Vos, *Endoreversible Thermodynamics of Solar Energy Conversion* (London: Oxford University Press, 1992), p. 103.

135. R. D. Schaller and V. I. Klimov, "High Efficiency Carrier Multiplication in PbSe Nanocrystals: Implications for Solar Energy Conversion," *Physical Review Letters* 92.18 (May 7, 2004): 186601.

136. National Academies Press, Commission on Physical Sciences, Mathematics, and Applications, *Harnessing Light: Optical Science and Engineering for the 21st Century*, (Washington, D.C.: National Academy Press, 1998), p. 166, http://books.nap.edu/books/0309059917/html/166.html.

137. Matt Marshall, "World Events Spark Interest in Solar Cell Energy Start-ups," *Mercury News*, August 15, 2004, http://www.konarkatech.com/news_articles_082004/b-silicon_valley.php and http://www.nanosolar.com/cache/merc0815 04.htm.

138. John Gartner, "NASA Spaces on Energy Solution," *Wired News,* June 22, 2004, http://www.wired.com/news/technology/0,1282,63913,00.html. See also Arthur Smith, "The Case for Solar Power from Space," http://www.lispace.org/articles/SSPCase.html.

139. "The Space Elevator Primer," Spaceward Foundation, http://www.elevator2010.org/site/primer.html.

140. Kenneth Chang, "Experts Say New Desktop Fusion Claims Seem More Credible," *New York Times,* March 3, 2004, http://www.rpi.edu/web/News/nytlahey3.html, reporting on R. P. Taleyarkhan, "Additional Evidence of Nuclear Emissions During Acoustic Cavitation," *Physical Review E: Statistical, Nonlinear, and Soft Matter Physics* 69.3, pt. 2 (March 2004): 036109.

141. The original Pons and Fleischman method of desktop cold fusion using palladium electrodes is not dead. Ardent advocates have continued to pursue the technology, and the Department of Energy announced in 2004 that it was conducting a new formal review of the recent research in this field. Toni Feder, "DOE Warms to Cold Fusion," *Physics Today* (April 2004), http://www.physicstoday.org/vol-57/iss-4/p27.html.

142. Akira Fujishima, Tata N. Rao, and Donald A. Tryk, "Titanium Dioxide Photocatalysis," *Journal of Photochemistry and Photobiology C: Photochemistry Review* 1 (June 29, 2000): 1–21; Prashant V. Kamat, Rebecca Huehn, and Roxana Nicolaescu, "A 'Sense and Shoot' Approach for Photocatalytic Degradation of Organic Contaminants in Water," *Journal of Physical Chemistry B* 106 (January 31, 2002): 788–94.

143. A. G. Panov et al., "Photooxidation of Toluene and p-Xylene in Cation-Exchanged Zeolites X, Y, ZSM-5, and Beta: The Role of Zeolite Physicochemical Properties in Product Yield and Selectivity," *Journal of Physical Chemistry B* 104 (June 22, 2000): 5706–14.

144. Gabor A. Somorjai and Keith McCrea, "Roadmap for Catalysis Science in the 21st Century: A Personal View of Building the Future on Past and Present Accomplishments," *Applied Catalysis* A:General 222.1–2 (2001): 3–18, Lawrence Berkeley National Laboratory number 3.LBNL-48555, http://www.cchem.berkeley.edu/~gasgrp/2000.html (publication 877). See also Zhao, Lu, and Millar, "Advances in mesoporous molecular sieve MCM-41," *Industrial & Engineering Chemistry Research* 35 (1996): 2075–90, http://cheed.nus.edu.sg/~chezxs/Zhao/publication/1996_2075.pdf.

145. NTSC/NSET report, *National Nanotechnology Initiative: The Initiative and Its Implementation Plan,* July 2000, http://www.nano.gov/html/res/nni2.pdf.

146. Wei-xian Zhang, Chuan-Bao Wang, and Hsing-Lung Lien, "Treatment of Chlorinated Organic Contaminants with Nanoscale Bimetallic Particles," *Catalysis Today* 40 (May 14, 1988): 387–95.

147. R. Q. Long and R. T. Yang, "Carbon Nanotubes as Superior Sorbent for Dioxin Removal," *Journal of the American Chemical Society* 123.9 (2001): 2058–59.

148. Robert A. Freitas, Jr. "Death Is an Outrage!" presented at the Fifth Alcor Conference on Extreme Life Extension, Newport Beach, California, November 16, 2002, http://www.rfreitas.com/Nano/DeathIsAnOutrage.htm.

149. For example, the fifth annual BIOMEMS conference, June 2003, San Jose, http://www.knowledgepress.com/events/11201717.htm.

150. First two volumes of a planned four-volume series: Robert A. Freitas Jr., *Nanomedicine*, vol. I, *Basic Capabilities* (Georgetown, Tex.: Landes Bioscience, 1999); *Nanomedicine*, vol. IIA, *Biocompatibility* (Georgetown, Tex.: Landes Bioscience, 2003); http://www.nanomedicine.com.

151. Robert A. Freitas Jr., "Exploratory Design in Medical Nanotechnology: A Mechanical Artificial Red Cell," *Artificial Cells, Blood Substitutes, and Immobilization Biotechnology* 26 (1998): 411–30, http://www.foresight.org/Nanomedicine/Respirocytes.html.

152. Robert A. Freitas Jr., "Microbivores: Artificial Mechanical Phagocytes using Digest and Discharge Protocol," Zyvex preprint, March 2001, http://www.rfreitas.com/Nano/Microbivores.htm; Robert A. Freitas Jr., "Microbivores: Artificial Mechanical Phagocytes," *Foresight Update* no. 44, March 31, 2001, pp. 11–13, http://www.imm.org/Reports/Rep025.html; see also microbivore images at the Nanomedicine Art Gallery, http://www.foresight.org/Nanomedicine/Gallery/Species/Microbivores.html.

153. Robert A. Freitas Jr., *Nanomedicine*, vol. I, *Basic Capabilities*, section 9.4.2.5 "Nanomechanisms for Natation" (Georgetown, Tex.: Landes Bioscience, 1999), pp. 309–12, http://www.nanomedicine.com/NMI/9.4.2.5.htm.

154. George Whitesides, "Nanoinspiration: The Once and Future Nanomachine," *Scientific American* 285.3 (September 16, 2001): 78–83.

155. "According to Einstein's approximation for Brownian motion, after 1 second has elapsed at room temperature a fluidic water molecule has, on average, diffused a distance of ~50 microns (~400,000 molecular diameters) whereas a 1-micron nanorobot immersed in that same fluid has displaced by only ~0.7 microns (only ~0.7 device diameter) during the same time period. Thus Brownian motion is at most a minor source of navigational error for motile medical nanorobots." See K. Eric Drexler et al., "Many Future Nanomachines: A Rebuttal to Whitesides' Assertion That Mechanical Molecular Assemblers Are Not Workable and Not a Concern," a Debate about Assemblers, Institute for Molecular Manufacturing, 2001, http://www.imm.org/SciAmDebate2/whitesides.html.

156. Tejal A. Desai, "MEMS-Based Technologies for Cellular Encapsulation," *American Journal of Drug Delivery* 1.1 (2003): 3–11, abstract available at http://www.ingentaconnect.com/search/expand?pub=infobike://adis/add/2003/00000001/00000001/art00001.

157. As quoted by Douglas Hofstadter in *Gödel, Escher, Bach: An Eternal Golden Braid* (New York: Basic Books, 1979).

158. The author runs a company, FATKAT (Financial Accelerating Transactions by

Kurzweil Adaptive Technologies), which applies computerized pattern recognition to financial data to make stock-market investment decisions, http://www.FatKat.com.

159. See discussion in chapter 2 on price-performance improvements in computer memory and electronics in general.

160. Runaway AI refers to a scenario where, as Max More describes, "superintelligent *machines*, initially harnessed for *human* benefit, soon leave us behind." Max More, "Embrace, Don't Relinquish, the Future," http://www.KurzweilAI.net/articles/art0106.html?printable=1. See also Damien Broderick's description of the "Seed AI": "A self-improving seed AI could run glacially slowly on a limited machine substrate. The point is, so long as it has the capacity to improve itself, at some point it will do so convulsively, bursting through any architectural bottlenecks to design its own improved hardware, maybe even build it (if it's allowed control of tools in a fabrication plant)." Damien Broderick, "Tearing Toward the Spike," presented at "Australia at the Crossroads? Scenarios and Strategies for the Future" (April 31–May 2, 2000), published on KurzweilAI.net May 7, 2001, http://www.KurzweilAI.net/meme/frame.html?main=/articles/art0173.html.

161. David Talbot, "Lord of the Robots," *Technology Review* (April 2002).

162. Heather Havenstein writes that the "inflated notions spawned by science fiction writers about the convergence of humans and machines tarnished the image of AI in the 1980s because AI was perceived as failing to live up to its potential." Heather Havenstein, "Spring Comes to AI Winter: A Thousand Applications Bloom in Medicine, Customer Service, Education and Manufacturing," *Computerworld*, February 14, 2005, http://www.computerworld.com/softwaretopics/software/story/0,10801,99691,00.html. This tarnished image led to "AI Winter," defined as "a term coined by Richard Gabriel for the (circa 1990–94?) crash of the wave of enthusiasm for the AI language Lisp and AI itself, following a boom in the 1980s." Duane Rettig wrote: ". . . companies rode the great AI wave in the early 80's, when large corporations poured billions of dollars into the AI hype that promised thinking machines in 10 years. When the promises turned out to be harder than originally thought, the AI wave crashed, and Lisp crashed with it because of its association with AI. We refer to it as the AI Winter." Duane Rettig quoted in "AI Winter," http://c2.com/cgi/wiki?AiWinter.

163. The General Problem Solver (GPS) computer program, written in 1957, was able to solve problems through rules that allowed the GPS to divide a problem's goals into subgoals, and then check if obtaining a particular subgoal would bring the GPS closer to solving the overall goal. In the early 1960s Thomas Evan wrote ANALOGY, a "program [that] solves geometric-analogy problems of the form A:B::C:? taken from IQ tests and college entrance exams." Boicho Kokinov and Robert M. French, "Computational Models of Analogy-Making," in L. Nadel, ed., *Encyclopedia of Cognitive Science,* vol. 1 (London: Nature Publishing Group, 2003), pp. 113–18. See also A. Newell, J. C. Shaw, and H. A.

Simon, "Report on a General Problem-Solving Program," *Proceedings of the International Conference on Information Processing* (Paris: UNESCO House, 1959), pp. 256–64; Thomas Evans, "A Heuristic Program to Solve Geometric-Analogy Problems," in M. Minsky, ed., *Semantic Information Processing* (Cambridge, Mass.: MIT Press, 1968).

164. Sir Arthur Conan Doyle, "The Red-Headed League," 1890, available at http://www.eastoftheweb.com/short-stories/UBooks/RedHead.shtml.

165. V. Yu et al., "Antimicrobial Selection by a Computer: A Blinded Evaluation by Infectious Diseases Experts," *JAMA* 242.12 (1979): 1279–82.

166. Gary H. Anthes, "Computerizing Common Sense," *Computerworld,* April 8, 2002, http://www.computerworld.com/news/2002/story/0,11280,69881,00.html.

167. Kristen Philipkoski, "Now Here's a Really Big Idea," *Wired News,* November 25, 2002, http://www.wired.com/news/technology/0,1282,56374,00.html, reporting on Darryl Macer, "The Next Challenge Is to Map the Human Mind," *Nature* 420 (November 14, 2002): 121; see also a description of the project at http://www.biol.tsukuba.ac.jp/~macer/index.html.

168. Thomas Bayes, "An Essay Towards Solving a Problem in the Doctrine of Chances," published in 1763, two years after his death in 1761.

169. SpamBayes spam filter, http://spambayes.sourceforge.net.

170. Lawrence R. Rabiner, "A Tutorial on Hidden Markov Models and Selected Applications in Speech Recognition," *Proceedings of the IEEE* 77 (1989): 257–86. For a mathematical treatment of Markov models, see http://jedlik.phy.bme.hu/~gerjanos/HMM/node2.html.

171. Kurzweil Applied Intelligence (KAI), founded by the author in 1982, was sold in 1997 for $100 million and is now part of ScanSoft (formerly called Kurzweil Computer Products, the author's first company, which was sold to Xerox in 1980), now a public company. KAI introduced the first commercially marketed large-vocabulary speech-recognition system in 1987 (Kurzweil Voice Report, with a ten-thousand-word vocabulary).

172. Here is the basic schema for a neural net algorithm. Many variations are possible, and the designer of the system needs to provide certain critical parameters and methods, detailed below.

Creating a neural-net solution to a problem involves the following steps:
- Define the input.
- Define the topology of the neural net (i.e., the layers of neurons and the connections between the neurons).
- Train the neural net on examples of the problem.
- Run the trained neural net to solve new examples of the problem.
- Take your neural-net company public.

These steps (except for the last one) are detailed below:

The Problem Input

The problem input to the neural net consists of a series of numbers. This input can be:

- In a visual pattern-recognition system, a two-dimensional array of numbers representing the pixels of an image; or
- In an auditory (e.g., speech) recognition system, a two-dimensional array of numbers representing a sound, in which the first dimension represents parameters of the sound (e.g., frequency components) and the second dimension represents different points in time; or
- In an arbitrary pattern-recognition system, an n-dimensional array of numbers representing the input pattern.

Defining the Topology

To set up the neural net, the architecture of each neuron consists of:

- Multiple inputs in which each input is "connected" to either the output of another neuron, or one of the input numbers.
- Generally, a single output, which is connected either to the input of another neuron (which is usually in a higher layer), or to the final output.

Set Up the First Layer of Neurons

- Create N_0 neurons in the first layer. For each of these neurons, "connect" each of the multiple inputs of the neuron to "points" (i.e., numbers) in the problem input. These connections can be determined randomly or using an evolutionary algorithm (see below).
- Assign an initial "synaptic strength" to each connection created. These weights can start out all the same, can be assigned randomly, or can be determined in another way (see below).

Set Up the Additional Layers of Neurons

Set up a total of M layers of neurons. For each layer, set up the neurons in that layer.

For $layer_i$:

- Create N_i neurons in $layer_i$. For each of these neurons, "connect" each of the multiple inputs of the neuron to the outputs of the neurons in $layer_{i-1}$ (see variations below).
- Assign an initial "synaptic strength" to each connection created. These weights can start out all the same, can be assigned randomly, or can be determined in another way (see below).
- The outputs of the neurons in $layer_M$ are the outputs of the neural net (see variations below).

The Recognition Trials
How Each Neuron Works

Once the neuron is set up, it does the following for each recognition trial:
- Each weighted input to the neuron is computed by multiplying the output of the other neuron (or initial input) that the input to this neuron is connected to by the synaptic strength of that connection.
- All of these weighted inputs to the neuron are summed.
- If this sum is greater than the firing threshold of this neuron, then this neuron is considered to fire and its output is 1. Otherwise, its output is 0 (see variations below).

Do the Following for Each Recognition Trial

For each layer, from $layer_0$ to $layer_M$:

For each neuron in the layer:
- Sum its weighted inputs (each weighted input = the output of the other neuron [or initial input] that the input to this neuron is connected to multiplied by the synaptic strength of that connection).
- If this sum of weighted inputs is greater than the firing threshold for this neuron, set the output of this neuron = 1, otherwise set it to 0.

To Train the Neural Net
- Run repeated recognition trials on sample problems.
- After each trial, adjust the synaptic strengths of all the interneuronal connections to improve the performance of the neural net on this trial (see the discussion below on how to do this).
- Continue this training until the accuracy rate of the neural net is no longer improving (i.e., reaches an asymptote).

Key Design Decisions
In the simple schema above, the designer of this neural-net algorithm needs to determine at the outset:
- What the input numbers represent.
- The number of layers of neurons.
- The number of neurons in each layer. (Each layer does not necessarily need to have the same number of neurons.)
- The number of inputs to each neuron in each layer. The number of inputs (i.e., interneuronal connections) can also vary from neuron to neuron and from layer to layer.
- The actual "wiring" (i.e., the connections). For each neuron in each layer, this consists of a list of other neurons, the outputs of which constitute the

inputs to this neuron. This represents a key design area. There are a number of possible ways to do this:

(i)   Wire the neural net randomly; or
(ii) Use an evolutionary algorithm (see below) to determine an optimal wiring; or
(iii) Use the system designer's best judgment in determining the wiring.

• The initial synaptic strengths (i.e., weights) of each connection. There are a number of possible ways to do this:

(i)   Set the synaptic strengths to the same value; or
(ii) Set the synaptic strengths to different random values; or
(iii) Use an evolutionary algorithm to determine an optimal set of initial values; or
(iv) Use the system designer's best judgment in determining the initial values.

• The firing threshold of each neuron.
• The output. The output can be:

(i)   the outputs of layer$_M$ of neurons; or
(ii) the output of a single output neuron, the inputs of which are the outputs of the neurons in layer$_M$; or
(iii) a function of (e.g., a sum of) the outputs of the neurons in layer$_M$; or
(iv) another function of neuron outputs in multiple layers.

• How the synaptic strengths of all the connections are adjusted during the training of this neural net. This is a key design decision and is the subject of a great deal of research and discussion. There are a number of possible ways to do this:

(i)   For each recognition trial, increment or decrement each synaptic strength by a (generally small) fixed amount so that the neural net's output more closely matches the correct answer. One way to do this is to try both incrementing and decrementing and see which has the more desirable effect. This can be time-consuming, so other methods exist for making local decisions on whether to increment or decrement each synaptic strength.
(ii) Other statistical methods exist for modifying the synaptic strengths after each recognition trial so that the performance of the neural net on that trial more closely matches the correct answer.

      Note that neural-net training will work even if the answers to the training trials are not all correct. This allows using real-world training data that may have an inherent error rate. One key to the success of a neural net–based recognition system is the amount of data used for training. Usually a very substantial amount is needed to obtain satisfactory results. Just like human students, the amount of time that a neural net spends learning its lessons is a key factor in its performance.

Variations

Many variations of the above are feasible:

- There are different ways of determining the topology. In particular, the interneuronal wiring can be set either randomly or using an evolutionary algorithm.
- There are different ways of setting the initial synaptic strengths.
- The inputs to the neurons in $layer_i$ do not necessarily need to come from the outputs of the neurons in $layer_{i-1}$. Alternatively, the inputs to the neurons in each layer can come from any lower layer or any layer.
- There are different ways to determine the final output.
- The method described above results in an "all or nothing" (1 or 0) firing called a nonlinearity. There are other nonlinear functions that can be used. Commonly a function is used that goes from 0 to 1 in a rapid but more gradual fashion. Also, the outputs can be numbers other than 0 and 1.
- The different methods for adjusting the synaptic strengths during training represent key design decisions.

The above schema describes a "synchronous" neural net, in which each recognition trial proceeds by computing the outputs of each layer, starting with $layer_0$ through $layer_M$. In a true parallel system, in which each neuron is operating independently of the others, the neurons can operate "asynchronously" (that is, independently). In an asynchronous approach, each neuron is constantly scanning its inputs and fires whenever the sum of its weighted inputs exceeds its threshold (or whatever its output function specifies).

173. See chapter 4 for a detailed discussion of brain reverse engineering. As one example of the progression, S. J. Thorpe writes: "We have really only just begun what will certainly be a long term project aimed at reverse engineering the primate visual system. For the moment, we have only explored some very simple architectures, involving essentially just feed-forward architectures involving a relatively small numbers of layers. . . . In the years to come, we will strive to incorporate as many of the computational tricks used by the primate and human visual system as possible. More to the point, it seems that by adopting the spiking neuron approach, it will soon be possible to develop sophisticated systems capable of simulating very large neuronal networks in real time." S. J. Thorpe et al., "Reverse Engineering of the Visual System Using Networks of Spiking Neurons," *Proceedings of the IEEE 2000 International Symposium on Circuits and Systems* IV (IEEE Press), pp. 405–8, http://www.sccn.ucsd.edu/~arno/mypapers/thorpe.pdf.

174. T. Schoenauer et al. write: "Over the past years a huge diversity of hardware for artificial neural networks (ANN) has been designed. . . . Today one can choose from a wide range of neural network hardware. Designs differ in terms of architectural approaches, such as neurochips, accelerator boards and multi-board neurocomputers, as well as concerning the purpose of the system, such as the ANN algorithm(s) and the system's versatility. . . . Digital neurohardware can be classified by the:[sic] system architecture, degree of parallelism, typical neural network partition per processor, inter-processor communication network and numerical representation." T. Schoenauer, A. Jahnke, U. Roth, and H. Klar, "Digital Neurohardware: Principles and Perspectives," in *Proc. Neuronale Netze in der Anwendung*—Neural Networks in Applications NN'98, Magdeburg, invited paper (February 1998): 101–6, http://bwrc.eecs.berkeley.edu/People/kcamera/neural/papers/schoenauer98digital.pdf. See also Yihua Liao, "Neural Networks in Hardware: A Survey" (2001), http://ailab.das.ucdavis.edu/~yihua/research/NNhardware.pdf.

175. Here is the basic schema for a genetic (evolutionary) algorithm. Many variations are possible, and the designer of the system needs to provide certain critical parameters and methods, detailed below.

THE EVOLUTIONARY ALGORITHM
Create N solution "creatures." Each one has:

- A genetic code: a sequence of numbers that characterize a possible solution to the problem. The numbers can represent critical parameters, steps to a solution, rules, etc.

For each generation of evolution, do the following:

- Do the following for each of the N solution creatures:

    (i) Apply this solution creature's solution (as represented by its genetic code) to the problem, or simulated environment.
    (ii) Rate the solution.

- Pick the L solution creatures with the highest ratings to survive into the next generation.
- Eliminate the (N − L) nonsurviving solution creatures.
- Create (N − L) new solution creatures from the L surviving solution creatures by:

    (i) Making copies of the L surviving creatures. Introduce small random variations into each copy; or
    (ii) Creating additional solution creatures by combining parts of the genetic code (using "sexual" reproduction, or otherwise combining portions of the chromosomes) from the L surviving creatures; or
    (iii) Doing a combination of (i) and (ii).

- Determine whether or not to continue evolving:

Improvement = (highest rating in this generation) – (highest rating in the previous generation).

If Improvement < Improvement Threshold, then we're done.

- The solution creature with the highest rating from the last generation of evolution has the best solution. Apply the solution defined by its genetic code to the problem.

Key Design Decisions

In the simple schema above, the designer needs to determine at the outset:

- Key parameters:
  N
  L
  Improvement threshold
- What the numbers in the genetic code represent and how the solution is computed from the genetic code.
- A method for determining the N solution creatures in the first generation. In general, these need only be "reasonable" attempts at a solution. If these first-generation solutions are too far afield, the evolutionary algorithm may have difficulty converging on a good solution. It is often worthwhile to create the initial solution creatures in such a way that they are reasonably diverse. This will help prevent the evolutionary process from just finding a "locally" optimal solution.
- How the solutions are rated.
- How the surviving solution creatures reproduce.

Variations

Many variations of the above are feasible. For example:

- There does not need to be a fixed number of surviving solution creatures (L) from each generation. The survival rule(s) can allow for a variable number of survivors.
- There does not need to be a fixed number of new solution creatures created in each generation (N − L). The procreation rules can be independent of the size of the population. Procreation can be related to survival, thereby allowing the fittest solution creatures to procreate the most.
- The decision as to whether or not to continue evolving can be varied. It can consider more than just the highest-rated solution creature from the most recent generation(s). It can also consider a trend that goes beyond just the last two generations.

176. Sam Williams, "When Machines Breed," August 12, 2004, http://www.salon.com/tech/feature/2004/08/12/evolvable_hardware/index_np.html.
177. Here is the basic scheme (algorithm description) of recursive search. Many varia-

tions are possible, and the designer of the system needs to provide certain critical parameters and methods, detailed below.

THE RECURSIVE ALGORITHM

Define a function (program) "Pick Best Next Step." The function returns a value of "SUCCESS" (we've solved the problem) or "FAILURE" (we didn't solve it). If it returns with a value of SUCCESS, then the function also returns the sequence of steps that solved the problem.

PICK BEST NEXT STEP does the following:

- Determine if the program can escape from continued recursion at this point. This bullet, and the next two bullets deal with this escape decision.

    First, determine if the problem has now been solved. Since this call to Pick Best Next Step probably came from the program calling itself, we may now have a satisfactory solution. Examples are:

    (i) In the context of a game (for example, chess), the last move allows us to win (such as checkmate).

    (ii) In the context of solving a mathematical theorem, the last step proves the theorem.

    (iii) In the context of an artistic program (for example, a computer poet or composer), the last step matches the goals for the next word or note.

If the problem has been satisfactorily solved, the program returns with a value of "SUCCESS" and the sequence of steps that caused the success.

- If the problem has not been solved, determine if a solution is now hopeless. Examples are:

    (i) In the context of a game (such as chess), this move causes us to lose (checkmate for the other side).

    (ii) In the context of solving a mathematical theorem, this step violates the theorem.

    (iii) In the context of an artistic creation, this step violates the goals for the next word or note.

If the solution at this point has been deemed hopeless, the program returns with a value of "FAILURE."

- If the problem has been neither solved nor deemed hopeless at this point of recursive expansion, determine whether or not the expansion should be abandoned anyway. This is a key aspect of the design and takes into consideration the limited amount of computer time we have to spend. Examples are:

(i) In the context of a game (such as chess), this move puts our side sufficiently "ahead" or "behind." Making this determination may not be straightforward and is the primary design decision. However, simple approaches (such as adding up piece values) can still provide good results. If the program determines that our side is sufficiently ahead, then Pick Best Next Step returns in a similar manner to a determination that our side has won (that is, with a value of "SUCCESS"). If the program determines that our side is sufficiently behind, then Pick Best Next Step returns in a similar manner to a determination that our side has lost (that is, with a value of "FAILURE").

(ii) In the context of solving a mathematical theorem, this step involves determining if the sequence of steps in the proof is unlikely to yield a proof. If so, then this path should be abandoned, and Pick Best Next Step returns in a similar manner to a determination that this step violates the theorem (that is, with a value of "FAILURE"). There is no "soft" equivalent of success. We can't return with a value of "SUCCESS" until we have actually solved the problem. That's the nature of math.

(iii) In the context of an artistic program (such as a computer poet or composer), this step involves determining if the sequence of steps (such as the words in a poem, notes in a song) is unlikely to satisfy the goals for the next step. If so, then this path should be abandoned, and Pick Best Next Step returns in a similar manner to a determination that this step violates the goals for the next step (that is, with a value of "FAILURE").

• If Pick Best Next Step has not returned (because the program has neither determined success nor failure nor made a determination that this path should be abandoned at this point), then we have not escaped from continued recursive expansion. In this case, we now generate a list of all possible next steps at this point. This is where the precise statement of the problem comes in:

(i) In the context of a game (such as chess), this involves generating all possible moves for "our" side for the current state of the board. This involves a straightforward codification of the rules of the game.

(ii) In the context of finding a proof for a mathematical theorem, this involves listing the possible axioms or previously proved theorems that can be applied at this point in the solution.

(iii) In the context of a cybernetic art program, this involves listing the possible words/notes/line segments that could be used at this point.

For each such possible next step:

(i) Create the hypothetical situation that would exist if this step were implemented. In a game, this means the hypothetical state of the board. In a mathematical proof, this means adding this step (for example, axiom) to the proof. In an art program, this means adding this word/note/line segment.

(ii) Now call Pick Best Next Step to examine this hypothetical situation. This is, of course, where the recursion comes in because the program is now calling itself.

(iii) If the above call to Pick Best Next Step returns with a value of "SUC-CESS," then return from the call to Pick Best Next Step (that we are now in) also with a value of "SUCCESS." Otherwise consider the next possible step.

If all the possible next steps have been considered without finding a step that resulted in a return from the call to Pick Best Next Step with a value of "SUC-CESS," then return from this call to Pick Best Next Step (that we are now in) with a value of "FAILURE."

End of PICK BEST NEXT STEP
If the original call to Pick Best Next Move returns with a value of "SUCCESS," it will also return the correct sequence of steps:

(i) In the context of a game, the first step in this sequence is the next move you should make.

(ii) In the context of a mathematical proof, the full sequence of steps is the proof.

(iii) In the context of a cybernetic art program, the sequence of steps is your work of art.

If the original call to Pick Best Next Step returns with a value of "FAILURE," then you need to go back to the drawing board.

Key Design Decisions
In the simple schema above, the designer of the recursive algorithm needs to determine the following at the outset:

• The key to a recursive algorithm is the determination in Pick Best Next Step of when to abandon the recursive expansion. This is easy when the program has achieved clear success (such as checkmate in chess or the requisite solution in a math or combinatorial problem) or clear failure. It is more difficult when a clear win or loss has not yet been achieved. Abandoning a line of inquiry before a well-defined outcome is necessary because otherwise the program might run for billions of years (or at least until the warranty on your computer runs out).

• The other primary requirement for the recursive algorithm is a straightforward codification of the problem. In a game like chess, that's easy. But in other situations, a clear definition of the problem is not always so easy to come by.

178. See Kurzweil CyberArt, http://www.KurzweilCyberArt.com, for further description of Ray Kurzweil's Cybernetic Poet and to download a free copy of the program. See U.S. Patent No. 6,647,395, "Poet Personalities," inventors: Ray Kurzweil and John Keklak. Abstract: "A method of generating a poet personality including reading poems, each of the poems containing text, generating analysis models, each of the analysis models representing one of the poems and storing the analysis models in a personality data structure. The personality data structure further includes weights, each of the weights associated with each of the analysis models. The weights include integer values."

179. Ben Goertzel: *The Structure of Intelligence* (New York: Springer-Verlag, 1993); *The Evolving Mind* (Gordon and Breach, 1993); *Chaotic Logic* (Plenum, 1994); *From Complexity to Creativity* (Plenum, 1997). For a link to Ben Goertzel's books and essays, see http://www.goertzel.org/work.html.

180. KurzweilAI.net (http://www.KurzweilAI.net) provides hundreds of articles by one hundred "big thinkers" and other features on "accelerating intelligence." The site offers a free daily or weekly newsletter on the latest developments in the areas covered by this book. To subscribe, enter your e-mail address (which is maintained in strict confidence and is not shared with anyone) on the home page.

181. John Gosney, Business Communications Company, "Artificial Intelligence: Burgeoning Applications in Industry," June 2003, http://www.bccresearch.com/comm/G275.html.

182. Kathleen Melymuka, "Good Morning, Dave . . . ," *Computerworld*, November 11, 2002, http://www.computerworld.com/industrytopics/defense/story/0,10801,75728,00.html.

183. JTRS Technology Awareness Bulletin, August 2004, http://jtrs.army.mil/sections/technicalinformation/fset_technical.html?tech_aware_2004-8.

184. Otis Port, Michael Arndt, and John Carey, "Smart Tools," Spring 2003, http://www.businessweek.com/bw50/content/mar2003/a3826072.htm.

185. Wade Roush, "Immobots Take Control: From Photocopiers to Space Probes, Machines Injected with Robotic Self-Awareness Are Reliable Problem Solvers," *Technology Review* (December 2002–January 2003), http://www.occm.de/roush1202.pdf.

186. Jason Lohn quoted in NASA news release "NASA 'Evolutionary' Software Automatically Designs Antenna," http://www.nasa.gov/lb/centers/ames/news/releases/2004/04_55AR.html.

187. Robert Roy Britt, "Automatic Astronomy: New Robotic Telescopes See and Think," June 4, 2003, http://www.space.com/businesstechnology/technology/automated_astronomy_030604.html.

188. H. Keith Melton, "Spies in the Digital Age," http://www.cnn.com/SPECIALS/cold.war/experience/spies/melton.essay.

189. "United Therapeutics (UT) is a biotechnology company focused on developing

chronic therapies for life-threatening conditions in three therapeutic areas: cardiovascular, oncology and infectious diseases" (http://www.unither.com). Kurzweil Technologies is working with UT to develop pattern recognition–based analysis from either "Holter" monitoring (twenty-four-hour recordings) or "Event" monitoring (thirty days or more).

190. Kristen Philipkoski, "A Map That Maps Gene Functions," *Wired News,* May 28, 2002, http://www.wired.com/news/medtech/0,1286,52723,00.html.

191. Jennifer Ouellette, "Bioinformatics Moves into the Mainstream," *The Industrial Physicist* (October–November 2003), http://www.sciencemasters.com/bioinformatics.pdf.

192. Port, Arndt, and Carey, "Smart Tools."

193. "Protein Patterns in Blood May Predict Prostate Cancer Diagnosis," National Cancer Institute, October 15, 2002, http://www.nci.nih.gov/newscenter/Prostate Proteomics, reporting on Emanuel F. Petricoin et al., "Serum Proteomic Patterns for Detection of Prostate Cancer," *Journal of the National Cancer Institute* 94 (2002): 1576–78.

194. Charlene Laino, "New Blood Test Spots Cancer," December 13, 2002, http://my.webmd.com/content/Article/56/65831.htm; Emanuel F. Petricoin III et al., "Use of Proteomic Patterns in Serum to Identify Ovarian Cancer," *Lancet* 359.9306 (February 16, 2002): 572–77.

195. For information of TriPath's FocalPoint, see "Make a Diagnosis," *Wired,* October 2003, http://www.wired.com/wired/archive/10.03/everywhere.html?pg=5. Mark Hagland, "Doctors' Orders," January 2003, http://www.healthcare-informatics.com/issues/2003/01_03/cpoe.htm.

196. Ross D. King et al., "Functional Genomic Hypothesis Generation and Experimentation by a Robot Scientist," *Nature* 427 (January 15, 2004): 247–52.

197. Port, Arndt, and Carey, "Smart Tools."

198. "Future Route Releases AI-Based Fraud Detection Product," August 18, 2004, http://www.finextra.com/fullstory.asp?id=12365.

199. John Hackett, "Computers Are Learning the Business," *Collections World,* April 24, 2001, http://www.creditcollectionsworld.com/news/042401_2.htm.

200. "Innovative Use of Artificial Intelligence, Monitoring NASDAQ for Potential Insider Trading and Fraud," AAAI press release, July 30, 2003, http://www.aaai.org/Pressroom/Releases/release-03-0730.html.

201. "Adaptive Learning, Fly the Brainy Skies," *Wired News,* March 2002, http://www.wired.com/wired/archive/10.03/everywhere.html?pg=2.

202. "Introduction to Artificial Intelligence," EL 629, Maxwell Air Force Base, Air University Library course, http://www.au.af.mil/au/aul/school/acsc/ai02.htm. Sam Williams, "Computer, Heal Thyself," *Salon.com,* July 12, 2004, http://www.salon.com/tech/feature/2004/07/12/self_healing_computing/index_np.html.

203. See http://www.Seegrid.com. Disclosure: The author is an investor in Seegrid and a member of its board of directors.

204. No Hands Across America Web site, http://cart.frc.ri.cmu.edu/users/hpm/

project.archive/reference.file/nhaa.html, and "Carnegie Mellon Researchers Will Prove Autonomous Driving Technologies During a 3,000 Mile, Hands-off-the-Wheel Trip from Pittsburgh to San Diego," Carnegie Mellon press release, http://www-2.cs.cmu.edu/afs/cs/user/tjochem/www/nhaa/official_press_release.html; Robert J. Derocher, "Almost Human," September 2001, http://www.insight-mag.com/insight/01/09/col-2-pt-1-ClickCulture.htm.

205. "Search and Rescue Robots," Associated Press, September 3, 2004, http://www.smh.com.au/articles/2004/09/02/1093939058792.html?oneclick=true.

206. "From Factoids to Facts," *Economist,* August 26, 2004, http://www.economist.com/science/displayStory.cfm?story_id=3127462.

207. Joe McCool, "Voice Recognition, It Pays to Talk," May 2003, http://www.bcs.org/BCS/Products/Publications/JournalsAndMagazines/ComputerBulletin/OnlineArchive/may03/voicerecognition.htm.

208. John Gartner, "Finally a Car That Talks Back," *Wired News,* September 2, 2004, http://www.wired.com/news/autotech/0,2554,64809,00.html?tw=wn_14techhead.

209. "Computer Language Translation System Romances the Rosetta Stone," Information Sciences Institute, USC School of Engineering (July 24, 2003), http://www.usc.edu/isinews/stories/102.html.

210. Torsten Reil quoted in Steven Johnson, "Darwin in a Box," *Discover* 24.8 (August 2003), http://www.discover.com/issues/aug-03/departments/feattech/.

211. "Let Software Catch the Game for You," July 3, 2004, http://www.newscientist.com/news/news.jsp?id=ns99996097.

212. Michelle Delio, "Breeding Race Cars to Win," *Wired News,* June 18, 2004, http://www.wired.com/news/autotech/0,2554,63900,00.html.

213. Marvin Minsky, *The Society of Mind* (New York: Simon & Schuster, 1988).

214. Hans Moravec, "When Will Computer Hardware Match the Human Brain?" *Journal of Evolution and Technology* 1 (1998).

215. Ray Kurzweil, *The Age of Spiritual Machines* (New York: Viking, 1999), p. 156.

216. See chapter 2, notes 22 and 23, on the International Technology Roadmap for Semiconductors.

217. "The First Turing Test," http://www.loebner.net/Prizef/loebner-prize.html.

218. Douglas R. Hofstadter, "A Coffeehouse Conversation on the Turing Test," May 1981, included in Ray Kurzweil, *The Age of Intelligent Machines* (Cambridge, Mass.: MIT Press, 1990), pp. 80–102, http://www.KurzweilAI.net/meme/frame.html?main=/articles/art0318.html.

219. Ray Kurzweil, "Why I Think I Will Win," and Mitch Kapor, "Why I Think I Will Win," rules: http://www.KurzweilAI.net/meme/frame.html?main=/articles/art0373.html; Kapor: http://www.KurzweilAI.net/meme/frame.html?main=/articles/art0412.html; Kurzweil: http://www.KurzweilAI.net/meme/frame.html?main=/articles/art0374.html; Kurzweil "final word": http://www.KurzweilAI.net/meme/frame.html?main=/articles/art0413.html.

220. Edward A. Feigenbaum, "Some Challenges and Grand Challenges for Computa-

tional Intelligence," *Journal of the Association for Computing Machinery* 50 (January 2003): 32–40.

221. According to the serial endosymbiosis theory of eukaryotic evolution, the ancestors of mitochondria (the structures in cells that produce energy and have their own genetic code comprising thirteen genes in humans) were originally independent bacteria (that is, not part of another cell) similar to the *Daptobacter* bacteria of today. "Serial Endosymbiosis Theory," http://encyclopedia.thefree dictionary.com/Serial%20endosymbiosis%20theory.

Chapter Six: The Impact . . .

1. Donovan, "Season of the Witch," *Sunshine Superman* (1966).
2. Reasons for the reduction in farm workforce include the mechanization that lessened the need for animal and human labor, the economic opportunities that were created in urban areas during World War II, and the development of intensive farming techniques that required less land for comparable yields. U.S. Department of Agriculture, National Agricultural Statistics Service, Trends in U.S. Agriculture, http://www.usda.gov/nass/pubs/trends/farmpopulation.htm. Computer-assisted production, just-in-time production (which results in lower inventory), and offshoring manufacturing to reduce costs are some of the methods that have contributed to the loss of factory jobs. See U.S. Department of Labor, *Futurework: Trends and Challenges of Work in the 21st Century*, http://www.dol.gov/asp/programs/history/herman/reports/futurework/report.htm.
3. For example, see Natasha Vita-More, "The New [Human] Genre Primo [First] Posthuman," paper delivered at Ciber@RT Conference, Bilbao, Spain, April 2004, http://www.natasha.cc/paper.htm.
4. Rashid Bashir summarizes in 2004:

> Much progress has also been made in therapeutic micro- and nanotechnology. . . . Some specific examples include (i) silicon-based implantable devices that can be electrically actuated to open an orifice from which preloaded drugs can be released, (ii) silicon devices functionalized with electrically actuated polymers which can act as a valve or muscle to release preloaded drugs, (iii) silicon-based micro-capsules with nano-porous membranes for the release of insulin, (iv) all polymer (or hydrogel) particles which can be preloaded with drugs and then forced to expand upon exposure to specific environmental conditions such as change in pH and release the loaded drug, (v) metal nano-particles coated with recognition proteins, where the particles can be heated with external optical energy and can locally heat and damage unwanted cells and tissue, etc.

R. Bashir, "BioMEMS: State-of-the-Art in Detection, Opportunities and Prospects," *Advanced Drug Delivery Reviews* 56.11 (September 22, 2004): 1565–86.

Reprint available at https://engineering.purdue.edu/LIBNA/pdf/publications/ BioMEMS%20review%20ADDR%20final.pdf. See also Richard Grayson et al., "A BioMEMS Review: MEMS Technology for Physiologically Integrated Devices," *IEEE Proceedings* 92 (2004): 6–21.

5. For activities of the International Society for BioMEMS and Biomedical Nano-technology, see http://www.bme.ohio-state.edu/isb. BioMEMS conferences are also listed on the SPIE Web site, http://www.spie.org/Conferences.

6. Researchers used a gold nanoparticle to monitor blood sugar in diabetics. Y. Xiao et al., " 'Plugging into Enzymes': Nanowiring of Redox Enzymes by a Gold Nanoparticle," *Science* 299.5614 (March 21, 2003): 1877–81. Also see T. A. Desai et al., "Abstract Nanoporous Microsystems for Islet Cell Replacement," *Advanced Drug Delivery Reviews* 56.11 (September 22, 2004): 1661–73.

7. A. Grayson, et al., "Multi-pulse Drug Delivery from a Resorbable Polymeric Microchip Device," *Nature Materials* 2 (2003): 767–72.

8. Q. Bai and K. D. Wise, "Single-Unit Neural Recording with Active Microelectrode Arrays," *IEEE Transactions on Biomedical Engineering* 48.8 (August 2001): 911–20. See the discussion of Wise's work in J. DeGaspari, "Tiny, Tuned, and Unattached," *Mechanical Engineering* (July 2001), http://www.memagazine.org/backissues/ july01/features/tinytune/tinytune.html; K. D. Wise, "The Coming Revolution in Wireless Integrated MicroSystems," Digest International Sensor Conference 2001 (Invited Plenary), Seoul, October 2001. Online version (January 13, 2004): http://www.stanford.edu/class/ee392s/Stanford392S-kw.pdf.

9. " 'Microbots' Hunt Down Disease," BBC News, June 13, 2001, http://news. bbc.co.uk/1/hi/health/1386440.stm. The micromachines are based on cylindrical magnets; see K. Ishiyama, M. Sendoh, and K. I. Arai, "Magnetic Micromachines for Medical Applications," *Journal of Magnetism and Magnetic Materials* 242–45, part 1 (April 2002): 41–46.

10. See Sandia National Laboratories press release, "Pac-Man-Like Microstructure Interacts with Red Blood Cells," August 15, 2001, http://www.sandia.gov/media/ NewsRel/NR2001/gobbler.htm. For an industry trade article in response, see D. Wilson, "Microteeth Have a Big Bite," August 17, 2001, http://www.e4engineering. com/item.asp?ch=e4_home&type=Features&id=42543.

11. See Freitas's books *Nanomedicine,* vol. 1, *Basic Capabilities* (Georgetown, Tex.: Landes Bioscience, 1999), and *Nanomedicine,* vol. 2A, *Biocompatibility* (George-town, Tex.: Landes Bioscience, 2003), both freely available online at http://www. nanomedicine.com. Also see the Foresight Institute's "Nanomedicine" page by Robert Freitas, which lists his current technical works (http://www.foresight.org/ Nanomedicine/index.html#MedNanoBots).

12. Robert A. Freitas Jr., "Exploratory Design in Medical Nanotechnology: A Mechani-cal Artificial Red Cell," *Artificial Cells, Blood Substitutes, and Immobilization Biotech-nology* 26 (1998): 411–30, http://www.foresight.org/Nanomedicine/Respirocytes. html.

13. Robert A. Freitas Jr., "Clottocytes: Artificial Mechanical Platelets," *Foresight Update* no. 41, June 30, 2000, pp. 9–11, http://www.imm.org/Reports/Rep018.html.

14. Robert A. Freitas Jr., "Microbivores: Artificial Mechanical Phagocytes," *Foresight Update* no. 44, March 31, 2001, pp. 11–13, http://www.imm.org/Reports/Rep025.html or http://www.KurzweilAI.net/meme/frame.html?main=/articles/art0453.html.

15. Robert A. Freitas Jr., "The Vasculoid Personal Appliance," *Foresight Update* no. 48, March 31, 2002, pp. 10–12, http://www.imm.org/Reports/Rep031.html; full paper: Robert A. Freitas Jr. and Christopher J. Phoenix, "Vasculoid: A Personal Nanomedical Appliance to Replace Human Blood," *Journal of Evolution and Technology* 11 (April 2002), http://www.jetpress.org/volume11/vasculoid.html.

16. Carlo Montemagno and George Bachand, "Constructing Nanomechanical Devices Powered by Biomolecular Motors," *Nanotechnology* 10 (September 1999): 225–31; "Biofuel Cell Runs on Metabolic Energy to Power Medical Implants," *Nature* online, Nov. 12, 2002, http://www.nature.com/news/2002/021111/full/021111–1.html, reporting on N. Mano, F. Mao, and A. Heller, "A Miniature Biofuel Cell Operating in a Physiological Buffer," *Journal of the American Chemical Society* 124 (2002): 12962–63; Carlo Montemagno et al., "Self-Assembled Microdevices Driven by Muscle," *Nature Materials* 4.2 (February 2005): 180–84, published electronically (January 16, 2005).

17. See the Lawrence Livermore National Laboratory Web site (http://www.llnl.gov) for updated information about this initiative, along with the Medtronic MiniMed Web site, http://www.minimed.com/corpinfo/index.shtml.

18. "Direct brain-to-brain communication . . . seem[s] more like the stuff of Hollywood movies than of government reports—but these are among the advances forecast in a recent report by the U.S. National Science Foundation and Department of Commerce." G. Brumfiel, "Futurists Predict Body Swaps for Planet Hops," *Nature* 418 (July 25, 2002): 359.

   Deep brain stimulation, by which electric current from implanted electrodes influences brain function, is an FDA-approved neural implant for Parkinson's disease and is being tested for other neurological disorders. See Al Abbott, "Brain Implants Show Promise Against Obsessive Disorder," *Nature* 419 (October 17, 2002): 658, and B. Nuttin et al., "Electrical Stimulation in Anterior Limbs of Internal Capsules in Patients with Obsessive-Compulsive Disorder," *Lancet* 354.9189 (October 30, 1999): 1526.

19. See the Retinal Implant Project Web site (http://www.bostonretinalimplant.org), which contains a range of resources including recent papers. One such paper is: R. J. Jensen et al., "Thresholds for Activation of Rabbit Retinal Ganglion Cells with an Ultrafine, Extracellular Microelectrode," *Investigative Ophthalmology and Visual Science* 44.8 (August 2003): 3533–43.

20. The FDA approved the Medtronic implant for this purpose in 1997 for only one side of the brain; it was approved for both sides of the brain on January 14, 2002.

S. Snider, "FDA Approves Expanded Use of Brain Implant for Parkinson's Disease," U.S. Food and Drug Administration, *FDA Talk Paper*, January 14, 2002, http://www.fda.gov/bbs/topics/ANSWERS/2002/ANS01130.html. The most recent versions provide for software upgrades from outside the patient.

21. Medtronic also makes an implant for cerebral palsy. See S. Hart, "Brain Implant Quells Tremors," ABC News, December 23, 1997, http://nasw.org/users/hart/sub html/abcnews.html. Also see the Medtronic Web site, http://www.medtronic.com.

22. Günther Zeck and Peter Fromherz, "Noninvasive Neuroelectronic Interfacing with Synaptically Connected Snail Neurons Immobilized on a Semiconductor Chip," *Proceedings of the National Academy of Sciences* 98.18 (August 28, 2001): 10457–62.

23. See R. Colin Johnson, "Scientists Activate Neurons with Quantum Dots," *EE Times*, December 4, 2001, http://www.eetimes.com/story/OEG20011204S0068. Quantum dots can also be used for imaging; see M. Dahan et al., "Diffusion Dynamics of Glycine Receptors Revealed by Single-Quantum Dot Tracking," *Science* 302.5644 (October 17, 2003): 442–45; J. K. Jaiswal and S. M. Simon, "Potentials and Pitfalls of Fluorescent Quantum Dots for Biological Imaging," *Trends in Cell Biology* 14.9 (September 2004): 497–504.

24. S. Shoham et al., "Motor-Cortical Activity in Tetraplegics," *Nature* 413.6858 (October 25, 2001): 793. For the University of Utah news release, see "An Early Step Toward Helping the Paralyzed Walk," October 24, 2001, http://www.utah.edu/news/releases/01/oct/spinal.html.

25. Stephen Hawking's remarks, which were mistranslated by *Focus*, were quoted in Nick Paton Walsh, "Alter Our DNA or Robots Will Take Over, Warns Hawking," *Observer*, September 2, 2001, http://observer.guardian.co.uk/uk_news/story/0,6903,545653,00.html. The widely reported mistranslation implied that Hawking was warning against developing smarter-than-human machine intelligence. In fact, he was advocating that we hasten to close links between biological and nonbiological intelligence. Hawking provided the exact quotes to KurzweilAI.net ("Hawking Misquoted on Computers Taking Over," September 13, 2001, http://www.KurzweilAI.net/news/frame.html?main=news_single.html?id%3D495).

26. See note 34 in chapter 1.

27. One example, Nomad for Military Applications, has been produced by Microvision, a company based in Bothell, Washington. See http://www.microvision.com/nomadmilitary/index.html.

28. Olga Kharif, "Your Lapel Is Ringing," *Business Week*, June 21, 2004.

29. Laila Weir, "High-Tech Hearing Bypasses Ears," *Wired News*, September 16, 2004, http://www.wired.com/news/technology/0,1282,64963,00.html?tw=wn_tophead_4.

30. Hypersonic Sound technology, http://www.atcsd.com/tl_hss.html; Audio Spotlight, http://www.holosonics.com/technology.html.

31. Phillip F. Schewe and Ben Stein, *American Institute of Physics Bulletin of Physics News* 236 (August 7, 1995), http://www.aip.org/enews/physnews/1995/physnews.

236.htm. See also R. Weis and P. Fromherz, "Frequency Dependent Signal-Transfer in Neuron-Transistors," *Physical Review E* 55 (1997): 877–89.

32. See note 18 above. Also see J. O. Winter et al., "Recognition Molecule Directed Interfacing Between Semiconductor Quantum Dots and Nerve Cells," *Advanced Materials* 13 (November 2001): 1673–77; I. Willner and B. Willner, "Biomaterials Integrated with Electronic Elements: En Route to Bioelectronics," *Trends in Biotechnology* 19 (June 2001): 222–30; Deborah A. Fitzgerald, "Bridging the Gap with Bioelectronics," *Scientist* 16.6 (March 18, 2002): 38.

33. Robert Freitas provides an analysis of this scenario: Robert A. Freitas Jr., *Nanomedicine*, vol. 1, *Basic Capabilities*, section 7.4.5.4, "Cell Message Modification" (Georgetown, Tex.: Landes Bioscience, 1999), pp. 194–96, http://www. nanomedicine.com/NMI/7.4.5.4.htm#p5, and section 7.4.5.6, "Outmessaging to Neurons," pp. 196–97, http://www.nanomedicine.com/NMI/7.4.5.6.htm#p2.

34. For descriptions of the Ramona project, including videos of the virtual-reality presentation at the TED conference and a behind-the-scenes "Making of Ramona" video, see "All About Ramona," http://www.KurzweilAI.net/meme/ frame.html?m=9.

35. I. Fried et al., "Electric Current Stimulates Laughter," *Nature* 391.6668 (February 12, 1998): 650. See Ray Kurzweil, *The Age of Spiritual Machines* (New York: Viking, 1999).

36. Robert A. Freitas Jr., *Nanomedicine*, vol. 1, *Basic Capabilities*, section 7.3, "Communication Networks" (Georgetown, Tex.: Landes Bioscience, 1999), pp. 186–88, http://www.nanomedicine.com/NMI/7.3.htm.

37. Allen Kurzweil, *The Grand Complication: A Novel* (New York: Hyperion, 2002); Allen Kurzweil, *A Case of Curiosities* (New York: Harvest Books, 2001). Allen Kurzweil is my first cousin.

38. As quoted in Aubrey de Grey, "Engineering Negligible Senescence: Rational Design of Feasible, Comprehensive Rejuvenation Biotechnology," Kronos Institute Seminar Series, February 8, 2002. PowerPoint presentation available at http://www.gen.cam.ac.uk/sens/sensov.ppt.

39. Robert A. Freitas Jr., "Death Is an Outrage!" presentation at the fifth Alcor Conference on Extreme Life Extension, Newport Beach, Calif., November 16, 2002, http:// www.rfreitas.com/Nano/DeathIsAnOutrage.htm, published on KurzweilAI.net January 9, 2003: http://www.KurzweilAI.net/articles/art0536.html.

40. Cro-magnon, "30 years or less, often much less . . .": http://anthro.palomar.edu/ homo2/sapiens_culture.htm.

Egypt: Jac J. Janssen quoted in Brett Palmer, "Playing the Numbers Game," in *Skeptical Review,* published online May 5, 2004, at http://www.theskeptical review.com/palmer/numbers.html.

Europe 1400: Gregory Clark, *The Conquest of Nature: A Brief Economic History of the World* (Princeton University Press, forthcoming, 2005), chapter 5, "Mortality in the Malthusian Era," http://www.econ.ucdavis.edu/faculty/gclark/GlobalHistory/ Global%20History-5.pdf.

1800: James Riley, *Rising Life Expectancy: A Global History* (Cambridge, U.K.: Cambridge University Press, 2001), pp. 32–33.

1900: http://www.cdc.gov/nchs/data/hus/tables/2003/03hus027.pdf.

41. The museum was originally located in Boston and is now in Mountain View, Calif. (http://www.computerhistory.org).

42. Lyman and Kahle on long-term storage: "While good paper lasts 500 years, computer tapes last 10. While there are active organizations to make copies, we will keep our information safe, we do not have an effective mechanism to make 500 year copies of digital materials. . . ." Peter Lyman and Brewster Kahle, "Archiving Digital Cultural Artifacts: Organizing an Agenda for Action," *D-Lib Magazine*, July–August 1998.

Stewart Brand writes: "Behind every hot new working computer is a trail of bodies of extinct computers, extinct storage media, extinct applications, extinct files. Science fiction writer Bruce Sterling refers to our time as 'the Golden Age of dead media, most of them with the working lifespan of a pack of Twinkies.' " Stewart Brand, "Written on the Wind," *Civilization Magazine*, November 1998 ("01998" in Long Now terminology), available online at http://www.longnow.org/10klibrary/library.htm.

43. DARPA's Information Processing Technology Office's project in this vein is called LifeLog, http://www.darpa.mil/ipto/Programs/lifelog; see also Noah Shachtman, "A Spy Machine of DARPA's Dreams," *Wired News*, May 20, 2003, http://www.wired.com/news/business/0,1367,58909,00.html; Gordon Bell's project (for Microsoft) is MyLifeBits, http://research.microsoft.com/research/barc/Media Presence/MyLifeBits.aspx; for the Long Now Foundation, see http://longnow.org.

44. Bergeron is assistant professor of anesthesiology at Harvard Medical School and the author of such books as *Bioinformatics Computing, Biotech Industry: A Global, Economic, and Financing Overview*, and *The Wireless Web and Healthcare*.

45. The Long Now Foundation is developing one possible solution: the Rosetta Disk, which will contain extensive archives of text in languages that may be lost in the far future. They plan to use a unique storage technology based on a two-inch nickel disk that can store up to 350,000 pages per disk, with an estimated life expectancy of 2,000 to 10,000 years. See the Long Now Foundation, Library Ideas, http://longnow.org/10klibrary/10kLibConference.htm.

46. John A. Parmentola, "Paradigm Shifting Capabilities for Army Transformation," invited paper presented at the SPIE European Symposium on Optics/Photonics in Security and Defence, October 25–28, 2004; available electronically at *Bridge* 34.3 (Fall 2004), http://www.nae.edu/NAE/bridgecom.nsf/weblinks/MKEZ-65 RLTA?OpenDocument.

47. Fred Bayles, "High-tech Project Aims to Make Super-soldiers," *USA Today*, May 23, 2003, http://www.usatoday.com/news/nation/2003-05-22-nanotech-usat_x.htm; see the Institute for Soldier Nanotechnologies Web site, http://web.mit.edu/isn; Sarah Putnam, "Researchers Tout Opportunities in Nanotech," MIT News Office, October 9, 2002, http://web.mit.edu/newsoffice/2002/cdc-nanotech-1009.html.

48. Ron Schafer, "Robotics to Play Major Role in Future Warfighting," http://www.jfcom.mil/newslink/storyarchive/2003/pa072903.htm; Dr. Russell Richards, "Unmanned Systems: A Big Player for Future Forces?" Unmanned Effects Workshop at the Applied Physics Laboratory, Johns Hopkins University, Baltimore, July 29–August 1, 2003.

49. John Rhea, "NASA Robot in Form of Snake Planned to Penetrate Inaccessible Areas," *Military and Aerospace Electronics,* November 2000, http://mae.pennnet.com/Articles/Article_Display.cfm?Section=Archives&Subsection=Display&ARTICLE_ID=86890.

50. Lakshmi Sandhana, "The Drone Armies Are Coming," *Wired News,* August 30, 2002, http://www.wired.com/news/technology/0,1282,54728,00.html. See also Mario Gerla, Kaixin Xu, and Allen Moshfegh, "Minuteman: Forward Projection of Unmanned Agents Using the Airborne Internet," IEEE Aerospace Conference 2002, Big Sky, Mont., March 2002: http://www.cs.ucla.edu/NRL/wireless/uploads/mgerla_aerospace02.pdf.

51. James Kennedy and Russell C. Eberhart, with Yuhui Shi, *Swarm Intelligence* (San Francisco: Morgan Kaufmann, 2001), http://www.swarmintelligence.org/SIBook/SI.php.

52. Will Knight, "Military Robots to Get Swarm Intelligence," April 25, 2003, http://www.newscientist.com/news/news.jsp?id=ns99993661.

53. Ibid.

54. S. R. White et al., "Autonomic Healing of Polymer Composites," *Nature* 409 (February 15, 2001): 794–97, http://www.autonomic.uiuc.edu/files/NaturePaper.pdf; Kristin Leutwyler, "Self-Healing Plastics," ScientificAmerican.com, February 15, 2001, http://www.sciam.com/article.cfm?articleID=000B307F-C71A-1C5AB882809EC588ED9F.

55. Sue Baker, "Predator Missile Launch Test Totally Successful," *Strategic Affairs,* April 1, 2001, http://www.stratmag.com/issueApr-1/page02.htm.

56. See the OpenCourseWare course list at http://ocw.mit.edu/index.html.

57. Brigitte Bouissou quoted on MIT OpenCourseWare's additional quotes page at http://ocw.mit.edu/OcwWeb/Global/AboutOCW/additionalquotes.htm and Eric Bender, "Teach Locally, Educate Globally," *MIT Technology Review,* June 2004, http://www.techreview.com/articles/04/06/bender0604.asp?p=1.

58. Kurzweil Educational Systems (http://www.Kurzweiledu.com) provides the Kurzweil 3000 reading system for people with dyslexia. It can read any book to the user while highlighting what is being read on a high-resolution image of the page. It incorporates a range of features to improve the reading skills of users.

59. As quoted by Natasha Vita-More, "Arterati on Ideas," http://64.233.167.104/search?q=cache:QAnJsLcXHXUJ:www.extropy.com/ideas/journal/previous/1998/02-01.html+Arterati+on+ideas&hl=en and http://www.extropy.com/ideas/journal/previous/1998/02-01.html.

60. Christine Boese, "The Screen-Age: Our Brains in our Laptops," CNN.com, August 2, 2004.

61. Thomas Hobbes, *Leviathan* (1651).

62. Seth Lloyd and Y. Jack Ng, "Black Hole Computers," *Scientific American*, November 2004.

63. Alan M. MacRobert, "The Allen Telescope Array: SETI's Next Big Step," *Sky & Telescope*, April 2004, http://skyandtelescope.com/printable/resources/seti/article_256.asp.

64. Ibid.

65. Ibid.

66. C. H. Townes, "At What Wavelength Should We Search for Signals from Extraterrestrial Intelligence?" *Proceedings of the National Academy of Sciences USA* 80 (1983): 1147–51. S. A. Kingsley in *The Search for Extraterrestrial Intelligence in the Optical Spectrum*, vol. 2, S. A. Kingsley and G. A. Lemarchand, eds. (1996) Proc. WPIE 2704: 102–16.

67. N. S. Kardashev, "Transmission of Information by Extraterrestrial Civilizations," *Soviet Astronomy* 8.2 (1964): 217–20. Summarized in Guillermo A. Lemarchand, "Detectability of Extraterrestrial Technological Activities," *SETIQuest* 1:1, pp. 3–13, http://www.coseti.org/lemarch1.htm.

68. Frank Drake and Dava Sobel, *Is Anyone Out There?* (New York: Dell, 1994); Carl Sagan and Frank Drake, "The Search for Extraterrestrial Intelligence," *Scientific American* (May 1975): 80–89. A Drake-equation calculator can be found at http://www.activemind.com/Mysterious/Topics/SETI/drake_equation.html.

69. Many of the descriptions of the Drake equation express f_L as the fraction of the *planet's* life during which radio transmission takes place, but this should properly be expressed as a fraction of the life of the universe, as we don't really care how long that planet has been around; rather, we care about the duration of the radio transmissions.

70. Seth Shostak provided "an estimate of between 10,000 and one million radio transmitters in the galaxy." Marcus Chown, "ET First Contact 'Within 20 Years,'" *New Scientist* 183. 2457 (July 24, 2004). Available online at http://www.newscientist.com/article.ns?id=dn6189.

71. T. L. Wilson, "The Search for Extraterrestrial Intelligence," *Nature*, February 22, 2001.

72. Most recent estimates have been between ten and fifteen billion years. In 2002 estimates based on data from the Hubble Space Telescope were between thirteen and fourteen billion years. A study published by Case Western Reserve University scientist Lawrence Krauss and Dartmouth University's Brian Chaboyer applied recent findings on the evolution of stars and concluded that there was a 95 percent level of confidence that the age of the universe is between 11.2 and 20 billion years. Lawrence Krauss and Brian Chaboyer, "Irion, the Milky Way's Restless Swarms of Stars," *Science* 299 (January 3, 2003): 60–62. Recent research from NASA has narrowed down the age of the universe to 13.7 billion years plus or minus 200 million, http://map.gsfc.nasa.gov/m_mm/mr_age.html.

73. Quoted in Eric M. Jones, "'Where Is Everybody?': An Account of Fermi's Ques-

tion," Los Alamos National Laboratories, March 1985, http://www.bayarea.net/~kins/AboutMe/Fermi_and_Teller/fermi_question.html.

74. First, consider the estimate of 10^{42} cps for the ultimate cold laptop (as in chapter 3). We can estimate the mass of the solar system as being approximately equal to the mass of the sun, which is 2 x 10^{30} kilograms. One twentieth of 1 percent of this mass is 10^{27} kilograms. At 10^{42} cps per kilogram, 10^{27} kilograms would provide 10^{69} cps. If we use the estimate of 10^{50} cps for the ultimate hot laptop, we get 10^{77} cps.

75. Anders Sandberg, "The Physics of Information Processing Superobjects: Daily Life Among the Jupiter Brains," *Journal of Evolution and Technology* 5 (December 22, 1999), http://www.jetpress.org/volume5/Brains2.pdf.

76. Freeman John Dyson, "Search for Artificial Stellar Sources of Infrared Radiation," *Science* 131 (June 3, 1960): 1667–68.

77. Cited in Sandberg, "Physics of Information Processing Superobjects."

78. There were 195.5 billion units of semiconductor chips shipped in 1994, 433.5 billion in 2004. Jim Feldhan, president, Semico Research Corporation, http://www.semico.com.

79. Robert Freitas has been a leading advocate of using robotic probes, especially self-replicating ones. See Robert A. Freitas Jr., "Interstellar Probes: A New Approach to SETI," *J. British Interplanet. Soc.* 33 (March 1980): 95–100, http://www.rfreitas.com/Astro/InterstellarProbesJBIS1980.htm; Robert A. Freitas Jr., "A Self-Reproducing Interstellar Probe," *J. British Interplanet. Soc.* 33 (July 1980): 251–64, http://www.rfreitas.com/Astro/ReproJBISJuly1980.htm; Francisco Valdes and Robert A. Freitas Jr., "Comparison of Reproducing and Nonreproducing Starprobe Strategies for Galactic Exploration," *J. British Interplanet. Soc.* 33 (November 1980): 402–8, http://www.rfreitas.com/Astro/ComparisonReproNov1980.htm; Robert A. Freitas Jr., "Debunking the Myths of Interstellar Probes," *AstroSearch* 1 (July–August 1983): 8–9, http://www.rfreitas.com/Astro/ProbeMyths1983.htm; Robert A. Freitas Jr., "The Case for Interstellar Probes," *J. British Interplanet. Soc.* 36 (November 1983): 490–95, http://www.rfreitas.com/Astro/TheCaseForInterstellarProbes1983.htm.

80. M. Stenner et al., "The Speed of Information in a 'Fast-Light' Optical Medium," *Nature* 425 (October 16, 2003): 695–98. See also Raymond Y. Chiao et al., "Superluminal and Parelectric Effects in Rubidium Vapor and Ammonia Gas," *Quantum and Semiclassical Optics* 7 (1995): 279.

81. I. Marcikic et al., "Long-Distance Teleportation of Qubits at Telecommunication Wavelengths," *Nature* 421 (January 2003): 509–13; John Roach, "Physicists Teleport Quantum Bits over Long Distance," *National Geographic News*, January 29, 2003; Herb Brody, "Quantum Cryptography," in "10 Emerging Technologies That Will Change the World," *MIT Technology Review*, February 2003; N. Gisin et al., "Quantum Correlations with Moving Observers," *Quantum Optics* (December 2003): 51; Quantum Cryptography exhibit, ITU Telecom World 2003, Geneva, Switzerland, October 1, 2003; Sora Song, "The Quantum Leaper," *Time*, March 15,

2004; Mark Buchanan, "Light's Spooky Connections Set New Distance Record," *New Scientist*, June 28, 1997.

82. Charles H. Lineweaver and Tamara M. Davis, "Misconceptions About the Big Bang," *Scientific American*, March 2005.

83. A. Einstein and N. Rosen, "The Particle Problem in the General Theory of Relativity," *Physical Review* 48 (1935): 73.

84. J. A. Wheeler, "Geons," *Physical Review* 97 (1955): 511–36.

85. M. S. Morris, K. S. Thorne, and U. Yurtsever, "Wormholes, Time Machines, and the Weak Energy Condition," *Physical Review Letters* 61.13 (September 26, 1988): 1446–49; M. S. Morris and K. S. Thorne, "Wormholes in Spacetime and Their Use for Interstellar Travel: A Tool for Teaching General Relativity," *American Journal of Physics* 56.5 (1988): 395–412.

86. M. Visser, "Wormholes, Baby Universes, and Causality," *Physical Review D* 41.4 (February 15, 1990): 1116–24.

87. Sandberg, "Physics of Information Processing Superobjects."

88. David Hochberg and Thomas W. Kephart, "Wormhole Cosmology and the Horizon Problem," *Physical Review Letters* 70 (1993): 2665–68, http://prola.aps.org/abstract/PRL/v70/i18/p2665_1; D. Hochberg and M. Visser, "Geometric Structure of the Generic Static Transversable Wormhole Throat," *Physical Review D* 56 (1997): 4745.

89. J. K. Webb et al., "Further Evidence for Cosmological Evolution of the Fine Structure Constant," *Physical Review Letters* 87.9 (August 27, 2001): 091301; "When Constants Are Not Constant," *Physics in Action* (October 2001), http://physicsweb.org/articles/world/14/10/4.

90. Joao Magueijo, John D. Barrow, and Haavard Bunes Sandvik, "Is It e or Is It c? Experimental Tests of Varying Alpha," *Physical Letters B* 549 (2002): 284–89.

91. John Smart, "Answering the Fermi Paradox: Exploring the Mechanisms of Universal Transcension," http://www.transhumanist.com/Smart-Fermi.htm. See also http://singularitywatch.com and his biography at http://www.singularitywatch.com/bio_johnsmart.html.

92. James N. Gardner, *Biocosm: The New Scientific Theory of Evolution: Intelligent Life Is the Architect of the Universe* (Maui: Inner Ocean, 2003).

93. Lee Smolin in "Smolin vs. Susskind: The Anthropic Principle," *Edge* 145, http://www.edge.org/documents/archive/edge145.html; Lee Smolin, "Scientific Alternatives to the Anthropic Principle," http://arxiv.org/abs/hep-th/0407213.

94. Kurzweil, *Age of Spiritual Machines*, pp. 258–60.

95. Gardner, *Biocosm*.

96. S. W. Hawking, "Particle Creation by Black Holes," *Communications in Mathematical Physics* 43 (1975): 199–220.

97. The original bet is located at http://www.theory.caltech.edu/people/preskill/info_bet.html. Also see Peter Rodgers, "Hawking Loses Black Hole Bet," *Physics World*, August 2004, http://physicsweb.org/articles/news/8/7/11.

98. To arrive at those estimates Lloyd took the observed density of matter—about one hydrogen atom per cubic meter—and computed the total energy in the universe. Dividing this figure by the Planck constant, he got about 10^{90} cps. Seth Lloyd, "Ultimate Physical Limits to Computation," *Nature* 406.6799 (August 31, 2000): 1047–54. Electronic versions (version 3 dated February 14, 2000) available at http://arxiv.org/abs/quant-ph/9908043 (August 31, 2000). The following link requires a payment to access: http://www.nature.com/cgi-taf/DynaPage.taf?file=/nature/journal/v406/n6799/full/4061047a0_fs.html&content_filetype=PDF.

99. Jacob D. Bekenstein, "Information in the Holographic Universe: Theoretical Results about Black Holes Suggest That the Universe Could Be Like a Gigantic Hologram," *Scientific American* 289.2 (August 2003): 58–65, http://www.sciam.com/article.cfm?articleID=000AF072-4891-1F0A-97AE80A84189EEDF.

Chapter Seven: *Ich bin ein Singularitarian*

1. In Jay W. Richards et al., *Are We Spiritual Machines? Ray Kurzweil vs. the Critics of Strong A.I.* (Seattle: Discovery Institute, 2002), introduction, http://www.KurzweilAI.net/meme/frame.html?main=/articles/art0502.html.

2. Ray Kurzweil and Terry Grossman, M.D., *Fantastic Voyage: Live Long Enough to Live Forever* (New York: Rodale Books, 2004).

3. Ibid.

4. Ibid.

5. Max More and Ray Kurzweil, "Max More and Ray Kurzweil on the Singularity," February 26, 2002, http://www.KurzweilAI.net/articles/art0408.html.

6. Ibid.

7. Ibid.

8. Arthur Miller, *After the Fall* (New York: Viking, 1964).

9. From a paper read to the Oxford Philosophical Society in 1959 and then published as "Minds, Machines and Gödel," *Philosophy* 36 (1961): 112–27. It was reprinted for the first of many times in Kenneth Sayre and Frederick Crosson, eds., *The Modeling of Mind* (Notre Dame: University of Notre Dame Press, 1963), pp. 255–71.

10. Martine Rothblatt, "Biocyberethics: Should We Stop a Company from Unplugging an Intelligent Computer?" September 28, 2003, http://www.KurzweilAI.net/meme/frame.html?main=/articles/art0594.html (includes links to a Webcast and transcripts).

11. Jaron Lanier, "One Half of a Manifesto," *Edge*, http://www.edge.org/3rd_culture/lanier/lanier_index.html; see also Jaron Lanier, "One-Half of a Manifesto," *Wired News*, December 2000, http://www.wired.com/wired/archive/8.12/lanier.html.

12. Ibid.

13. Norbert Wiener, *Cybernetics: or, Control and Communication in the Animal and the Machine* (Cambridge, Mass.: MIT Press, 1948).

14. "How Do You Persist When Your Molecules Don't?" *Science and Consciousness Review* 1.1 (June 2004), http://www.sci-con.org/articles/20040601.html.

15. David J. Chalmers, "Facing Up to the Problem of Consciousness," *Journal of Consciousness Studies* 2.3 (1995): 200–219, http://jamaica.u.arizona.edu/~chalmers/papers/facing.html.

16. Huston Smith, *The Sacred Unconscious,* videotape (The Wisdom Foundation, 2001), available for sale at http://www.fonsvitae.com/sacredhuston.html.

17. Jerry A. Fodor, *RePresentations: Philosophical Essays on the Foundations of Cognitive Science* (Cambridge, Mass.: MIT Press, 1981).

Chapter Eight: The Deeply Intertwined Promise and Peril of GNR

1. Bill McKibben, "How Much Is Enough? The Environmental Movement as a Pivot Point in Human History," Harvard Seminar on Environmental Values, October 18, 2000.

2. In the 1960s, the U.S. government conducted an experiment in which it asked three recently graduated physics students to build a nuclear weapon using only publicly available information. The result was successful; the three students built one in about three years (http://www.pimall.com/nais/nl/n.nukes.html). Plans for how to build an atomic bomb are available on the Internet and have been published in book form by a national laboratory. In 2002, the British Ministry of Defence released measurements, diagrams, and precise details on bomb building to the Public Record Office, since removed (http://news.bbc.co.uk/1/hi/uk/1932702.stm). Note that these links do not contain actual plans to build atomic weapons.

3. "The John Stossel Special: You Can't Say That!" ABC News, March 23, 2000.

4. There is extensive information on the Web, including military manuals, on how to build bombs, weapons, and explosives. Some of this information is erroneous, but accurate information on these topics continues to be accessible despite efforts to remove it. Congress passed an amendment (the Feinstein Amendment, SP 419) to a Defense Department appropriations bill in June 1997, banning the dissemination of instructions on building bombs. See Anne Marie Helmenstine, "How to Build a Bomb," February 10, 2003, http://chemistry.about.com/library/weekly/aa021003a.htm. Information on toxic industrial chemicals is widely available on the Web and in libraries, as are information and tools for cultivating bacteria and viruses and techniques for creating computer viruses and hacking into computers and networks. Note that I do not provide specific examples of such information, since it might be helpful to destructive individuals and groups. I realize that even stating the availability of such information has this potential, but I feel that the benefit of open dialogue about this issue outweighs this concern. Moreover, the availability of this type of information has been widely discussed in the media and other venues.

5. Ray Kurzweil, *The Age of Intelligent Machines* (Cambridge, Mass.: MIT Press, 1990).

6. Ken Alibek, *Biohazard* (New York: Random House, 1999).

7. Ray Kurzweil, *The Age of Spiritual Machines* (New York: Viking, 1999).

8. Bill Joy, "Why the Future Doesn't Need Us," *Wired,* April 2000, http://www.wired.com/wired/archive/8.04/joy.html.

9. Handbooks on gene splicing (such as A. J. Harwood, ed., *Basic DNA and RNA Protocols* [Totowa, N.J.: Humana Press, 1996]) along with reagents and kits that enable gene splicing are generally available. Even if access to these materials were limited in the West, there are a large number of Russian companies that could provide equivalent materials.

10. For a detailed summary site of the "Dark Winter" simulation, see "DARK WINTER: A Bioterrorism Exercise June 2001": http://www.biohazardnews.net/scen_smallpox.shtml. For a brief summary, see: http://www.homelandsecurity.org/darkwinter/index.cfm.

11. Richard Preston, "The Specter of a New and Deadlier Smallpox," *New York Times,* October 14, 2002, available at http://www.ph.ucla.edu/epi/bioter/specterdeadlier smallpox.html.

12. Alfred W. Crosby, *America's Forgotten Pandemic: The Influenza of 1918* (New York: Cambridge University Press, 2003).

13. "Power from Blood Could Lead to 'Human Batteries,'" *Sydney Morning Herald,* August 4, 2003, http://www.smh.com.au/articles/2003/08/03/1059849278131.html. See note 129 in chapter 5. See also S. C. Barton, J. Gallaway, and P. Atanassov, "Enzymatic Biofuel Cells for Implantable and Microscale Devices," *Chemical Reviews* 104.10 (October 2004): 4867–86.

14. J. M. Hunt has calculated that there are 1.55×10^{19} kilograms (10^{22} grams) of organic carbon on Earth. Based on this figure, and assuming that all "organic carbon" is contained in the biomass (note that the biomass is not clearly defined, so we are taking a conservatively broad approach), we can compute the approximate number of carbon atoms as follows:

    Average atomic weight of carbon (adjusting for isotope ratios) = 12.011.

    Carbon in the biomass = 1.55×10^{22} grams / 12.011 = 1.3×10^{21} mols.

    $1.3 \times 10^{21} \times 6.02 \times 10^{23}$ (Avogadro's number) = 7.8×10^{44} carbon atoms.

    J. M. Hunt, *Petroleum Geochemistry and Geology* (San Francisco: W. H. Freeman, 1979).

15. Robert A. Freitas Jr., "The Gray Goo Problem," March 20, 2001, http://www.KurzweilAI.net/articles/art0142.html.

16. "Gray Goo Is a Small Issue," Briefing Document, Center for Responsible Nanotechnology, December 14, 2003, http://crnano.org/BD-Goo.htm; Chris Phoenix and Mike Treder, "Safe Utilization of Advanced Nanotechnology," Center for Responsible Nanotechnology, January 2003, http://crnano.org/safe.htm; K. Eric Drexler, *Engines of Creation,* chapter 11, "Engines of Destruction" (New York: Anchor Books, 1986), pp. 171–90, http://www.foresight.org/EOC/EOC_Chapter_11.html; Robert A. Freitas Jr. and Ralph C. Merkle, *Kinematic Self-Replicating Machines,* section 5.11, "Replicators and Public Safety" (Georgetown, Tex.: Landes Bioscience, 2004),

pp. 196–99, http://www.MolecularAssembler.com/KSRM/5.11.htm, and section 6.3.1, "Molecular Assemblers Are Too Dangerous," pp. 204–6, http://www.Molecular Assembler.com/KSRM/6.3.1.htm; Foresight Institute, "Molecular Nanotechnology Guidelines: Draft Version 3.7," June 4, 2000, http://www.foresight.org/guidelines/.

17. Robert A. Freitas Jr., "Gray Goo Problem" and "Some Limits to Global Ecophagy by Biovorous Nanoreplicators, with Public Policy Recommendations," Zyvex preprint, April 2000, section 8.4 "Malicious Ecophagy" and section 6.0 "Ecophagic Thermal Pollution Limits (ETPL)," http://www.foresight.org/NanoRev/Ecophagy. html.

18. Nick D. Bostrom, "Existential Risks: Analyzing Human Extinction Scenarios and Related Hazards," May 29, 2001, http://www.KurzweilAI.net/meme/frame.html? main=/articles/art0194.html.

19. Robert Kennedy, *13 Days* (London: Macmillan, 1968), p. 110.

20. In H. Putnam, "The Place of Facts in a World of Values," in D. Huff and O. Prewitt, eds., *The Nature of the Physical Universe* (New York: John Wiley, 1979), p. 114.

21. Graham Allison, *Nuclear Terrorism* (New York: Times Books, 2004).

22. Martin I. Meltzer, "Multiple Contact Dates and SARS Incubation Periods," *Emerging Infectious Diseases* 10.2 (February 2004), http://www.cdc.gov/ncidod/EID/ vol10no2/03-0426-G1.htm.

23. Robert A. Freitas Jr., "Microbivores: Artificial Mechanical Phagocytes Using Digest and Discharge Protocol," Zyvex preprint, March 2001, http://www.rfreitas.com/ Nano/Microbivores.htm, and "Microbivores: Artificial Mechanical Phagocytes," *Foresight Update* no. 44, March 31, 2001, pp. 11–13, http://www.imm.org/Reports/ Rep025.html.

24. Max More, "The Proactionary Principle," May 2004, http://www.maxmore.com/ proactionary.htm and http://www.extropy.org/proactionaryprinciple.htm. More summarizes the proactionary principle as follows:

1. People's freedom to innovate technologically is valuable to humanity. The burden of proof therefore belongs to those who propose restrictive measures. All proposed measures should be closely scrutinized.

2. Evaluate risk according to available science, not popular perception, and allow for common reasoning biases.

3. Give precedence to ameliorating known and proven threats to human health and environmental quality over acting against hypothetical risks.

4. Treat technological risks on the same basis as natural risks; avoid underweighting natural risks and overweighting human-technological risks. Fully account for the benefits of technological advances.

5. Estimate the lost opportunities of abandoning a technology, and take into account the costs and risks of substituting other credible options, carefully considering widely distributed effects and follow-on effects.

6. Consider restrictive measures only if the potential impact of an activity has both significant probability and severity. In such cases, if the activity also generates benefits, discount the impacts according to the feasibility of adapting to the adverse effects. If measures to limit technological advance do appear justified, ensure that the extent of those measures is proportionate to the extent of the probable effects.

7. When choosing among measures to restrict technological innovation, prioritize decision criteria as follows: Give priority to risks to human and other intelligent life over risks to other species; give non-lethal threats to human health priority over threats limited to the environment (within reasonable limits); give priority to immediate threats over distant threats; prefer the measure with the highest expectation value by giving priority to more certain over less certain threats, and to irreversible or persistent impacts over transient impacts.

25. Martin Rees, *Our Final Hour: A Scientist's Warning: How Terror, Error, and Environmental Disaster Threaten Humankind's Future in This Century—on Earth and Beyond* (New York: Basic Books, 2003).

26. Scott Shane, *Dismantling Utopia: How Information Ended the Soviet Union* (Chicago: Ivan R. Dee, 1994); see also the review by James A. Dorn at http://www.cato.org/pubs/journal/cj16n2-7.html.

27. See George DeWan, "Diary of a Colonial Housewife," *Newsday*, 2005, for one account of the difficulty of human life a couple of centuries ago: http://www.newsday.com/community/guide/lihistory/ny-history-hs331a,0,6101197.story.

28. Jim Oeppen and James W. Vaupel, "Broken Limits to Life Expectancy," *Science* 296.5570 (May 10, 2002): 1029–31.

29. Steve Bowman and Helit Barel, *Weapons of Mass Destruction: The Terrorist Threat*, Congressional Research Service Report for Congress, December 8, 1999, http://www.cnie.org/nle/crsreports/international/inter-75.pdf.

30. Eliezer S. Yudkowsky, "Creating Friendly AI 1.0, The Analysis and Design of Benevolent Goal Architectures" (2001), The Singularity Institute, http://www.singinst.org/CFAI/; Eliezer S. Yudkowsky, "What Is Friendly AI?" May 3, 2001, http://www.KurzweilAI.net/meme/frame.html?main=/articles/art0172.html.

31. Ted Kaczynski, "The Unabomber's Manifesto," May 14, 2001, http://www.KurzweilAI.net/meme/frame.html?main=/articles/art0182.html.

32. Bill McKibben, *Enough: Staying Human in an Engineered Age* (New York: Times Books, 2003).

33. Kaczynski, "The Unabomber's Manifesto."

34. Foresight Institute and IMM, "Foresight Guidelines on Molecular Nanotechnology," February 21, 1999, http://www.foresight.org/guidelines/current.html; Christine Peterson, "Molecular Manufacturing: Societal Implications of Advanced Nanotechnology," April 9, 2003, http://www.KurzweilAI.net/meme/frame.html?

main=/articles/art0557.html; Chris Phoenix and Mike Treder, "Safe Utilization of Advanced Nanotechnology," January 28, 2003, http://www.KurzweilAI.net/meme/frame.html?main=/articles/art0547.html; Robert A. Freitas Jr., "The Gray Goo Problem," KurzweilAI.net, 20 March 2002, http://www.KurzweilAI.net/meme/frame.html?main=/articles/art0142.html.

35. Robert A. Freitas Jr., private communication to Ray Kurzweil, January 2005. Freitas describes his proposal in detail in Robert A. Freitas Jr., "Some Limits to Global Ecophagy by Biovorous Nanoreplicators, with Public Policy Recommendations."

36. Ralph C. Merkle, "Self Replicating Systems and Low Cost Manufacturing," 1994, http://www.zyvex.com/nanotech/selfRepNATO.html.

37. Neil King Jr. and Ted Bridis, "FBI System Covertly Searches E-mail," *Wall Street Journal Online* (July 10, 2000), http://zdnet.com.com/2100-11-522071.html?legacy=zdnn.

38. Patrick Moore, "The Battle for Biotech Progress—GM Crops Are Good for the Environment and Human Welfare," *Greenspirit* (February 2004), http://www.greenspirit.com/logbook.cfm?msid=62.

39. "GMOs: Are There Any Risks?" European Commission (October 9, 2001), http://europa.eu.int/comm/research/biosociety/pdf/gmo_press_release.pdf.

40. Rory Carroll, "Zambians Starve As Food Aid Lies Rejected," *Guardian* (October 17, 2002), http://www.guardian.co.uk/gmdebate/Story/0,2763,813220,00.html.

41. Larry Thompson, "Human Gene Therapy: Harsh Lessons, High Hopes," *FDA Consumer Magazine* (September–October 2000), http://www.fda.gov/fdac/features/2000/500_gene.html.

42. Bill Joy, "Why the Future Doesn't Need Us."

43. The Foresight Guidelines (Foresight Institute, version 4.0, October 2004, http://www.foresight.org/guidelines/current.html) are designed to address the potential positive and negative consequences of nanotechnology. They are intended to inform citizens, companies, and governments, and provide specific guidelines to responsibly develop nanotechnology-based molecular manufacturing. The Foresight Guidelines were initially developed at the Institute Workshop on Molecular Nanotechnology Research Policy Guidelines, sponsored by the institute and the Institute for Molecular Manufacturing (IMM), February 19–21, 1999. Participants included James Bennett, Greg Burch, K. Eric Drexler, Neil Jacobstein, Tanya Jones, Ralph Merkle, Mark Miller, Ed Niehaus, Pat Parker, Christine Peterson, Glenn Reynolds, and Philippe Van Nedervelde. The guidelines have been updated several times.

44. Martine Rothblatt, CEO of United Therapeutics, has proposed replacing this moratorium with a regulatory regime in which a new International Xenotransplantation Authority inspects and approves pathogen-free herds of genetically engineered pigs as acceptable sources of xenografts. Rothblatt's solution also helps stamp out rogue xenograft surgeons by promising each country that joins the IXA, and helps to enforce the rules within its borders, a fair share of the pathogen-free xenografts for its own citizens suffering from organ failure. See Martine Roth-

blatt, "Your Life or Mine: Using Geoethics to Resolve the Conflict Between Public and Private Interests," in *Xenotransplantation* (Burlington, Vt.: Ashgate, 2004). Disclosure: I am on the board of directors of United Therapeutics.

45. See Singularity Institute, http://www.singinst.org. Also see note 30 above. Yudkowsky formed the Singularity Institute for Artificial Intelligence (SIAI) to develop "Friendly AI," intended to "create cognitive content, design features, and cognitive architectures that result in benevolence" before near-human or better-than-human AIs become possible. SIAI has developed The SIAI Guidelines on Friendly AI: "Friendly AI," http://www.singinst.org/friendly/. Ben Goertzel and his Artificial General Intelligence Research Institute have also examined issues related to developing friendly AI; his current focus is on developing the Novamente AI Engine, a set of learning algorithms and architectures. Peter Voss, founder of Adaptive A.I., Inc., has also collaborated on friendly-AI issues: http://adaptive ai.com/.

46. Integrated Fuel Cell Technologies, http://ifctech.com. Disclosure: The author is an early investor in and adviser to IFCT.

47. *New York Times,* September 23, 2003, editorial page.

48. The House Committee on Science of the U.S. House of Representatives held a hearing on April 9, 2003, to "examine the societal implications of nanotechnology and H.R. 766, the Nanotechnology Research and Development Act of 2002." See "Full Science Committee Hearing on the Societal Implications of Nanotechnology," http://www.house.gov/science/hearings/full03/index.htm, and "Hearing Transcript," http://commdocs.house.gov/committees/science/hsy86340.000/hsy 86340_0f.htm. For Ray Kurzweil's testimony, see also http://www.KurzweilAI.net/ meme/frame.html?main=/articles/art0556.html. Also see Amara D. Angelica, "Congressional Hearing Addresses Public Concerns About Nanotech," April 14, 2003, http://www.KurzweilAI.net/articles/art0558.html.

Chapter Nine: Response to Critics

1. Michael Denton, "Organism and Machine," in Jay W. Richards et al., *Are We Spiritual Machines? Ray Kurzweil vs. the Critics of Strong A.I.* (Seattle: Discovery Institute Press, 2002), http://www.KurzweilAI.net/meme/frame.html?main=/articles/ art0502.html.

2. Jaron Lanier, "One Half of a Manifesto," *Edge* (September 25, 2000), http://www.edge.org/documents/archive/edge74.html.

3. Ibid.

4. See chapters 5 and 6 for examples of narrow AI now deeply embedded in our modern infrastructure.

5. Lanier, "One Half of a Manifesto."

6. An example is Kurzweil Voice, developed originally by Kurzweil Applied Intelligence.

7. Alan G. Ganek, "The Dawning of the Autonomic Computing Era," *IBM Systems*

Journal (March 2003), http://www.findarticles.com/p/articles/mi_m0ISJ/is_1_42/ai_98695283/print.

8. Arthur H. Watson and Thomas J. McCabe, "Structured Testing: A Testing Methodology Using the Cyclomatic Complexity Metric," NIST special publication 500–35, Computer Systems Laboratory, National Institute of Standards and Technology, 1996.

9. Mark A. Richards and Gary A. Shaw, "Chips, Architectures and Algorithms: Reflections on the Exponential Growth of Digital Signal Processing Capability," submitted to *IEEE Signal Processing*, December 2004.

10. Jon Bentley, "Programming Pearls," *Communications of the ACM* 27.11 (November 1984): 1087–92.

11. C. Eldering, M. L. Sylla, and J. A. Eisenach, "Is There a Moore's Law for Bandwidth," *IEEE Communications* (October 1999): 117–21.

12. J. W. Cooley and J. W. Tukey, "An Algorithm for the Machine Computation of Complex Fourier Series," *Mathematics of Computation* 19 (April 1965): 297–301.

13. There are an estimated 100 billion neurons with an estimated interneuronal connection "fan out" of about 1,000, so there are about 100 trillion (10^{14}) connections. Each connection requires at least 70 bits to store an ID for the two neurons at either end of the connection. So that's approximately 10^{16} bits. Even the uncompressed genome is about 6 billion bits (about 10^{10}), a ratio of at least 10^6: 1. See chapter 4.

14. Robert A. Freitas Jr., *Nanomedicine*, vol. I, *Basic Capabilities*, section 6.3.4.2, "Biological Chemomechanical Power Conversion" (Georgetown, Tex.: Landes Bioscience, 1999), pp. 147–48, http://www.nanomedicine.com/NMI/6.3.4.2.htm#p4; see illustration at http://www.nanomedicine.com/NMI/Figures/6.2.jpg.

15. Richard Dawkins, "Why Don't Animals Have Wheels?" *Sunday Times*, November 24, 1996, http://www.simonyi.ox.ac.uk/dawkins/WorldOfDawkins-archive/Dawkins/Work/Articles/1996-11-24wheels.shtml.

16. Thomas Ray, "Kurzweil's Turing Fallacy," in Richards et al., *Are We Spiritual Machines?*

17. Ibid.

18. Anthony J. Bell, "Levels and Loops: The Future of Artificial Intelligence and Neuroscience," *Philosophical Transactions of the Royal Society of London B* 354 (1999): 2013–20, http://www.cnl.salk.edu/~tony/ptrsl.pdf.

19. Ibid.

20. David Dewey, "Introduction to the Mandelbrot Set," http://www.ddewey.net/mandelbrot.

21. Christof Koch quoted in John Horgan, *The End of Science* (Reading, Mass.: Addison-Wesley, 1996).

22. Roger Penrose, *Shadows of the Mind: A Search for the Missing Science of Consciousness* (New York: Oxford University Press, 1996); Stuart Hameroff and Roger Penrose, "Orchestrated Objective Reduction of Quantum Coherence in Brain

Microtubules: The 'Orch OR' Model for Consciousness," *Mathematics and Computer Simulation* 40 (1996): 453–80, http://www.quantumconsciousness.org/penrosehameroff/orchOR.html.

23. Sander Olson, "Interview with Seth Lloyd," November 17, 2002, http://www.nanomagazine.com/i.php?id=2002_11_17.

24. Bell, "Levels and Loops."

25. See the exponential growth of computing graphs in chapter 2 (pp. 67, 70).

26. Alfred N. Whitehead and Bertrand Russell, *Principia Mathematica*, 3 vols. (Cambridge, U.K.: Cambridge University Press, 1910, 1912, 1913).

27. Gödel's incompleteness theorem first appeared in his "Uber formal unenscheiderbare Satze der *Principia Mathematica* und verwandter Systeme I," *Monatshefte für Mathematik und Physik* 38 (1931): 173–98.

28. Alan M. Turing, "On Computable Numbers with an Application to the Entscheidungsproblem," *Proceedings of the London Mathematical Society* 42 (1936): 230–65. The "Entscheidungsproblem" is the decision or halting problem—that is, how to determine ahead of time whether an algorithm will halt (come to a decision) or continue in an infinite loop.

29. Church's version appeared in Alonzo Church, "An Unsolvable Problem of Elementary Number Theory," *American Journal of Mathematics* 58 (1936): 345–63.

30. For an entertaining introductory account of some of the implications of the Church-Turing thesis, see Douglas R. Hofstadter, *Gödel, Escher, Bach: An Eternal Golden Braid* (New York: Basic Books, 1979).

31. The busy-beaver problem is one example of a large class of noncomputable functions, as seen in Tibor Rado, "On Noncomputable Functions," *Bell System Technical Journal* 41.3 (1962): 877–84.

32. Ray, "Kurzweil's Turing Fallacy."

33. Lanier, "One Half of a Manifesto."

34. A human, that is, who is not asleep and not in a coma and of sufficient development (that is, not a prebrain fetus) to be conscious.

35. John R. Searle, "I Married a Computer," in Richards et al., *Are We Spiritual Machines?*

36. John R. Searle, *The Rediscovery of the Mind* (Cambridge, Mass.: MIT Press, 1992).

37. Hans Moravec, Letter to the Editor, *New York Review of Books,* http://www.kurzweiltech.com/Searle/searle_response_letter.htm.

38. John Searle to Ray Kurzweil, December 15, 1998.

39. Lanier, "One Half of a Manifesto."

40. David Brooks, "Good News About Poverty," *New York Times* November 27, 2004, A35.

41. Hans Moravec, Letter to the Editor, *New York Review of Books,* http://www.kurzweiltech.com/Searle/searle_response_letter.htm.

42. Patrick Moore, "The Battle for Biotech Progress—GM Crops Are Good for the

Environment and Human Welfare," *Greenspirit* (February 2004), http://www. greenspirit.com/logbook.cfm?msid=62.

43. Joel Cutcher-Gershenfeld, private communication to Ray Kurzweil, February 2005.

44. William A. Dembski, "Kurzweil's Impoverished Spirituality," in Richards et al., *Are We Spiritual Machines?*

45. Denton, "Organism and Machine."

Epilogue

1. As quoted in James Gardner, "Selfish Biocosm," *Complexity* 5.3 (January–February 2000): 34–45.

2. In the function $y = 1/x$, if $x = 0$, then the function is literally undefined, but we can show that the value of y exceeds any finite number. We can transform $y = 1/x$ into $x = 1/y$ by flipping the nominator and denominator of both sides of the equation. So if we set y to a large finite number, then we can see that x becomes very small but not zero, no matter how big y gets. So the value of y in $y = 1/x$ can be seen to exceed any finite value for y if $x = 0$. Another way to express this is that we can exceed any possible finite value of y by setting x to be greater than 0 but smaller than 1 divided by that value.

3. With estimates of 10^{16} cps for functional simulation of the human brain (see chapter 3) and about 10^{10} (under ten billion) human brains, that's 10^{26} cps for all biological human brains. So 10^{90} cps exceeds this by a factor of 10^{64}. If we use the more conservative figure of 10^{19} cps, which I estimated was necessary to simulate each nonlinearity in each neuron component (dendrite, axon, and so on), we get a factor of 10^{61}. A trillion trillion trillion trillion trillion is 10^{60}.

4. See the estimates in the preceding note; 10^{42} cps exceeds this by a factor of ten thousand trillion (10^{16}).

5. Stephen Jay Gould, "Jove's Thunderbolts," *Natural History* 103.10 (October 1994): 6–12; chapter 13 in *Dinosaur in a Haystack: Reflections in Natural History* (New York: Harmony Books, 1995).

Index

Page numbers in *italics* refer to illustrations.